SEMICONDUCTOR DEVICES AND CIRCUITS

NEW TITLES IN ELECTRONIC TECHNOLOGY

Theodore Bogart
LAPLACE TRANSFORMS AND CONTROL SYSTEMS THEORY FOR TECHNOLOGY (1982)

Rodney B. Faber
APPLIED ELECTRICITY AND ELECTRONICS FOR TECHNOLOGY, 2nd edition (1982)

Luces M. Faulkenberry
AN INTRODUCTION TO OPERATIONAL AMPLIFIERS, 2nd edition (1982)

Joseph D. Greenfield
PRACTICAL DIGITAL DESIGN USING ICs, 2nd edition (1983)

Joseph D. Greenfield and William C. Wray
USING MICROPROCESSORS AND MICROCOMPUTERS: THE 6800 FAMILY (1981)

Curtis Johnson
PROCESS CONTROL INSTRUMENTATION TECHNOLOGY, 2nd edition (1982)

Larry Jones and A. Foster Chin
ELECTRONIC INSTRUMENTS AND MEASUREMENTS (1983)

Thomas Young
LINEAR INTEGRATED CIRCUITS (1981)

William Starr
ELECTRICAL WIRING AND DESIGN: A PRACTICAL APPROACH (1983)

Henry Zanger
SEMICONDUCTOR DEVICES AND CIRCUITS (1984)

SEMICONDUCTOR DEVICES AND CIRCUITS

Henry Zanger
QUEENSBOROUGH COMMUNITY COLLEGE

JOHN WILEY & SONS.
New York   Chichester   Brisbane   Toronto   Singapore

*To my wife, Cynthia,
and my daughters,
Diane, Ariane and
Jordana.*

Copyright © 1984, by John Wiley & Sons, Inc.

All rights reserved. Published simultaneously in Canada.

Reproduction or translation of any part of
this work beyond that permitted by Sections
107 and 108 of the 1976 United States Copyright
Act without the permission of the copyright
owner is unlawful. Requests for permission
or further information should be addressed to
the Permissions Department, John Wiley & Sons.

Library of Congress Cataloging in Publication Data:

Zanger, Henry, 1931–
   Semiconductor devices and circuits.

   (Electronic technology series)
   Includes index.
   1. Semiconductors.   2. Transistor circuits.
I. Title.   II. Series.
TK7871.85.Z36   1983   621.3815   83-1190
ISBN 0-471-05323-6

Printed in the United States of America

10 9 8 7 6 5 4 3 2 1

PREFACE

Transistor technology has undergone major developments in the last ten years. Most of the disadvantages of the device, such as instability with temperature (the effects of leakage current) and low gain at high frequencies have either been eliminated completely or drastically reduced. These much-improved device characteristics have enabled the theory of the operation of transistor circuits to be substantially simplified by approximations that introduce relatively small errors. (The meaning and magnitude of these errors are discussed in this text.)

This book uses modern analysis techniques coupled with simplified equivalences. It relies heavily on basic circuit theory, KVL, KCL, and Thevenin's equivalence. Voltage and current laws are used directly in the examination of devices, giving more insight into circuit performance. The need to memorize formulas is practically eliminated.

The book presents these essential concepts and techniques:

1. Modern analysis (i.e., dynamic resistance) techniques are used.
2. A uniform technique is applied to all transistor configurations.
3. Simplified equivalent circuits are used, together with the direct application of basic electric circuit theorems. (The first chapter reviews basic theorems and elaborates on the use of equivalent circuits.)
4. Practical, though simple, examples are used throughout the book to clarify and reinforce the theoretical discussions.
5. To enhance the practical aspects of the discussion, manufacturers' specifications are analyzed, focusing on important parameters and their effects on circuit performance in particular applications.

6. A summary of important terms is presented at the end of each chapter. It highlights the main topics of the chapter and is a good review.
7. The problems accompanying each chapter are largely analysis-oriented and for the most part numerical in nature. They are designed to serve as a reinforcement and require original thought, not just the simple application of formulas.

The presentation of the material goes from the simple to the complex. An attempt has been made to present only material that correctly belongs in an electronics curriculum. Although one may argue that other material should be included, trying to satisfy everyone usually results in unnecessary complexity and renders the text cumbersome and too encyclopedic. In any case, the student should be encouraged to use additional reference books.

Chapter 1, as noted, is essentially a review, and may be appropriately assigned for homework in those curricula in which a circuit theory course precedes this material in the syllabus.

Chapters 2 and 3 discuss standard diode and Zener diode theory and applications, in particular, the use of diodes in power supplies and Zeners in voltage regulation.

Although it may be argued that these chapters should also cover other diodes, such as photodiodes and 4-layer diodes, I felt that these devices are not nearly as common as the diode, and are not generally covered by many schools. Consequently, these special devices are discussed in a separate chapter (Chapter 12). Similarly, all special 3-terminal devices (e.g., phototransistors and thyristors) are dealt with in separate chapters (Chapters 12 and 13).

Chapters 4 through 6 cover BJT and FET devices, as well as applications in single stage circuits. A special section is devoted to VMOSFETs, because in the near future these devices may replace the transistor in many circuit applications. Emphasis is placed on the manufacturer's specifications and detailed discussions of a number of data sheets are included.

Chapter 7 extends the circuit analysis to multistage circuits containing both BJTs and FETs. It leads logically to problems of frequency response covered in Chapter 8.

Chapter 9 discusses large signal circuits. Power, as well as distortion, is considered here. The application of VMOS to power circuits is detailed with practical examples taken from manufacturers' suggested applications.

Chapters 10 and 11 deal with differential (DA) and operational (OA) amplifiers. These chapters are somewhat introductory in nature since the applications of the OA are so varied, and a full presentation of the material would require a separate text. The purpose here is to make sure that the student who may not choose to take a separate course in the subject has some familiarity with a few practical applications, and exposure to a number of commercially available DA circuits of the IC or hybrid construction types. The concept of negative feedback is covered in Chapter 11 in connection with the OA.

As noted, Chapters 12 and 13 are devoted to "special" semiconductor devices such as thyristors, photodevices, and LEDs.

Throughout this book, an effort has been made to minimize the mathematical complexities of the material covered. In many cases, simple algebraic developments have been substituted for more sophisticated calculus. For completeness, however, the advanced analysis is presented in the appendices. This means that students do not need a calculus background to use this text. The book is therefore particularly suitable for the two-year technology curriculum. The inclusion of the more sophisticated materials in the appendices makes the text also useful for the four-year technology and engineering schools.

Writing a textbook requires persistence and hard work. It was made easier by the strong support and encouragement I received from two editors at Wiley, Judy Green and Susan Weiss. Their tireless efforts during the reviewing process have earned my sincere thanks. I also thank my wife and daughters for their understanding during the long writing period.

Henry Zanger

CONTENTS

1	**BASIC CIRCUIT THEOREMS REEXAMINED**	**1**
1-1	Chapter Objectives	1
1-2	Introduction	1
1-3	Linearity	2
1-4	Kirchhoff's Current Law (KCL)	3
1-5	Kirchhoff's Voltage Law (KVL)	5
1-6	Thevenin's and Norton's Theorems	7
1-7	Maximum Power Theorem	12
1-8	The Superposition Theorem	14
1-9	Equivalent Circuits	18
1-10	Dependent Sources	22
	Summary of Important Terms	23
	Problems	24
2	**SEMICONDUCTOR DIODES: PHYSICS AND CHARACTERISTICS**	**28**
2-1	Chapter Objectives	28
2-2	Introduction	28
2-3	The Physics of Semiconductors	29
2-4	Diode Characteristics	42
2-5	Equivalent Circuits of the Diode	51
	Summary of Important Terms	59
	Problems	61

x CONTENTS

3 DIODE APPLICATIONS — 63

- 3-1 Chapter Objectives — 63
- 3-2 Introduction — 63
- 3-3 Wave Shaping — 64
- 3-4 dc Power Supplies — 72
- 3-5 Signal Switching — 97
- 3-6 Zener Diode Applications — 98
- 3-7 Diode Specifications—The Data Sheet — 107
- Summary of Important Terms — 113
- Problems — 114

4 THE BIPOLAR TRANSISTOR — 119

- 4-1 Chapter Objectives — 119
- 4-2 Introduction — 119
- 4-3 Transistor Physics — 120
- 4-4 Transistor Characteristics — 127
- 4-5 Transistor Biasing—Basic Concepts — 130
- 4-6 Biasing Stability — 132
- 4-7 Biasing Circuits — 134
- 4-8 dc Load Line—Classes of Operation — 152
- 4-9 Biasing for Switching Applications — 156
- 4-10 NPN vs. PNP — 157
- 4-11 Miscellaneous Biasing Circuits — 158
- Summary of Important Terms — 160
- Problems — 161

5 THE TRANSISTOR AS AN AMPLIFIER (SMALL SIGNAL) — 167

- 5-1 Chapter Objectives — 167
- 5-2 Introduction — 167
- 5-3 Important Amplifier Parameters — 168
- 5-4 Transistor Equivalent Circuits — 171
- 5-5 Transistor Circuits—ac Analysis — 178
- 5-6 Summary of Single Stage Transistor Amplifiers — 209
- 5-7 Graphical Analysis and Distortion — 211
- 5-8 The ac Load Line — 216
- 5-9 The Transistor as a Switch — 220
- Summary of Important Terms — 223
- Problems — 224

6 FIELD EFFECT TRANSISTORS (FET) — 232

- 6-1 Chapter Objectives — 232
- 6-2 Introduction — 232
- 6-3 The Fundamentals of the JFET — 233

6-4	JFET dc Parameters and Characteristics	236
6-5	JFET Biasing	240
6-6	JFET ac Analysis (Small Signal)	251
6-7	CS and SF Applications	276
6-8	MOSFET (Metal Oxide Semiconductor FET)	284
6-9	MOSFET Biasing	289
6-10	MOSFET ac Analysis	295
6-11	FETS in Switching Applications	298
6-12	VMOS FET	301
Summary of Important Terms		304
Problems		306

7 MULTITRANSISTOR AMPLIFIERS — 318

7-1	Chapter Objectives	318
7-2	Introduction	319
7-3	Biasing	319
7-4	Small Signal ac Analysis	327
Summary of Important Terms		342
Problems		343

8 FREQUENCY RESPONSE — 349

8-1	Chapter Objectives	349
8-2	Introduction	349
8-3	Bels and Decibels	350
8-4	Low Frequency Response	353
8-5	High Frequency Response	360
8-6	The Frequency Response Plot	368
8-7	Transistor Parameters, f_β, f_T	373
Summary of Important Terms		374
Problems		375

9 LARGE SIGNAL AMPLIFIERS — 380

9-1	Chapter Objectives	380
9-2	Introduction	381
9-3	r_e and β for Large Signals	381
9-4	Linearity and Distortion	383
9-5	Power Considerations—Efficiency and Maximum Power Rating	385
9-6	Classes of Operations	389
9-7	Class A Power Amplifiers	391
9-8	Class B Audio Amplifiers	397
9-9	VMOS Power Amplifiers	407
9-10	Class C Amplifiers	410
9-11	Power Ratings and Device Temperature	413

xii CONTENTS

9-12	Heat Sinks	417
9-13	Transistor Specifications	419
Summary of Important Terms		426
Problems		427

10 DIFFERENTIAL AMPLIFIERS 430

10-1	Chapter Objectives	430
10-2	Introduction	430
10-3	Differential Amplifier (DA) Functional Description	431
10-4	DA Configurations	435
10-5	The Differential Amplifier (DA) Circuit Biasing	435
10-6	Current Sources	436
10-7	The DA Circuit—ac Analysis	440
10-8	Input and Output Resistance	445
10-9	The Typical I.C. DA	446
Summary of Important Terms		446
Problems		448

11 OPERATIONAL AMPLIFIERS 450

11-1	Chapter Objectives	450
11-2	Introduction	450
11-3	Negative Feedback	451
11-4	Voltage Gains of an I.C. Operational Amplifier (OA)	456
11-5	z_{in} or r_{in} and $z_o(r_o)$ of the Inverting OA	459
11-6	Noninverting OA	459
11-7	Adders, Subtractors, and Multipliers	461
11-8	Differentiator and Integrator	466
11-9	OA-DA Applications	473
11-10	Practical OA-DA—The Data Sheet	478
Summary of Important Terms		491
Problems		494

12 THYRISTORS 499

12-1	Chapter Objectives	499
12-2	Introduction	499
12-3	The Silicon Controlled Rectifier (SCR)	500
12-4	The TRIAC	505
12-5	SCR and TRIAC Specifications	506
12-6	SCR and TRIAC Circuit	511
12-7	Miscellaneous Thyristors	515
Summary of Important Terms		517
Problems		518

13 MISCELLANEOUS SEMICONDUCTOR DEVICES — 521

13-1	Chapter Objectives	521
13-2	Introduction	521
13-3	The Unijunction Transistor (UJT)	522
13-4	UJT Applications	526
13-5	Other Types of UJT	530
13-6	Photosensitive Devices	534
13-7	Semiconductor Light Sources	546
13-8	Other Semiconductor Devices	549
Summary of Important Terms		551
Problems		552

APPENDIX A	Forward Resistance of the Diode, r_f	556
APPENDIX B	Derivation of $S(\beta)$ for the Fixed Bias Circuit with R_E	557
APPENDIX C	$S(\beta)$ for the Voltage Divider Bias Circuit	559
APPENDIX D	The Derivation of g_m	560
APPENDIX E	Exact Solution for Equivalent Circuit of Fig. 5-26b	561
APPENDIX F	The Logarithm (Base 10)	564
APPENDIX G	Antilogarithm (Base 10)	566
APPENDIX H	Largest Possible Power Dissipation in the Transistors of a Class B Two Transistor Circuit	567
APPENDIX I	Integrator and Differentiator General Expressions	569

ANSWERS TO MISCELLANEOUS PROBLEMS — 572

INDEX — 585

SEMICONDUCTOR DEVICES AND CIRCUITS

CHAPTER

1
BASIC CIRCUIT THEOREMS REEXAMINED

1-1 CHAPTER OBJECTIVES

This chapter reviews some of the basic circuit theorems that will be used throughout the text. The student is reintroduced to the various applications of superposition, Thevenin's and Norton's theorems, KCL, KVL, etc. With the extensive use of equivalent circuits in this text anticipated, the student is given an introduction to the idea of circuit equivalence and some simple examples of two- and three-terminal equivalent circuits.

The meaning and significance of linearity is reviewed, with a particular emphasis on the errors introduced when nonlinear circuits are *assumed* to be linear in order to allow the use of the various linear circuit theorems.

1-2 INTRODUCTION

How often, when trying to describe something, have you used such terms as "it is like" or "it is similar to"? How often have you used known objects or images in describing a new and different phenomenon? Instead of dealing with a problem directly, you may often find it useful to first establish an "equivalent" (or analogous) situation and solve the problem in its *equivalent* form.

The idea of "equivalence" occupies an important place in circuit analysis. A complex circuit may be solved by first replacing major portions of the circuit with acceptable, yet very simple, equivalents, and then finding the unknown quantity (current or voltage) in the simplified circuit. The student may recall finding the

2 BASIC CIRCUIT THEOREMS REEXAMINED

parallel equivalent of a *series* R-L circuit or the delta (Δ) equivalent of a wye (Y) connected circuit.

Thevenin's and Norton's Theorems are two more very important examples of equivalent circuit techniques. The equivalent circuit is the basis for most of the analysis and design approaches to active circuits (circuits containing transistors, diodes, or vacuum tubes). It is, therefore, important to review the various equivalent circuit techniques before proceeding to a detailed analysis of circuits and devices. This chapter includes such a review and a reexamination of other network theorems that will prove very useful in the following chapters. Kirchhoff's current and voltage laws as well as the superposition theorems will be reviewed in anticipation of the extensive use of these techniques in later chapters.

1-3 LINEARITY

Most techniques of circuit analysis used in this text are restricted to linear circuits. For that matter, a large portion of the study of electrical-electronic circuits and systems, and their analysis relies heavily on various linear techniques. We will first review the *definition* of *linearity* as applied to circuit theory and then examine the restrictions that the requirement of linearity places on circuits and devices.

Given a circuit with an input E_i and output I_o, see Fig. 1-1, we say that the circuit is *linear* if the *output* is *directly proportional to the input*. Stated another way, the circuit is linear if the relation between input and output is graphically represented by a *straight line*, see Fig. 1-2. This means that a known percent change in the *driving source* (the input) results in an *identical* percent change in the *response* signal (the output). The resistor is a linear component since the current through the resistor is directly proportional to the voltage across it. ($I = V \times 1/R$) Similarly, capacitors and air-core inductors are linear components. At a specific frequency, the voltage and current associated with these components are linearly related. It is clear that circuits containing only resistors, capacitors, and air-core inductors are linear circuits.

Most of the electronic devices that will be analyzed in this text are *not strictly* linear. Current-voltage relations in the vacuum tube, transistor, or diode are *not*, in reality, linear. To make linear network analysis methods applicable to such devices, we "linearize" them. "Linearization" may best be understood when we examine the process of "linearizing" a circle. As shown in Fig. 1-3, the circle may be drawn as an accumulation of straight lines. The few large linear segments shown in Fig. 1-3a result in a poor approximation of the circle. In Fig. 1-3b, a larger

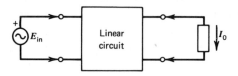

FIGURE 1-1 Linear circuit.

LINEARITY 3

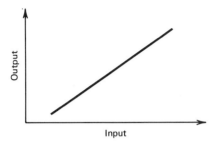

FIGURE 1-2 Linear graph (relating output to input).

number of smaller linear segments more closely and accurately approximates the circle.

A voltage current relation such as shown in Fig. 1-4 may be considered linear over a small portion of the curve, that is, between points A, B. A larger segment of the graph (points A', B') is certainly not linear. Linear analysis is valid only when we can assume the circuit to be linear, and are still satisfied with the accuracy

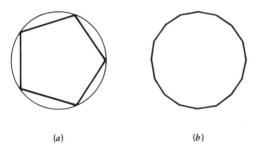

FIGURE 1-3 "Linearizing" the circle. (a) Poor linearization accuracy (few large linear segments). (b) Good linearization accuracy (many small linear segments).

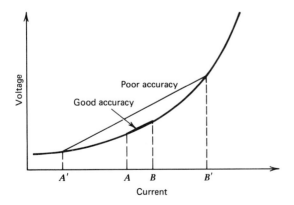

FIGURE 1-4 Nonlinear current-voltage relations linearized.

4 BASIC CIRCUIT THEOREMS REEXAMINED

achieved by our "linearization." It must always be recognized that the assumption of linearity, that is, the "linearization" of circuits that are essentially nonlinear, introduces an error in the analysis. Moreover, the analysis holds only for the operating range for which the linearity assumption was made. We are usually more than willing to accept the inaccuracies involved in order to be able to use the various linear network theorems.

1-4 KIRCHHOFF'S CURRENT LAW (KCL)

KCL states that the algebraic sum of all currents entering and leaving a node equals zero. We define *entering* currents as *positive* currents and *leaving* currents as *negative*. Figure 1-5 shows an application of KCL. Here, I_1 and I_3 are entering currents, hence positive, while I_2, I_4, and I_5 are leaving the node, hence negative. Note that the direction of the unknown current I_5 has been assumed. This is necessary so that we may assign to it the appropriate sign.

Applying KCL

$$I_1 + I_3 - I_2 - I_4 - I_5 = 0$$

solving for I_5

$$I_5 = I_1 + I_3 - I_2 - I_4$$

substituting the values shown in Fig. 1-5, we get

$$I_5 = 2 + 4 - 1 - 7 = -2 \text{ A}$$

The negative sign of I_5 means that the assumed direction of I_5 was incorrect. I_5 is actually a current of 2 A entering the node.

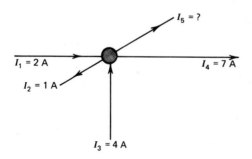

FIGURE 1-5 KCL at a node.

KIRCHHOFF'S VOLTAGE LAW (KVL)

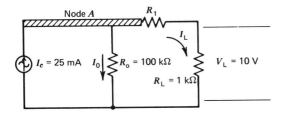

FIGURE 1-6 Application of KCL.

EXAMPLE 1-1

In Fig. 1-6 find I_o

SOLUTION

First find the current I_L

$$I_L = \frac{V_L}{R_L} = \frac{10 \text{ V}}{1 \text{ k}\Omega} = 10 \text{ mA}$$

Applying KCL to node A, we get

$$I_c - I_o - I_L = 0$$

Solving for I_o and substituting the given values, we get

$$I_o = I_c - I_L$$
$$I_o = 25 - 10 = 15 \text{ mA}$$

1-5 KIRCHHOFF'S VOLTAGE LAW (KVL)

Kirchhoff's voltage law (KVL) is to voltage what KCL is to current. KVL states that the algebraic sum of all voltages in a closed loop, summed in one direction, is zero.

When summing voltages, we use the following polarity convention: if the negative terminal of the voltage is encountered *before* the positive terminal, in a particular *summing direction*, that voltage is assigned a positive sign (this is a *plus* voltage or voltage *rise*). If the positive terminal is encountered first, that voltage is assigned a negative sign (this is a *minus* voltage or a voltage *drop*).

In Fig. 1-7 the voltages are E_s, V_1, and V_2, in the closed loop *abca*. Summing in the *clockwise* direction, we get

$$E_s - V_1 + V_2 = 0$$

V_1 has a negative sign since, in the chosen summing direction, the positive end of V_1 is encountered before the negative end. Summing in the *counter-clockwise* direction, we get

$$-E_s - V_2 + V_1 = 0$$

6 BASIC CIRCUIT THEOREMS REEXAMINED

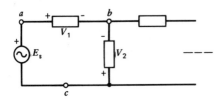

FIGURE 1-7   KVL in a closed circuit.

which can be rearranged as

$$E_s - V_1 + V_2 = 0$$

yielding the same expression as in the *clockwise* summation. Summing voltages around any closed loop, therefore, may be done in *either* direction.

EXAMPLE 1-2

In Fig. 1-8 the directions of all currents is as shown and

$$R_B = 9 \text{ k}\Omega; \quad R_E = 1 \text{ k}\Omega; \quad I_C = 1 \text{ mA} \quad \text{and} \quad E_{in} = 3 \text{ V}$$

Find I_B by the use of KVL and KCL.

SOLUTION

Applying KVL to the closed loop consisting of E_{in}, V_B, and V_E, we get

$$E_{in} - V_B - V_E = 0$$

but since $V_B = I_B \times R_B$ and $V_E = I_E \times R_E$ we get, by substitution,

$$E_{in} - I_B \times R_B - I_E \times R_E = 0$$

Applying KCL to the circuit of Fig. 1-8, we get

$$I_E = I_C + I_B \quad \text{(KCL at the node joining the three currents)}$$

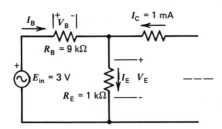

FIGURE 1-8   Application of KVL.

Hence, by substitution in the KVL equation:

$$E_{in} - I_B \times R_B - (I_C + I_B) \times R_E = 0$$
$$E_{in} - I_B \times R_B - I_C \times R_E - I_B \times R_E = 0$$

Solving the above for I_B yields

$$I_B = \frac{E_{in} - I_C \times R_E}{R_B + R_E}$$

substituting the given values yields

$$I_B = \frac{3 - 10^{-3} \times 10^3}{9 \times 10^3 + 10^3} = \frac{3 - 1}{10 \times 10^3} = 0.2 \text{ mA}$$

Note that in Example 1-2 both KCL and KVL were used.

1-6 THEVENIN'S AND NORTON'S THEOREMS

Thevenin's and Norton's theorems allow complex linear circuitry to be replaced by simple equivalent circuits. Both the Thevenin's and Norton's equivalents are 2-*terminal* equivalents. The original complex circuit (which may consist of inductors, capacitors, resistors and voltage and current sources, all linear elements) has only *two* output terminals that are connected to a load or other circuitry (see Fig. 1-9a).

Thevenin's theorem states that the complex circuit may be replaced by an equivalent circuit that consists of a *single voltage* source E_{TH} in *series* with an impedance Z_{TH} (Fig. 1-9b).

Norton's theorem states that the complex circuit may be replaced by an equivalent circuit consisting of a *single current* source I_N in *parallel* with an impedance Z_N (Fig. 1-9c).

In Fig. 1-9, the three circuits (shown boxed-in) are said to be *equivalent* of each other. The term "equivalent" means that any measurement performed on terminal pairs A, B, or C, D, or E, F yields exactly the same values. For example, the voltages measured across all terminal pairs are equal, that is, $V_{AB} = V_{CD} = V_{EF}$. Similarly, the current through load Z is the same assuming the value of Z is identical in each case. It is important to realize that the three circuits inside the boxes are indeed different and have completely different *internal* voltages and currents.

Realizing the meaning of *terminal equivalence*, we will now proceed to find the particular Thevenin's and Norton's values (E_{TH}, Z_{TH}, I_N, and Z_N) satisfying the equivalence. The Thevenin's equivalent, E_{TH} and Z_{TH}, can be obtained from two measurements (or calculations) on the original circuit, the *open circuit voltage* V_{OC}

8 BASIC CIRCUIT THEOREMS REEXAMINED

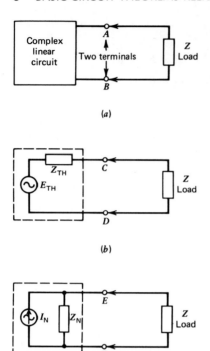

FIGURE 1-9 Two-terminal circuits. (a) Original circuit. (b) Thevenin's equivalent circuit. (c) Norton's equivalent circuit.

and the terminal impedance Z_{AB}. Open circuit voltage is defined as the voltage between terminals A, B with the load completely removed (Fig. 1-10a). The *terminal impedance* is defined as the impedance between A, B with the load removed and all internal ideal *current* sources replaced by *open* circuits and all ideal *voltage* sources replaced by *shorts*. (We refer to this last operation as "source elimination.") We can show that

$$E_{TH} = V_{OC} \quad \text{volts (V)} \tag{1-1}$$

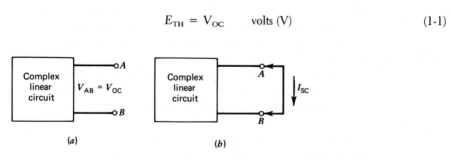

FIGURE 1-10 V_{OC} and I_{SC} computations. (a) Circuit showing V_{OC}. (b) Circuit showing I_{SC}.

THEVENIN'S AND NORTON'S THEOREMS 9

The open circuit voltage (load removed) measured at terminals C, D (Fig. 1-11a) is equal to E_{TH}. There is no voltage drop across Z_{TH}. This E_{TH} is also equal to the open circuit voltage across terminals A, B of the original circuit. To obtain E_{TH} it is only necessary to measure or calculate the open circuit voltage of the original circuit. (The significance of Z_{AB} is discussed later.)

To obtain the Norton's equivalent, namely, the values I_N and Z_N (Fig. 1-9c), two measurements (or calculations) on the original circuit are required. It is necessary to find the short circuit current I_{SC} and the terminal impedance Z_{AB}. The short circuit current I_{SC} is the current through a short connected between terminals A, B (Fig. 1-10b). The terminal impedance Z_{AB} is the *same* impedance discussed in connection with the Thevenin's equivalent. It can be shown that

$$I_N = I_{SC} \quad \text{amperes (A)} \tag{1-2}$$

The short circuit current (load replaced by a short) measured on the Norton's equivalent cirucit, I_{EFSC} (Fig. 1-11b) is equal to I_N. This is also equal to the short circuit current in the original circuit. Since all three circuits in Fig. 1-9 are equivalent, the terminal impedance of these circuits must be equal. For the Norton's circuit, the terminal impedance is Z_N, since the current source I_N is replaced by an open circuit. (Note that the load is removed.) For the Thevenin's circuit, the terminal impedance is Z_{TH} (since E_{TH} is replaced by a short); hence,

$$Z_{AB} = Z_N = Z_{TH} \quad \text{ohms } (\Omega) \tag{1-3}$$

The *same value impedance is used in each* equivalent circuit. Since the Thevenin's and Norton's circuits are both equivalents of the original circuit they must all be

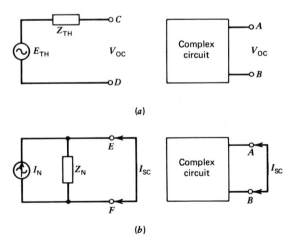

FIGURE 1-11 Open circuit and short circuit equivalents. (a) Open circuit voltages. (b) Short circuit currents.

10 BASIC CIRCUIT THEOREMS REEXAMINED

equivalent of each other,

$$I_N = \frac{E_{TH}}{Z_N} = \frac{E_{TH}}{Z_{TH}} \tag{1-4}$$

Equation (1-4) shows that the Norton's equivalent can be obtained from the Thevenin's equivalent, and vice versa. Z_{TH} (or Z_N) can be expressed in terms of the open circuit voltage and the short circuit current. From Eq. (1-4)

$$Z_N = Z_{TH} = \frac{E_{TH}}{I_N}$$

Since $E_{TH} = V_{OC}$, and $I_N = I_{SC}$

$$Z_N = Z_{TH} = \frac{V_{OC}}{I_{SC}} \tag{1-5}$$

Equation (1-5) allows the calculation of Z_N or Z_{TH} from I_{SC} and V_{OC}.

To obtain either the Thevenin's or Norton's equivalent circuit, we need evaluate only two of the above three parameters: Z_{AB}, the terminal impedance; V_{OC}, the open circuit voltage; I_{SC}, the short circuit current. In the following two examples, resistive circuits are used for the sake of simplicity. The theorems hold, as stated above, for all the linear circuits (R, L, and C).

EXAMPLE 1-3

Calculate the current I_L in Fig. 1-12a for the following values of R_L
 (a) $R_L = 10\ \Omega$
 (b) $R_L = 20\ \Omega$
 (c) $R_L = 30\ \Omega$

SOLUTION

This problem can be solved without the use of Thevenin's or Norton's equivalent. But this means that it would be necessary to compute a three-way current division for each value of R_L in 3 different cases. (The 1.5 A source current divides three ways at the junction of R_1 and R_L.) Instead, we find the Thevenin's equivalent for terminals A, B (R_L removed) and then compute the three I_L currents.

To find the Thevenin's equivalent, we chose to compute Z_{AB} and I_{SC}. I_{SC} can be found by inspection.

$$I_{SC} = 1.5\ \text{A}$$

When A, B in Fig. 1-12a are shorted, the source current flows only in the short (since R_2 and R_3 are paralleled by a short circuit). Z_{AB} is given by

$$Z_{AB} = R_2 \| R_3\ (R_2 \text{ in parallel with } R_3)$$

THEVENIN'S AND NORTON'S THEOREMS

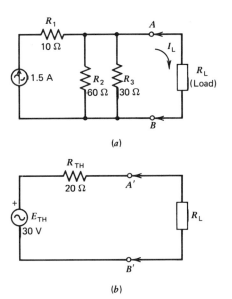

FIGURE 1-12 Application of Thevenin's theorem. (a) Original circuit. (b) Thevenin's equivalent circuit.

This is obvious since the 1.5 A ideal current source must be "eliminated" (open circuit the current source). Therefore,

$$Z_{AB} = \frac{60 \times 30}{60 + 30} = 20 \; \Omega$$

From Eq. (1-3)

$$Z_{TH} = Z_N = Z_{AB} = 20 \; \Omega$$

From Eq. (1-5)

$$E_{TH} = Z_N \times I_{SC} = 20 \; \Omega \times 1.5 \; A = 30 \; V$$

Figure 1-12b shows the Thevenin's equivalent of the circuit of Fig. 1-12a.

The current I_L for all the different values of R_L may now be readily found using simple series circuits analysis.

(a) $R_L = 10 \; \Omega; \; I_L = \dfrac{E_{TH}}{R_T} = \dfrac{30 \; V}{(20 + 10) \; \Omega} = 1 \; A$

(b) $R_L = 20 \; \Omega; \; I_L = \dfrac{E_{TH}}{R_T} = \dfrac{30 \; V}{(20 + 20) \; \Omega} = 0.75 \; A$

(c) $R_L = 30 \; \Omega; \; I_L = \dfrac{E_{TH}}{R_T} = \dfrac{30 \; V}{(20 + 30) \; \Omega} = 0.6 \; A$

12 BASIC CIRCUIT THEOREMS REEXAMINED

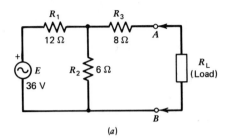

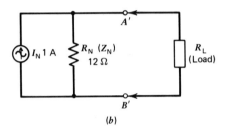

FIGURE 1-13 Application of Norton's theorem. (a) Original circuit. (b) Norton's equivalent circuit.

EXAMPLE 1-4

Find the Norton's equivalent circuit for terminals A, B in Fig. 1-13a.

SOLUTION

We chose to find V_{OC} and Z_{AB}. If we use the voltage divider rule

$$V_{OC} = E \times \frac{R_2}{R_1 + R_2} = 36 \text{ V} \times \frac{6 \text{ }\Omega}{(12 + 6) \text{ }\Omega} = 12 \text{ V}$$

V_{OC} equals the voltage across R_2 since no voltage drop occurs across R_3 (no current through R_3 when R_L is removed).

$$Z_{AB} = R_3 + R_1 \| R_2 \quad (E \text{ is replaced by a short})$$

$$Z_{AB} = R_3 + \frac{R_1 \times R_2}{R_1 + R_2} = 8 + \frac{12 \times 6}{12 + 6} = 12 \text{ }\Omega = Z_N$$

$$I_N = \frac{V_{OC}}{Z_N} = \frac{12 \text{ V}}{12 \text{ }\Omega} = 1 \text{ A}$$

1-7 MAXIMUM POWER THEOREM

It is often desirable to maximize the power delivered to a load connected to a complex circuit. The maximum power theorem states that maximum power is delivered to the load if the load impedance is the *conjugate* of the Thevenin's (or

Norton's) impedance. If, for example, $Z_N = 10 - j6$ (a capacitive driving source), the load must be selected so that $Z_{LOAD} = 10 + j6$. Z_L is the conjugate of Z_N if the real part of Z_L equals the real part of Z_N and the imaginary part of Z_L equals the negative (opposite sign) of the imaginary part of Z_N. In the case of pure resistive circuits, this means $R_L = R_N$ (load resistance equals the equivalent source resistance).

EXAMPLE 1-5

V_{OC} and I_{SC} were measured on the complex circuit of Fig. 1-14 at terminals A, B, yielding

$$V_{OC} = 50\angle 0° \text{ V} \quad \text{and} \quad I_{SC} = 10\angle -53° \text{ A}.$$

Calculate
(a) Z_L for maximum power.
(b) The real power dissipated in Z_L.

SOLUTION

We first get the Thevenin's equivalent

$$E_{TH} = V_{OC} = 50\angle 0° \text{ V} \tag{1-1}$$

$$Z_{TH} = \frac{V_{OC}}{I_{SC}} = \frac{50\angle 0° \text{ V}}{10 - \angle 53° \text{ A}} = 5\angle 53° \text{ }\Omega \tag{1-5}$$

To find the desired Z_L, we must obtain Z_{TH} in rectangular form

$$Z_{TH} = 5\angle 53° = 5(\cos 53°) + j5 (\sin 53°) = 3 + j4 \text{ }\Omega$$

For maximum power, therefore, Z_L is the conjugate of Z_{TH}.
(a) $Z_L = 3 - j4 = 5\angle -53° \text{ }\Omega$
(b) To find the power, we find the current in the selected Z_L (Fig. 1-14b)

$$I_L = \frac{E}{Z_N + Z_L} = \frac{50\angle 0° \text{ V}}{(3 + j4 + 3 - j4)} = \frac{50}{6} = 8.33 \text{ A}$$

$$P_{real} = I^2 \times (\text{real part of } Z_L) = 8.33^2 \times 3 = 208.3 \text{ W}$$

(Real power is dissipated by resistive elements only.)

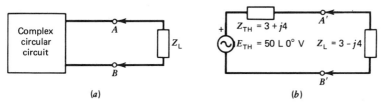

FIGURE 1-14 Maximum power theorem. (a) Original circuit. (b) Thevenin's equivalent with selected load.

1-8 THE SUPERPOSITION THEOREM

An important and powerful theorem (so often used in circuit analysis that one does not even realize it is being used) is the *superposition theorem*. The *superposition theorem*, when applied to a multisource circuit, permits the computation of voltage across (or current through) any single circuit branch (or circuit component). This individual voltage is an algebraic sum of partial voltages, each computed separately, for each source while all others are "eliminated" (open ideal current sources replace ideal voltage source by short circuits). If we use this technique, each partial voltage (or current) across the particular branch is computed as if only a single source is present in the circuit. The technique is best demonstrated by an example. As in previous examples, we demonstrate the superposition theorem on purely resistive circuitry for the sake of clarity and simplicity. But the theorem holds for *any* linear circuit.

EXAMPLE 1-6

Find the voltage V_{BN} (across R_2 in parallel with the current source) in Fig. 1-15a. Note that all sources are ac sources and of the same frequency.

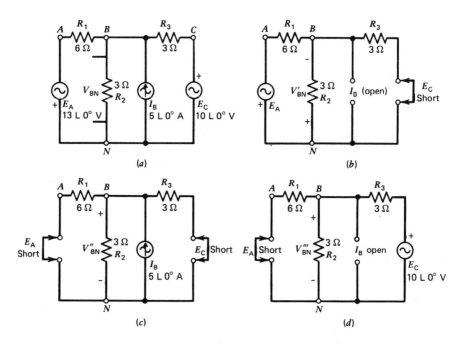

FIGURE 1-15 Application of superposition. (a) Original circuit. (b) E_A acting alone (compute V'_{BN}). (c) I_B acting alone (compute V''_{BN}). (d) E_C acting alone (compute V'''_{BN}).

THE SUPERPOSITION THEOREM 15

SOLUTION

The three steps involved in the application of superposition are shown in Figs. 1-15b,c,d. Figure 1-15b shows E_A alone, all other sources having been "eliminated." For this figure we compute V'_{BN}, the voltage due to E_A operating alone. Similarly in Fig. 1-15c, I_B is operating alone and we compute V''_{BN}. In Fig. 1-15d, E_C is operating alone and we compute V'''_{BN}. The actual voltage V_{BN} is the algebraic sum of the three partial voltages.

$$V_{BN} = V'_{BN} + V''_{BN} + V'''_{BN}$$

1. Calculation of V'_{BN} (Fig. 1-15b) due to E_A acting alone:
The total resistance R_{TA} "seen" by the source E_A is

$$R_1 + R_2 \| R_3 = 6 + \frac{3 \times 3}{3 + 3} = 7.5 \, \Omega$$

The voltage V'_{BN} is the voltage across $R_2 \| R_3$ or

$$V'_{BN} = I_T(R_2 \| R_3) = \frac{EA}{R_{TA}} \times (R_2 \| R_3)$$

Substituting values, we get

$$V'_{BN} = \frac{-13\angle 0°}{7.5} \left(\frac{3 \times 3}{3 + 3} \right) = \frac{-13\angle 0°}{7.5 \, \Omega} \times 1.5 = -2.6\angle 0 \text{ V}$$

2. Calculation of V''_{BN} (Fig. 1-15c) due to I_B acting alone:
The voltage V''_{BN} is the voltage across the parallel combination of R_1, R_2, R_3 (Fig. 1-15c). The total current through this total resistance is the source current $I_B = 5\angle 0°$ A. Hence,

$$V''_{BN} = I_B \times (R_1 \| R_2 \| R_3) = 5\angle 0 \times (6\|3\|3)$$
$$= (5\angle 0° \text{ A})(1.2 \, \Omega) = 6\angle 0 \text{ V}$$

3. Calculation of V'''_{BN} (Fig. 1-15d) due to E_C acting alone:
The total resistance R_{TC} seen by E_C is

$$(R_3 + R_1 \| R_2) = 3 + \frac{6 \times 3}{6 + 3} = 5 \, \Omega$$

V'''_{BN} is the voltage across $R_1 \| R_2$ or

$$V'''_{BN} = I_T(R_1 \| R_2) = \frac{E_C}{R_{TC}} \times (R_1 \| R_2)$$

16 BASIC CIRCUIT THEOREMS REEXAMINED

Substituting, we get

$$V'''_{BN} = \frac{10\angle 0° \text{ V}}{5 \text{ }\Omega} \times \frac{6 \times 3}{6 + 3} = 4\angle 0° \text{ V}$$

Using the superposition theorem, the actual voltage V_{BN} is the algebraic sum or:

$$V_{BN} = V'_{BN} + V''_{BN} + V'''_{BN}$$
$$= -2.6\angle 0° + 6\angle 0° + 4\angle 0° = 7.4\angle 0° \text{ V}$$

In Ex. 1-6, for simplicity, it was assumed that all sources have the same frequency. The superposition theorem can be applied to circuits that include sources of different frequencies—in particular, circuits that include both ac and dc sources. The final solution in these cases, however, will consist of a sum of voltages such as a dc voltage added to an ac voltage. There is *no single* numerical answer, since voltages (or currents) of different frequencies cannot be added to yield a single numerical result. In practice, two separate computations are performed, an ac computation (at a specific frequency or range of frequencies) and a dc computation. (This mixture of ac and dc voltages and currents is present in most transistor circuits.)

EXAMPLE 1-7

Find the total curent I_3 (a complex waveform consisting of I_{dc} and I_{ac}) through R_3 in Fig. 1-16a. For the ac calculations use $f = 1000$ Hz (frequency of 1 kHz).

SOLUTION

1. I_{dc} calculations due to E_{CC} acting alone:

The equivalent circuit for dc purposes is shown in Fig. 1-16b. Note that superposition is being used. E_{in} is replaced by a short and only E_{CC} is in the circuit. However, the shorted E_{in} has no effect on the equivalent circuit. At dc (frequency of zero) the capacitive reactance X_c is *infinite*. ($X_c = \infty$; the capacitor is an open circuit.) I_{dc} in R_3 is found as follows:
The total resistance connected to E_{CC} is:

$$R_{dc} = R_1 + R_2 \| R_3$$

$$= 95 \text{ k}\Omega + \frac{10 \text{ k}\Omega \times 10 \text{ k}\Omega}{10 \text{ k}\Omega + 10 \text{ k}\Omega} = 100 \text{ k}\Omega$$

$$I_t = \frac{20 \text{ V}}{100 \text{ k}\Omega} = \frac{20}{10^5} = 200 \times 10^{-6} = 200 \text{ }\mu\text{A}$$

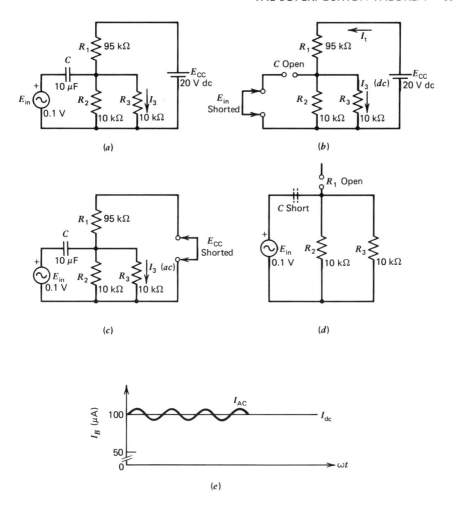

FIGURE 1-16 ac and dc circuit analysis. (a) Original circuit. (b) dc equivalent circuit. (c) ac equivalent circuit. (d) *Approximate* ac equivalent circuit (R_1 neglected). (e) Resultant current in R_3 versus time.

Using the current divider rule for simple parallel circuits, we get

$$I_{dc} = I_t \frac{R_2}{R_2 + R_3}$$

$$= 200 \ \mu A \ \frac{10 \ k\Omega}{(10 \ k\Omega + 10 \ k\Omega)} = 100 \ \mu A$$

$I_{dc} = 100 \ \mu A$ (current in R_3 due to E_{CC} acting alone).

2. I_{ac} calculations due to E_{in} acting alone:

The ac equivalent circuit is shown in Fig. 1-16c. By superposition, only E_{in} is in the circuit. X_C is evaluated at the given frequency:

$$|X_c| = \frac{1}{\omega_C} = \frac{1}{6.28 \times 1000 \times 10 \times 10^{-6}} = 15.9 \ \Omega$$

Inspecting the circuit (Fig. 1-16c) carefully, we note that R_1, R_2, R_3 are all in parallel, and

$$R_t = R_1 \| R_2 \| R_3 = 10 \text{ k} \| 10 \text{ k} \| 95 \text{ k} = 4.75 \text{ k}\Omega$$

The series combination of X_C and R_t act as a voltage divider for E_{in}, where $|X_C|$ = 15.9 Ω and R_t = 4.75 kΩ. Since $|X_C| \ll R_t$ (X_C is much smaller than R_t), the drop across X_C is negligible. X_C may be approximated, therefore, by a short. (This is a good approximation *only* because $X_C \ll R_t$.) Shorting X_C yields the approximate equivalent circuit of Fig. 1-16d. (R_1 is shown as an open circuit, hence neglected, since $R_1 \gg R_2 \| R_3$, $R_1 \gg 10$ kΩ$\|$10 kΩ.) Since E_{in} is connected directly across R_3, the current in R_3 is:

$$I_{ac} = \frac{E_{in}}{R_3} = \frac{0.1}{10 \text{ k}} = 10 \ \mu A$$

The total current through R_3 is the algebraic sum of the superposition currents: $I_3 = I_{dc} + I_{ac} = 100 \ \mu A$ (dc) and 10 μA (ac). This is a 10 μA ac current superimposed on a 100 μA dc current, as shown in Fig. 1-16e.

1-9 EQUIVALENT CIRCUITS

The use of equivalent circuits has two basic purposes:

(a) attain simplicity;
(b) permit the use of standard network analysis methods, such as superposition, KVL, and KCL.

A fringe benefit often resulting from the equivalent circuit representation is an understanding of the operation of the device and/or circuit.

1-9.1 Passive Equivalents

Before applying linear network methods, such as equivalent circuits using Thevenin's and Norton's theorems or superposition, it is first necessary to make sure that the network being considered is indeed linear. Circuits containing no devices such as diodes, transistors, or vacuum tubes are linear. Consequently, the equivalent circuits for this class of circuits have no inherent representation errors. Any such circuit can be represented by a precise Thevenin's or Norton's equivalent. The representation depends solely on our desired numerical accuracy. In many cases,

EQUIVALENT CIRCUITS

the use of an *approximate* equivalent is justified. Example 1-8 should clarify this technique.

EXAMPLE 1-8

Find the approximate impedance Z_{AB} of the circuit of Fig. 1-17a at the following frequencies:
(a) $f = 1.59$ Hz;
(b) $f = 159$ kHz.
(c) Find the exact value of Z_{AB} for the two frequencies of parts (a) and (b).

SOLUTION

For all parts it is first necessary to find the reactance X_C at the given frequency.

(a) $X_C = \dfrac{1}{2\pi f_c} = \dfrac{1}{2\pi \times 1.59 \times 0.1 \times 10^{-6}} = 1$ MΩ, at 1.59 Hz

The 1 MΩ reactance (X_C) is in parallel with a 1 kΩ resistance in Fig. 1-17a. The total impedance Z_{AB} of this parallel combination at 1.59 Hz may be approx-

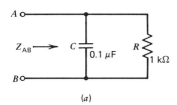
(a)

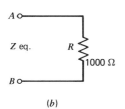
(b)

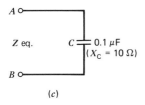
(c)

FIGURE 1-17 Simple approximation. (a) Original circuit. (b) Approximate equivalent, $f = 1.59$ H$_z$. (c) Approximate equivalent, $f = 159$ kH$_z$.

20   BASIC CIRCUIT THEOREMS REEXAMINED

imated by an equivalent impedance Z_{eq}; $Z_{eq} \cong 1000 \; \Omega$. When this approximation was made, an error of about 0.1% was introduced (or the ratio of R/X_c). The equivalent circuit is a 1000 Ω resistor (Fig. 1-17b).

(b) $X_C = \dfrac{1}{2\pi \times 159 \times 10^3 \times 0.1 \times 10^{-6}} = 10 \; \Omega$, at 159 kHz

Here, the approximate equivalent impedance is X_C since $|X_C| \ll R$ (Fig. 1-17c). The error involved in this approximation is about 1% ($|X_C|/R \times 100 = 10/1000 \times 100 = 1\%$).

(c) The exact value of Z_{AB} is $Z_{AB} = \dfrac{R \times (-jX_C)}{R - jX_C}$

For part (a)

$$Z_{AB} = \frac{1000 \times (-j10^6)}{1000 - j10^6} = \frac{10^9 \angle -90}{(10^6 + 0.5 \angle -89.9°)} \approx 10^3 \angle -0.1° \; \Omega$$

For part (b)

$$Z_{AB} = \frac{1000(-j10)}{1000 - j10} = \frac{10^4 \angle -90°}{1000 \times .05 \angle -5°} \approx 10 \angle -89.5° \; \Omega$$

Note, the approximate values found in (a) and (b) above are reasonably accurate for our purposes and much simpler to obtain.

As Example 1-8 shows, the equivalent circuit depends on the frequency used. It is necessary to obtain a new equivalent circuit every time the frequency is changed. In particular, circuits that use both ac and dc will yield two distinct equivalent circuits: the ac and dc equivalents.

1-9.2   Device Equivalent

The devices discussed in this text are all inherently *nonlinear*. Since most of the previous analysis methods (to be applied to these devices) are applicable to *linear* circuits *only*, we must first "linearize" the device (see Section 1-1). Such "linearization" introduces an error that can be estimated. For the most part, we can only estimate the maximum error involved. If the nonlinearity of the device (over the range of operations involved) is, say, no more than 1%, it is safe to assume that the linearization error does not exceed 1% (nonlinearity is percent deviation from a straight line). Most devices can be "linearized" over a desired *small* operating range with a very small error. In Chapter 3, which deals with transistors, the term "*small*" will be quantified.

In addition to the "linearization" error (over which we have little control), we may choose to use an *approximate* device equivalent, rather than a precise one. In doing this, one may choose to neglect some device characteristics, which are

EQUIVALENT CIRCUITS   21

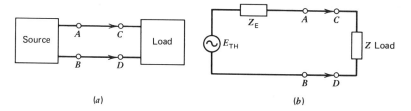

FIGURE 1-18 Source and load equivalents. (a) Circuit block diagram. (b) Equivalent circuit (Thevenin's equivalent for source).

considered unimportant, in a particular application. This may lead to an even simpler approximate equivalent circuit.

Detailed equivalent circuits for devices are discussed in the next two chapters.

1-9.3 Terminal Equivalence

Section 1-6 presented the Norton's and Thevenin's equivalents. These can be used only for a particular *pair* of terminals. Most circuits can usually be classified as either driving circuits (sources) or loads. Both source and load may be complex circuits. Figure 1-18a shows a load (terminals C, D) connected to a source (terminals A, B). The load may be a passive circuit or the two input terminals of an active circuit (amplifier). The source may be an ac or dc source or an amplifier used to drive the load. No matter what the complexity of the source and load circuits we will endeavor to replace both with appropriate equivalent circuits (Fig. 1-18b). For the source, the terminals A, B will serve as the equivalent terminals, either for obtaining a Norton's equivalent or a Thevenin's equivalent. The load equivalent will be obtained by measurements or calculations associated with terminals C, D. (For the above equivalence calculations, the source and load are disconnected.)

While the source is replaced by its Norton's or Thevenin's equivalent, the load is replaced by its equivalent impedance. This means that the load is assumed to be *passive* with no sources. In cases where the load is the input terminals of the next amplifier stage, the equivalent impedance is referred to as the *input impedance* of the next stage. Again, two terminal equivalents only are being considered here. Both the source and the load are accessible via two terminals only.

Many devices have three or more terminals. The equivalent circuits of such devices will contain as many terminals as the actual device contains. A transistor that has three terminals will be represented by an equivalent that has three terminals. The behavior of the equivalent representation must be the same as that of the device itself under all specified operating conditions. In Fig. 1-19a the voltage V_{CE} measured at terminals C, E of the device must be the same as $V_{C'E'}$ measured at corresponding terminals of the equivalent circuit (Fig. 1-19b). Clearly, in Fig. 1-19, the corresponding terminals B, E and B', E' must be connected to identical sources. It should be understood that the equivalent representation must be one

22 BASIC CIRCUIT THEOREMS REEXAMINED

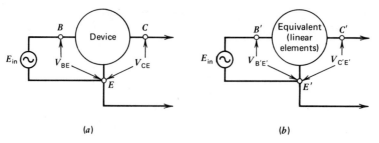

FIGURE 1-19 Three-terminal equivalent. (a) Device connections. (b) Equivalent representation (utilizing linear elements).

that can be analyzed by standard network techniques. (The $\Delta - Y$ conversion is another three terminal equivalence.) The device (Fig. 1-20a) could have been viewed as having *four* terminals or *two terminal pairs*: BE and CE. The terminal E plays a double role. It is part of the input terminals (to which the source is connected) and the output terminals (which may be connected to the next stage or the load). With this approach conventional 2-terminal analysis may be used to obtain Thevenin's and/or Norton's equivalent circuits for each pair (Fig. 1-20b). An important aspect of the equivalent circuits of Fig. 1-20b is that both E_{TH} and I_N are *dependent* sources. They depend on current or voltage in other parts of the circuit.

1-10 DEPENDENT SOURCES

A "dependent" source is one whose specific value of voltage or current depends on a current or voltage in another part of the circuit. In Fig. 1-21a, E is an independent *voltage* source and I_1 is an independent *current* source. However, I_2 depends on I_1; hence, it is shown as a *dependent* source. When superposition is applied to this circuit, we cannot "eliminate" I_1 without eliminating I_2 as well. Consequently, either both I_1 and I_2 are present or both "eliminated." Clearly, the only way I_2 can be assumed to be zero is if I_1 is assumed to be zero. Conversely, if we assume I_1

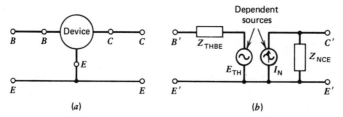

FIGURE 1-20 Four-terminal equivalent. (a) Device. (b) Equivalent circuit (E' common terminal).

SUMMARY OF IMPORTANT TERMS 23

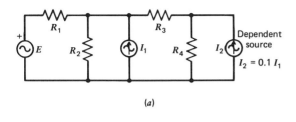

(a)

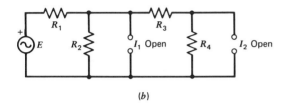

(b)

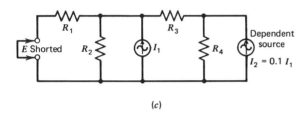

(c)

FIGURE 1-21 Dependent sources. (a) Original circuit. (b) Superposition applied: E acting alone. (c) Superposition: I_1 and I_2 acting alone.

$= 0$, this implies that $I_2 = 0$. The two circuits that will result when superposition is applied to the circuit of Fig. 1-21a are shown in Figs. 1-21b and c. The circuit of Fig. 1-21c must be solved with both I_1 and I_2 present.

SUMMARY OF IMPORTANT TERMS

Dependent Source A voltage or current source whose value depends on a voltage or current somewhere else in the circuit.

KCL Kirchhoff's Current Law. The algebraic sum of all the currents *entering* a node (with proper polarities for *entering* and *leaving* currents) equals zero.

KVL Kirchhoff's Voltage Law. The algebraic sum of all the voltages around a closed loop (with proper polarities) is zero.

Linear (relation) A relation between two variables (for example, voltage and current) such that the plot of this relation is a straight line.

Norton's circuit A two-terminal equivalent circuit consisting of a current source in parallel with an impedance Z_N (or R_N).

Open circuit voltage (V_{OC}) The voltage across two terminals with the load driven by these terminals removed ($V_{OC} = E_{TH}$).

24 BASIC CIRCUIT THEOREMS REEXAMINED

Short circuit current (I_{SC}) The current through a short placed between two terminals ($I_{SC} = I_N$).

Superposition A theorem that permits calculation (or measurement) of voltages and currents in a multisource circuit by considering the effects of each source acting alone.

Thevenin's circuit A two-terminal equivalent circuit that consists of a voltage source E_{TH} in series with an impedance Z_{TH} (or R_{TH}).

I_{SC} See short circuit current.

V_{OC} See open circuit voltage.

PROBLEMS

1-1. Figure 1-22 shows a voltage-current relationship and two straight line approximations.

    (a) Find the maximum error introduced by approximation A (in volts and amps).
    (b) Find the maximum error in approximation B (in volts and amps).
    (c) Draw a better *two segment* approximation and calculate its maximum error.

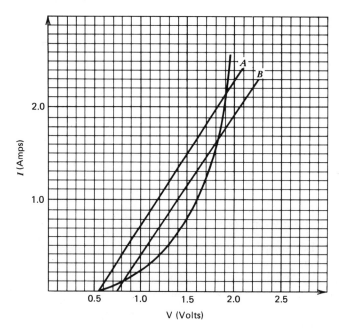

FIGURE 1-22

1-2. Plot the following mathematical V-I relationships and determine which are linear.

    (a) $I = V/R$ $R = 10$
    (b) $I = 5(1 - V/3)^2$

(c) $I = 0.1 \times 3$ V
(d) $I = 6(1 - V)$
(e) $V = IR + 6 \qquad R = 100$
(f) $V \times I = 20$

1-3. Find I_5 in Fig. 1-23. $I_1 = 20$ mA, $I_2 = 6$ mA, $I_3 = 8$ mA.

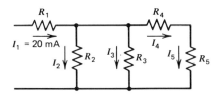

FIGURE 1-23

1-4. Find I_x in the diagram of Fig. 1-24.

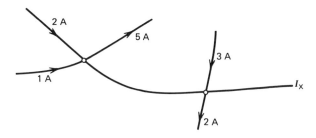

FIGURE 1-24

1-5. Find V_o in Fig. 1-25.

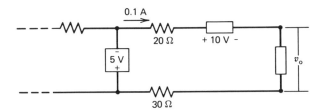

FIGURE 1-25

1-6. Use KVL to find I for the circuits of Fig. 1-26. (The resistive voltage drops are $I \times 10$ and $I \times 20$.)

26 BASIC CIRCUIT THEOREMS REEXAMINED

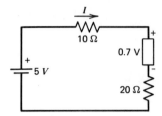

FIGURE 1-26

1-7. Find the Thevenin's and Norton's equivalents for points A-B in the circuit of Fig. 1-27 for the following source voltages (remove R_L):

    (a)   e = 10 sin 10t
    (b)   e = 10 sin 100t
    (c)   e = 10 sin 1000t

You may use approximations wherever appropriate.

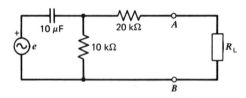

FIGURE 1-27

1-8. Find I_D in Fig. 1-28. Convert the voltage divider into its Thevenin equivalent and use KVL.

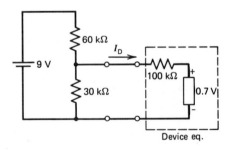

FIGURE 1-28

1-9. Find V_3 in Fig. 1-29 using superposition.

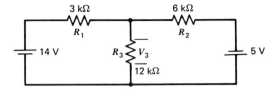

FIGURE 1-29

1-10. Plot V_o for Fig. 1-30 using superposition. (Use ωt as the abcissa.) Consider the capacitor a short for the ac signal.

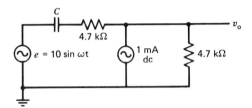

FIGURE 1-30

1-11. A three-terminal equivalent circuit with a dependent current source is shown in Fig. 1-31. Find V_o. (Hint: Find I_1 in loop 1; then find I. Find V_o using Ohm's Law.)

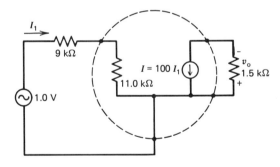

FIGURE 1-31

CHAPTER

2

SEMICONDUCTOR DIODES: PHYSICS AND CHARACTERISTICS

2-1 CHAPTER OBJECTIVES

From this chapter, the student will obtain the necessary background in semiconductor physics to be able to understand the operation of the diode. This consists of a brief description of the atomic structure and the energy considerations relating to this structure.

Following this brief physics review, a description of the operation of the diode is given. This part includes a discussion of some of the important characteristics of the diode and their representations, as well as the manufacturer's specifications of diode performance. This gives the student an appreciation of the significance of these parameters in various applications.

The student, having mastered this chapter, will be able to analyze most simple diode circuits. Three methods of analysis are presented: an approximate method, neglecting diode resistance; a graphical method, relying on the load-line technique; and an iterative method, which requires repeated recomputation of parameters.

2-2 INTRODUCTION

The diode plays a most important role in modern technology. Practically every electronic system, whether it be the hi-fi amplifier or the computer, uses diodes in some form or another.

The diode may be described as a two-terminal device, which is *polarity* sensitive. That is, the current through the diode can flow only in one direction. (This is a simplified and idealized description.)

The first *vacuum* diode (based on the phenomenon of thermionic emission—the emission of electrons from a heated metallic wire) dates back to the early 1900s. About 30 years later, the semiconductor diode was introduced commercially (the first such diode, the "cat whiskers" diode, was actually demonstrated in 1905. The technology of *germanium* and *silicon* semiconductors came into being in the 1930s). The *semiconductor* diode, sometimes called the *solid state* diode, has many important advantages over the vacuum diode. The solid state diode is much smaller in size, less expensive, and higly reliable. At present, vacuum diodes are used on very rare occasions.

2-3 THE PHYSICS OF SEMICONDUCTORS

The behavior of the solid state diode (and later the transistor) can be understood through the analysis of the atomic structure of the materials used in their construction. It is necessary, before proceeding with this analysis, that some aspects of general atomic theory be understood.

2-3.1 Atomic Structure of Material[1]

The atom consists of a nucleus and "orbiting" electrons (negative electrical charges). The nucleus contains protons (positive electrical charges) and a number of other particles that are not relevant to our discussion. To maintain a neutral electrical state the number of electrons orbiting the nucleus is the same as the number of protons inside the nucleus.

The electron orbits are organized in "shells." A "shell" may consist of many circular paths that may be occupied by electrons. Figure 2-1 shows the structure of a *germanium* atom. The electron orbits or shells are named K, L, M, and N. Note that the K orbit contains 2 electrons, the L contains 8, the M shell has 18 electrons, and the outer shell, the N shell, contains 4 electrons. The outer shell contains the *valence* electrons. Our interest is going to center around the outer shell; hence it is expanded in Fig. 2-2. Here, we see clearly that each electron is orbiting the nucleus in a unique path. The radius of each orbit is different. No two electrons may occupy the same orbit. (If, for example, the first-lowest-orbit has a radius of 5×10^{-9} cm, the next orbit will have a radius of approximately 20×10^{-9} cm.)

Different elements have a different number of protons and electrons and hence may have one, two, or up to seven shells, named K, L, M, N, O, P, Q. Theoretical calculations have shown that there is a maximum number of electrons that may occupy a shell. The K shell may have a maximum of two electrons, the L-8, M-18, N-32, O-50, P-72, and Q-98. That is not to say that a shell is necessarily fully occupied. Experiments so far have not found an element with more than 31 electrons in the O shell, more than 10 in the P shell, or more than 2 in a Q shell. This clearly indicates that shells may be fully or partially occupied by electrons.

---

[1] Postulated by N. Bohr in 1913.

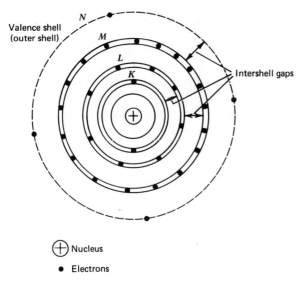

FIGURE 2-1 Two-dimensional view of the structure of a germanium atom, showing four valence electrons.

The shells are separated from each other by radial gaps in which no orbits may exist. (See Fig. 2-1.)

2-3.2 Electron Energy—Energy Bands

The structure of the atom is maintained by a balance between a number of forces: the force of *attraction* between electrons in orbit and protons in the nucleus, the force involved in the motion of the electrons in their path, etc. These forces vary with distance from the nucleus. These forces imply that there is a certain amount of energy associated with each electron. This energy, like the forces acting on the electrons, varies with the radius of the orbit of the electron. We may then speak of specific electron energy, the value of which is unique for each orbit. The L shell, for example, contains eight possible orbits, which may not be occupied by electrons, for a particular element. The L shell is said to consist of eight discrete energy levels. (See Fig. 2-3.)

As the radius of orbit is increased, the energy level is also increased. The outer shells have higher energy levels than the inner shell. Furthermore, the valence band of energies (the outer shell of an atom of a particular element) are the highest for the particular element.

We may talk about energy bands, whether they are occupied by electrons or not, and we can calculate, theoretically, the energy levels associated with each band.

As noted before, electron orbits exist at specific radii and an empty space—no orbits—separates the different shells. The gap in which no orbits are possible is called a *forbidden* energy gap. Thus, electron shells are separated by forbidden

THE PHYSICS OF SEMICONDUCTORS   31

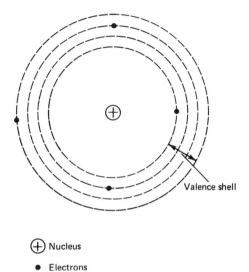

FIGURE 2-2   Expanded view of the valence shell of a germanium atom.

energy gaps. Clearly, if an electron changes its orbit, it also changes its energy level. A *reduction* in the orbit radius of the electron will cause the electron energy to *decrease* and the difference of energies will be released in the form of *radiated* energy. To move the electron into a larger orbit (say from the K shell to the L shell) requires a discrete amount of energy. This may be done by providing heat

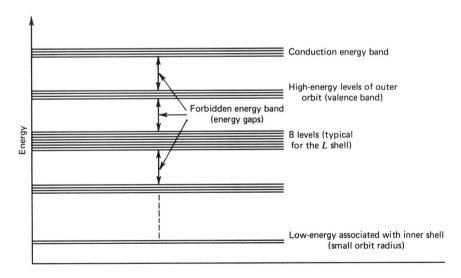

FIGURE 2-3   Energy band representation.

energy or voltage-electrical energy (recall that the volt is defined in joules per coulomb, or energy per 6.28×10^{18} electrons) to the atom.

2-3.3 Conduction and Valence Bands

The outer shell, as already noted, is called the valence band or valence energy band. The valence band may be any shell, the K, L, M, etc., as long as it is the outermost shell. In copper, with 29 electrons, the valence band is the N shell, while in silicon with 14 electrons the M shell contains the valence electrons. The valence electrons occupying this outer band are most likely to be removed completely from the atomic structure and become free electrons—electrons that may be moved with the application of additional energy, from one atom to the other. These are the electrons that, when voltage is applied to the material, produce the electrical current. They are conduction electrons also referred to as *carriers*.

In order to remove an electron from the valence shell, its energy must be increased. Like the movement of electrons from orbit to orbit, a discrete amount of energy is required to move the electron from the valence band into the conduction band, converting the electron in the outer atomic shell into a free electron. The conduction electron has higher energy levels than the valence electron. Here again, there is an energy gap between the valence energy band and the conduction band (Fig. 2-3). The energy gap concept may be best demonstrated by a numerical example. In Fig. 2-4 it is assumed that the valence band electrons may have energies between 0.5 eV[1] and 0.6 eV, while the conduction band electrons may have energies between 1.8 and 1.9 eV. No electron can have an energy level of, say, 1.2 eV. The forbidden gap extends from 0.6 to 1.8 eV. It is clear that in this case an electron can be moved from the valence band to the conduction band by imparting to it a minimum of 1.2 eV; less than 1.2 eV imparted to the electron will *not* move the electron to the conduction band.

2-3.4 Conductors, Semiconductors, and Insulators

Electric conductivity of a material is directly related to the density of free electrons. For example, a good conductor would have a free electron density of about 10^{23}/cm^3 (10^{23} free electrons per each cm^3 of volume). An insulator might have free electron densities as low as 10/cm^3. The conductor has 10^{22} as many free electrons (per unit volume) as the insulator. Semiconductors fall somewhere between the last two, with densities ranging between 10^8/cm^3 to 10^{14}/cm^3. The free electron density is closely related to the atomic structure, in particular to the energy gap between valence electrons and conduction (or free) electrons. The free electrons are valence electrons that have been removed from their orbit by an increase in

---

[1] eV, electron-volt. 1 eV = 1.6×10^{-19} joule. Recall that 1 joule is 1 coulomb-volt (volt = joule/coulomb). Hence, 1 joule = 6.28×10^{18} electron-volt and 1 eV = $1/6.28 \times 10^{18}$ J.

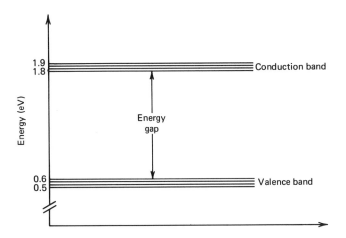

FIGURE 2-4 Numerical example of energy bands.

their energy. This increase, with the absence of any electrical energy (voltage), is due largely to temperature. (Each electron that becomes a free electron leaves behind a hole. The process is referred to as hole-electron pair generation. The reverse process, whereby electrons drop into holes, is called recombination.) Consequently, one expects that with a higher temperature, carrier concentration (free electrons in this context) will increase. Indeed, silicon at 25°C has a free electron concentration of $2 \times 10^{10}/cm^3$, while at 100°C the concentration is about $2 \times 10^{12}/cm^3$. In order to maintain consistency, we will restrict our interest to a fixed temperature of about 27°C (80°F or 300°K). If we assume a fixed temperature, it can be concluded that the density of conduction electrons must depend on the "ease" with which the valence electrons can be "bounced" into the conduction bands. This, of course, depends on the energy gap between the valence and conduction bands. A large gap means that a relatively large energy is involved in the transfer of an electron from valence to conduction. A lower gap means less energy is required for the task. Conductors, semiconductors, and insulators may be classified by their energy gap (the term energy gap will be used to denote the valence to conduction gap exclusively). A conductor may have a gap of 0.05 eV or less; the gap for semiconductors is about 0.7 eV to 1.4 eV, while insulators usually exhibit energy gaps of 8 eV or more. (The silicon semiconductor has a gap of 1.1 eV, germanium 0.7 eV; hence, within this semiconductive range, germanium is a better conductor with a free electron concentration 1,000 times that of silicon.)

2-3.5 Covalent Bonds

As noted (see Fig. 2-1), germanium (Ge), which is classified as a semiconductor, has *four* valence electrons (in the N shell that is the outermost shell for germanium). Silicon (Si), another semiconductor, also has *four* valence electrons (in the M shell that is the outermost shell for silicon). The valence electrons are those involved in

34 SEMICONDUCTOR DIODES: PHYSICS AND CHARACTERISTICS

the atomic bonds (bonds between atoms) that produce crystal-like atomic structures. A *stable* atomic structure requires eight electrons in the outer shell. Since both germanium (Ge) and silicon (Si) have only four valence electrons, the atoms are structured in such a way that electrons are *shared* with neighboring atoms. Figure 2-5 gives a two-dimensional, somewhat simplified representation of the crystalline atomic structure. Each nucleus is surrounded by a total of eight electrons: four *shared* electrons and four of its own valence electrons. (The number next to the electrons in Fig. 2-5 indicates the nucleus to which they belong.) This atomic bond, called a *covalent bond*, involves the *sharing* of valence electrons. Note that the *neutrality* of the atom is preserved, since the number of electrons around the nucleus and protons inside the nucleus are the same.

At room temperature (27°C) there is enough heat energy available to "bounce" some electrons out of their covalent bond, out of their valence band, into the conduction band. These conduction electrons are now free to move about and contribute to the conductivity of the material. (This is referred to as "thermal ionization.")

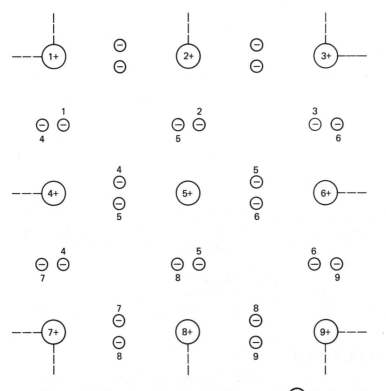

FIGURE 2-5 Covalent bonds in a germanium crystal. (1+) Nucleus (the number used for reference). 1⊖ Valence electron (the number next to the electron indicates which nucleus the electron "belongs to").

When an electron leaves its valence shell, a "hole" or the *absence* of an electron is left behind (Fig. 2-6). The plus sign in Fig. 2-6 replaces the space occupied by electron 5 before it became a free electron. Clearly, the absence of an electron is a positive charge—hence, the plus sign. The "hole" (positive charge) can move about very much like the electron. Consider, in Fig. 2-6, that electron 7 between nuclei 7 and 8 "jumps" into the hole shown. We then have a hole replacing this electron, while the previous hole is now eliminated. The hole has effectively moved from its original location to the new location between nuclei 7 and 8.

We may now speak of two types of carriers, electrons-negative charge carriers and holes-positive charge carriers. In the following discussion, as well as in Chapter 3, holes and electrons are treated in the same manner; the flow or movement of either will be considered as an electrical current. The flow of electrons from right to left has the same effect as the flow of holes from left to right.

2-3.6 Doping

In the construction of semiconductors, the free electron concentration must be carefully controlled and it cannot be allowed to be so very dependent upon temperature. This control is achieved by a careful introduction of controlled amounts of *impurities* into the semiconductor atomic structure. The impurities are introduced into the crystalline atomic structure and not simply "mixed" with the semiconductor. The process is referred to as "doping." Depending on the type of impurities used, the doping process results in the increased concentration of holes or free electrons in the original pure semiconductor.

If we inject into the atomic structure atoms that have *five* valence electrons (phosphorus, arsenic, antimony), the crystalline structure will contain an electron (for each injected impurity atom) that is *not* part of the covalent bond (Fig. 2-7). This electron is *easily* moved from its orbit into the conduction band and becomes

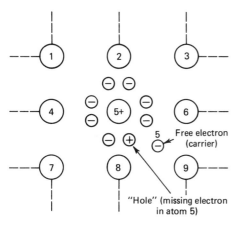

FIGURE 2-6 Thermal ionization—creation of hole and electron carriers (electron-hole pair generation).

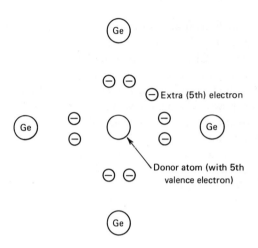

FIGURE 2-7   Injection of donor atoms.

a free electron. Stated differently, the energy gap has been substantially reduced by this doping process. Each impurity atom injected has donated an electron. This type of impurity is called a *donor impurity*. By controlling the number of donor impurity atoms injected, the free-electron concentration can be controlled. By introducing a relatively large number of impurity atoms, the original free electrons in the intrinsic (pure) material become negligible. A semiconductor material, which has been injected with donor impurities, is called N-type material (negative because of the large number of free electrons).

By introducing impurities with *three* valence electrons (boron, aluminum, gallium, indium), we may control the concentration of holes, since every trivalent impurity atom injected adds one hole to the atomic structure. These impurities are called *acceptor* atoms, since they produce an excess of "free" holes that can "accept" electrons. This kind of doping produces a so-called P-type material (positive because of the excess of positive carriers).

The doping process modifies substantially the electron energy distribution in the solid. With no dopant added, each electron that leaves the valence band and becomes a free electron (high energy electron) leaves behind a hole (absence of an electron). Thus, in pure (intrinsic) semiconductor material, the concentration of holes equals that of electrons. The average-electron energy[1] (or the energy of an average electron) is precisely in the middle of the energy gap (see Fig. 2-8). The N-type material has an excess of free electrons. Hence, the average energy is *closer* to the *conduction* energy level, while in the P-type material, the electron average energy is *closer* to the *valence* energy level. This energy distribution, in particular the difference between the N and P material average energy, gives rise to the diode barrier potential discussed in Section 2-3.9.

---

[1] This is a simplification of the Fermi level, the precise definition of which is beyond the scope of this book.

THE PHYSICS OF SEMICONDUCTORS   37

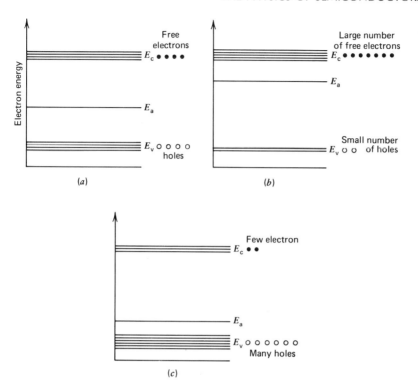

FIGURE 2-8   Energy levels and average energy. E_a—Average energy. E_c—Energy of conduction electrons (conduction band). E_v—Energy of valence band. (a) Pure semiconductor. (b) N-Type. (c) P-Type.

2-3.7   Minority and Majority Carriers

The P material has a relatively large number of holes (mobile hole, free to move and carry electric current) as a result of the impurities injected. At the same time, the process of thermal ionization, which produces free electrons, is also present. As a result, in P material, we have a large number of mobile positive charges (holes) and a small number of mobile negative charges (free electrons). Since the holes in this case constitute the *majority* of the available current carriers, we call them *majority carriers*. The free electrons in the P-type material, are the minority carriers. It is clear then that in the N-type material, where the majority of available current carriers are electrons, the *free electrons* are the majority carriers and the holes the minority carriers. Since minority and majority carriers are always of opposite electric charge, they carry current in opposite direction. In Fig. 2-9 the minority and majority carriers are traveling in the same direction. The net current is $(5 \times 10^{12} - 10^6)$ majority carriers as shown. The minus sign is due to the opposing effect minority and majority carriers have.

In our discussion of the semiconductor diode, the effect of both carriers will be

38 SEMICONDUCTOR DIODES: PHYSICS AND CHARACTERISTICS

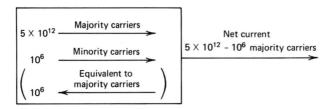

FIGURE 2-9 Net current flow.

considered. The concept of minority and majority carriers will be even more important in the analysis of the behavior of the transistor (Chapter 3).

2-3.8 The *P-N* Junction

The semiconductor diode is constructed by "bringing together" P and N type semiconductor materials. The *junction* of the P and N materials is metallurgical, involving the atoms of the two oppositely doped materials. The intimate P-N contact causes a redistribution of holes and electrons in the vicinity of the junction. The high concentration of holes on the P side and electrons on the N side produces a *diffusion* of electrons from N to P material and holes from P to N material. (Diffusion is the natural process through which charges flow from higher to lower concentrations and differences in concentration are eliminated. Sugar, for example, diffuses from a highly concentrated solution through a membrane into a low concentration solution, producing equal concentrations on both sides of the membrane. Electrons will diffuse from a high concentration area in an attempt to equalize concentrations.) The process of diffusion is shown in Fig. 2-10. Figure 2-10*a* shows the concentration distribution *before* the diffusion takes place, while Fig. 2-10*b* shows the equilibrium reached after diffusion.

Each electron that diffuses from N to P leaves a charged atom behind, an atom which is missing one electron, that is, a *positive ion*. Similarly, the diffusing holes leave behind, in the P material, *negative ions*. As these charges build up, they produce a potential difference across the junction, the *junction barrier voltage*. Two processes now take place: the diffusion of holes and electrons and the drift (the flow of holes and electrons due to the presence of a voltage) of holes and electrons. Equilibrium is reached when the net current is zero (Fig. 2-11); that is, the drift of holes equals the diffusion of holes (they are moving in opposite directions) and the drift of electrons equals the diffusion of electrons (Fig. 2-11).

When equilibrium is reached, the region in the vicinity of the junction consists largely of charged atoms—positive ions on the N side and negative ions on the P side. These ions themselves are immobile and locked in the atomic lattice. They *cannot* serve as current carriers. The carriers, originally present in this region, have traversed the junction to combine with atoms on the other side. The free electrons

THE PHYSICS OF SEMICONDUCTORS 39

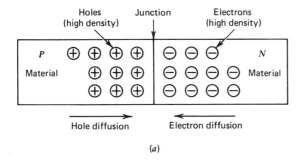

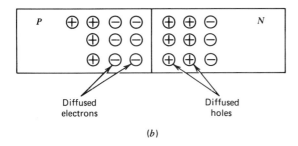

FIGURE 2-10 Electron and hole distribution in vicinity of junction. (a) Before diffusion took place. (b) Showing effects of diffusion.

in the N material traveled to the P region and "filled" the holes in that region producing negative ions (an excess electron over and above the number of protons in the nucleus). Similarly, the mobile holes produced positive ions in the N material. As a result, the junction region is almost completely *depleted* of carriers (very few mobile electrons or holes are left in the region). The region near (on either side of) the junction is simply called the *depletion region* (Fig. 2-11). To keep the

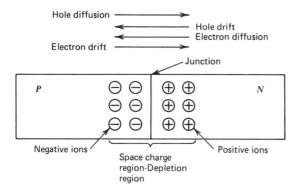

FIGURE 2-11 Equilibrium and space charge region—depletion region.

40 SEMICONDUCTOR DIODES: PHYSICS AND CHARACTERISTICS

discussion simple, it is assumed that the change in carrier concentrations is abrupt: no carriers in the depletion region and the *normal* hole and electron concentration outside the region, in the P and N regions respectively.

The voltage produced across the junction depends on the *average electron energy* of the P and N regions. This becomes clear when we examine the *energy diagram* of the junction. The equilibrium state of the junction dictates that the average electron energy must be the same throughout both materials, across the junction. The result of the intimate P-N contact is shown in Fig. 2-12. The magnitude of the energy "shift," the "hill," is essentially $E_{aP} - E_{aN}$, the difference between the P and N average electron energy. The barrier voltage then depends not only upon $E_{aP} - E_{aN}$ but also upon the hole and electron concentrations in the P and N materials. This barrier potential difference is approximately 0.7 V for silicon and 0.35 V for germanium. (The exact value, of course, depends upon precise junction construction and temperature.)

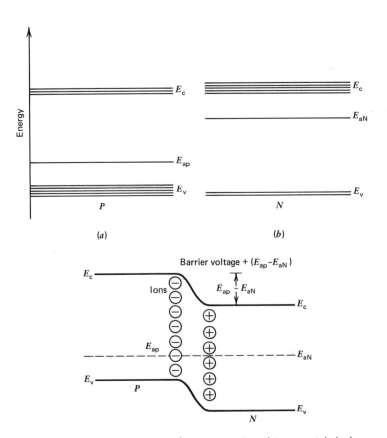

FIGURE 2-12 Junction energy diagram. (a) P and N materials before contact. (b) Energy level realignment after contact. E_c—Conduction band energy; E_{aP}—Average energy, P material E_v—Valence band energy; E_{aN}—Average energy, N material.

2-3.9 Forward-Biased Junction (Conducting Diode)

In order to produce forward current through the junction, it is necessary to overcome the barrier voltage. Since the barrier voltage has a polarity as shown in Fig. 2-12b, an external voltage that *exceeds* the barrier voltage and that is of *opposite* polarity must be applied (Fig. 2-13). The effect of this applied voltage is to reduce and ultimately eliminate the depletion region. Electrons injected by the voltage source into the N region eliminate (if a high enough voltage is applied) the positive ions. Similarly, the negative ions in the P side are eliminated by the injection of holes. The depletion region is completely eliminated, and there is no barrier to forward current flow (other than the resistance of the semiconductor material and the external circuit resistance R). The current under these conditions is largely a diffusion current. Recall that in equilibrium, the drift currents balance out the diffusion current. Since the externally applied voltage reduces, or eliminates, the barrier voltage, which causes the drift current, only the diffusion current is present. Because of the large hole and electron concentrations on the two sides of the junction, the diffusion current can be very high. Note that the majority carriers (holes) from the P region diffuse through the junction, into the N region. Holes, however, are minority carriers in the N region. In the process of carrier transport the concentration of minority carriers near the junction is increased. This phenomenon is of major importance in the operation of the transistor (Chapter 4).

2-3.10 Reverse Bias Junction

The effect of *forward bias* is to practically eliminate the drift current of minority carriers across the junction. *Reverse bias* aids the internal barrier voltage and, consequently, reduces diffusion current and increases drift current. Figure 2-14 shows the circuit connections and the currents involved. As already noted, the drift current consists of minority carriers, electrons from the P region and holes from

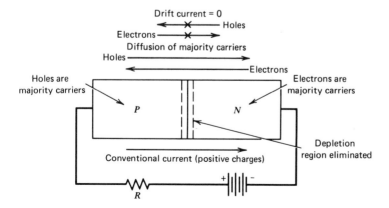

FIGURE 2-13   Forward-biased diode.

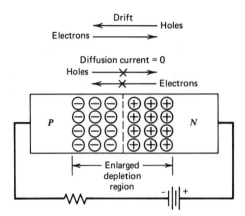

FIGURE 2-14 Reverse-biased junction.

the N region. As the reverse voltage is increased, a point is reached where the diffusion current is zero and the current consists of the drift current of minority carriers only. The direction of this current is negative (as compared with the forward majority carrier current; see Figs. 2-13, 2-14). At the point where *all* minority carriers contribute to the reverse current (sometimes called leakeage current) the *saturation* of the reverse current (I_s) is reached. A further increase in the reverse voltage (within limits) will have no effect on the reverse current.

Clearly, I_s depends largely (almost completely) on the concentration of minority carriers, or the total availability of these carriers. Since minority carriers have a much lower concentration (in doped material) than majority carriers, as a result, I_s is many orders of magnitude smaller than the forward current (forward bias). Typically, I_s is only a few μA, while the forward current is 100 to 200 mA (for a typical germanium diode). For silicon diodes, I_s is typically only a few pA.

2-4 DIODE CHARACTERISTICS

Based on a largely qualitative analysis, we have, so far, established two modes of operation for the P-N junction: forward bias, with substantial conduction, and reverse bias, with negligible conduction. A full description of the behavior of the commercial *semiconductor diode* (a P-N junction in its final sealed package with leads attached to the P and N sides) follows. For the remainder of this text, the standard symbol for a diode (Fig. 2-15) is used. In the conducting state (forward-

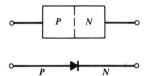

FIGURE 2-15 Diode symbols.

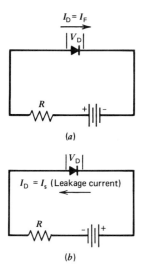

FIGURE 2-16 Diode biasing.

bias) there is a forward current (in the direction of conventional current) from P to N (Fig. 2-16a). Figure 2-16b shows a reverse bias connection with a reverse or leakage current.

2-4.1 V-I plot (Voltage-Current Plot)

A measurement of forward and reverse diode current (I_D) as a function of forward and reverse diode voltage (V_D) would yield the typical graph shown in Fig. 2-17. The operation of the diode may be divided into three regions[1] (see Fig. 2-17). Region A is forward biased, largely the diffusion current of majority carriers—holes from P to N(I_h), electrons from N to P(I_e). Note that holes are majority carriers in the P region and electrons are majority carriers in the N region. The two currents, flowing in opposite directions, *add* to yield the diode forward current. In region B, the forward voltage applied is less than the diode barrier voltage. The result is a mix of diffusion and drift currents (of majority and minority carriers, respectively). The net current (very small) is the difference between majority and minority currents. In region C, reverse bias is applied, producing an almost purely minority carrier current largely drift current (diode leakage current). Note that different current and voltage scales have been used in the positive and negative current axis in Fig. 2-17: mA and volts for the positive side, μA and 20 V steps for the negative side. (The avalanche occurring at 50 V is representative only.)

---

[1] The three regions overlap somewhat with regard to drift and diffusion currents. The overall description is somewhat simplified. Currents other than diffusion and drift are also present. Diffusion and drift, however, are, for the most part, the more important current components.

44 SEMICONDUCTOR DIODES: PHYSICS AND CHARACTERISTICS

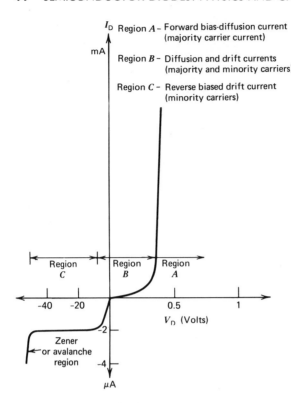

FIGURE 2-17 Diode characteristics.

The plot shown in Fig. 2-17 can be approximated by the equation

$$I_D = I_s \left(e^{V_D / \frac{kT}{q}} - 1 \right) \text{ A}^{(2)} \tag{2-1}$$

I_s is the maximum (or saturation) reverse current. (In Fig. 2-17 this value is 2 μA.)

T is the absolute temperature of the junction in degrees Kelvin (which can be calculated by adding 273 to the temperature in degrees Celsius °C[3]).

k is a constant (Boltzmann's constant, $k = 1.38 \times 10^{-23}$ joules/°K).

q is the charge of an electron, $q = 1.6 \times 10^{-19}$ coulomb.

V_D is the voltage applied across the diode. V_D may be positive for forward bias (regions A and B, Fig. 2-17) or negative for reverse biasing (region C, Fig. 2-17).

---

[2] The symbol e has the numerical value 2.71828. . . . . It has major significance in differential and integral calculus.

[3] For example, 27°C (80.6°F) equals 27 + 273 = 300°K. This temperature is usually referred to as room temperature.

Equation (2-1) has been derived by the application of semiconductor theories. It agrees closely with measured results. The agreement between the theoretical equation and experimental data can be improved by using $2kT/q$ in the reverse bias region, and kT/q in the forward biased region. For the region between (region B in Fig. 2-17) values ranging between kT/q and $2kT/q$ may be used. Equation (2-1) describes the behavior of the diode in all regions, forward bias and reverse bias (excluding avalanche or zener effects. These effects will be discussed in Section 2-4.4).

The diode current, as shown in Eq. (2-1), is dependent on the ratio $V_D/kT/q$. Let us evaluate the denominator kT/q for a specific junction temperature, $T = 300°K$ (27°C).

$$\frac{kT}{q} = \frac{1.38 \times 10^{-23} \times 300}{1.6 \times 10^{-19}} = 259 \times 10^{-4}$$
$$= 0.0259 \text{ V} = 25.9 \text{ mV} \approx 26 \text{ mV}$$

Note that the unit of kT/q is volts. This can be shown by dimensional analysis as follows: Since [k] = joules/°K, [q] = coulomb; [T] = °K, joules/°K x °K/coulomb = joules/coulomb = volts. Equation (2-1) can be rewritten for $T = 300°K$ as

$$I_D = I_s (e^{V_D/0.026} - 1)$$

For V_D in mV, this becomes

$$I_D = I_s(e^{V_D/26} - 1) \text{ (for } T = 300°K)$$

When the ratio $V_D/26$ is negative and $|V_D/26| \gg 1$ (V_D in mV), then

$$I_D = -I_s \quad (2\text{-}1a)$$

This means that for a negative bias voltage (reverse bias), relatively large with respect to 26 mV (at room temperature), the current through the diode is the reverse saturation current I_s. (The minus sign associated with I_s indicates reverse current, from N to P.) In Fig. 2-17 (region C) I_s is given as -2 μA and independent of V_D, which is in agreement with the result above, that for this region $I_D = -I_s$.

For example, if $V_D = -130$ mV (V_D is 5 times kT/q at 300°K) and $I_s = 2$ μA, by substitution in Eq. (2-1)

$$I_D = I_s (e^{-130/26} - 1) = I_s (e^{-5} - 1) \approx I_s(-1)$$
$$= -I_s = -2.10^{-6} = -2 \text{ μA}$$

namely, $I_D = -I_s$.

In a similar way, Eq. (2-1) may be simplified for the forward bias region (region A in Fig. 2-17), provided $V_D/kT/q \gg 1$. The simplified equation becomes

$$I_D = I_s e\left(e^{V_D/\frac{kT}{q}}\right) \quad (2\text{-}1b)$$

46 SEMICONDUCTOR DIODES: PHYSICS AND CHARACTERISTICS

This equation results from neglecting the 1 in Eq. (2-1). This approximation is quite reasonable since

$$e^{V_D/\frac{kT}{q}} \gg 1$$

For operation at $T = 300°K$ where $kT/q = 26$ mV, Eq. (2-1b) becomes

$$I_D = I_s e^{V_D/26} \quad (V_D \text{ in mV}) \tag{2-2}$$

$$I_D = I_s e^{V_D/0.026} \quad (V_D \text{ in V}) \tag{2-2a}$$

For example, for $T = 300°K$, $V_D = 260$ mV and $I_s = 2$ μA from Eq. (2-1)

$$I_D = I_s(e^{260/26} - 1) = I_s(e^{10} - 1)$$
$$\approx I_s e^{10} \text{ (here, } e^{10} \gg 1\text{)}$$
$$I_D = 2 \times 10^{-6} \times e^{10} = 0.044 \text{ A} = 44 \text{ mA}$$

Figure 2-18 shows V-I plots (forward bias region) for silicon and germanium diodes. The actual diode characteristics (commercial diodes) conform reasonably well to Eq. (2-1). The difference between the plots for silicon and germanium is due largely, to the marked difference in I_s for the two diodes. The following two examples will clarify the effect of I_s on the forward current in the diodes.

EXAMPLE 2-1

Given:

$I_s = 1$ μA (typical for germanium diode)
$V_D = 0.312$ V (forward voltage applied to diode)
$T = 300°K$ (junction temperature)

Find I_D (forward current) [Use Eq. (2-1)].

SOLUTION

At 300° K junction temperature

$$kT/q = 26 \text{ mV}$$

and

$$\frac{V_D}{kT/q} = 312/26 = 12$$

DIODE CHARACTERISTICS

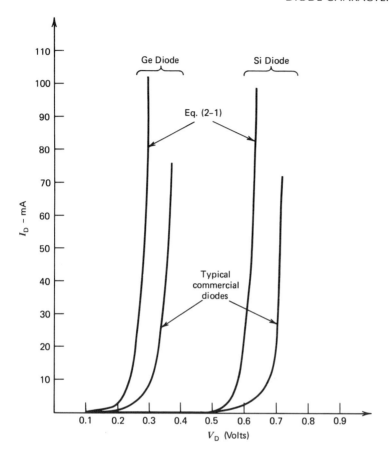

FIGURE 2-18 Silicon and germanium diode characteristics (forward region).

Using Eq. (2-1b) (since $V_D/kT/q = 12 \gg 1$)

$$I_D = I_s e^{(V_D/kT/q)} = I_s e^{12} = 1 \times 10^{-6} \times e^{12} = 163 \text{ mA}$$

(Here, $I_D = I_F$, the forward current under forward bias conditions.)

EXAMPLE 2-2

$$I_s = 2 \text{ pA (typical for silicon diode)}$$
$$\left.\begin{array}{l} V_D = 0.312 \text{ V} \\ T = 300°\text{K} \end{array}\right\} \text{ as in Ex. 2-1}$$

Find I_D (forward current) [use (Eq. 2-1)].

48 SEMICONDUCTOR DIODES: PHYSICS AND CHARACTERISTICS

SOLUTION

Equation (2-1b) may be used

$$I_D = I_s e^{12} = 2 \times 10^{-12} \times e^{12} = 2.10^{-12} \times 1.63 \times 10^5 = 0.33 \ \mu A$$

The forward currents calculated in Examples 2-1 and 2-2 indicate that conduction in the germanium diode starts at a much lower forward-bias, as compared to the silicon diode. The "knee" of the germanium diode is at about $V_D = 0.35$ V and for a silicon diode at about $V_D = 0.7$ V. [Note that these values have been taken from the V-I plot for the commercial diodes, rather than from the plot of Eq.(2-1).] It must be emphasized that these "knee" values are only *approximations*. Different types of diodes and, for that matter, different samples of the same diode type may exhibit knee voltages varying from about 0.3 V to 0.4 V for germanium and about 0.6 to 0.8 V for silicon.

Although Eq. (2-1) is not a precise representation of the actual diode, it is nevertheless used in our computations, since it provides a mathematical analysis tool with a reasonable accuracy. (This text will use 0.35 V and 0.7 V as the barrier potentials for germanium and silicon, respectively.)

As Fig. 2-18 shows, the forward current, for both diodes, at a particular forward voltage, depends on the saturation leakage current I_s. As a result, diodes of a particular type (Ge or Si) designed to operate at high forward currents (power diodes) exhibit high leakage currents, while low forward current diodes (signal diodes) have low leakage currents. A silicon diode designed to operate at 3 A may have a leakage current of about 0.5 to 1 nA, while a silicon diode designed to operate, say, at a 200 mA forward current may have a leakage current of about 20 pA. (A similar pair of germanium diodes may have leakage currents substantially larger, 50 μA and 2 μA, respectively.) Table 2-1 gives pertinent data for typical silicon and germanium diodes.

TABLE 2-1 TYPICAL DIODE CURRENTS AT DIFFERENT TEMPERATURES (T)

	at $T = 27°C$		at $T = 100°C$
	I_F	I_s	I_s
Silicon (low current diode)	200 mA ($V_D \simeq 0.67$ V)	20 pA	4 nA
Silicon (high current)	5 A ($V_D \simeq 0.67$ V)	0.5 nA	100 nA
Germanium (low current diode)	200 mA ($V_D \simeq 0.3$ V)	2 μA	1.5 mA
Germanium (high current)	5 A ($V_D \simeq 0.3$ V)	50 μA	37 mA

2-4.2 Temperature Effects

It is obvious, from basic semiconductor theory, that temperature variations severely alter the diode characteristics. An increase in temperature results in an increase in available carriers, which directly affects the currents in the diode.

Measurements have shown that the leakage current in a diode approximately doubles for every 10°C rise in temperature.[1] A germanium diode with $I_s = 5$ μA at 27°C junction temperature will have a leakage current of about 1 mA at 100°C. Such high leakage current presents a serious problem. The reverse-biased diode can no longer be considered an open circuit (at least not in low current, high temperature applications). The same temperature change in silicon, however, will result in $I_s = 1$ μA or less, since I_s at 27°C is substantially smaller, in the order of a few picoamperes. This characteristic of silicon diodes makes it predominant in both commercial and industrial applications. As a matter of fact, germanium diodes are presently used largely in power applications and in circuits where the lower barrier voltage (0.35 V as compared to 0.7 V of silicon) is important.

The effects of temperature on the forward diode current I_F, are controlled by two opposing factors. As we noted, I_s (which directly affects I_F) increases with temperature. The exponent in Eq. (2-1b), howevver, decreases as the temperature increases. The net effect is nevertheless a rise in I_F as the temperature rises. (Note that I_D is the general current in the diode, while I_F denotes the diode current when forward biased only.)

EXAMPLE 2-3

Given

$$I_s = 0.1 \text{ μA}$$
$$T = 300°K (27°C)$$
$$V_D = 416 \text{ mV}$$

assume I_s doubles for every increase of 10°C. Calculate:
(a) I_F (forward current) for $T = 300°K$.
(b) I_F for $T = 330°K$ (a 30°K or 30°C rise).
(c) Ratio of (b) to (a) above.

SOLUTION

Since $V_D \gg kT/q$ Eq. (2-1b) may be used in our computations.

---

[1] Experiments have shown some differences between temperature sensitivities of silicon and germanium diodes. It is impossible, however, to generalize since the temperature behavior of a diode depends strongly on construction details and doping levels and temperature range. The figure given (double for every 10°C) is a reasonably accurate approximation around 25°C.

(a) For $T = 300°K$, $kT/q = 26$ mV

$$I_F = 0.1 \times 10^{-6} \times e^{416/26} = 0.89 \text{ A}$$

(b) At $T = 330°K$ $kT/q = 28.5$ mV
I_s at this temperature is 0.8 µA Note that I_s has been doubled 3 times (doubling for each 10°C) from 0.1 to 0.2 µA, 0.2 to 0.4 µA, and then to 0.8 µA.

$$I_F = 0.8 \times 10^{-6} \times e^{416/28.5} = 0.8 \times 10^{-6} \times e^{14.6} = 0.175 \text{ A}$$

(c) $\dfrac{I_{Fb}}{I_{Fa}} = \dfrac{0.175}{0.89} \simeq 2$

As Example 2-3 shows, I_F doubled for the 30°C change. This demonstrates that the variations of I_s with temperature are dominant. It was previously mentioned that in order to turn a diode ON (ON means to produce substantial current) a voltage of about 0.35 V for germanium and 0.7 V for silicon is required (see Fig. 2-18). This ON voltage is affected by temperature. Since conduction is enhanced with increases in temperature, as shown by Example 2-3, the diode turns ON at a *lower* voltage as the temperature is increased. (We will come back to this phenomenon in our discussion of temperature effects on transistors.)

2-4.3 "Zener" and "Avalanche" Breakdown

Figure 2-17 shows a point, in the reverse bias portion of the graph, at which the reverse current increases sharply—the "zener" or "avalanche" region. The sudden increase in reverse current is the result of a very large reverse electric field. (Since electric field is usually defined in terms of volts/cm, or volts per unit distance, a reverse voltage of 100 V may be sufficient to produce this effect, if the physical dimension of the diode depletion region is about 10^{-4} cm, or 1 micron, yielding a field of $100/1 \times 10^{-4} = 10^6$ V/cm.)

The high reverse field accelerates the conduction band (free) electrons. As these collide with valence electrons, the energy of collision may be sufficient to transfer the valence electrons into the conduction band. This increases the number of minority carriers, in turn increasing the reverse (minority) current. As the field is increased, this "multiplication" in minority carries is accelerated and very high reverse currents are produced. This process is called "avalanche."

The minority carrier concentration may be increased by directly "tearing" electrons out of the valence band (removing them from the atomic orbit) into the conduction band. This phenomenon is called *tunneling*. It requires a substantially larger field than required to produce avalanche (for moderate doping concentrations). Tunneling may lead to a rapid increase in diode reverse current. The process is called "zener breakdown." At very high doping levels the zener breakdown occurs at lower fields than the avalanche breakdown. It is not surprising, therefore, that diodes designed to exhibit breakdown characteristics at lower voltages are heavily doped and the process involved is the zener breakdown. (For voltages from 2.4 to

6 V the process is zener breakdown, and from 6 V up the process is avalanche breakdown.) These processes are not necessarily destructive, that is, the diode can operate in this region with no damage to the diode, provided other parameters, such as power dissipation, diode temperature, etc., are kept within bounds.

The voltage across the diode when avalanche, or zener, occurs is almost completely independent of the current through the diode; hence its use as a voltage reference or voltage regulator. (See Sec. 3-6 for zener diode applications.)

It is interesting to note that these "reference" diodes are all called *zener* diodes even though they may actually operate in the *avalanche* region.

2-5 EQUIVALENT CIRCUITS OF THE DIODE

The analysis of diode circuits is substantially simplified by replacing the diode in any circuit by an *equivalent circuit*, consisting of simple circuit components. This substitution is, however, applicable to linear systems only (see Section 1-9). Clearly, the diode is *not* a linear device. One has only to examine Eq. (2-1) or Fig. 2-17 to see the marked nonlinearities involved. It is nevertheless possible to select regions for which the diode behavior (its voltage versus current characteristics) can be approximated by a straight line. Figure 2-19 shows straight line approximations of the diode characteristics. As shown, the approximations are made for two separate

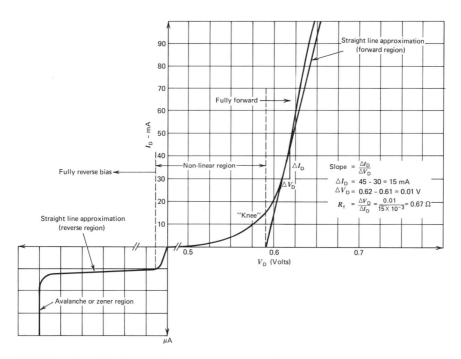

FIGURE 2-19 Diode operating regions and straight-line approximations.

52 SEMICONDUCTOR DIODES: PHYSICS AND CHARACTERISTICS

regions: A, the *fully forward-biased* region, beyond the diode knee; B, the *fully reverse-biased* region (reverse current equals I_s). The *nonlinear* region between these two cannot be linearized without introducing large approximation errors. Fortunately, this region is rarely used in diode circuits (although often used in harmonic generation).

2-5.1 Forward Biased Diode Equivalent (ON Diode)

Figure 2-20 shows a straight line approximation of a typical diode characteristic curve (*V-I* plot) in the forward region only. (It is the same as the straight line approximation of the forward region in Fig. 2-19.) This approximation assumes that $I_D = 0$ until $V_D \simeq 0.7$ V (0.6 to 0.7 V for silicon and 0.3 to 0.35 V for germanium). For $V_D \geq 0.7$ V, the diode behaves like a resistor with resistance r_f. The value of r_f is determined by the slope of the straight line (Fig. 2-20).

$$r_f = \Delta V_D / \Delta I_D \; \Omega \tag{2-3}$$

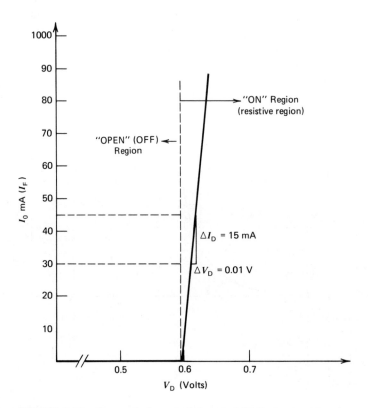

FIGURE 2-20 Forward characteristics straight line approximation.

EQUIVALENT CIRCUITS OF THE DIODE

In Fig. 2-20, the value r_f is found by

$$r_f = \frac{\Delta V_D}{\Delta I_D} = 0.01/(15 \times 10^{-3}) = 0.67 \; \Omega$$

A more rigorous analysis (see below) would show that r_f is not constant and depends on the particular point on the graph. For example, between $V_D = 0.64$ V and $V_D = 0.65$ V, $\Delta I_D = 144 - 98 = 46$ mA; hence, $r_f = (0.65 - 0.64)/(46 \times 10^{-3}) = 0.22 \; \Omega$, substantially different from the r_f calculated near $V_D = 0.62$ V. Fortunately, both these values are so small (compared with other circuit component values) that either value can be used. (In many cases we even "idealize" the diode by assuming $r_f = 0$.)

The exact value of r_f can be obtained by use of calculus as shown in Appendix A which is based on

$$r_f = dV_D/dI_D \tag{2-4}$$

[r_f is the derivative of V_D with respect to I_D evaluated at a given I_D. This expression is the derivative form of Eq. (2-3).]

This analysis yields the expression for r_f.

$$r_f = \frac{Kt/q[V]}{I_D[A]} \; \Omega \tag{2-5}$$

Equation (2-5) shows that the value of r_f depends on the diode operating point, that is, the diode current I_D.

EXAMPLE 2-4

Given a forward biased diode operating at 27°C with $I_D = 260$ mA, find r_f.

SOLUTION

We find $T = 27 + 273 = 300°K$. kT/q at 300°K equals 0.026 V. Hence,

$$r_f = \frac{0.026[V]}{I_D[A]} = \frac{0.026[V]}{0.26[A]} = \frac{26 \; mV}{260 \; mA} = 0.1 \; \Omega$$

Equation (2-5) rewritten for $T = 300°K$ is known as the Shockley relation (recall that $kT/q = 26$ mV for $T = 300°K$):

$$r_f(\Omega) = \frac{0.026[V]}{I_D[A]} = \frac{26[mV]}{I_D[mA]} \tag{2-5a}$$

Our equivalent circuit may now be shown as a resistor r_f in series with a voltage of 0.7 V (Fig. 2-21). The 0.7 V represents the *voltage drop*, not a voltage source, produced by the application of an external forward biasing voltage. The equivalent

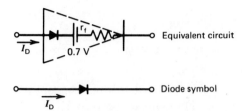

FIGURE 2-21 Diode equivalent (silicon). *Forward biased.*

circuit (Fig. 2-21) also contains an ideal diode (diode symbol encircled). This ideal diode is assumed to behave like a switch producing a short when the voltage across the diode (anode to cathode) is 0, or positive, and an open for negative anode to cathode voltages. The ideal diode contributes *nothing* to the circuit; it does, however, remind us that we are dealing with a diode *equivalent* and not with a simple battery and resistance. It also reminds us that a positive (forward) voltage is needed to produce the battery-resistor behavior.

Before using this equivalent circuit, it is necessary first to establish that the diode is forward biased and, if using the exact expression for r_f [Eq. (2-5)], I_D must be known. Note, however, that to obtain I_D, r_f must be known. Is it a trap? Not really. Example 2-5 shows why it is not.

EXAMPLE 2-5

Given the circuit shown in Fig. 2-22, find r_f and I. (Ge diode.)

SOLUTION

$$1 - 0.35 - I \times r_f - I \times R = 0 \quad \text{KVL}$$

and

$$I = \frac{1 - 0.35}{R + r_f} = \frac{0.65}{R + r_f} \text{ A}$$

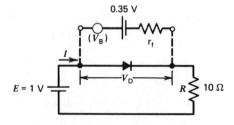

FIGURE 2-22 Diode circuit.

EQUIVALENT CIRCUITS OF THE DIODE

as a first step assume $r_f = 0$, and since $R = 10\ \Omega$,

$$I \approx \frac{1 - 0.35}{10} = 65\ \text{mA}$$

Now for $I_D = 65\ \text{mA}$

$$r_f = \frac{26}{65} = 0.4\ \Omega\ \text{[from Eq. (2-5a)]}$$

Using this value of r_f in the original expression for I, we get

$$I = \frac{1 - 0.35}{10 + 0.4} = 62.5\ \text{mA}$$

The iteration process in Example 2-5 may be repeated. Find r_f for $I_D = 62.5\ \text{mA}$, then find I. The effect of another iteration is, however, small (I becomes 62.4 mA). This iterative process is effective for relatively low values of r_f (high I_D) as compared to the resistance in the circuit (R). For a circuit with $E = 0.4\ \text{V}$, for example, it would be necessary to perform four iterations to obtain 5% accuracy.

The main advantages of the iterative solution are its simplicity, its not having to refer to device characteristics, and its suitability for automated or computerized solution.

2-5.2 Load Line Technique

The problem of Example 2-5 can be solved, graphically, by the use of the "load line" technique.

In circuits that contain diodes, transistors, or other electronic devices, currents and voltages are often dependent on two major factors: the passive circuit components used (application of KVL or KCL) and the characteristics of the device incorporated in the circuit.

In other words, these currents and/or voltages must satisfy two constraints. Circuit constraints in that they must conform to the applicable circuit law (KVL and KCL) and device constraints in that they must agree with currents and voltages typical for the device used. The circuit and device constraints can be represented by appropriate equations, for example, KVL and the diode equation [Eq. (2-1)], or by graphs, for example, a plot of KVL and the diode V-I plot (Fig. 2-23). The line representing KVL in Example 2-5 is called the "load line" (it is indeed a straight line). The solution of Example 2-5 may be obtained by solving the two simultaneous equations involved, algebraically or graphically. The "load line" technique is a graphic solution.

EXAMPLE 2-6

Repeat Example 2-5 using the "load line" technique.

56 SEMICONDUCTOR DIODES: PHYSICS AND CHARACTERISTICS

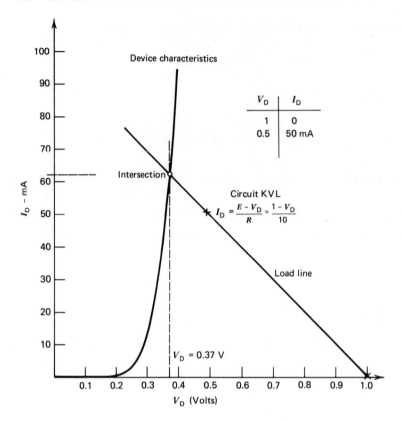

FIGURE 2-23 Diode operating point.

SOLUTION

The two equations (or curves) involved are the KVL equation, $E - V_D - I_D \times R = 0$, and Eq. (2-1b), the device equation. The KVL equation can be rewritten as

$$I_D = \frac{E - V_D}{R} = \frac{1 - V_D}{10} \quad (E = 1\text{ V}, R = 10\text{ }\Omega)$$

This equation, the load line equation, is plotted in Fig. 2-23, superimposed on the V-I plot of the diode. The two curves, the load line and the V-I characteristics, intersect at $I_D = 62$ mA, $V_D = 0.37$ V. The solution is $I = 62$ mA (compared with 62.5 mA in Example 2-5 using the iteration method).

In many cases the assumption $r_f = 0$ may be made without introducing excessive error (for example, consider $R = 100\text{ }\Omega$ and $E = 10$ V). This assumption (always a good starting point) yields somewhat "idealized" behavior, namely that of a perfect

EQUIVALENT CIRCUITS OF THE DIODE

switch (Fig. 2-24). At 0.35 V the diode is exhibiting zero resistance. Below 0.35 V the diode is OFF, OPEN, exhibiting infinite resistance. In many cases a further idealization is possible, setting the barrier voltage (V_B) to zero. This assumption is reasonable as long as $V_B \ll E$. (The percent error introduced by this assumption is approximately $V_B/E \times 100\%$.) It yields the ideal diode previously mentioned.

EXAMPLE 2-7

In Fig. 2-22 use $R = 100 \ \Omega$, $E = 10$ V and find I.

SOLUTION

It is clear from the source polarity that the diode is forward biased. First, find an approximate solution based on the ideal diode assumptions, $r_f = 0$, $V_B = 0$ (both the forward resistance and the barrier voltage are assumed to be zero),

$$I = \frac{10}{100} = 100 \text{ mA}$$

Now examine the validity of the assumptions: $r_f = 0$? r_f in this problem, with $I = 100$ mA, is $r_f = 26/100 = 0.26 \ \Omega$. Since $r_f \ll 100 \ \Omega$, the assumption $r_f = 0$ is valid (it introduces only a 0.26% error). $V_B = 0$? V_B is usually approximated as $V_B = 0.35$ V (for germanium). Again, $V_B \ll 10$ V (V_B is much smaller than the driving voltage in the circuit). Hence, the assumption is valid (it introduces an error of about $0.35/10 \times 100 = 3.5\%$). In this problem, the forward biased diode was represented by a short circuit (the most idealized and simplest equivalent-circuit, "ideal" diode).

2-5.3 Dynamic and Static Diode Resistance

The value of r_f discussed in Section 2-5.1 was based on incremental changes in V_D and I_D ($\Delta V_D/\Delta I_D$). This r_f is referred to as the *dynamic resistance*.

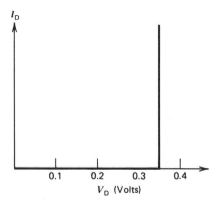

FIGURE 2-24 Idealized diode characteristics ($r_f = 0$). Germanium diode.

58 SEMICONDUCTOR DIODES: PHYSICS AND CHARACTERISTICS

The *static* resistance is defined as V_D/I_D. For example, for $V_D = 0.3$ V, $I_D = 102$ mA (Fig. 2-18 germanium diode); the static resistance is $R_{FS} = 0.3/0.102 \simeq 3\ \Omega$. Another point on the same curve, $V_D = 0.28$ V, $I_D = 47$ mA, yields $R_{FS} = 0.28/0.047 = 6\ \Omega$. In both cases, the forward current is considerable; however, R_{FS} is distinctly different. When this method is used, the effect of V_B is included in the computation of R_{FS}. Because the value of the static resistance is relatively large (compared to the dynamic r_f) and varies so dramatically within the forward bias region ("fully" forward biased), it is advisable to avoid using this method. (It is presented here for reference purposes only.)

2-5.4 Reverse Bias Diode Equivalent Circuit

In the *reverse bias region* the diode conducts a current, I_s, which is not directly related to the magnitude of the reverse voltage applied. The magnitude of this current is measured in picoamperes (pA 10^{-12} A) for silicon diodes and in μA for germanium. For practical purposes the diode may be considered OPEN ($I_s = 0$) if I_s is small compared to the currents in the circuit. In some instances, for greater accuracy, it may be useful to replace the back biased diode with its "leakage" resistance (R_s).

The value of R_s is usually in the megohm range and depends primarily on I_s and only secondarily on the back bias voltage.

EXAMPLE 2-8

In Fig. 2-25 find R_s for the given reverse voltages with the following diodes:
  (a) $E = 1$ V, silicon diode ($I_s = 2$ pA)
  (b) $E = 10$ V, silicon diode ($I_s = 2$ pA)
  (c) $E = 1$ V, germanium diode ($I_s = 2$ μA)
  (d) $E = 10$ V germanium diode ($I_s = 2$ μA)

SOLUTION

The diode in Fig. 2-25 is obviously reverse biased.
  (a) $V_D = -1$ V (the negative sign is used to indicate reverse bias conditions).

$$V_R = I_s \times R = 2 \times 10^{-12} \times 100 = 2 \times 10^{-10}$$

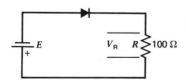

FIGURE 2-25 Reverse-biased diode.

By Ohm's Law and KVL

$$R_s = \frac{E - V_R}{I_s} = \frac{1 - 2 \times 10^{-10}}{2 \times 10^{-12}}$$

$$\approx \frac{1}{2 \times 10^{-12}} = 500{,}000 \text{ M}\Omega$$

Here, V_R was neglected since its magnitude is very small relative to E. It will also be neglected in parts (b), (c), and (d).

(b) $V_D = -10$ V

$$R_s = \frac{10 \text{ V}}{2 \times 10^{-12} \text{ A}} = 5 \times 10^{12} \text{ }\Omega$$

(c) $R_s = \dfrac{1 \text{ V}}{2 \times 10^{-6} \text{ A}} = 0.5 \times 10^6 = 0.5 \text{ M}\Omega$

(d) $R_s = \dfrac{10 \text{ V}}{2 \times 10^{-6} \text{ A}} = 5 \times 10^6 = 5 \text{ M}\Omega$

The approximation $R_s = \infty$ is certainly valid for the silicon diode and in most cases for the germanium diode. Thus, when reverse-biased, the diode may be considered an OPEN switch, as a first approximation.

SUMMARY OF IMPORTANT TERMS

Acceptor atoms Atoms with (typically) three electrons in the outer shell (valence electrons) capable of accepting a fourth electron into the shell (for example, atoms of boron, aluminum, gallium, indium).

Avalanche A process that leads to very high reverse (minority carrier) current in a p-n junction. It is present under a high reverse electrical field that produces a high rate of collision between electrons and valence electrons.

Conduction band; conduction electrons An energy band (electron energy level) associated with free electrons. (Electrons that constitute the electrical current.) See free electrons.

Covalent bond A bond between atoms that is based on the sharing of valence electrons between the atoms.

Depletion region The region between the P and N materials of a p-n junction where the concentration of free electrons has been substantially reduced, depleted, by the formation of the p-n junction. This region increases with increased reverse voltage.

Diffusion A process by which molecular or electron concentration is equalized by the movement of molecules from areas of high to areas of low concentration.

Diode resistance—dynamic (r_f) Defined as $\Delta V_D/\Delta I_D$. The change (very small) of diode voltage over the change of diode current.

Diode resistance—static (R_f) Defined as V_D/I_D. Diode dc voltage divided by diode dc current.

Donor atom An atom with (typically) five electrons in its outer shell capable of releasing an electron from the shell (for example, phosphorus, arsenic, antimony).

Doping A process by which impurity atoms, with three or five valance electrons, are introduced into the semiconductor material.

Electron shell A group of electrons orbiting the nucleus with their distinct orbits in very close proximity to each other.

Electron volt A unit of energy based on the product of charge times voltage. 1 eV = 1.6 × 10^{-19} joules.

Energy gap Electron energy levels that do not exist in the atomic structure (related to the special gap between the electron shells).

Forward bias Voltage applied to the diode in a polarity to induce conduction.

Free electrons Electrons that have been removed from the atomic structure (usually valence electrons leaving their orbit) and are free to be part of the electrical conduction. Also called conduction electrons.

Germanium—Ge A semiconductor material from which diodes and transistors are made.

Junction barrier voltage A voltage internal to the *p-n* junction. To induce conduction, an external voltage larger in magnitude and of appropriate polarity must be connected to the *p-n* junction to overcome the barrier voltage. Typically, 0.35 V for germanium and 0.7 V for silicon.

Junction—*p-n* The metalurgical (atomic) junction of P- and N-type materials.

Load line A graphic method used to find I_D—the diode forward current.

Majority carriers Electrical charges, electrons (−), or holes (+) that constitute the majority charges in a particular region. Usually contribute to forward diode current.

Minority carriers Charges (electrons or holes) that are in the minority in a particular region. Usually contribute to reverse (leakage) diode current.

N—material Semiconductor material that is doped with donor atoms. High concentration of free negative charges (free electrons).

P—material Semiconductor material doped with acceptor atoms. High concentration of free positive charges (holes).

Reverse bias External voltage applied to a diode with a polarity opposing forward conduction.

Semiconductor Material with four electrons in the valence band. Medium concentration of free (conduction) electrons. Classified in terms of its electrical conductivity between conductors and insulators (silicon, germanium).

Solid state diode A diode (*p-n* junction) constructed of semiconductor materials.

Tunneling The process that leads to zener breakdown. Electrons are torn from the valence shell by high electrical fields, thus becoming free electrons and very sharply increasing reverse (minority carrier) conduction.

Zener breakdown See *tunneling*.

I_D Forward diode current

I_s Saturation reverse current. In the order of μA for Ge and pA for Si.

k Boltzman's constant = 1.38 × 10^{-23} joules/°K, used in the expression for I_D.

°K Degrees kelvin (see T).

q The charge of an electron = 1.6 × 10^{-19} coulomb (figures in the expression for I_D.

r_f See *diode resistance—dynamic*.

R_F See *diode resistance—static*.

T Absolute temperature in degrees kelvin. T = °C + 273 (degrees kelvin = degrees celsius + 273).
V_D Diode voltage.
v-i plot A plot of diode voltage versus diode current.

PROBLEMS

2-1. The forward voltage across a diode is held constant at 0.52 V. The reverse saturation current I_s = 2 pA (at 27°C) and is assumed constant.

(a) Find I_D at 27°C (300°K).
(b) Find I_D at 47°C.
(c) Find the average change in I_D per °C ($\Delta I_D / \Delta T$).

2-2. Assume that I_s doubles for every 10°C increase in temperature.

(a), (b), (c) Repeat (a), (b), and (c) of Problem 2-1.
(d) Compare the rate of change of I_D in Problems 2-1(c) and 2-2(c).

2-3. In the circuit of Fig. 2-26, E = 1 V. R = 5 Ω
Find I, using:

(a) The idealized silicon diode characteristics (V_D = 0, r_f = 0),
(b) The approximation V_D = 0.7 V, r_f = 0 (draw the equivalent circuit).
(c) The iterative method (V_D = 0.7 V).
(d) The load line method (use the v-i characteristics for a commercial silicon diode given in Fig. 2-18).

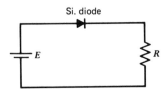

FIGURE 2-26

2-4. Repeat Problem 2-3 for E = 20 V, R = 100 Ω.

2-5. For Problem 2-1 find:

(a) The static diode resistance at 27°C.
(b) The dynamic diode resistance at 27°C.
(c) Repeat (a) for 47°C.
(d) Repeat (b) for 47°C.
(e) Compare the static and dynamic resistances.

2-6. In the diode circuit of Fig. 2-27 a measurement yielded V_R = 0.2 V. Find:

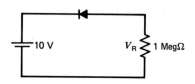

FIGURE 2-27

    (a) The diode leakage current.
    (b) The diode reverse resistance.

2-7. In Problem 2-6 the temperature of the diode (diode only) was increased by 20°C. Find:

    (a) The diode reverse resistance (leakage resistance).
    (b) The voltage V_R.

CHAPTER

3
DIODE APPLICATIONS

3-1 CHAPTER OBJECTIVES

This chapter presents some basic diode applications. These include circuits that modify the waveform of an ac signal, wave-shaping circuits, and circuits that are used to convert ac to dc. In all applications, the student will be given the ability to analyze the circuits both qualitatively and quantitatively. Rectifier circuits, whose major function is the conversion of ac to dc are analyzed, somewhat in detail, giving the student some understanding of the design process as well, even though the basic emphasis is on analysis. The discussion of rectifiers is followed by a number of filtering circuits, circuits that eliminate or reduce any leftover ac component present in the dc voltage. The meaning of voltage regulation, the extent to which the dc voltage varies or does not vary with changes in load conditions, is discussed and the use of zener diodes in regulating (stabilizing) the dc voltage analyzed. Throughout this chapter, the student will learn the significance of the device specifications as given in the manufacturer's data sheet.

3-2 INTRODUCTION

Diodes are used in a large variety of applications. In many instances, the diodes are constructed to match the applications—that is, their characteristics are designed by the manufacturer to best fit specific applications, such as high power diodes, high-speed low current diodes, and so on (see Sec. 3-7). Only a few simple and popular diode applications are presented here.

3-3 WAVE SHAPING

The diode circuits covered in this section involve *time-varying* voltages and currents. It becomes essential, as a first step in circuit analysis, to establish the "state" of the diode, either forward or reverse biased, for specific instances in time. Only after this determination has been made, can the appropriate diode equivalent circuits be selected (for the particular time segment). In most instances, the diode is considered OPEN ($R_s = \infty$) in the reverse bias region and the equivalent shown in Figs. 2-21 and/or 2-22 is used for the forward region.

3-3.1 Clipper and Clamper Circuits

It is often desired to keep a voltage below a certain level. For example, in many amplifier circuits the input signal may not exceed a certain amplitude. The *clipper* circuit serves this purpose; it "clips" the signal at a selected amplitude, keeping the voltage below the voltage limit. This usually results in a waveshape distortion as is demonstrated below. The distortion of the waveshape in itself may sometimes be the primary purpose of the clipper (harmonic generation).

The clamper is often used to shift the dc operating voltage of a waveform. It may be desired to change a square wave having an amplitude of 0 to $+10$ V to one with a 10 V amplitude from -10 to 0 V. These two circuits are best explained by means of examples.

EXAMPLE 3-1

In Fig. 3-1a, $e^{(1)}$ is a sinusoidal voltage with period $T = 20$ ms ($f = 50$ Hz) with an amplitude of 3 V or

$$e = 3 \sin 314\, t. \qquad \text{(Fig. 3-1b)}$$

Draw i and v_o.

SOLUTION

To determine for what value of e the diode is forward-biased and for what values the diode is reverse biased, replace the diode with its forward-biased equivalent circuit. When forward biased, the current i must flow as indicated by the arrow. We may now write KVL for the loop, neglecting r_f.

$$e - 0.7 - i \times R = 0.$$

Solving for i yields

$$i = \frac{e - 0.7}{R}$$

---

[1] The use of lowercase symbols is reserved for ac or dynamic quantities, while uppercase symbols are used for dc parameters.

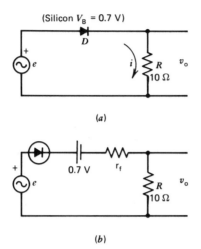

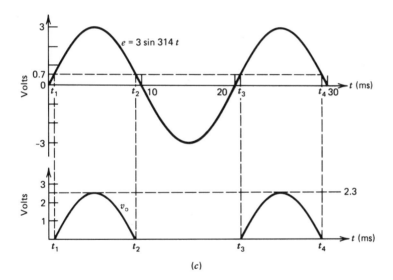

FIGURE 3-1 Series "clipper" circuit. (a) Actual circuit. (b) Forward biased equivalent circuit. (c) Plots of e and v_o.

i must be positive (a negative value for i implies a current in the reverse direction, that is, a reverse diode current). But i can be positive only if $e - 0.7$ is positive; hence, e must be larger than 0.7 V. e takes values *above* 0.7 V between t_1 and t_2 and between t_3 and t_4, etc. (Fig. 3-1c). For these time intervals $i \neq 0$ (i is nonzero).

At $t = t_1$, $t = t_2$ etc., $i = 0$ because the voltage in the loop is zero ($e - 0.7 = 0$). If reverse currents are neglected, $i = 0$ for all intervals outside $t_1 - t_2$, t_3

66 DIODE APPLICATIONS

$- t_4$, etc. In the forward biased intervals, Eq. (2-8) is valid and may be used to compute i as follows:

$$\text{at} \quad t = t_1, \quad i = 0$$
$$\text{at} \quad t = 5 \text{ ms}, \quad e = 3 \text{ V}$$

hence,

$$i = \frac{3 - 0.7}{R + r_F} = \frac{3 - 0.7}{10 + r_F}$$

Using the iterative method, we set $r_F = 0$

$$i = \frac{3 - 0.7}{10} = 230 \text{ mA}$$

and

$$r_F = \frac{26}{230} = 0.11 \, \Omega$$

A second iteration would change i very little (since $r_F \ll 10$)

$$i = \frac{3 - 0.7}{10 + 0.11} = 227.5 \text{ mA} \approx 230 \text{ mA}$$
$$i = 230 \text{ mA}.$$

Between t_1 and t_2, i will follow the sinusoidal shape of e. There is some distortion of the sinusoidal shape because of the substantial increase in r_F as e approaches 0.7 V. The student should find r_F for various values of e near $e = 0.7$ V and find i for these points (in this range i will largely depend on r_F rather than R). In our analysis this effect is neglected; it is assumed that r_F is very low (for the most part it can be assumed that $r_F = 0$). Since $v_o = i \times R = i \times 10$, we get the plot shown in Fig. 3-1c. If we assume $r_F = 0$, then

$$v_o = e - 0.7$$

for the forward biased intervals. This agrees with the plot of v_o shown. For the reverse biased intervals, the diode is open; hence, $v_o = 0$. (This circuit is called a "series clipping circuit.")

EXAMPLE 3-2

In Fig. 3-2a, $e = 3 \sin \omega t$ ($T = 20$ ms). Plot i and v_o.

WAVE SHAPING 67

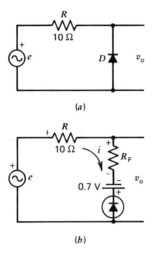

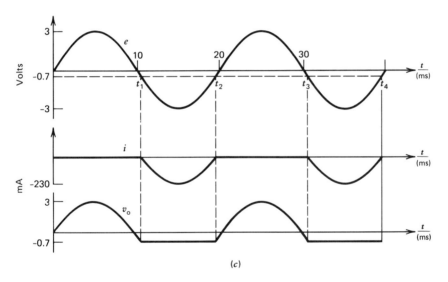

FIGURE 3-2 Parallel "clipper" circuit. (a) Actual circuit. (b) Equivalent circuit. (c) Plots of e, i, and v_o.

SOLUTION

Applying KVL to the equivalent circuit (Fig. 3-2b) yields

$$e - iR - ir_F + 0.7 = 0$$

$$i = \frac{3 + 0.7}{R + r_F}$$

68 DIODE APPLICATIONS

The chosen direction of i implies a *reverse* current through the diode (cathode to anode). For the current to be a forward current, i must be negative (anode to cathode), effectively reversing the direction of the current. This leads to $e + 0.7$ negative, or $e + 0.7 < 0$; hence, $e < -0.7$. For all values of e less than -0.7 V, the diode is forward biased. This is true for the intervals $t_1 - t_2$, $t_3 - t_4$, etc. At t_1, t_2, t_3, and t_4, $i = 0$. For the intervals $t_1 - t_2$ and $t_3 - t_4$, etc., i can be calculated by the expression for i given above. At $t = 15$ ms (where $e = -3$ V)

$$i = \frac{-3 + 0.7}{R + r_F}$$

If we assume $r_F = 0$,

$$i = \frac{-3 + 0.7}{10} = \frac{-2.3}{10} = -0.23 \text{ A} = -230 \text{ mA}$$

For this value of i, $r_F = 26/230 = 0.11\ \Omega$ which can be neglected ($r_F \ll R$). The assumption $r_F = 0$, therefore, is quite valid.

A plot of i is shown in Fig. 3-2c. v_o can be found by realizing that in the intervals $t_1 - t_2$, $t_3 - t_4$, etc., for which forward bias conditions exist,

$$v_o = i \times r_F - 0.7$$

(the appropriate KVL is $v_o - i \times r_F + 0.7 = 0$). If we assume $r_F = 0$, we have, for these intervals, $v_o = 0.7$ V (see Fig. 3-2c). For the reverse bias intervals (outside $t_1 - t_2$, $t_3 - t_4$, etc.), $v_o = e$ since no current flows and there is no voltage drop across R.

The complete waveform for v_o is shown in Fig. 3-2c. This circuit is called a "parallel clipping circuit."

EXAMPLE 3-3

Find i and v_o for circuit of Fig. 3-3a using the "ideal" diode equivalent.

SOLUTION

This solution, assuming an "ideal" diode, is approximate only. It is simple to obtain and leads the way to a more accurate solution, should one be desired. It assumes the diode is an open switch when reverse-biased and a closed switch when forward-biased.

As before, we have to establish forward bias conditions. The idealized equivalent circuit is shown in Fig. 3-3b. For forward bias, the anode (positive terminal) potential of the diode must be equal or more positive than the cathode. This implies anode potential of -2 V or larger (more positive). Consequently, for all values of $e > -2$ V, the diode is forward-biased (interval $t_1 - t_2$ in Fig. 3-3c). When the diode is ON, $v_o = -2$ V (the -2 V battery is effectively connected across v_o).

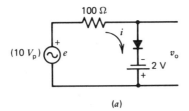

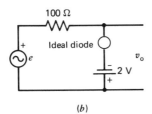

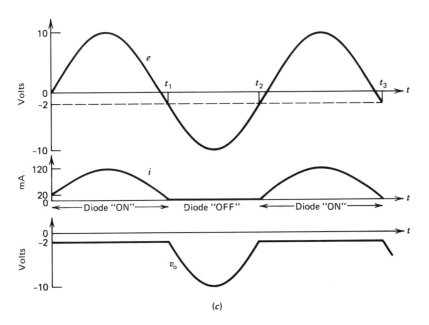

FIGURE 3-3 Parallel "clipper" with bias. (a) Circuit. (The subscript "p" is an abbreviation for "peak.") (b) "Idealized" equivalent circuit. (c) Plots of e, i, and v_o.

For this interval

$$e - i \times R + 2 = 0$$

$$i = \frac{e + 2}{R}$$

70   DIODE APPLICATIONS

For intervals outside $t_1 - t_2$, etc., the diode is OFF, $i = 0$ and $v_o = e$ (see plots in Fig. 3-3c).

The circuits in Examples 3-1 and 3-2 may be modified by the addition of dc biasing sources. The solution of such circuits would proceed as before. The additional voltage sources are simply included in the applicable KVL equation.

EXAMPLE 3-4

Plot v_o for Fig. 3-4a (v_{in} is shown in Fig. 3-4b).

SOLUTION

If the ideal diode equivalent ($r_F = 0$, $V_B = 0$) is used, it is clear that V_{BN} (the voltage across the diode) *cannot* be positive. That is, V_B (with respect to the GND point—N) can only be zero or negative. (V_{BN} positive turns the diode on, shorting voltage V_{BN}. V_{BN} becomes zero when the diode is on.)

At time t_1, a 25 V negative transition takes place at the input (node A) (v_{in} swings negative by 25 V). This transition is fully transmitted to node B (the capacitor behaves like a short for fast varying voltages). V_B becomes -25 V.

During the interval t_1 to t_2, the voltage V_B (node B) is charging up toward GND (0 V). The rate of this charging process depends on the product $R \times C$ (R in ohms, C in farads). For $R \times C \gg t_2 - t_1$, V_B changes very little during the interval t_1 to t_2 (Fig. 3-4c). At time t_2, a 25 V positive transition occurs, bringing V_B back to 0 V.

For $R \times C \approx t_2 - t_1$ ($R \times C$ having about the same value as $t_2 - t_1$), V_B rises toward GND during the interval t_1 to t_2. At t_2, the 25 V positive input swing, coupled directly to node B, brings V_B up to GND (Fig. 3-4d). Recall that V_B cannot be positive. The diode essentially clips off the dashed portion of the waveform in Fig. 3-4d.

Figure 3-4e gives v_o for the case $R \ll t_2 - t_1$. For this condition, V_B charges "all the way" up to GND during the interval t_1 to t_2.

For the case $R \times C \gg t_2 - t_1$, the circuit effects a shift in the signal dc levels. The input signal swings between $+10$ V and -15 V, while v_o swings between 0 and -25 V.

3-3.2   Peak Detecting

In many measuring systems, we may wish to measure the *peak* amplitude of a periodic or nonperiodic waveform. Since a regular meter cannot respond fast enough to measure the short duration pulse, a peak detector is used to convert the peak value into a dc voltage. (The peak detector "holds" the amplitude for a large enough duration to allow the use of dc meters.) A typical *peak detector* (or peak sampler) is shown in Fig. 3-5a. The circuit is somewhat idealized (various source

WAVE SHAPING 71

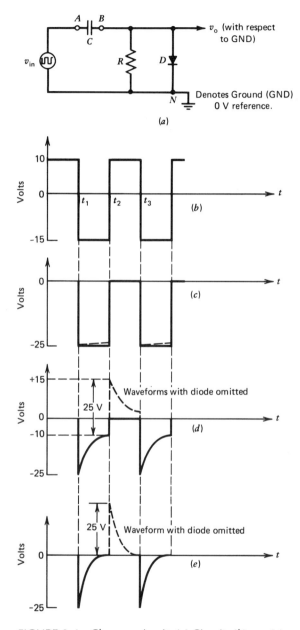

FIGURE 3-4 Clamper circuit. (a) Circuit. (b) v_{in}. (c) v_o for the case. $R \times C \gg t_2 - t_1$. (d) v_o for $R \times C \approx t_2 - t_1$. (e) v_o for $R \times C \ll t_2 - t_1$.

and load resistances have been neglected) and our explanation, again, assumes an ideal diode.

For any e_{in} positive, the diode is ON and the capacitor is charged to the value of e_{in}. This value is maintained across the capacitor (since there is no discharge

72 DIODE APPLICATIONS

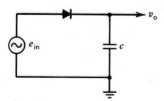

FIGURE 3-5 Peak detector (ideal, no load).

path) until a more positive e_{in} voltage occurs. v_o retains the value of the largest positive value of e_{in}, that is, the peak value of e_{in}. With the diode terminals *reversed*, a *negative* peak is obtained.

In practical circuits the forward resistance of the diode, r_F and the load resistance R_L, representing an instrument or display connected across the capacitor must be considered. Two conditions must be met:

1. $r_F \times C \ll T$ pulse (the duration of the pulse—T pulse)
2. $R_L \times C \gg T$ no pulse (the interval between pulses—T no pulse)

In words, condition 1 means that we must provide a *fast* charge-up of the capacitor during any short duration pulse presence (the diode is on). Condition 2 means that the charge must be maintained unaltered during the waiting time, when no larger pulse is present (the diode is off).

3-4 DC POWER SUPPLIES

The dc power supply finds use in many of our applications and instruments. The radio, stereo amplifier, electronic calculator, the computer, and many other systems operate on dc voltage, while our main electrical power comes from a 120 V 60 Hz ac power system. It is not surprising, therefore, that ac to dc conversion is extremely useful and very widely used. The dc power supply serves to accomplish this conversion.

The power supply consists of separate elements as shown in Fig. 3-6. The transformer is used to step up or down the ac voltage when necessary, to suit the desired dc voltage. It also serves to isolate the dc ground from the ac line neutral terminal. The rectifier converts ac with zero average voltage to pulsating dc. The filter removes the ac component still present in the waveform. The regulator is used to keep the dc voltage constant in spite of load and input voltage variations. The regulator is used only when constant—regulated—dc voltage is required.

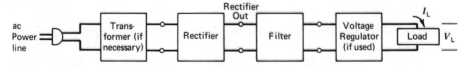

FIGURE 3-6 Power supply block diagram.

3-4.1 Rectifiers

Rectifiers belong to the class of wave-shaping circuits whose primary purpose is to convert ac voltage (and current) to dc.

The majority of rectifier circuits use diodes as the central component. These circuits convert an input sine wave with a zero average voltage (long-term average of a sine wave = 0) to waveforms that have nonzero averages. In other words, the pure ac voltage is converted into a combination of ac and dc voltage. *The average value of a waveform is its dc components.* Let us investigate a simple (and somewhat idealized) sine wave rectifier circuit (where the input is a sinewave) in an attempt to provide a numerical relationship between the ac and dc quantities.

3-4.2 Half-Wave Rectifier

The circuit of Fig. 3-7a has a sinusoidal input signal $e_{in} = E_m \sin \omega t$ and produces an output v_o as shown in Fig. 3-7b. (Assuming an ideal diode.) While the average of e_{in} is zero, the average of v_o can be found by dividing the area (above the axis over the period T) by the period T

$$V_{av} = \frac{\text{area}}{T}$$

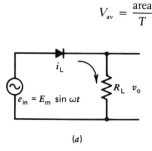

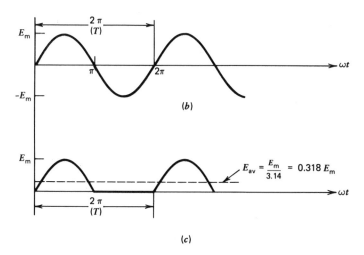

FIGURE 3-7 Half-wave rectifier. (a) Circuit. (b) Input voltage (e_{in}). (c) Output voltage (V_o).

74 DIODE APPLICATIONS

Note that the period T is equivalent to $\omega t = 2\pi$ radians (when plotted on an ωt axis). The average value of one-half cycle or π radians is

$$V_{av} = V_{dc} = \frac{2 \times E_m}{2\pi} = \frac{E_m}{\pi}$$
$$= 0.318\ E_m\ [V]$$
(3-1)

(The area $= 2E_m$ is found by integration methods that are not essential to this presentation.) The average of the v_o waveform is a voltage of magnitude $E_m/3.14$ (E_m/π). Recall that E_m is the amplitude of the input sine wave. This establishes a relation between the ac amplitude of the input and the dc component at the output.

The waveform v_o shown in Fig. 3-7b consists of only *half* of the original input signal (hence, the name *half-wave rectifier*). The negative portion of the input signal makes no contribution to the output v_o. This indicates a rather inefficient ac to dc conversion. (The full wave and bridge rectifier circuits discussed in Sections 3-4.3 and 3-4.4 remedy this deficiency.) Furthermore, v_o contains a very large ac component and cannot be used in applications in which a very steady (constant) dc voltage is required. (Should you choose, for example, to use this voltage to power your pocket radio, you will hear a very loud 60 Hz "hum" in your speakers.) As discussed later, various "filtering" methods can be used to reduce the ac components in v_o.

The dc current ($I_{av} = I_{dc}$) in the circuit of Fig. 3-7a is

$$I_{dc} = \frac{V_{dc}}{R_L}$$

$$I_{dc} = \frac{E_m/\pi}{R_L} = I_m/\pi\ [A]$$
(3-2)

$$(I_m = E_m/R_L)$$

The relationship $I_{dc} = I_m/\pi$ or $I_m = I_{dc} \times \pi$ points to another disadvantage of this circuit—namely, that the *maximum* current through the diode is 3.14 times the usable dc current. The diode must be capable of carrying this large peak current. Its *peak current rating* must be high compared to the average dc current used.

Another important consideration in diode applications, in general, is the maximum voltage that appears across the diode in the reverse-biased condition ($I_D = 0$). In the case of the half-wave rectifier circuit (Fig. 3-7a), that peak reverse voltage is $-E_m$. During the negative half of the cycle, the current in the circuit is 0. The full half cycle of input waveform, therefore, appears across the diode. At its peak, this waveform attains the value E_m. When specifying a diode for use in a half-wave rectifier, one must select a peak inverse voltage rating for the diode that is larger than E_m, or PIV $> E_m$.

The PIV is the *maximum reverse voltage* the diode can withstand. Clearly, the PIV specification must be larger than the actual maximum reverse voltage across the diode.

EXAMPLE 3-5

Design a half-wave rectifier circuit to produce a 1 A dc current (average current) in a 10 Ω load resistor (neglect r_F).
  (a) Select the input e_{in}.
  (b) Specify diode PIV and peak current.
  (c) With line voltage of 120 V 60 Hz, select a suitable transformer.

SOLUTION

The basic circuit is shown in Fig. 3-8a.
  (a) By Ohm's law

$$V_{dc} = I_{dc} \times R_L = 1 \times 10 = 10 \text{ V}$$

From Eq. (3-1)

$$E_m = \pi \times V_{dc} = \pi \times 10 = 31.4 \text{ V}$$

hence,

$$e_{in} = 31.4 \sin \omega t$$

The amplitude of e_{in} must be 31.4 V.
  (b) Since PIV > E_m, we can specify PIV = 40 V. This provides some safety margin. PIV need only be larger than 31.4 V. From Eq. (3-2)

$$I_m = 3.14 \times I_{dc} = 3.14 \times 1 = 3.14 \text{ A}.$$

Hence, the diode must be rated for a peak current larger than 3.14 A. Choosing a diode with a peak current rating of 5 A provides some safety margin.

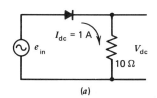

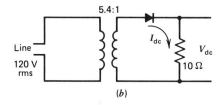

FIGURE 3-8 Half-wave rectifier design. (a) Basic circuit. (b) Circuit using 120 V line.

(c) The various diode parameters and circuit currents and voltages discussed in the preceding sections are independent of frequency. Indeed, if the half-wave rectifier is used in conjunction with a 400 Hz ac system (as compared with our common 60 Hz system), the relations between I_m and I_{dc}, E_m and V_{dc}, etc. will remain the same. In many common applications, the available input voltage is 120 V at 60 Hz rms line voltage. Consequently, a *transformer* has to be used to obtain the desired e_{in}.

From part (a) the required source is $e_{in} = 31.4 \sin \omega t$ with rms value of $E_{rms} = 0.707 \times 31.4 = 22.2$ V. Since the available line voltage is 120 V rms, a transformer with a step-down ratio of $120/22.2 = 5.4/1$ must be used. The complete circuit is shown in Fig. 3-8b.

3-4.3 Full-Wave Rectifier

The full-wave rectifier (Fig. 3-9) can be viewed as the combination of two half-wave rectifiers operating on alternate half-cycles of the input waveform. The two voltages e_{AN} and e_{BN} are shown in Fig. 3-9b. During the interval t_0 to t_1, t_2 to t_3, etc., the diode D_1 is conducting (ON) while D_2 is OPEN (OFF). For this time interval, the circuit is essentially a half-wave rectifier consisting of e_{AN}, D_1, and R_L. Similarly, during the intervals t_1 to t_2, t_3 to t_4, etc., we have a half-wave rectifier consisting of e_{BN}, D_2, and R_L. As far as R_L is concerned, the two half-wave rectifiers produce a *sum* effect in R_L, with v_o as shown in Fig. 3-9d. Here, both halves of the waveform contribute to the dc average. The dc output voltage ($V_{dc} = V_{av}$) becomes

$$V_{dc} = V_{av} = 2 \times E_m/\pi = 0.636 \, E_m \text{ V} \tag{3-3}$$

E_m is the amplitude of *half* of the secondary voltage. Equation (3-3) shows that the average output of a full-wave rectifier is twice the value for the half-wave rectifier. Here, $e_{AN} = E_m \sin \omega t$ and $e_{BN} = E_m \sin(\omega t \pm 180°)$. The total current provided by *both* diodes is

$$I_{dc} = I_{av} = 0.636 \, I_m \tag{3-3a}$$

where $I_m = E_m/R_L$ (assuming no diode or source resistance).

From Fig. 3-9 it can easily be seen that the reverse diode voltage (for each diode) is $2E_m$. During the t_0 to t_1 interval, D_2 is open. The voltage across D_2 (at time t_A) is the potential difference between points C and B; hence, $V_{CB} = V_{CN} - e_{BN} = E_m - (-E_m) = 2E_m$. A similar case can be made for D_1 at time t_B.

EXAMPLE 3-6

In Fig. 3-9, let the line voltage be 120 V rms and the transformer ratio (total primary to total secondary) is 5:1. The secondary is center-tapped; that is, e_{AN} and e_{BN} are each exactly half the total secondary voltage. $R_L = 20 \, \Omega$. Find:

DC POWER SUPPLIES 77

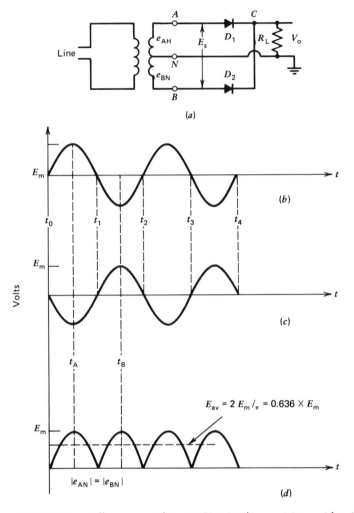

FIGURE 3-9 Full-wave rectifier. (a) Circuit. (b) e_{AN}. (c) e_{BN}. (d) $v_o(V_{CN})$.

(a) V_{dc} (dc voltage across R_L)
(b) I_{dc} (dc current in R_L)
(c) Specify PIV and I_{peak} for the diodes.

SOLUTION

(a) From the turns ratio and line voltage we can find E_m

$$V_{line} = 120 \text{ V rms}$$
$$E_s = 120/5 = 24 \text{ V rms}$$
$$E_s/2 = 12 \text{ V rms};$$
$$|E_m| = |E_{AN}| = |E_{BN}| = 12 \times 1.414 \text{ V peak}$$
$$= 17 \text{ V peak}$$

78   DIODE APPLICATIONS

This is the amplitude of the driving ac source (e_{AN} or e_{BN}).

$$V_{dc} = \frac{2}{\pi} \times E_m = 0.636 \times E_m = 0.636 \times 17 \qquad (3\text{-}3)$$
$$= \underline{10.8 \text{ V dc}}$$

(b) $I_{dc} = V_{dc}/R_L = 10.8/20 = \underline{0.54 \text{ A dc}}$
(c) $I_m = I_{dc}/0.636 = 0.54/0.636 = \underline{0.85 \text{ A peak}}$

$$\left(I_m = \frac{E_m}{20} = \frac{17}{20} = \underline{0.85 \text{ A peak}} \right)$$

PIV > $2E_m$, PIV > 34 V. Select $PIV = 50$ V.

The peak current in each diode is 0.85 A. It is reasonable to select a I_{max} rating of 1 A (or higher). Note that while the peak current through each diode is $I_{dc}/0.636$ = 0.85 A, the average current through each diode is half the total dc current. (Each diode conducts for only half the time.) For better safety margins it is advisable to select PIV ≈ 100 V and I_{peak} ≈ 2 A.

In the above calculations it was assumed that $V_B = 0$, no diode barrier voltage. The actual E_m would be 0.7 V lower when a silicon diode is used, about 0.35 V lower for germanium diodes. r_F was also neglected. The diode resistance r_F, however, is indeed very small since each diode dc current is 0.27 A ($I_{dc}/2$) that would yield an $r_F = \dfrac{26 \text{ mV}}{270 \text{ mA}} \simeq 0.1 \text{ }\Omega$ that is very small compared to R_L.

3-4.4   Bridge Rectifier

It is often useful to be able to produce a full-wave rectified waveform without using a center-tapped transformer. The bridge rectifier accomplishes this task (Fig. 3-10). The transformer shown in Fig. 3-10a is for purposes of obtaining the desired ac voltage to be rectified (from the standard 120 V ac line). In the bridge rectifier circuit, the output terminals are completely independent. There are two dc terminals; neither is *common* to the ac voltage (see, in contrast, the half-wave rectifier of Fig. 3-7 where one terminal is common to *both* ac and dc voltages). With this circuit both negative or positive dc supply voltages (with respect to GND) can be produced. Connecting the +dc terminal to GND yields a *negative* supply, while connecting the −dc terminal to ground yields a positive supply.

Figure 3-10c shows the operation of the circuit for the time segment t_0 to t_1. During this interval the polarity of the ac voltage is such that it makes AC_1 positive and AC_2 negative. This polarity turns D_1 and D_3 ON (forward biased) and D_2 and D_4 OFF (reverse biased). As a result, the current I_L, through R_L, has the direction shown, from +dc to −dc.

A similar situation exists for the interval t_1 to t_2 (Fig. 3-10d). This time AC_1 is

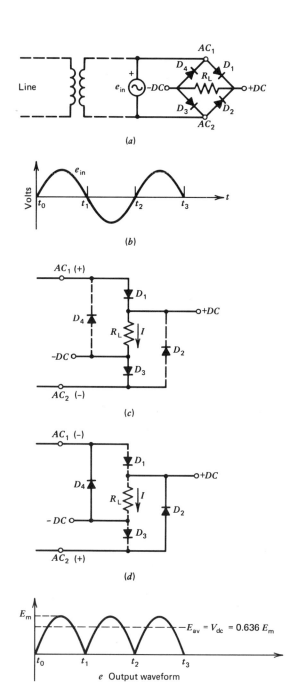

FIGURE 3-10 Bridge rectifier. (a) Circuit (transformer shown as option when needed). (b) ac input voltage—e_{in}. (c) Equivalent circuit for interval t_0 to t_1. D_1 and D_3—ON. D_2 and D_4—OFF. (d) Equivalent circuit for interval t_1 to t_2. D_2 and D_4—ON. D_1 and D_3—OFF. (e) Output waveform.

80 DIODE APPLICATIONS

negative and AC_2 positive. Hence, D_2 and D_4 are ON and D_1 and D_3 are OFF. The current through the load I_L again has the direction +dc to −dc.

For either half of the ac voltage, the output has the same polarity, resulting in a full-wave-rectified output waveform (Fig. 3-10e). It is easy to see that in this configuration the maximum reverse voltage applied to each diode (when in the OFF state) is E_m, and

$$E_{av} = E_{dc} = 0.636 \times E_m \tag{3-4}$$

where E_m is the amplitude of e_{in}. Each pair of diodes D_1, D_3 and D_2, D_4 carry half the total average current through the load. Each pair conducts for only half the cycle.

EXAMPLE 3-7

A bridge rectifier is used to deliver 5 A at 30 V dc.
 (a) Find the PIV rating of the diodes.
 (b) Find the max current rating of the diodes.

SOLUTION

 (a) To obtain 30 V dc (30 V average) with a full-wave rectifier, the peak ac voltage must be, from Eq. (3-4),

$$E_m = V_{dc}/0.636 = 30/0.636 = 47 \text{ V}$$

PIV > 47 V (choose a PIV rating of 50 V or higher).
 (b) $I_m = 5/0.636 = 7.86$ A peak
Select I_{max} rating of 10 A (or higher).

3-4.5 Filtering

The rectifiers discussed in the previous sections produced an average voltage (dc component). However, the waveform still contains a large ac component. An inspection of the rectified waveforms (Figs. 3-7c and 3-9d) reveals that, indeed, the voltage at the output terminals keeps varying with time. One can calculate the frequency of these waveforms. The *half-wave* rectifier produces an output with a frequency *equal* to the original ac source frequency. The *full-wave* rectifier output consists of a waveform with *double* the original ac source frequency.

The power supply used to power most of our dc appliances (radio, hi-fi, etc.) can tolerate only a very small ac component. For example, the hi-fi power supply (inside your hi-fi cabinet) is typically 15 V dc with an ac component of *less* than 15 mV (a ratio of 1000:1, dc to ac). To achieve this low ac content we use *filters, whose primary purpose is to eliminate ac voltages while not affecting the dc voltage.* Two basic criteria are used in the evaluation of filter performance: First, how

DC POWER SUPPLIES   81

effective is the filter in reducing ac, i.e., *attenuating* ac? Second, how little does the filter affect the dc voltage?

The filter circuit is inserted between the rectifier and the load or a regulator if one is used (Fig. 3-6). Many different circuits may be used to filter the ac; three types (Fig. 3-11) are explored here. Since the capacitor filter (Fig. 3-11a) is common to all three filter circuits (in all three circuits, there is a capacitor across the rectifier terminals), this circuit will be analyzed first.

THE CAPACITOR FILTER

The capacitor filter is much like the peak detector of Section 3-3.2. With the absence of a load resistor R_L ($R_L = \infty$), the capacitor is charged to the peak voltage (E_m) and holds the charge indefinitely. Consequently, $V_{dc} = E_m$ and $V_R = 0$ (where V_{dc} is the dc output voltage, V_R is the ripple or ac component of the output voltage).

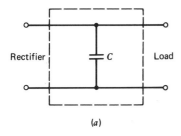

(a)

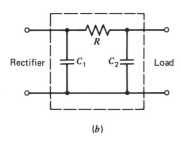

(b)

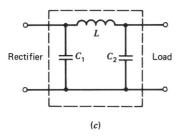

(c)

FIGURE 3-11 Three types of filters. (a) Capacity filter. (b) R-C filter. (c) L-C filter (π).

82 DIODE APPLICATIONS

The introduction of R_L causes the capacitor to discharge between pulses (Fig. 3-12). The quantitive analysis of the circuit is based on a number of assumptions and observations.

a. $V_R \ll E_m$. V_R is the ac component of the output voltage; the ripple voltage is small compared with E_m.
b. The ripple voltage (the ac component) is a *triangular* wave (all segments are straight lines). This is a reasonable approximation, in particular when V_R is small.
c. V_{dc} is midway between the peaks of V_R. (This is a result of the fact that V_R is symmetrical around its center axis.)
d. $T_1 + T_2 = T$, the period of the rectified waveform.

The waveform shown in Fig. 3-12 is that of the output of a half-wave rectifier. The analysis is for either half-wave or full-wave circuits. In the case of a full-wave rectifier, the period T would be half of the original ac source period, while for the half-wave shown, the period is that of the original ac source. It is clear that as T increases, T_2 increases and the capacitor C discharges further; hence, V_R increases.

V_R can be evaluated by applying the relationship $C \times V = Q = I \times t$. For incremental changes in V, ΔV,

$$\Delta V \times C = \Delta Q = I \times \Delta t \text{ coulombs} \tag{3-5}$$

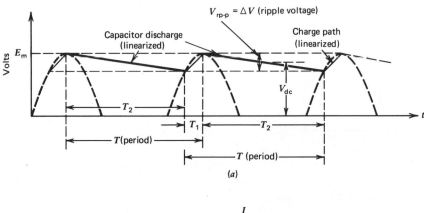

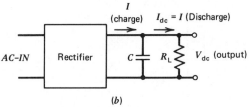

FIGURE 3-12 Capacitor filter (with half-wave rectifier). (a) Output waveform. (b) Rectifier, filter capacitor, and load.

If we apply Eq. (3-5) to our circuit and realize that for the discharge portion (T_2) $I = I_{dc}$, $\Delta V = V_{Rp\text{-}p}$, and $\Delta T = T_2$, we get

$$V_{Rp\text{-}p} \times C = I_{dc} \times T_2 \tag{3-6}$$

I_{dc} is assumed constant. This is quite valid for small V_R. Since $I_{dc} = V_{dc}/R_L$,

$$V_{Rp\text{-}p} \times C = (V_{dc}/R_L) \times T_2; \tag{3-6a}$$
$$V_{Rp\text{-}p} = [V_{dc}/(R_L \times C)] \times T_2$$

For *triangular* waveforms, $V_{rms} = V_{p\text{-}p}/2\sqrt{3}$ (by application of the root-mean-square procedure). Rewriting Eq. (3-6a) for V_{Rrms} instead of $V_{Rp\text{-}p}$ yields

$$V_{Rrms} = [V_{dc}/(2\sqrt{3}\, R_L \times C)] \times T_2 \quad [V] \tag{3-7}$$

A careful inspection of Fig. 3-12a reveals that as $V_{Rp\text{-}p}$ decreases, T_2 approaches T (T_1 decreases) and V_{dc} approaches E_m. It was initially assumed that V_R is very small. As a result, $V_{dc} \simeq E_m$ and $T_2 \simeq T$. Equation (3-7) becomes, therefore,

$$V_{Rrms} \simeq [E_m/(2\sqrt{3}\, R_L \times C)] \times T \quad [V] \tag{3-7a}$$

T is substituted for T_2 and V_{dc} for E_m

$$V_{Rrms} \simeq E_m/(2\sqrt{3} \times f \times R_L \times C) \quad [V] \tag{3-7b}$$

where $f = 1/T$ (or $1/f = T$) is the frequency of the rectified waveforms (not necessarily the same as the ac source frequency).

The ripple may now be computed from circuit component values (R_L, C), and from source voltage parameters (E_m, T).

It is useful to define the ripple factor r (relative ripple) or the percent ripple (%r) as

$$\text{Ripple factor} \quad (r) = V_{Rrms}/V_{dc} \tag{3-8}$$
$$\text{Percent ripple} \quad (\%\, r) = (V_{Rrms}/V_{dc}) \times 100 \tag{3-8a}$$

From Eq. (3-7) and the assumption $T_2 = T$

$$r = V_{Rrms}/V_{dc}$$
$$= \frac{V_{dc} \times T_2}{2\sqrt{3} \times R_L \times C} / V_{dc} \simeq T/(2\sqrt{3}\, R_L \times C) \tag{3-8b}$$
$$= 0.29/(f \times R_L \times C)$$

where $1/2\sqrt{3} = 0.29$ and $T = 1/f$

$$\%r = [0.29/(f \times R_L \times C)] \times 100\% \tag{3-8c}$$

84 DIODE APPLICATIONS

In most instances, V_{dc} and I_{dc} are given. R_L can be obtained from

$$R_L = V_{dc}/I_{dc}.$$

Since our calculations are based on the assumption $V_R \ll V_{dc}$, it is advisable that when analyzing a circuit, we first find the ripple factor (which gives the relative amplitude of V_R with respect to V_{dc}). For ripple factors of 0.03 (3%) or less, the error due to our assumption is 5% or less.

EXAMPLE 3-8

A power supply provides 100 mA at 20 V dc. It uses capacitive filtering and is driven from a 60 Hz source. Calculate the ripple factor for the following:
(a) half-wave rectification and $C = 1000$ μF
(b) full-wave rectification and $C = 1000$ μF
(c) full-wave rectification and $C = 500$ μF

SOLUTION

$$r = V_{Rrms}/V_{dc} = \frac{0.29}{f \times R_L \times C} \qquad (3\text{-}8b)$$

and

$$R_L = \frac{V_{dc}}{I_{dc}} = \frac{20\text{ V}}{100 \times 10^{-3}\text{ A}} = 200\ \Omega$$

(a) In this case $f = 60$ Hz (the frequency of the rectified waveform is the same as that of the ac source)

$$r = \frac{0.29}{60 \times 200 \times 1000 \times 10^{-6}} = \underline{0.024 \quad (2.4\%)} \qquad (3\text{-}8b)$$

(b) For full-wave rectification $f = 120$ Hz (double the source frequency)

$$r = \frac{0.29}{120 \times 200 \times 1000 \times 10^{-6}} = \underline{0.012 \quad (1.2\%)}$$

(c) $f = 120$ Hz

$$r = \frac{0.29}{120 \times 200 \times 500 \times 10^{-6}} = \underline{0.024 \quad (2.4\%)}$$

To calculate V_{dc}, we observe in Fig. 3-12 that

$$V_{dc} = E_m - \frac{1}{2} V_{Rp\text{-}p} \quad \text{(assumption (c) above)} \qquad (3\text{-}9)$$

Combining Eq. (3-9) and Eq. (3-6a), we set

$$V_{dc} = E_m - \frac{1}{2}\left(\frac{V_{dc}}{R_L \times C} \times T_2\right) \qquad (3\text{-}9a)$$

Rearranging (3-9a), we set

$$E_m = V_{dc}\left(1 + \frac{1}{2R_L \times C} \times T_2\right) \qquad (3\text{-}10)$$

By substituting $R_L = V_{dc}/I_{dc}$, Eq. 3-10 becomes

$$E_m = V_{dc} + \frac{I_{dc}}{2 \times C} \times T_2 \qquad (3\text{-}10a)$$

Equations (3-10) and (3-10a) establish a relation between the dc parameters V_{dc}, I_{dc} and the ac voltage E_m. The capacitor C is usually calculated from ripple factor requirements [Eq. (3-8)].

Note that T_2 is the *discharge* time between charges (Fig. 3-12a). An exact evaluation of T_2 is rather complex and, for the most part, the approximation $T_2 = T$ is adequate. Equations (3-10) and (3-10a) become, respectively,

$$E_m \simeq V_{dc}\left(1 + \frac{1}{2R_L \times C} \times T\right) \qquad (3\text{-}10b)$$

$$\simeq V_{dc}\left(1 + \frac{1}{2f \times R_L \times C}\right)$$

$$E_m \simeq V_{dc} + \frac{I_{dc}}{2 \times C} \times T \simeq V_{dc} + \frac{I_{dc}}{2f \times C} \qquad (3\text{-}10c)$$

The value E_m obtained by either equation (given V_{dc}, I_{dc}, and C) is somewhat larger than the actual E_m because of the assumption $T_2 = T$ (since $T > T_2$).

EXAMPLE 3-9

A dc power supply uses full-wave rectification from a 120 V-60 Hz source. The following are the applicable specifications:
   Ripple factor not to exceed 0.04 (4%);
   Maximum dc current $I_{dc} = 0.5$ A;
   dc voltage $V_{dc} = +30$ V.
The circuit uses a center tapped transformer and a simple capacitor filter.
   (a) Draw the circuit.
   (b) Calculate the values of all components.
   (c) Find the required transformer ratio.

86 DIODE APPLICATIONS

SOLUTION

(a) The circuit is shown in Fig. 3-13.
(b) As a first step, R_L and f are found.

$$R_L = \frac{V_{dc}}{I_{dc}} = \frac{30}{0.5} = 60 \, \Omega \quad (R_L \text{ for full load})$$

$f = 120$ Hz (2×60 Hz or double the ac source frequency).
The ripple factor specified is used to calculate C. From Eq. (3-8b),

$$0.04 = \frac{0.29}{120 \times 60 \times C}; \quad C = \frac{0.29}{120 \times 60 \times 0.04} = \underline{1006.9 \, \mu F}$$

Use $C = 1000 \, \mu F$. (Most capacitors of this size have an actual capacity exceeding the nominal value.)
E_m is found from Eq. (3-10c).

$$E_m \simeq 30 + \frac{0.5}{2 \times 120 \times 1000 \times 10^{-6}} = 32.08 \text{ V}$$

The total secondary ac rms voltage is:

$$E_{srms} = 2 \times E_m \times 0.707 = 2 \times 32.08 \times 0.707 = 45.36 \text{ V}$$

(c) The transformer turns ratio N (primary to total secondary) is

$$N = \frac{120}{45.36} = \underline{2.64}$$

The majority of capacitive filtered rectifier circuits can be designed, and/or analyzed, using two basic expressions. The equation giving the ripple voltage, directly obtained from Eq. 3-6:

$$V_{Rp\text{-}p} = \frac{I_{dc}}{C} \times T = \frac{I_{dc}}{C \times f} \tag{3-6b}$$

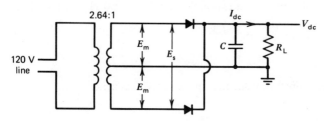

FIGURE 3-13 Full-wave rectifier power supply.

DC POWER SUPPLIES

and the expression for V_{dc}, recopied here for convenience:

$$V_{dc} = E_m - \frac{1}{2} V_{Rp\text{-}p} \tag{3-9}$$

Equation 3-6b can be derived intuitively, realizing that a larger I_{dc} will cause a faster discharge of the capacitor, and, hence, a larger ripple voltage, I_{dc} in the numerator. A larger period, T, would have a similar effect, since it would allow more time for discharge, T in the numerator. An increase in C would reduce ripple voltage, since it would store more charge and thus 'hold' the voltage, reduce voltage discharge rate. A *larger* C means a *lower* ripple voltage, C in the denominator. Equation 3-9 can be obtained directly from the waveform shown in Fig. 3-12. (To complete the analysis, it is, of course, necessary to understand the meaning of each of the various parameters, in particular: ripple factor, V_{Rrms}, etc.)

PIV, PEAK CURRENTS, AND CONDUCTION ANGLE

Two of the most important diode parameters that must be specified for rectifier applications are the PIV and the peak current. These parameters have been discussed in connection with the rectifier circuits. The addition of a filter capacitor C, however, changes these quantities substantially. With the filter capacitor $V_{dc} \approx E_m$. The dc output voltage is close to the peak ac. (Most rectifiers have very small ripple voltages.) Since $V_{dc} = E_m - \frac{1}{2} V_{Rp\text{-}p}$ [Eq. (3-9)], the assumption $V_{dc} \approx E_m$ is quite valid. The diodes in both half-wave and full-wave rectifiers are subject to a PIV of $2E_m$ or *double the peak input voltage*. (E_m for the full wave is given for one half the transformer secondary as shown in Fig. 3-13.) The cathode of the diode (in a rectifier with a positive output voltage) is connected to V_{dc}, while the voltage applied to the anode varies between $-E_m$ and $+E_m$. The peak reverse voltage is $V_{dc} - (-E_m) = V_{dc} + E_m \approx 2E_m$. For those types of rectifiers a diode PIV of $2E_m$ or larger must be specified.

In the case of the *bridge* rectifier, the actual maximum reverse voltage applied to each of the diodes is E_m; hence, all four bridge rectifier diodes must be capable of withstanding a reverse voltage equal to or larger than E_m.

The peak forward current through the diode for a capacitive filter rectifier circuit can be calculated by equating the total charge used by the load to the charge supplied through the diode. The diode conducts for only a portion of the full period (T_1 in Fig. 3-12a). If we assume that the diode current is constant for the period T_1, the peak current can be obtained by equating charges ($Q = I \times t$)

$$I_{peak} \times T_1 = I_{dc} \times T \quad \text{coulombs} \tag{3-11}$$

and

$$I_{peak} = I_{dc} \times \frac{T_1}{T} \quad [A] \tag{3-11a}$$

88 DIODE APPLICATIONS

Equation (3-11a) shows that the total peak current gets *larger*, as the duration of conduction gets smaller. (T used here is the period of the rectified waveform. It is equal to the source period for half-wave rectifiers and to half that value for the full-wave or bridge rectifiers.)

T_1, when translated into degrees T_1 (degrees) $= \phi_c = (360°/T) \times T_1$, where ϕ_c is called the *angle of conduction*. If all quantities are referred to the ac source frequency, $\phi_c = (360°/T) \times T_1$ for half-wave and $\phi_c = (180°/T) \times T_1$ for full-wave. (This is due to the fact that the half-wave rectifier T corresponds to 360° of the line voltage while for the full-wave rectifier T corresponds to 180° of the line voltage. These are two different values of T.) It is clear that for small V_R, where a large capacitor is used, the conduction angle is small, and hence, the short-term (pulse) peak current through the diode is many times the dc current. Since, in the full-wave rectifier, two current pulses (one through each set of diodes) are used to charge C, the peak current required would be half that required for the half-wave rectifier for the same ripple voltage. Table 3-1 gives some typical I_{peak}/I_{dc} values for various % ripple values:

R-C FILTER

The R-C filter is a simple resistor capacitor circuit (Fig. 3-14) that serves primarily to reduce ripple voltage. The circuit behaves like a voltage divider for ac voltages; hence, (for ac only)

$$v_o = \frac{v_{in} \times (-jX_C)}{R - jX_C} = \frac{v_{in}\{-(j/\omega C)\}}{R - (j/\omega C)} \qquad (3\text{-}12)$$

where v_{in} and v_o are the ac component (ripple voltage) at the input and output of the R-C filter, respectively, ω is the angular frequency ($2\pi f$) of the ripple waveform, and j comes from complex number notation. In terms of magnitude only, the ripple reduction can be expressed as a ratio of $|v_o/v_{in}|$. This ratio is less than unity indicating that the ripple voltage is reduced by the R-C network.

TABLE 3-1 % RIPPLE VERSUS I_{peak}/I_{dc} FOR FULL-WAVE (FW) AND HALF-WAVE (HW) RECTIFIERS

Percentage r	I_{peak}/I_{dc}	
	FW	HW
0.5	16.6	33.1
1	11.7	23.4
2	8.3	16.6
3	6.8	13.6
4	5.8	11.6
5	5.2	10.4

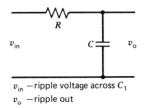

v_{in} — ripple voltage across C_1
v_o — ripple out

FIGURE 3-14 R-C filter. v_{in} and v_o are ac (ripple) voltages. v_{in}—ripple voltage across C_1. v_o—ripple voltage at output.

$$\left|\frac{v_o}{v_{in}}\right| = \left|\frac{-j\omega C}{R - (j/\omega C)}\right| = \frac{(1/\omega C)}{\sqrt{R^2 + (1/\omega C)^2}} \quad (3\text{-}12a)$$

$$= \frac{1}{\sqrt{\{R/(1/\omega C)\}^2 + 1}}$$

This expression can be simplified for the case in which $R \gg (1/\omega C)$ or $R/(1/\omega C) \gg 1$ (the reactance of the capacitor is much smaller than the resistance R). By ignoring the 1 compared to $[R/(1/\omega C)]^2$, we get

$$\left|\frac{v_o}{v_{in}}\right| \approx \frac{1}{\sqrt{(R/(1/\omega C))^2}} \quad (3\text{-}12b)$$

$$= \frac{(1/\omega C)}{R} = \frac{X_C}{R}$$

For example, for the case in which $R/(1/\omega C) = 5$, the *exact* relation gives a ripple reduction of $1/\sqrt{5^2 + 1} = 0.196$, while the *approximate* relationship yields a reduction of $1/5 = 0.2$ (a 4% difference).

The R-C circuit is usually connected to the output of a capacitor filter power supply, such as shown in Fig. 3-13. The complete circuit is shown in Fig. 3-15. R-C_2 is the added filter network connected to the simple capacitive filter C_1.

The analysis of the circuit proceeds in two steps. First, the simple capacitive filter circuit is analyzed (see Example 3-9). We then proceed to calculate the effects of the added R-C network. In terms of Fig. 3-15, Eq. (3-12b) becomes

$$v_{ro}/v_{r1} \approx X_C/R \quad (3\text{-}12c)$$

where v_{ro} and v_{r1} are the ripple voltages shown in Fig. 3-15.

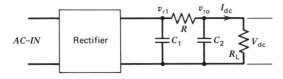

FIGURE 3-15 Power supply with R-C filter.

EXAMPLE 3-10

Connect an R-C section to the power supply of Example 3-9 (Fig. 3-13) so that the ripple factor is 0.004 (a 10 to one improvement over the circuit of Fig. 3-13).
  (a) Find R if we select $C_2 = 2000$ μF.
  (b) Find V_{dc} (at terminals A-N Fig. 3-16) for the maximum current.

SOLUTION

(a) To obtain a 1:10 reduction in ripple factor, the ratio X_C/R must be such that $X_C/R \leq 1/10$ ($R/X_C \geq 10$), based on the approximation given by Eq. (3-12c). Since the frequency is $2 \times 60 = 120$ Hz (ripple frequency of a full-wave rectifier), R can be obtained by:

$$R \geq 10 X_{C_2} = 10 \times 1/\omega C_2$$
$$= 10 \times 1/(2\pi \times 120 \times 2000 \times 10^{-6}) = 6.6 \, \Omega$$

$R \geq 6.6 \, \Omega$. Use $R = 6.6 \, \Omega$.

The maximum I_{dc} was specified as 0.5 A. The value of R_L for this current is $30/0.5 = 60$ ohms. The actual ripple voltage depends on R and the parallel combination of $X_{C_2} \| R_L$ and not simply R and X_{C_2}. It is, however, sufficient to consider X_C only since in most cases $R_L \gg X_{C_2}$ so that $X_{C_2} \| R_L \simeq X_{C_2}$. In this example $X_{C_2} = 0.66 \, \Omega$ and $R_L = 60 \, \Omega$.

(b) To obtain V_{dc} (worst case), observe that for $I_{dc} = 0.5$ A, the voltage drop across R is $6.6 \times 0.5 = 3.3$ V and $V_{dc} = 30 - 3.3 = 26.7$ V. Note that should $I_{dc} = 0$, no-load conditions, V_{dc} becomes 30 V. V_{dc} will change as the current is drawn (more on this subject in Section 3-4.6).

Note that the ripple factor has not exactly been improved by a factor of 10, as expected, since V_{dc} is no longer 30 V. $r = 0.004$ for $V_{dc} = 30$ V, yielding $V_{Rrms} = 0.004 \times 30 = 0.12$ V [from Eq. (3-8)]. The actual ripple factor is then

$$r = \frac{V_{Rrms}}{V} = \frac{0.12}{26.7} = 0.0045$$

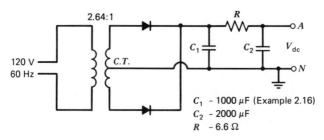

FIGURE 3-16 Power supply for Example 3-10.

Here, 26.7 V is the actual V_{dc} at the output terminal. (Using R or C_2 larger by about 10% will correct this discrepancy.)

L-C FILTER

As noted before, the R-C filter has one major drawback. It causes V_{dc} to change as I_{dc} changes (in Example 3-10 a drop from 30 to 26.7 V was caused by the introduction of R). This problem can be substantially reduced by replacing R in the R-C filter with an inductance Z_L. The ratio R/X_C, which determined the ripple factor improvement for the R-C filter, may now be replaced by the ratio Z_L/X_C. (Note that $Z_L = R_{dc} + jX_L$.) The reactance of the inductor X_L can be made very large, while keeping the internal dc resistance of the inductor R_{dc} relatively small. This results in a large Z_L/X_C ratio with a correspondingly large improvement in the ripple factor. At the same time, since R_{dc} is small, the dc voltage drop introduced by the inductor is minimal (Fig. 3-17). For $X_L \gg R_{dc}$, the ratio Z_L/X_C may be replaced by X_L/X_C. Rewriting Eq. (3-12c) for the L-C filter

$$\frac{v_{ro}}{v_{rl}} = \frac{X_{C_2}}{X_L} \tag{3-12d}$$

EXAMPLE 3-11

In Example 3-10, use an L-C filter in place of the R-C filter. Assume that for the inductor used the ratio $X_L/R_{dc} = 50$ (the ratio between the inductive reactance at the ripple frequency and the dc resistance of the inductor).
(a) Find L.
(b) Calculate V_{dc} at $I_{dc} = 0.5$ A.

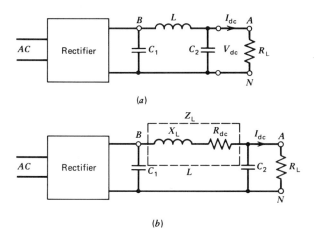

FIGURE 3-17   L-C filter. (a) Basic circuit. (b) Showing X_L and R_{dc} of inductor.

92 DIODE APPLICATIONS

SOLUTION

(a) To get a 1:10 reduction in ripple factor, the ratio $X_{C_2}/X_L \leq 10$ or $X_L/X_{C_2} \geq 10$ [from Eq. (3-12d)]. Hence,

$$X_L \geq 10 \times X_{C_2} = 10 \times \frac{1}{2\pi \times 120 \times 2000 \times 10^{-6}} = 6.6 \, \Omega$$

$$X_L = \omega L \geq 6.6 \, \Omega \quad (\text{use } X_L = 6.6 \, \Omega)$$

$$L = \frac{6.6}{2\pi \times 120} = 8.7 \text{ mH}$$

Using $L = 10$ mH (larger than the 8.7 mH required) will guarantee that $X_L/X_C > 10$.

(b) Based on the assumption $X_L/R_{dc} = 50$,

$$R_{dc} = \frac{X_L}{50} = \frac{6.6}{50} = 0.13 \, \Omega$$

V_{dc} at $I_{dc} = 0.5$ A can now be calculated.

$$V_{dc} = 30 - 0.13 \times 0.5 = 29.935 \text{ V}$$

(The value 30 V is the voltage of the capacitor filter, terminals BN.)

The larger capacitor C_2 (2000 μF) is not necessary in this case, since a 100 mH inductor (10 times larger than before) is readily available so that $X_L = 66 \, \Omega$. To get a 1:10 ripple reduction, the capacitor should be $C_2 = 200$ μF (not 2000 μF). The accompanying increase in R_{dc} is not significant (R_{dc} becomes 1.3 Ω, still quite small compared to the equivalent R in the R-C filter).

When using inductors in power supply filter applications care must be taken to select an inductor that has the required inductance under dc current conditions, in particular, full dc current. The inductance of most inductors used in power supplies decreases with an increase in the dc current through the inductor.

The last two filter configurations are referred to as Π (pi) filters (their structure resembles a Π). In the analysis of these filters it was assumed that $R_L \gg X_{C_2}$ and $R_L \| X_{C_2} \approx X_{C_2}$. For examples 3-9 and 3-10, $R_L = 60$ ohms, $X_{C_2} = 0.66$ ohm (2000 μF) the assumption above is indeed valid. In cases for which this assumption *cannot* be made, the analysis must include the effects of R_L, the load. Figure 3-18 shows a general circuit representing either an L-C or R-C filter section.

The ripple improvement v_{ro}/v_{r1} can be found by applying voltage division to the ripple voltage,

$$\frac{v_{ro}}{v_{r1}} = \frac{R_L \| X_{C_2}}{Z_s + R_L \| X_{C_2}}$$

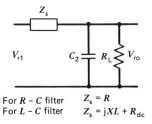

For R - C filter $Z_s = R$
For L - C filter $Z_s = jXL + R_{dc}$

FIGURE 3-18 General filter section (L section).

The evaluation of this quotient follows standard complex techniques and will not be carried out here because most applications can be analyzed by the approximations used above.

3-4.6 Voltage Regulators

In Examples 3-10 and 3-11 the dc voltage output of a power supply (rectifier and filter combination) is, to some extent, dependent upon the current drawn from the supply. The larger the current, the lower the output voltage. This reduction is certainly *not* a desired effect. If we need a 30 V supply, we expect the voltage to be 30 V for *all* conditions. In practice, it is almost impossible to guarantee such behavior. To demonstrate the causes for the voltage variations, the power source may be replaced by its Thevenin's equivalent (Fig. 3-19). R_{Th} is made up of the total *internal resistance*, the transformer resistive loss, the diode resistance, and filter resistance. Clearly, it is practically impossible to construct a power supply for which $R_{Th} = 0$. As a consequence, the voltage drop $R_{Th} \times I_{dc}$ increases as I_{dc} increases, causing a commensurate reduction in V_{dc}. This dependence of V_{dc} on I_{dc} is given as the regulation or % regulation of the source, where

$$VR = \text{Voltage Regulation} = \frac{V_{dcNL} - V_{dcFL}}{V_{dcFL}} \quad (3\text{-}13)$$

$$\%VR = VR \times 100\% \quad (3\text{-}13a)$$

where

V_{dcNL} = dc voltage for $I_{dc} = 0$ (no load current)
V_{dcFL} = dc voltage for $I_{dc} = I_{dcmax}$ (maximum load current)

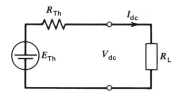

FIGURE 3-19 Thevenin's equivalent of power supply (connected to load).

(Note that I_{dcmax} is the maximum available dc current as specified by the manufacturer.) The power supply of Example 3-10 (neglecting transformer and diode losses as was indeed done in the example) had a regulation of $VR = \dfrac{30 - 26.7}{26.7}$ = 0.1236 or $\%VR = 100 \times VR = 12.36\%$. This value implies a very poorly regulated supply. (The maximum current specified for the supply was only 0.5 A.) Figures such as 0.1% VR or better are quite common. Such power supplies, however, use *special voltage regulating* circuitry.

Two basic voltage regulation circuits are available: the *series* type and the *shunt* type (Fig. 3-20). The series regulator behaves like a series resistance that varies inversely with the current. A current *increase* is accompanied by a series resistance *decrease*, compensating for the expected increase in the $R_S \times I_{dc}$ voltage drop. Essentially, the series regulator attempts to keep the total series voltage drop (across R_{Th} and the regulator) constant, regardless of the current drawn. Most modern series regulators use a transistor as the series control element.

The shunt regulator (Fig. 3-20a) attempts to keep I_t (the current through R_{Th}) constant. Since $I_t = I_{dc} + I_s$, clearly, I_s has to increase when I_{dc} decreases, and vice versa. For a constant I_t, the voltage drop $R_{Th} \times I_t$ is constant, and hence, V_{dc} is constant. (The details of a shunt regulator will be analyzed in the discussion of Zener diode application in Section 3-6.2.)

EXAMPLE 3-12

A half-wave rectifier is used in a power supply. An *L-C* section (in addition to the initial capacitor) is used for filtering (Fig. 3-21). $E_m = 50$ V, $I_{dcmax} = 1$ A, line frequency = 60 Hz, $\%r = 5\%$ across C_1, $\%r = 0.1\%$ across C_2 (output), and the power loss of diode and transformer when I_{dcmax} is flowing = 2 W.

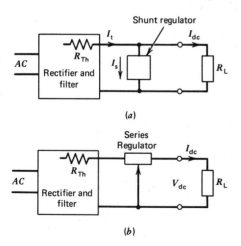

FIGURE 3-20 Regulation circuits. (a) Shunt type. (b) Series type.

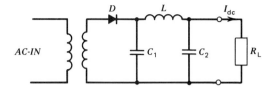

FIGURE 3-21 Half-wave rectifier for Example 3-12.

(a) Find C_1, L, C_2 (assume for L that $X_L/R_{dc} = 25$).
(b) Specify the diode (PIV, I_{peak}, I_{dc}).
(c) Find % regulation.

SOLUTION

(a) First, R_L is found

$$R_L = \frac{V_{dc}}{I_{dc}} = \frac{50}{1} = 50 \, \Omega \quad \text{(maximum load)}$$

C_1 is obtained from Eq. (3-8c).

$$\% \text{ ripple} = 5 = \frac{0.29}{60 \times 15 \times C_1} \times 100 \quad \text{(ripple across } C_1\text{)},$$

from which

$$C_1 = \frac{0.29}{60 \times 50 \times 0.05} = \underline{1933 \, \mu F}$$

Use $C_1 = \underline{2000 \, \mu F}$
Since a 50:1 ripple improvement is called for, X_L and X_{C2} can be found using the approximations

$$\frac{X_L}{X_{C2}} = \frac{5}{0.1} = 50 \qquad (3\text{-}12d)$$

C_2 is chosen, somewhat arbitrarily,

$$C_2 = 2000 \, \mu F$$

hence,

$$X_{C2} = \frac{1}{2\pi f C_2} = \frac{1}{2\pi \times 60 \times 2000 \times 10^{-6}} = 1.3 \, \Omega$$

Since $X_L = 50 \times X_{C2}$, (3-12d)

$$X_L = 50 \times 1.3 = 65 \, \Omega \quad (= \omega L)$$

$$L = \frac{X_L}{\omega} = \frac{65 \, \Omega}{2\pi 60} = \frac{65 \, \Omega}{2\pi \times 60} = 0.17 \, H$$

Note that $R_L \gg X_{C2}$; hence, our approximations are valid. $R_{dc} = 65/25 = 2.6 \, \Omega$, based on the stated assumption that the inductor will have a $Q = 25$, $(X_L/R_{dc} = 25)$.

Summarizing, we see that

$$C_1 = 2000 \, \mu F, \quad C_2 = 2000 \, \mu F, \quad L = 0.17 \, H$$

(b) To specify the diode, we evaluate the actual PIV and I_{peak}. The PIV is $2 \times E_m$. Because of the small ripple (5%) $E_m \simeq V_{dc}$ and PIV = $2 \times 50 = 100$ V. The diode must be able to withstand a PIV of at least 100 V. To provide a safety margin, a diode with a higher PIV should be specified.

PIV = 150 V (1.5 safety margin)

I_{peak} can be found from Table 3-1 for 5%.

$$I_{peak} = 10.4 \times I_{dc} = 10.4 \times 1 = 10.4 \, A$$

To provide a safety margin, specify $I_{peak} = 15$ A. This is the peak current the diode can withstand. $I_{dc} = 1.0$ A; hence, we specify the average current carrying capacity as 1.5 A (a 150% safety margin).

(c) To calculate regulation, replace the power supply by a Thevenin's equivalent. The R_{Th} is the total resistance of the source. This includes the transformer, diode and inductor resistance. The latter was found to be 2.6 Ω.
The transformer and diode resistance can be calculated from their power loss. At 1 A (full load current) this loss is given as 2 W.

$$2(W) = I_{dc}^2 \times R; \quad R = \frac{2}{1^2} = 2 \, \Omega.$$

The Thevenin resistance $R_{Th} = 2 + 2.6 = 4.6 \, \Omega$. The voltage drop across $R_{Th} = 4.6 \times 1 = 4.6$ V and the voltage at C_1 must be $50 + 4.6 = 54.6$ V to yield 50 V across the load. With $I_{dc} = 0$, no load, the output voltage is 54.6 V (no drop across R_{Th}) while for full load it is 50 V. Hence,

$$\% \, VR = \frac{54.6 - 50}{50} \times 100 = 9.2\% \quad \text{(from 3-13)}$$

Note that the voltage across C_1 is 54.6 V; hence, $E_m \approx 54.6$ V and the PIV > 109.2 V (the selected PIV = 150 V is acceptable). The change from 50 V to 54.6

SIGNAL SWITCHING

V affects the ripple calculations and hence, the selection of C_1. The error, however, is too small to require recomputation. In addition, the standard value for C_1 selected, (2000 μF) was more than required (1933 μF).

Detailed specification:

$$E_m = 54.6 + \frac{V_{rpp}}{2}^1 = 54.6 + 4.3 = \underline{58.9 \text{ V}}$$

$$C_1 = C_2 = 2000 \text{ μF at } 100 \text{ V}$$

$$L = 0.17 \text{ H}; \quad R_{dc} = 2.6 \text{ Ω} \quad (Q = 25)$$

Diode:

$$\text{PIV} = 150 \text{ V}$$
$$I_{peak} = 15 \text{ A}$$
$$I_{av} = 1.5 \text{ A} \quad (I_{dc})$$

3-4.7 Voltage Doubler

It is sometimes desirable to produce a dc voltage that is larger than the peak ac voltage (E_m). This can be accomplished by a circuit as shown in Fig. 3-22. Here, the V_{dc} produced is double the ac line voltage. (For simplicity, assume no load and neglect diode voltage drops.) During the half cycle in which L_1 is negative and L_2 is positive, D_1 conducts charging C_1 to E_m, as shown in Fig. 3-22b. During the second half cycle (L_1 positive, L_2 negative) diode D_2 is conducting (Fig. 3-22c). The voltage applied to D_2 and the output is the series combination (series aiding) $E_m + V_{C1} = E_m + E_m$. C_2 is charged to this value $2E_m$. The output voltage $V_{dc} = 2E_m$. Note that C_2 is charged only during the second half cycle, indicating a half-wave rectifier behavior ($f = 60$ Hz).

The diode capacitor circuit used in the voltage doubler can be cascaded to produce voltage triplers and quadruplers. These circuits are mostly used in very high voltage applications for which it is desirable to keep the transformer voltage as low as possible and obtain a dc output voltage as high as possible. Most of these circuits are suitable for low current applications only and usually require a very large capacitor to keep the ripple voltage low.

3-5 SIGNAL SWITCHING

The switching characteristics of the diode (switches from OFF to ON as the applied voltage crosses the diode threshold) can be used to control a signal voltage. Consider

---

[1] $V_{Rpp} = V_{R(rms)} \times 2\sqrt{3} = \overbrace{(5/100) \times 50}^{V_{R(rms)}} \times 3.46 = 8.6$ V. The 0.7 V diode drop should be added to give $E_m = 59.6$ V.

98 DIODE APPLICATIONS

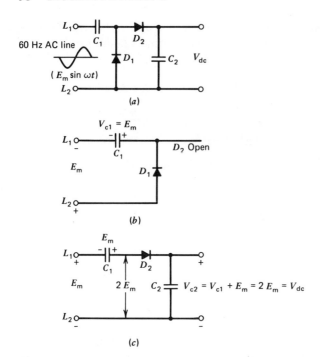

FIGURE 3-22 Voltage doubler. (a) Complete circuit. (b) Firrst half cycle equivalent $(L_1 \to -, L_2 \to +)$. (c) Second half cycle equivalent $(L_1 \to +, L_2 \to -)$.

the circuit of Fig. 3-23. (For simplicity, assume ideal diodes.) When the switch Sw is in the $+5$ V position, D_2 is OFF (no voltage applied to D_2 to turn it ON). As a result, D_1 is on as long as e_{in} is less than 5 V (D_1 voltage is $5 - e_{in}$) and e_{in} appears at the output: $v_o = e_{in}$. When Sw is in the GND position, D_2 is ON and $v_o = 0$ V (effectively shorted to GND by D_2 and Sw). The circuit is used to "gate" e_{in} (open the gate $v_o = e_{in}$, close it $v_o = 0$ V). The gating operation is controlled by a 0 to 5 V control signal (shown here as a switch Sw).

3-6 ZENER DIODE APPLICATIONS

3-6.1 Zener Diode Parameters

Section 2-4.3 discussed, briefly, the "avalanche" and "zener" breakdown. While all diodes exhibit one or the other, the "zener diode" has been specifically designed to take advantage of the "avalanche" phenomenon. As can readily be seen in Figs. 2-17 and 2-19, the voltage across the diode when in the zener (or avalanche) region is nearly constant and independent of the current through the diode. Figure 3-24 is a typical current versus voltage plot of a 5.6 V zener diode. At a current of about 20 mA (the test current, I_{ZT}), the voltage across the diode is about 5.6 V. At a much larger current, say 75 mA, the voltage across the zener has become about 6

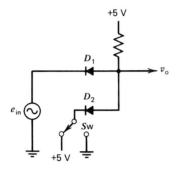

FIGURE 3-23 Simple diode "gate."

V. A change of 0.4 V or $(0.4/5.6) \times 100 = 7\%$ for a better than 3 to 1 current change. This indicates that the voltage V_Z across the diode is largely *independent* of the current through the zener. Here, $\Delta V_Z = 6 - 5.6 = 0.4$ V (change in V_Z) and $\Delta I_Z = 75 - 20 = 55$ mA (corresponding change in I_Z). The dynamic zener resistance R_Z can now be calculated:

$$R_Z = \frac{\Delta V_Z}{\Delta I_Z} = \frac{0.4}{55 \times 10^{-3}} = 7.3 \, \Omega$$

The zener diode when operating in the zener-avalanche region may now be represented by a Thevenin's equivalent (Fig. 3-25) where V_Z is the nominal zener

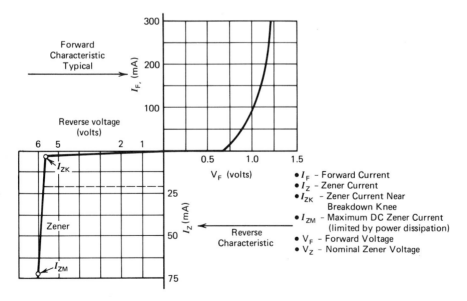

FIGURE 3-24 Zener diode characteristics ($V_Z = 5.6$ V) (courtesy Motorola Corp.).

voltage and R_Z is the dynamic resistance of the diode. The total voltage V_{AB} will change with the current through the zener only to the extent of the voltage drop across R_Z ($I_Z \times R_Z$). Different zener diodes have different values of R_Z. In high power zeners, for example, R_Z is as low as $0.12\ \Omega$ (the IN4553 at $I_Z = 2.25$ A).

Some of the other important design parameters of the zener diode are shown in Fig. 3-24. I_{ZK} is the smallest zener current that produces the zener effect. (We refer to the zener effect even though most zener diodes operate in the "avalanche" region. The distinction has been discussed in Section 2-4.3 and is of little importance in the applications involved.) It is referred to as the zener "knee" current I_{ZK} (the current at the "knee" of the zener plot). I_{ZK} is usually a small fraction of the maximum zener current I_{ZM} (no more than about 5% of I_{ZM}).

It must be remembered that the zener region is part of the reverse-biased diode operation. Hence, leakage current may be specified for the zener when the reverse voltage is insufficient to produce the zener effect and the diode acts as a back-biased diode. (Similarly, forward diode parameters are usually also given by the manufacturer.)

3-6.2 Zener dc Voltage Regulation

One of the most predominant uses of zener diodes is voltage regulation. The voltage regulator circuit is designed to minimize variations of a dc supply voltage. These changes are usually induced by either a load current change or a change in the ac source voltage (used to produce the rectified dc output).

A typical zener regulator is shown in Fig. 3-26. For the sake of clarity, it is assumed that the diode is operating in the zener region, and hence, $V_{dc} = V_Z$ (independent of I_Z). It is self-evident that when, for example, a 12 V dc voltage supply is required, a zener with $V_Z = 12$ V must be selected. (Standard zener diodes are available with V_Z ranging between 2.4 to 200 V with maximum power dissipation ranging between .25 to 50 W.) It is necessary to provide for a voltage drop across R_S; this means that $V_{in} > V_{dc}$. The zener current I_Z must be limited to *no less* than I_{ZK} and no more than I_{ZM}. As a practical design guide,

$$0.05\ I_{ZM} \leq I_Z \leq 0.95\ I_{ZM} \tag{3-14}$$

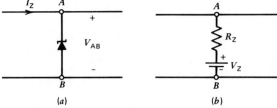

FIGURE 3-25 Zener diode. (a) Zener diode symbol. (b) Thevenin's equivalent of zener (operating in zener region).

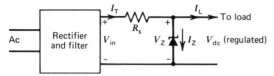

P_{max} — maximum power in zener (specified by manufacturer)
I_{ZM} — maximum permissable current in zener, (specified by manufacturer)
I_{ZMAX} — maximum current flowing in zener (determined by circuit parameters)
I_{ZMIN} — minimum current flowing in zener (determined by circuit)
I_{ZMIN} — minimum current to load (determined by circuit)
I_{ZMAX} — maximum current to load (determined by circuit)
I_{ZK} — zener diode "knee" current

FIGURE 3-26 Zener regulator (shunt type).

(For higher safety margins, I_Z in the range of $0.1 I_{ZM}$ to $0.9 I_{ZM}$ is somtimes used.) I_Z must be larger than (or equal to) $0.05 I_{ZM}$, to ensure operation in the zener region, above I_{ZK}, and less than (or equal to) about $0.95 I_{ZM}$ to prevent excessive power dissipation. Usually the data sheet lists both P_{max} and I_{ZM}. Nominally $I_{ZM} = P_{max}/V_Z$ (or $P_{max} = V_Z \times I_{ZM}$). In practice, I_{ZM} is about 10% less than P_{max}/V_Z. This guarantees that the maximum current will never result in an excessive power dissipation.

The following analysis of the zener regulator (Fig. 3-26) will also serve to explain its operation and the significance of the various zener specifications. The analysis uses notation as listed in Fig. 3-26 (some of it is given in Fig. 3-24). From Fig. 3-26,

$$I_T = I_Z + I_L; \quad I_Z = I_T - I_L \tag{3-15}$$

$$I_T = \frac{V_{in} - V_Z}{R_S} \quad \text{(the voltage across } R_S \text{ divided by } R_S\text{)} \tag{3-16}$$

$$R_S = \frac{V_{in} - V_Z}{I_T} \tag{3-16a}$$

Let us consider variations in I_L as they affect I_Z. Regulation is accomplished by keeping I_T constant. This requires that I_Z be the *smallest* (I_{Zmin}) when $I_L = I_{Lmax}$ (full load current), and I_Z be the *highest* (I_{Zmax}) when $I_L = 0$ (I_L minimum). Should there be variations in V_{in} (due to unfiltered ripple, ac voltage changes, or load current induced internal voltage drop), the minimum I_Z (I_{Zmin}) will occur when

$$I_L = I_{Lmax}$$

at full load current *and*

$$V_{in} = V_{in\,min}$$

when V_{in} is minimum. This condition is shown in Fig. 3-27a. Maximum I_Z (I_{Zmax}) will occur when

$$I_L = 0 \quad \text{(minimum load current)}$$

and

$$V_{in} = V_{in\,max} \quad \text{(maximum ac voltage)}$$

This condition is shown in Fig. 3-27b.

The basic criteria may now be restated as

$$I_{Zmin} = 0.05\, I_{ZM} \quad \text{(or, in general, } I_{Zmin} \geq 0.05\, I_{ZM}) \tag{3-17}$$

This guarantees operation above I_{ZK}, the knee current, well in the zener region.

$$I_{Zmax} = 0.95\, I_{ZM} \quad \text{(or } I_{Zmax} < 0.95\, I_{ZM}) \tag{3-18}$$

To make sure that the actual power dissipated in the zener never exceeds P_{max}, the actual power in the zener is simply $P_Z = V_Z \times I_Z$ for any zener current I_Z. Under all conditions, $P_Z \leq P_{max}$. From Fig. 3-27a,

$$I_T = I_{Zmin} + I_{Lmax} = 0.05\, I_{ZM} + I_{Lmax} \tag{3-19}$$

A useful rule of thumb requires that I_{ZM} (the current carrying capacity of the zener) be about $1.1\, I_{Lmax}$.

$$I_{ZM} \geq 1.1\, I_{Lmax} \quad \text{or} \quad I_{Lmax} \leq 0.9\, I_{ZM} \tag{3-20}$$

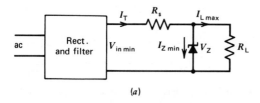

(a)

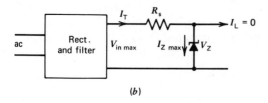

(b)

FIGURE 3-27 Zener regulator design. (a) Conditions for I_{Zmin}. (b) Conditions for I_{Zmax}.

ZENER DIODE APPLICATIONS 103

(This is derived from the fact that for the no load case, $I_L = 0$, I_Z consists of $I_{Zmin} = 0.05\, I_{ZM}$ plus I_{Lmax}.) When large variations in V_{in} (about $\pm 20\%$) are expected, I_{ZM} may have to be two or three times the full load current. From Eqs. (3-16a) and (3-19) for full load conditions,

$$R_S = \frac{V_{in\,min} - V_Z}{I_T} = \frac{V_{in\,min} - V_Z}{0.05\, I_{ZM} + I_{Lmax}}\ \Omega \qquad (3\text{-}21)$$

R_S given by Eq. (3-21) is a maximum value for R_S. Should you choose R_S larger than indicated, I_{Zmin} will become less than that required by Eq. (3-17). Equation (3-21) may be rewritten as

$$R_{S\,max} = \frac{V_{in\,min} - V_Z}{0.05\, I_{ZM} + I_{Lmax}}\ \Omega \qquad (3\text{-}21a)$$

From Fig. 3-27b, for I_{Zmax},

$$R_S = \frac{V_{in\,max} - V_Z}{I_T} = \frac{V_{in\,max} - V_Z}{I_{Zmax}} \qquad (3\text{-}22)$$

$$= \frac{V_{in\,max} - V_Z}{0.95\, I_{ZM}}\ \Omega$$

Here, R_S is a minimum value. R_S lower than indicated by Eq. (3-22) will cause excessive zener current (power dissipated exceeding P_{max}). The acceptable range of R_S is

$$\frac{V_{in\,max} - V_Z}{0.95\, I_{ZM}} \le R_S \le \frac{V_{in\,max} - V_Z}{0.05\, I_{ZM} + I_{Lmax}} \qquad (3\text{-}23)$$

EXAMPLE 3-13

Design a zener regulator with the following requirements.
  dc output voltage 12 V
  maximum load current 0.5 A
  unregulated voltage (rectifier output) $V_{in} = 20\ V \pm 20\%$. The 20% variation in V_{in} is due to all causes, ac input variations, ripple, and internal $I \times R$ drops.
  (a) Select a specific diode from the manufacturer's data sheets.
  (b) Find R_S.
  (c) Calculate the voltage regulation (compare to unregulated supply).

SOLUTION

Equation (3-20) indicates that the zener selected must have $I_{ZM} \ge 1.1 \times I_{Lmax}$; hence, $I_{ZM} \ge 1.1 \times 0.5 = 0.55\ A = 550\ mA$. For this current the power in the zener is $P_Z = 0.55 \times 12 = 6.6\ W$. Since standard zeners are usually available with P_{max} of 0.25 W, 0.4 W, 0.5 W, 0.75 W, 1 W, 1.5 W, 10 W, 50 W, the lowest P_{max} satisfying our requirement that $P_Z \leq P_{max}$ is $P_{max} = 10$ W ($P_{max} > 6.6$ W). We select a 10 W zener with $V_Z = 12$ V. *The type number is 1N2976* (Motorola). The important characteristics of the 1N2976 are listed below (from the Motorola catalog):

$V_Z = 12$ V

$I_{ZM} = 720$ mA

$R_Z = 3\ \Omega$ (maximum dynamic resistance at $I_Z = 250$ mA)

To obtain R_S, V_{inmin} and V_{inmax} must be obtained.

$$V_{inmin} = 20 - \frac{20}{100} \times 20 = 16 \text{ V} \qquad (20 \text{ V} - 20\%)$$

$$V_{inmax} = 20 + \frac{20}{100} \times 20 = 24 \text{ V} \qquad (20 \text{ V} + 20\%)$$

From Eq. (3-23)

$$\frac{24 - 12}{0.95 \times 0.75} \leq R_S \leq \frac{16 - 12}{0.05 \times 0.75 + 0.5}$$

$$16.8\ \Omega \leq R_S \leq 7.4\ \Omega$$

This solution is impossible, that is, 16.8 cannot be less than 7.4 Ω! The reason for this contradiction is the large variations in V_{in}. A zener diode with a much larger I_{ZM} must be selected and, therefore, a higher power rating. So we select a 50 W zener, type 1N2810 (or 1N3311). The specifications are:

$V_Z = 12$ V

$I_{ZM} = 3.6$ A

$R_Z = 1.0\ \Omega$

Again,

$$\frac{24 - 12}{0.95 \times 3.6} \leq R_S \leq \frac{16 - 12}{0.05 \times 3.6 + 0.5}$$

$$3.5\ \Omega \leq R_S \leq 5.9\ \Omega$$

(b) We select $R_S = 5.5\ \Omega$ (rated at 5 W $> I_{Zmax}^2 \times 5.5$) The circuit is shown in Fig. 3-28.

(c) The voltage regulation can be calculated by first obtaining the values for I_{Zmin} and I_{Zmax}. From Fig. 3-27a with $R_S = 5.5\ \Omega$, $V_Z = 12$ V

$$V_{inmin} = 16 \text{ V} \quad \text{and} \quad I_{Lmax} = 0.5 \text{ A} \quad \text{(full load current)}$$

$$I_T = \frac{V_{inmin} - V_Z}{R_S} = \frac{16 - 12}{5.5} = 0.727 \text{ A}$$

$$I_{Zmin} = I_T - I_{Lmax} = 0.727 - 0.5 = 0.227 \text{ A}$$

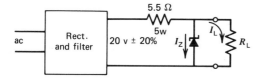

FIGURE 3-28 Regulator circuit for Example 3-13.

From Fig. 3-27b

$$I_{Zmax} = I_T = \frac{V_{in\,max} - V_Z}{R_S} = \frac{24 - 12}{5.5} = 2.18 \text{ A}$$

$$\Delta I_Z = 2.18 - 0.227 = 1.95 \text{ A}$$

with a dynamic resistance of 1.0 Ω

$$\Delta V_Z = 1.0 \times 1.95 = 1.95 \text{ V} \quad \text{or} \quad \pm 0.975 \text{ V}$$

This represents a regulation of

$$\pm \frac{0.975}{12} \times 100 = \pm 8.1\% \quad \text{(no load to full load)}$$

The 8.1% voltage regulation in Example 3-13 is rather poor. In practice, however, R_Z is usually substantially smaller than 1.0 Ω. (Note that the 1.0 Ω value given by the manufacturer is a *maximum* value.) Regulations of 5% or better are attainable with zener diodes. (For better regulation special amplifier controlled regulating circuits are used.) The ±8.1% regulation is better than a twofold improvement from the unregulated source (±20%).

The zener diode is often used instead of a voltage divider. If, for example, a 5 V supply is required and a well regulated 12 V supply is available (with sufficient current capacity), the zener can be used to obtain the 5 V supply as shown in Fig. 3-26 where V_{in} is the 12 V supply voltage available. The diode may be a 5.1 V zener with appropriate power specifications (a 1N3826 rated at 1 W or a 1N3996 rated at 10 W, etc.). In this case, since V_{in} is constant, R_S can be selected in the range

$$\frac{V_{in} - V_Z}{0.95 \times I_{ZM}} \leq R_S \leq \frac{V_{in} - V_Z}{0.05\, I_{ZM} + I_{Lmax}}$$

3-6.3 ac Regulation

The zener diode may be used to regulate ac voltages as well. The regulation is accomplished by "clipping" the sinusoidal wave at predetermined levels (Fig. 3-29a). During the positive half of the output cycle D_1 conducts in the zener mode

106 DIODE APPLICATIONS

(V_Z) while D_2 acts as a forward-biased diode (V_D). For the second half cycle, the negative half D_2 conducts in the zener mode and D_1 acts as an ON diode. The output voltage is shown in Fig. 3-29b. $V_{o(p-p)} = 2(V_Z + V_D)$ regardless of the input voltage amplitude. It should be noted that $V_{o(rms)}$ is *not* simply 0.707 ($V_Z + V_D$) since the output waveform is not sinusoidal. The regulator does not maintain a constant *rms* voltage. As V_{in} increases or as V_o becomes a smaller portion of V_{in}, (for example, $V_{in(p-p)} = 12$ V and $V_{o(p-p)} = 10$ V, compared to $V_{in(p-p)} = 100$ and $V_{o(p-p)} = 10$ V), $V_{o(rms)}$ does increase appreciably.

3-6.4 Temperature Characteristics of Zener Diodes

Power supplies, ac or dc, are often subject to large temperature variations. A field-operated power supply in the New York area may be expected to operate in the range between $-20°C$ to about $+50°C$. Some military applications may require an even wider operating range. The voltage variations of the power supply must be kept within a specified value for the complete operating temperature range. The manufacturer often specifies the "temperature coefficient" of the power supply.

The *temperature coefficient* (tc) is usually defined as "percent voltage variations in the output voltage per degree Celsius (°C)." Typically, temperature coefficients for power supplies are 0.01% per °C (0.01%/°C). This means that, for example, a 10 V (nominal) power supply with a tc of 0.01%/°C will vary by $0.01/100 \times 10 = 0.001$ V for each degree Celsius variation in temperature. For a change of 100 °C a ΔV of 0.1 V that is 1% of the supply voltage (10 V) will take place. The temperature coefficient may be positive, a rise in temperature causes a rise in voltage, or negative, a rise in temperature causes a drop in voltage.

It becomes essential, for wide temperature range applications, that components

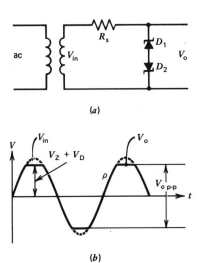

FIGURE 3-29 ac regulator. (a) Circuit. (b) Waveforms.

TABLE 3-2 TEMPERATURE COEFFICIENTS OF VARIOUS ZENER DIODES VERSUS ZENER VOLTAGE.[a]

Diode No.	V_Z - Zener Voltage (V)	Temperature Coefficient, tc (%/°C)
1N5221	2.4	−0.085
1N5225	3.0	−0.075
1N5230	4.7	±0.030
1N5232	5.6	+0.038
1N5235	6.8	+0.050
1N5238	8.7	+0.065
1N5240	10	+0.075
1N5250	20	+0.086
1N5259	39	+0.095

[a] Courtesy Motorola Corp.

with suitable temperature coefficients be selected. The key component in a zener regulated power supply, as far as temperature effects are concerned, is the zener diode itself.

The temperature coefficient of zener diodes is intrinsically related to the zener voltage V_Z. As we noted, zener diodes may operate in both the zener or avalanche regions. At voltages (V_Z) below about 6 V, the zener breakdown is predominant, while above 6 V, avalanche dominates. The zener phenomenon has a negative temperature coefficient, while the avalanche has a positive temperature coefficient. It is expected, then, that diodes with $V_Z \approx 6$ V will have a tc near zero. As V_Z decreases, the tc becomes larger with a negative sign, and as V_Z increases, the tc gets larger and is positive. Typical data is given in Table 3-2. One can, of course, combine diodes with V_Z of, say, 2.4 and 8.6 V to get a total V_Z of 10 V with a tc of close to zero (the sum of the two temperature coefficients). Manufacturers sometimes combine a forward diode with a zener diode to obtain a very low, compensated temperature coefficient. These packages are usually called "compensated zeners" and have extremely low temperature coefficients.

3-7 DIODE SPECIFICATIONS—THE DATA SHEET

Throughout this text, reference has been made to manufacturer's specifications, whether it was in relation to maximum power dissipation, peak current ratings, or maximum voltage limitations such as PIV and the like. This data is available from the specification sheet or data sheet issued by the manufacturer, usually as part of a data book. Since data sheets are complex documents utilizing many abbreviations and shortcuts, it is useful to learn how to interpret the data.

The format used for data sheets differs from one manufacturer to another. Nevertheless, most data sheets consist of a number of distinct parts (Fig. 3-30). One section contains the name, number, and a brief description of the device. It usually

Silicon Rectifiers

1N91 thru 1N93

$V_R = 100 - 300$ V
$I_O = 1.0$ A

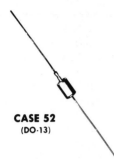

CASE 52
(DO-13)

Low-current germanium rectifiers for applications requiring extremely low forward voltage drop, low power dissipation, and high rectification efficiency, such as biasing and battery charging circuits.

MAXIMUM RATINGS*

Rating	Symbol	1N91	1N92	1N93	Unit
Peak Repetitive Reverse Voltage (Rated I_O, $T_A \leq 55°C$, see Figure 2)	$V_{RM(rep)}$	100	200	300	Volts
DC Blocking Voltage $T_A \leq 80°C$	V_R	100	200	300	Volts
$T_A \geq 80°C$		Derate V_R 6.7%/°C above 80°C			
Average Rectified Forward Current (Single Phase, Resistive Load, 60 Hz) $T_A = 55°C$, 100% V_{RM}	I_O	← 1.0 →			Amp
$T_A = 75°C$, 50% V_{RM}		← 0.25 →			
Non-Repetitive Peak Surge Current (Surge Applied at Rated Load Conditions, see Figure 4)	$I_{FM(surge)}$	← 30 →			Amp
Operating Junction Temperature Range	T_J	−65 to +95			°C
Storage Temperature Range	T_{stg}	−65 to +125			°C

THERMAL CHARACTERISTICS (ALL TYPES)

Characteristic	Symbol	Max Limit	Unit
Thermal Resistance, Junction to Ambient	θ_{JA}		°C/W
1 inch Leads		100	
1/4 inch Leads		70	

ELECTRICAL CHARACTERISTICS (ALL TYPES)

Characteristic	Symbol	Max Limit	Unit
Peak Forward Voltage Drop ($I_F = 150$ mAdc, $T_A = 25°C$, see Figure 1)	V_F	0.45	Volts
DC Reverse Current (Rated V_R, $T_A = 25°C$, see Figure 3)	I_R	0.22	mA

FIGURE 3-30 Diode data sheet. (Courtesy Motorola Inc.) (a) Maximum ratings and electrical characteristics. (b) Graphic description of characteristics (partial).

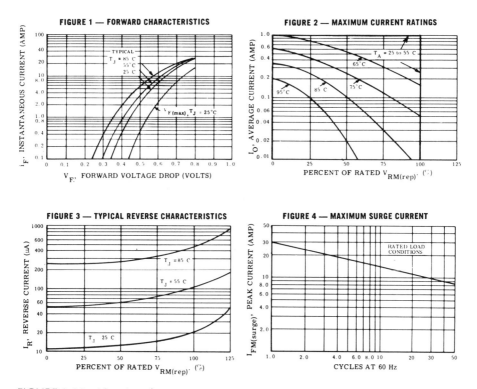

FIGURE 3-30 (Continued).

contains a brief summary of applications, which is intertwined with some promotional material. The purpose, besides increasing the sales of the product, is to make it easy for the designer to see whether the particular device is at all suitable for the intended application. The second part of the data sheet is the maximum ratings (or absolute maximum ratings). This section gives maximum limits for the various operating conditions of the device. Should the device be operated outside these limits, permanent damage may result, and the manufacturer will assume no responsibility for device failure. The thermal characteristics, normally given for power devices only, is essentially an extension of the maximum ratings. The last portion is the data concerning performance characteristics—electrical characteristics, usually given for operation at 25°C and "normal" environment (pressure, moisture, etc.). Most characteristics have a range, a minimum and maximum value, and a typical value. It is the practice of manufacturers to give minimum and maximum values *or* typical values, not both. This is largely done for the convenience of the manufacturer. Should additional data be required, it can be obtained by direct contact with the manufacturer. Mechanical data, such as the type of package and its dimensions, is generally given in a separate section in the data book.

110 DIODE APPLICATIONS

The sections containing maximum ratings and characteristics make extensive use of graphs to represent the various parameters over a wide range of operating conditions (Fig. 3-30b). It would be impossible to cover all these graphs and their significance in this section; consequently, we concentrate here on the two specifications tables: the maximum ratings and the characteristics.

It should be kept in mind that the data sheet is usually geared in the direction of some general device application. The data sheet for a high power *rectifier* diode contains substantial data relating to high power and high current performance characteristics, which are practically never given for the general purpose *signal* diode. It is reasonable to expect that an *absolutely complete* data sheet for any device will never be available. This is particularly important when attempting to use a device in a somewhat unconventional application. The source for any additional data is the manufacturer itself.

To a large extent, the tables in Fig. 3-30 are self-explanatory, particularly in light of the material presented in this chapter. Since nomenclature and symbols are rarely standardized, it becomes necessary to relate the terms used in the data sheet to those used in this text. Note that the parameters are given in the data sheet, together with some details of the operating conditions for which the ratings are given.

Maximum Rating:

$V_{RM(rep)}$ Maximum reverse voltage that may be applied to the diode repetitively (during the nonconducting half cycle). Figure 2 (of the data sheet) gives the temperature dependence of V_{RM}. (It is noteworthy that V_{RM} is dependent on the average forward current I_D.)

V_R *Continuous* maximum reverse voltage.

These two terms are essentially the maximum allowable PIV. V_R deteriorates as the temperature increases. It must be *derated* at 6.7%/°C above 80°C. This means that for every °C above 80°C, V_R must be reduced by 6.7%. For example, at 90°C V_R (for 1N91) becomes $100 - 10 \times 0.067 \times 100 = 33$ V. (At 80°C, V_R is given in the data sheet as 100 V.)

I_o Average forward current for the specified temperature and V_{RM} and the selected application (60 Hz rectifier).

$I_{FM(surge)}$ Peak forward current. Figure 4 (of the data sheet) shows that if the current is repetitive (more than a single cycle during the initial charging of the filter capacitor), we must keep the peak below the 30 A value given for a single cycle peak current. For 10 cycles, the peak current must be kept below about 15 A.

T_J Operating junction temperature (See Ch. 9).

T_{STG} Storage temperature.

Thermal Characteristics—See Chapter 9.

DIODE SPECIFICATIONS—THE DATA SHEET 111

——— Silicon Zener Diodes ———

MZ1000-1 thru MZ1000-37

1 W
3.3-100 V

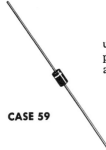

CASE 59

Miniature plastic encapsulated zener diodes for regulated power supply circuits, surge protection, arc suppression and other functions in television, automotive and other consumer product applications.

MAXIMUM RATINGS

Rating	Value	Unit
DC Power Dissipation @ T_L = 50°C	1.0	Watt
Derate above 50°C	8.33	mW/°C
Lead Temperature*	−65 to +175	°C

*Maximum lead temperature for 10 seconds at 1/16″ from case = 230°C

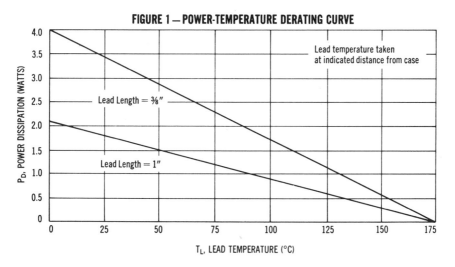

FIGURE 1 — POWER-TEMPERATURE DERATING CURVE

FIGURE 3-31 Zener diode data sheet. (Courtesy Motorola Inc.)

112 DIODE APPLICATIONS

────── Silicon Zener Diodes ──────

MZ1000-1 thru MZ1000-37 (continued)

MECHANICAL CHARACTERISTICS

CASE: Void free, transfer molded.

FINISH: All external surfaces are corrosion resistant. Leads are readily solderable.

POLARITY: Cathode indicated by color band. When operated in zener mode, cathode will be positive with respect to anode.

MOUNTING POSITION: Any.

WEIGHT: 0.42 gram (approximately).

ELECTRICAL CHARACTERISTICS (T_C = 25°C unless otherwise noted) V_F = 1.5 V max @ 200 mA on all types

Type No.	Zener Voltage V_Z @ I_{ZT} Volts*			Test Current I_{ZT} mA	Typical Z_{ZT} @ I_{ZT} Ohms	Max DC Zener Current I_{ZM} mA	Maximum Reverse Leakage Current		Temperature Coefficient %/°C
	Min	Nom	Max				I_R µA Max	@ V_R Volts	
MZ1000-1	2.97	3.3	3.63	76	15	276	150	1	−.070
MZ1000-2	3.24	3.6	3.96	69	15	252	150	1	−.065
MZ1000-3	3.51	3.9	4.29	64	13.5	234	75	1	−.060
MZ1000-4	3.87	4.3	4.73	58	13.5	217	20	1	−.050
MZ1000-5	4.23	4.7	5.17	53	12	193	20	1	−.043
MZ1000-6	4.59	5.1	5.61	49	10.5	178	20	1	±.030
MZ1000-7	5.04	5.6	6.16	45	7.5	162	20	2	±.028
MZ1000-8	5.58	6.2	6.82	41	3	146	20	3	+.045
MZ1000-9	6.12	6.8	7.48	37	5.25	133	20	4	.050
MZ1000-10	6.75	7.5	8.25	34	6	121	20	5	.058
MZ1000-11	7.38	8.2	9.02	31	6.75	110	20	5.9	.062
MZ1000-12	8.19	9.1	10.01	28	7.5	100	20	6.6	.068
MZ1000-13	9	10	11	25	10.5	91	20	7.2	.075
MZ1000-14	9.9	11	12.1	23	12	83	10	8.0	.076
MZ1000-15	10.8	12	13.2	21	13.5	76	10	8.6	.077
MZ1000-16	11.7	13	14.3	19	15	69	10	9.4	.079
MZ1000-17	13.5	15	16.5	17	21	61	10	10.8	.082
MZ1000-18	14.4	16	17.6	15.5	24	57	10	11.5	.083
MZ1000-19	16.2	18	19.8	14	30	50	10	13.0	.085
MZ1000-20	18	20	22	12.5	33	45	10	14.4	.086
MZ1000-21	19.8	22	24.2	11.5	34.5	41	10	15.8	.087
MZ1000-22	21.6	24	26.4	10.5	37.5	38	10	17.3	.088
MZ1000-23	24.3	27	29.7	9.5	52.5	34	10	19.4	.090
MZ1000-24	27	30	33	8.5	60	30	10	21.6	.091
MZ1000-25	29.7	33	36.3	7.5	67.5	27	10	23.8	.092
MZ1000-26	32.4	36	39.6	7	75	25	10	25.9	.093
MZ1000-27	35.1	39	42.9	6.5	90	23	10	28.1	.094
MZ1000-28	38.7	43	47.3	6	105	22	10	31.0	.095
MZ1000-29	42.3	47	51.7	5.5	120	19	10	33.8	.095
MZ1000-30	45.9	51	56.1	5	142.5	18	10	36.7	.096
MZ1000-31	50.4	56	61.6	4.5	165	16	10	40.3	.096
MZ1000-32	55.8	62	68.2	4	177.5	14	10	44.6	.097
MZ1000-33	61.2	68	74.8	3.7	225	13	10	49.0	.097
MZ1000-34	67.5	75	82.5	3.3	262.5	12	10	54.0	.098
MZ1000-35	73.8	82	90.2	3	300	11	10	59.0	.098
MZ1000-36	81.9	91	100.1	2.8	375	10	10	65.5	.099
MZ1000-37	90	100	110	2.5	525	9	10	72.0	.100

*1. Nominal voltages other than those stated above, matched sets, and tighter voltage tolerances are available. 1N4728 thru 1N4764 series (1M3.3ZS10 thru 1M100ZS10).

2. Voltages to 200 volts are available in other package configurations on request.

FIGURE 2 — TYPICAL ZENER DIODE CHARACTERISTICS and SYMBOL IDENTIFICATION

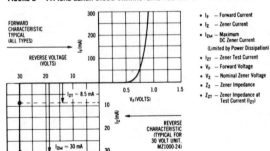

- I_F — Forward Current
- I_Z — Zener Current
- I_{ZM} — Maximum DC Zener Current (Limited by Power Dissipation)
- I_{ZT} — Zener Test Current
- V_F — Forward Voltage
- V_Z — Nominal Zener Voltage
- Z_Z — Zener Impedance
- Z_{ZT} — Zener Impedance at Test Current (I_{ZT})

FIGURE 3-31 (*Continued*).

SUMMARY OF IMPORTANT TERMS 113

Electrical Characteristics:

V_F Forward voltage. A maximum value of 0.45 V under a forward current of 150 mA at 25°C. This maximum value depends on the temperature as shown in Fig. 1 (of the data sheet), which also gives a more complete relation between I_F and the forward voltage drop.

I_R Reverse current (at the full allowable reverse voltage, for example, 100 V reverse voltage for the 1N91). Figure 3 (of the data sheet) shows that for lower V_{RM} (below rated), I_R is lower and I_R increases with an increase in temperature.

A zener diode data sheet with appropriate explanations of the terms used is given in Fig. 3-31. The power rating and its temperature derating is discussed in Chapter 9 in connection with power transistors.

V_Z in the data sheet is given for a particular test current I_{ZT}. At currents other than I_{ZT}, V_Z may be slightly different (Sec. 3-6.1). Note also that the range of V_Z is $\pm 10\%$. This means that the actual V_Z may vary by $\pm 10\%$ from its nominal value. Diodes with tighter tolerance are available at higher costs. The leakage current I_R is given for the condition when the diode is reverse biased but *not* in a zener (or avalanche) state. For a 33 V zener, I_R is given for a 1 V reverse voltage (well below the zener voltage). The temperature coefficient is discussed in Sec. 3-6.4.

SUMMARY OF IMPORTANT TERMS

Clipper circuit   Diode-resistor circuit used in waveshaping. (Clips the waveform at preselected levels.)

Clamper circuit   Diode—capacitor circuit used in waveshaping. Serves to shift the signal reference level. (Clamps the waveform to a selected level.)

Conduction angle   The interval, expressed in electrical degrees, during which there is a current flow through a rectifier diode.

Filtering   The circuit used to reduce the residual ac component (ripple voltage) in a rectified signal.

Filters—capacitor   Single capacitor used as a ripple voltage filter.

Filter: R-C   Resistor—capacitor circuit used in ripple voltage filtering.

Filter: L-C   Inductor—capacitor circuit used in ripple voltage filtering.

Peak current $I_{(peak)}$ The maximum current through a diode in a power supply (rectifier with or without filter).

Peak detector   A circuit whose output is dc level with the magnitude of the peak (largest) input signal amplitude.

Peak inverse voltage (PIV)   The maximum permissible diode reverse voltage. The term PIV is often used to denote both the actual reverse diode voltage in a rectifier circuit as well as the maximum permissible reverse voltage as specified by the manufacturer.

Power supply   A circuit, including rectifiers, filters, and regulators as required, used to convert ac to dc voltage.

Rectifier   A circuit (typically using diodes) used to convert ac (sinewaves) into pulsating dc (which is subsequently filtered to yield dc voltage).

114 DIODE APPLICATIONS

Rectifier—half-wave Rectifier circuit that utilizes half of the sinusoidal input voltage only (180° conduction).

Rectifier—full-wave Rectifier circuit that rectifies the full input waveform (360° conduction), *usually* requiring a center tapped transformer.

Rectifier—bridge A full-wave rectifier circuit not requiring the use of a transformer.

Ripple factor The ratio of rms ripple voltage to dc voltage at the output terminals of a dc power supply.

Ripple voltage The ac voltage component of a rectified signal.

Ripple % Ripple factor expressed in percent.

Voltage doubler (tripler, etc.) A circuit used to produce double (triple, etc.) the dc voltage that can be produced by a half-wave, full-wave, or bridge rectifier with the same input.

Voltage regulation Principles and methods used to reduce variations in the dc voltage of a power supply. (Variations may be due to changes in the load current or variations in input voltage.)

Voltage regulation % The percent dc voltage change from a no-load current (no external load connected to power supply) to a full load current (external load drawing specified maximum current).

Zener diode A diode designed to operate in the zener breakdown region (see Ch. 2).

I_Z Zener current (in zener diode).

I_{ZK} The zener "knee" current. Zener current at the point where zener breakdown occurs (approximately). Specified by the manufacturer.

I_{Zmax} Maximum allowable zener current (specified by the manufacturer).

I_{Zmin} Smallest zener current that still keeps the diode in the zener region.

V_Z Nominal zener voltage. (Varies for different diode types and is specified by the manufacturer.)

For the symbols I_{FM}, I_o, I_R, T_J, T_{STG}, V_R, V_{RM}, see Sec. 3-7—diode specifications.

PROBLEMS

3-1. Plot v_o for the circuit shown in Fig. 3-32.

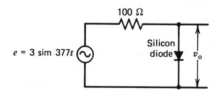

FIGURE 3-32

3-2. Plot v_o for the circuit in Fig. 3-33. Assume $V_D = 0.7$ V (forward diode drop). *Hint:* The output is shorted as long as either diode is conducting.

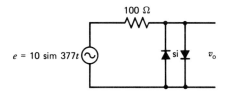

FIGURE 3-33

3-3. Plot v_o (approximate) for the circuit of Fig. 3-34 (assume $RC \ll T$). *Hint:* There is no output unless the diode is forward biased, conducting. Get the waveform at Point A first.

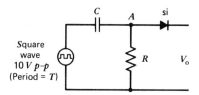

FIGURE 3-34

3-4. A peak detector circuit (Fig. 3-35) is connected to a load $R_L = 100$ KΩ. The shortest pulse width that must be considered is $T_p = 0.2$ μs and the largest interval between pulses is $T_w = 5$ μs. Assume that the charge time constant (RC) must be a tenth $(1/10)$ of T_p and the discharge time constant, ten times T_w. Find C and R_F (the forward diode resistance).

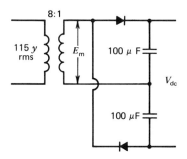

FIGURE 3-35

3-5. A half-wave rectifier is connected to a 10:1 transformer operating off the 115 V − 60 Hz power line. It is used to drive a 20 Ω load ($R_L = 20$ Ω).
  (a) Sketch the circuit.
  (b) Sketch the output voltage waveform.

116 DIODE APPLICATIONS

Find the following:

    (c) V_{dc} across the load.
    (d) I_{dc} through R_L.
    (e) I_{peak}, peak diode current.
    (f) The diode peak reverse voltage.

3-6.   A 5000 μF filter capacitor is added to the circuit of Problem 3-5.

    (a) Sketch (approximately) the waveform of the diode current.

Find:

    (b) The ripple voltage — rms.
    (c) V_{dc} (across the load).
    (d) I_{dc} in R_L.
    (e) The diode peak reverse voltage.

3-7.   A half-wave rectifier with a capacitor filter delivers 0.2 A (max) at 12 V dc. It uses the 115 V – 60 Hz line as the primary power source. The % ripple may not exceed 2%. Calculate the following:

    (a) The capacity of the filter capacitor (minimum).
    (b) The angle of conduction.
    (c) The diode peak current.
    (d) The transformer secondary voltage (rms) and the transformer turns ratio.

3-8.   A full-wave rectifier (no filter) is driven from a transformer (the primary is connected to a 115 – 60 Hz power line). It is designed to deliver 5 V dc at 0.2 A maximum.

    (a) Sketch the circuit. Show all related details.
    (b) Sketch the voltage output waveform.
    (c) Find the transformer secondary voltage and the turns ratio.
    (d) Find I_{peak} (diode peak current).
    (e) Find the diode peak reverse voltage.

3-9.   A full-wave rectifier with a capacitor filter of 1000 μF had $V_{dc} = 5$ V and $I_{dc} = 0.1$ A.

    (a) Sketch the voltage output waveform

Find:

    (b) % ripple.
    (c) I_{peak} (diode peak current).
    (d) The transformer secondary voltage.
    (e) The transformer turns ratio.
    (f) The diode PIV.

3-10.   A full-wave rectifier with a capacitive filter is to deliver 0.5 A at 24 V, with a ripple factor of 4% (maximum). It uses the 115 V – 60 Hz power line as the primary power source.

    (a) Find the transformer secondary voltage and its turns ratio.
    (b) Find C (the filter capacitor value).
    (c) Specify the diodes' PIV (minimum).

(d) Specify the peak current rating for the diodes.
(e) Sketch the circuit.

3-11. (a) Recompute the value of C above if the primary source is changed to 115 V – 400 Hz.
(b) Find the ripple voltage when using the capacitor found in Problem 3-10(b) and a 115 V – 400 Hz source. (The 400 Hz source is commonly used in boats and airplanes.)

3-12. A power supply (F.W) has a 2% ripple factor when it delivers 0.2 A at 20 V dc. (It operates off the 115 V – 60 Hz power system.)

(a) Design an R-C filter to improve the ripple factor to 0.2% (use reasonable capacitive values, not to exceed 1000 μF).
(b) Find the output voltage V_{dc} with the R-C filter in the circuit. (The current still is 0.2 A.)
(c) Design an L-C filter to improve the ripple factor to 0.02%, from the original 2%. (Assume the inductor $Q = 40$, $\omega L/R_{dc} = 40$.)
(d) Find V_{dc} with the L-C filter connected.
(e) Could a ripple factor of 0.02% be obtained with an R-C filter? Explain using numerical calculations.

3-13. For Problem 3-12 the voltage regulation of the original power supply is 1% (0 to 0.2 A full load). Find the regulation of the power supply with the added

(a) R-C filter
(b) L-C filter

Hint: Replace the power supply with a Thevenin's equivalent in which R_{Th} is computed to produce the voltage drop implied by the given regulation.

3-14. A bridge rectifier is used in conjunction with a capacitor filter. The voltage output is 15 V dc, the maximum load current is 0.5 A.

(a) Sketch the circuit.
(b) Find C, the filter capacitor required to keep the ripple factor at or below 1.0%. (Power source is 115 V–60 Hz.)
(c) Find the diodes' peak reverse voltage.

3-15. A zener regulator produces 12 V dc at 0.3 A full load.

(a) Sketch the circuit.
(b) If $R_s = 10\ \Omega$, what is the power dissipated in R_s?
(c) Specify the zener ratings (available power ratings are ½ W, 1 W, 2 W, 5 W, 10 W, 50 W).

3-16. A zener regulator is to deliver 5 V dc 1 A full load. The unregulated supply provides 9 V at full load. Its ripple voltage at full load is ±0.2 V.

(a) Select a suitable zener—power rating.
(b) Sketch the circuit.
(c) Find R_s and give its power rating.
(d) Give an estimate of the ripple voltage across the zener. (Consider the dynamic on resistance of the zener to be 0.1 Ω.)

118 DIODE APPLICATIONS

3-17. Sketch an ac zener regulator to deliver ≈ 25 V p-p output from the 115 V ac line.

3-18. Find R_s in Problem 3-17 if $I_L = 0.1$ A, $I_{Z1} = I_{Z2} = 20$ mA.

3-19. Calculate V_{dc} for the circuit shown in Fig. 3-35. *Hint:* Calculate the voltage across each capacitor (half-wave capacitor filter circuit).

CHAPTER

4

THE BIPOLAR TRANSISTOR

4-1 CHAPTER OBJECTIVES

In this chapter the student is introduced to the bipolar junction transistor (BJT), a semiconductor device that is used extensively in most of today's common electronic appliances. The student will gain an understanding of the physics of the device and of the basic current-voltage relations in the BJT. Before the uses of transistors in amplifier circuits are studied, the preliminaries of biasing the BJT are discussed. The student will learn of the need to bias the transistor (set up specific dc voltages and currents) and various circuits used to accomplish the biasing task. As part of the basic biasing discussion, the reader will also be presented with three basic types, configurations, of transistor amplifier circuits and how these different configurations affect the biasing circuit. Basic concepts relating to the need of stability in biasing and some quantitative biasing analysis are presented.

4-2 INTRODUCTION

Until 1947 all electronic circuits used vacuum tubes to obtain amplifications. The current in the vacuum tube was controlled by voltage applied to a metal grid placed between a thermionic emitter (the cathode emitted electrons when heated) and a conductive plate, the anode.

In 1947 the first transistor was demonstrated at Bell Laboratories. Amplification has been obtained in a semiconductor three-terminal device many times smaller in size than the vacuum tube and eventually much lower in cost than the tube.

120   THE BIPOLAR TRANSISTOR

The development of the transistor started one of the most outstanding revolutions in the electronic industry. Today, the microcomputer is a direct descendant of the transistor.

Since the theory of operation of the transistor relies heavily on the physics of P-N (and N-P) junctions, it is appropriate to review, briefly, the principles involved.

4-3   TRANSISTOR PHYSICS

In the *forward-biased* junction (where the externally applied voltage overcomes the diode barrier voltage) majority carriers are transported across the junction. Thus, the electrons from the N region (electrons are majority carriers in this region) are transferred to the P region and holes from the P region reach the N region. This conduction causes the P region to be inundated with electrons transferred from the N region. The N region is similarly flooded with holes. It is of the utmost importance to realize that this forward biasing results in a substantial increase in *minority* carriers in the two regions. The P region suddenly has a large number of electrons, and the N region a large number of holes. Both are minority carriers in their respective regions.

For the reverse-biased case (where an external voltage is applied to aid the junction barrier voltage and block majority current flow) the current consists of minority carriers. Since we have *very few* such carriers available, this reverse current in the diode is very small.

Let us now "combine" two junctions producing a single "grown" or metallurgically fused structure as shown in Fig. 4-1. Furthermore let us forward-bias junction A and reverse bias junction B, as shown. The forward-biasing of junction A causes electron flow across A and substantially increases the concentration of electrons (minority carriers) in the P region immediately next to junction A (Fig. 4-2). The P region now consists of two separate subregions: one with *high* minority carrier concentration (electrons near junction A) and a *low* electron concentration near junction B. (Recall that the reverse saturation current I_s consists of electrons moving across junction B from P to N, reducing electron concentration on the P side of junction B.) *Diffusion* (the natural phenomenon that equalizes concentrations in a region) takes place as a result of the large variation in electron concentration within the base region and the electrons diffuse through the base region (the P region) from the A junction toward the B junction.

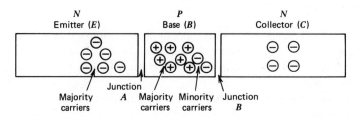

FIGURE 4-1   Back-to-back semiconductor junction: forming a transistor.

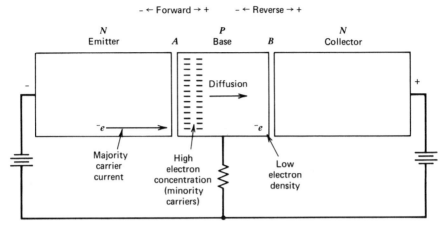

FIGURE 4-2 Charge concentrations in a biased transistor.

Next, the collection of the electrons (minority carriers) into the N region (the collector) across the B junction takes place. What was a very small reverse current (due to low electron concentration in the single P-N junction) is now a substantial current due to the injection of minority carriers into the P region by the forward-biased adjacent N-P junction. The current through the device, emitter to collector, is largely the current injected into the emitter by the forward bias applied to base-emitter junction. This injection current is controlled by the base-emitter voltage. To obtain collector current (in an unsaturated transistor) a reverse-bias condition must be set up in the base-collector junction. Once this has been done, the current to the collector is independent (to a large degree) of the exact value of the reverse bias. Since we assume that "all" electrons are swept into the collector, an increase in the C-B reverse voltage *cannot* increase this reverse saturation current. The result is a collector current independent of collector voltage. The emitter current can be viewed as the input current, and the collector current as the output current, with the exact relation between the two established by the characteristics of the device. It must be noted that not all of the emitter injected electrons reach the collector. Some recombine with holes in the base region and form part of the base current. Another source of base current is the hole current, base to emitter. In practical transistors the *base current is directly related to the emitter injected current*, and is a very small portion of it. *Base current is of the order of 1% of the emitter current* (more later on the ratio of base current to emitter current).

Figure 4-3 shows the basic currents involved in the transistor operation.

The terminal currents shown, I_E, I_B, I_C, are given in terms of electron current (solid lines) and hole current (dashed lines). The polarity of conventional current agrees with that of the hole current; hence, in later discussions, the direction of currents will be that of the hole currents (dashed lines). The above analysis was applied to a N-P-N structure, yielding what is classified as an NPN type transistor. The same analysis, using hole current, can be applied to a P-N-P structure (a PNP

122 THE BIPOLAR TRANSISTOR

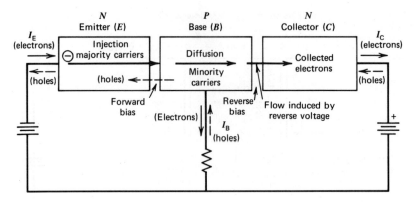

FIGURE 4-3 Currents in an NPN structure.

transistor) giving the identical general relationships. Figure 4-4 shows the symbols for NPN and PNP transistors and the directions of *conventional* dc currents in the terminals. The two transistor symbols differ only by the direction of the emitter current (conventional).

Figures 4-1, 4-2, and 4-3 are symmetrical, and one is tempted to conclude that the biasing arrangement can be reversed and still obtain valid transistor behavior. In other words, forward bias the collector-base junction and reverse bias the base emitter junction and, by previous analysis, obtain an identical operation. This would have been correct except for the physical structure of the P-N-P regions.

Figure 4-5 gives a simplified version of the transistor structure. The collector is constructed to surround the base and emitter regions so that it "collects" most of the current "emitted" by the emitter (injected current) that travels through the base. This guarantees a very low base current (relative to the emitter or collector current) and a very efficient operation. The effective width of the base region also affects the base current. A very small base region reduces side effects in that region (for example, electron-hole recombination) and hence, reduces the base current. The dimensions of the various regions are usually in the μm (10^{-6} m) range, so that structural strength restricts size reduction.

As noted before, I_B is a fraction of I_E. By controlling I_B (a very small current), we control I_E and hence I_C. Clearly, I_E is much larger than I_B, yielding an am-

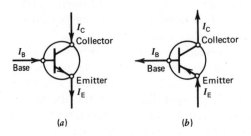

FIGURE 4-4 Transistor symbols and currents. (a) NPN. (b) PNP.

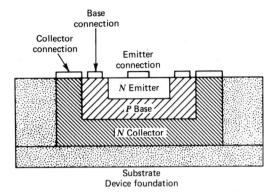

FIGURE 4-5 Physical transistor structure.

plification effect, a small I_B controlling a large I_E. If, for example, I_B is 1% of I_E, this implies that a 1 mA base current will produce a 100 mA emitter current. The application of KCL to Fig. 4-4 shows that

$$I_E = I_B + I_C \tag{4-1}$$

or

$$I_C = I_E - I_B \tag{4-1a}$$

If $I_B = 1$ mA and $I_E = 100$ mA, then $I_C = 100 - 1 = 99$ mA. The proportions I_C/I_B and I_C/I_E are characteristic of the particular device structure and doping levels and are usually given by the manufacturer. For convenience, the ratios are given literal symbols

$$I_C/I_B = \beta \tag{4-2}$$

(h_{FE} is often used in place of β.)

$$I_C/I_E = \alpha \tag{4-3}$$

β can be viewed as a current amplification factor, while α represents the efficiency of current transportation through the transistor (collected current relative to the injected current). β is usually large, in the order of 100 or more (I_B is 1% of I_c or less). The relations between β and α may be expressed as

$$\beta = \alpha/(1 - \alpha) \tag{4-4}$$
$$\alpha = \beta/(1 + \beta) \tag{4-5}$$

From Examples 4-1, 4-2, and 4-3, it is evident that a large β means α near unity. (As α approaches unity, β approaches infinity.)

EXAMPLE 4-1

The following currents were measured in a transistor:

$$I_E = 20 \text{ mA} \qquad I_C = 19.8 \text{ mA}$$

Calculate α, β, and I_B.

SOLUTION

From Eq. (4-3)

$$\alpha = I_C/I_E = \frac{19.8}{20} = 0.99$$

From Eq. (4-4)

$$\beta = 0.99/(1 - 0.99) = 0.99/0.01 = 99$$

From Eq. (4-2)

$$I_B = I_C/\beta = 19.8/99 = 0.2 \text{ mA} = 200 \text{ }\mu\text{A}$$

(The significance of β and α will become more evident from the discussion of transistor amplifiers in Chapter 5.)

4-3.1 Cutoff and Saturation

The previous section discussed the operation of the transistor in the "active" region, that is, base-emitter forward biased, base-collector reverse biased. Two other distinct operating conditions will now be investigated. First, the operation of the transistor under the conditions of base-emitter *not* forward biased, that is, $V_{BE} < 0.7$ (for silicon) and $I_B = 0$. In this state $I_C = 0$ since $I_C = \beta \times I_B$. When the base-emitter junction is *not* forward biased, we have no injected current and $I_E = 0$, $I_C = 0$. This is referred to as the "cutoff" region; the current is *cutoff*. Second, the state under which *both* junctions are *forward biased* represents the *saturation*-region. The current through the device (I_E or I_C) is essentially the forward current through two forward-biased diodes connected in series. Under this condition $V_{BE} \simeq 0.7$ V and $V_{BC} \simeq 0.7$ V (for silicon transistors). The polarities of these voltages are such (Fig. 4-6) that $V_{CE} \simeq 0$ V. ($V_{CE} = V_{CB} - V_{BE} \simeq 0.7 - 0.7$.) This means that to keep the operation in the active region, V_{CE} must not be permitted to be very low. (This is often caused by a high collector current and improper selection of circuit resistances as discussed in Chapter 5.)

TRANSISTOR PHYSICS

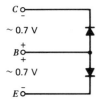

FIGURE 4-6   Saturated transistor representation (NPN).

4-3.2 Secondary Effects

So far, the behavior of the transistor was somewhat idealized. It was assumed, for example, that I_C is completely independent of V_{CE} (or V_{CB}) as long as the C-B junction is reverse biased. Any resistive voltage drop inside the transistor, such as that caused by bulk resistance and contact resistance, were ignored. While it is true that circuit analysis will largely ignore these factors (or represent their lumped effect), understanding them may give us a better insight into transistor operation.

In Chapter 2 we noted that a depletion region (a region in which there are very few, if any, mobile charges) exists in the P-N junction. The width of the region (usually measured in $\mu m = 10^{-6}$ m), depends on the voltage applied to the junction. Forward biasing the junction reduces this region to zero, producing conduction in the P-N device. Reverse bias increases the width of this region. The higher the reverse bias, the larger the depletion region. These variations in depletion region width have an effect on transistor operation.

As V_{CB}, the collector-base reverse voltage, is increased, the C-B depletion region increases in width and the base region itself shrinks (Figs. 4-7a and 4-7b). The result is that I_C increases somewhat as V_{CB} increases. A narrower base region causes an increase in the current swept through the base (I_C). This base width effect (named the Early effect after J.M. Early who first noted it in 1952) is most pronounced at high V_{CB} ($\simeq 50$ V). The Early effect results in an increase in β as V_{CB} increases. Recall that $\beta = I_C/I_B$, and since the Early effect causes an increase in I_C while I_B remains constant, it produces an increase in β. Since V_{CE} is directly related to V_{CB},

$$V_{CE} = V_{CB} + V_{BE} \simeq V_{CB} + 0.7 \text{ V (for silicon)} \tag{4-6}$$

variations in V_{CE} will also affect β while previously it was assumed that β is constant and I_C was assumed independent of V_{CE}. In circuit analysis (Ch. 5), the Early effect is usually negligible. (The Early effect is mentioned again in connection with transistor characteristics in Sec. 4-4.)

Because of some other internal effects, the β of the transistor decreases with an increase in I_C (assuming V_{CE} is constant) above a certain optimum value of I_C. The typical behavior of β with respect to I_C is shown in Fig. 4-8.

In the discussion of the forward diode resistance r_F (Sec. 2-5), we were concerned

126 THE BIPOLAR TRANSISTOR

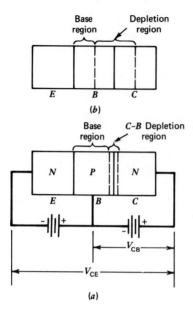

FIGURE 4-7 Base width variations. (a) E-B forward biased; C-B reverse biased. (b) E-B forward biased; C-B with larger reverse biased.

only with the physics of the device. r_F was calculated at 300°K (300 degrees Kelvin) junction temperature

$$r_F = 26/I_D \qquad (2\text{-}5a)$$

where I_D is in milliamperes, neglecting some important practical parameters. Different methods are used to connect the external diode terminals to the P and N

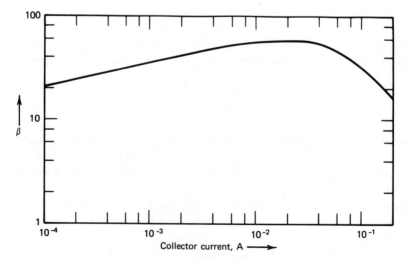

FIGURE 4-8 Current gain (β) as a function of the collector current for a typical transistor (log-log plot).

regions. This introduces contact resistance. In addition, there is bulk semiconductor material between the contacts and the internal P and N regions. This bulk (material) resistance was also neglected. In many cases Eq. (2-5a) is modified to incorporate these two effects.

$$r_F = 26/I_D + R \qquad (4\text{-}7)$$

where R is somewhat arbitrarily selected to represent contact and bulk resistance and may vary between 0.5 Ω to 2 Ω.

Contact and bulk resistance are both present in the transistor in both C-B and B-E junctions. Various techniques have been developed to minimize these resistive effects, which become extremely important in power (high current) transistors. In the analysis of signal transistors (low power) these effects may indeed be neglected. It is, however, worth bearing in mind that discrepancies between calculated and measured performance may be due in large measure to these neglected effects.

4-4 TRANSISTOR CHARACTERISTICS

The design of transistor circuits is based largely on the relationships among the various currents and voltages in the three terminals: emitter, base, and collector. Transistor manufacturers usually provide plots that give the relation between collector-emitter voltage (V_{CE}) and collector current (I_C) for particular base currents (I_B). Figure 4-9 shows two *typical* plots for $I_B = 10$ μA: one for low values of V_{CE} (0 to 15 V) and one for high values of V_{CE} (0 to 60 V). The region where V_{CE} is very small ($V_{CE} < 0.7$ V) produces saturation, since V_{CB} is not large enough to establish reverse bias across the C-B junction. Note that both plots show V_{CE} as the independent variable and not V_{CB}. From Eq. (4-6), however, it is clear that $V_{CB} \approx -0.7$ V results in $V_{CE} \approx 0$ V (for silicon). For values of V_{CE} between approximately 1 to 30 V, I_C is fairly constant (1 mA), as shown in Fig. 4-9b, and can be calculated from Eq. (4-2) as $I_C = \beta I_B$. β is a characteristic of the specific transistor and is given by the manufacturer. As V_{CE} increases beyond 30 V, the Early effect becomes more pronounced. I_C shows a marked increase with increases in V_{CE}. A point is reached (≈ 55 V) where the C-B junction enters into the avalanche region (Fig. 4-9b). Collector-base current increases drastically, producing a sharp increase in I_C. In this region, the collector-base diode behaves much like a conducting zener diode. As a result, the collector-emitter path consists of a forward-biased diode (base-emitter) in series with a conducting zener (collector-base), which clearly results in high collector current. It is important to note that (as with the zener diode), this avalanche behavior is not destructive in itself, as long as the power ratings of the transistor are not exceeded. (The high I_C may, however, result in power dissipation in excess of the rated power leading to permanent damage.)

Figure 4-10 is essentially a duplication of Fig. 4-9b for a number of different values of I_B. Such plots are called a family of collector characteristic curves. (The two variables plotted in Fig. 4-10 are *both* related to the collector: that is, collector current and collector-emitter voltage. I_B is held constant for each plot.)

128 THE BIPOLAR TRANSISTOR

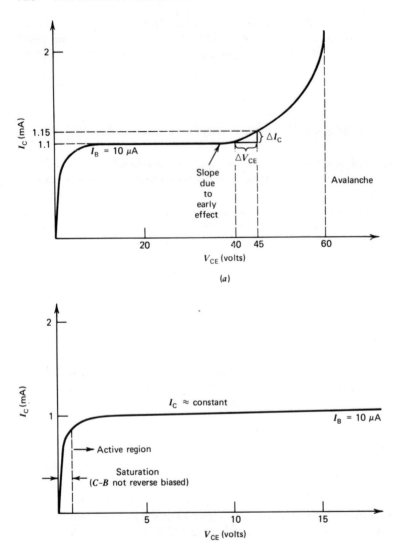

FIGURE 4-9 Typical transistor characteristics. (a) Low V_{CE}. (b) High V_{CE}.

Although Fig. 4-10 (and all typical collector characteristics) does not give plots for all possible base currents (for example, $I_B = 15$ μA is not given), these missing graphs can always be obtained by interpolation (or extrapolation) if necessary. For example, the plot for $I_B = 15$ μA is midway between the plots for $I_B = 10$ μA and $I_B = 20$ μA.

Figures 4-9 and 4-10 yield a number of important transistor parameters. The β, or current gain, of the transistor can be calculated from these plots. (Recall that β

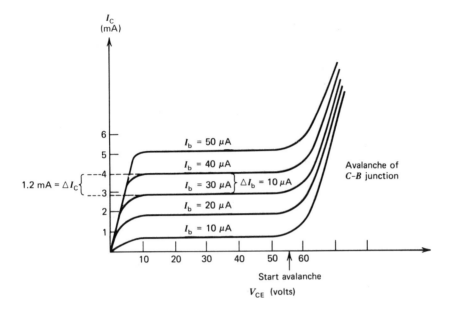

FIGURE 4-10 Collector characteristics I_C versus V_{CE}.

$= I_C/I_B$.) Figure 4-9 shows that for a wide range of V_{CE}, $I_C = 1$ mA for $I_B = 10$ μA; hence,

$$\beta_{dc} = (1 \times 10^{-3})/(10 \times 10^{-6}) = 100$$

The value for β above was based on dc currents, both I_B and I_C where dc values; hence, the subscript dc. The β_{ac}, the current gain for changing currents, can be obtained by introducing a change in I_B, ΔI_B and examining the corresponding change in I_C, ΔI_C. From Fig. 4-10, if $\Delta I_B = 10$ μA, then $\Delta I_C = 1.2$ mA and $\beta_{ac} = \Delta I_C/\Delta I_B = (1.2 \times 10^{-3})/(10 \times 10^{-6}) = 120$. Note that β_{ac} was computed for the particular case in which $V_{CE} = 30$ V and I_B changes from 30 to 40 μA. β_{ac} may vary with changes in transistor operating conditions as shown in Fig. 4-8. It is common to define β_{dc} as the ratio of I_C to I_B (or for β_{ac} as the ratio of ΔI_C to ΔI_B) for a constant value of V_{CE}. The definitions are:

$$\beta_{dc} = I_C/I_B|_{V_{CE} = \text{constant}} \quad (4\text{-}8)$$

$$\beta_{ac} = \Delta I_C/\Delta I_B|_{V_{CE} = \text{constant}} \quad (4\text{-}9)$$

Equation (4-9) is extended to (by differential calculus)

$$\beta_{ac} = dI_C/dI_B \quad (4\text{-}9a)$$

by selecting as small a ΔI_B as possible. Both Eqs. (4-8) and (4-9) give the current gain for a particular value of I_B. (As already noted, β may change with changes in

130 THE BIPOLAR TRANSISTOR

I_C, see Sec. 4-32 and Fig. 4-8). In most practical applications, however, a single value for β is used even when a wide range of V_{CE} and I_B is involved. While this approximation introduces an error, it is quite acceptable in most cases. (The function of good circuit design is to make the operation independent of β.)

Another important transistor parameter is the "collector-emitter resistance." This is defined as

$$R_{CE} = V_{CE}/I_C|_{I_B = \text{constant}} \quad (4\text{-}10)$$

$$r_{CE} = dV_{CE}/dI_C|_{I_B = \text{constant}} \quad \text{or} \quad r_{CE} = \Delta V_{CE}/\Delta I_C \quad (4\text{-}11)$$

Here, I_B is constant. R_{CE} and r_{CE} may be used to represent the collector to emitter resistive behavior of the transistor for dc and ac applications, respectively. Since r_{CE} is defined in differential terms, the measurement of r_{CE} involves the ratio of ΔV_{CE} to ΔI_C, as shown in Fig. 4-9b. Here, we get $R_{CE} = 40/(1.1 \times 10^{-3}) = 36.4$ kΩ, for $V_{CE} = 40$ V and $I_B = 10$ μA. $r_{CE} = (45 - 40)/(1.15 - 1.1) \times 10^{-3} = 5/(0.05 \times 10^{-3}) = 100$ kΩ. The ac collector resistance was evaluated for $_{CE}$ varying for 40 to 45 V and for $I_B = 10$ μA (I_B is kept constant at 10 μA). Note that r_{CE} is very large compared to R_{CE}. Had Fig. 4-9a (where we assumed $I_C \approx$ constant) been used to obtain r_{CE}, it would have seen that $r_{CE} = \infty$ (since $\Delta I_C = 0$, $\Delta V_{CE}/\Delta I_E \to \infty$). When considering the ac behavior of the transistor (Ch. 5), we will take notice of the fact that r_{CE} is very large and may be ignored in many transistor circuits. (The parameter r_{CE} is usually given by the manufacturer as h_{oe} where $r_{CE} = 1/h_{oe}$).

4-5 TRANSISTOR BIASING—BASIC CONCEPTS

The performance of the transistor in a circuit (and the transistor parameters to some extent) depend on the dc operating conditions established by the circuit. For example, specific dc collector currents are required for different applications. Typically, audio amplifier applications require a substantial dc collector current while other applications, such as switching circuits, call for zero dc collector current. The circuit that produces these desired operating *dc* currents is called the *biasing* circuit. The currents produced by this circuit are the bias currents. The dc bias currents in the transistor prepare the transistor for its assigned task. But bias currents are not *directly* involved in the task, no more than the setting up of chessmen on a checkerboard affects the game itself. You cannot play unless the board is set up. However, setting up the board does not guarantee further results.

The transistor circuit designer determines the desired operating conditions, called the *quiescent* operating point or the Q-point. Supply voltages must be selected, and specific circuits must be designed to provide the necessary bias currents. Only when properly biased can the circuit fulfill the assigned task, that is, amplify, perform a switching function, etc.

Figure 4-11 shows three choices for the operating point. Q_A represents operating

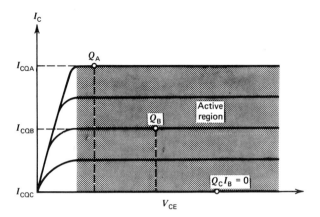

FIGURE 4-11 Operating points.

at or near saturation. Q_B represents operation approximately in the center of the "active" region. Q_C shows operation at the "cut off" point. Many applications require biasing in the active region. For such applications we attempt to design the circuit so that the operating point Q_B is somewhere near the center of the active region (more on this in Sec. 4-8). This requires a specific collector current (I_{CQB} in Fig. 4-11). The biasing circuit must be designed to produce the required value of I_{CQ} (the quiescent collector current). Changes in I_{CQ} result in the shifting of the operating point, possibly out of the active region. It is essential, therefore, that once the bias circuit has been designed, I_{CQ} remain constant regardless of changes in temperature of *replacement of the transistor*. The latter consideration is particularly important. It is inconceivable that every time a repair is made and a transistor replaced, it will be necessary to redesign the circuit to obtain the original I_{CQ}.

The β of transistors varies substantially, not only with the temperature, but also from transistor to transistor (even if the two transistors are of the *same type*). For example, the β of the 2N4265 transistor has a range of 100 to 400 at 25°C. (In many cases only minimum *or* maximum β is given by the manufacturer.) But β may change from 100 at 25°C to 70 at 15°C. It is very important, therefore, that the bias circuit is designed so that the operating point (I_{CQ}) is independent of β. Otherwise, large fluctuations in I_{CQ} can be expected to make the circuit nearly useless. In the detailed discussion of specific biasing circuits, the subject of bias stability (the stability of I_{CQ}) will be of major importance.

Another important point is that in the general analysis of transistor circuits, we distinguish between dc analysis (namely, the biasing considerations) and signal (or ac) analysis. In most instances, the results of the analysis will consist of two separate sets of values. One will specify dc currents and voltages. The other set will specify ac or signal amplitudes. (By superposition the *total* current, or voltage, at a point is the sum of the dc and ac currents or voltages. We will rarely concern ourselves with this *total* current or voltage.)

4-5.1 Three Basic Circuit Configurations

The transistor may be connected in several different ways. Since it has three terminals (collector, base, emitter), any terminal can serve as the signal input or output terminal. While this leads to six possible configurations, only three of these have extensive practical applications.

Table 4-1 gives the three practical configurations.
Note that the terminal *not* associated with either input or output is referred to as the common terminal. This gives rise to the name of the circuit.

The design of biasing circuitry is not concerned with *signal* flow at all. Therefore, the various biasing circuits are essentially the *same* for all three configurations. For example, the fixed bias circuit (discussed in detail in Sec. 4-7) is used for all three configurations. Figure 4-12 shows the fixed bias circuit as it is used for the three configurations. This makes the dc analysis of the biasing circuits completely independent of the "configuration" used. This is so as long as appropriate methods (capacitors or transformers) are used to isolate the dc quantities from the ac signal (dc amplifiers are excluded since no such isolation is provided). The ac considerations for a particular configuration affect the choice of resistor *values*, but not necessarily the *type* of biasing circuit that can be used.

4-6 BIASING STABILITY

All biasing circuits are designed to produce a *desired* dc collector current I_{CQ} at a desired Q-point. The specific value of I_{CQ} depends on the particular application (analog applications use the center of the active region of the collector characteristics. Digital applications usually require the saturation or cut-off operating point.) In choosing a Q-point, it is, therefore, necessary to use the collector characteristics of the specific transistor. These are either obtained from the manufacturer or, alternatively, from a transistor curve tracer.

As noted in Sec. 4-5, it is important to make the operating point (I_{CQ}, V_{CEQ}) as stable (constant) as possible, to withstand variations in temperature and β. The choice of a particular biasing circuit is based largely on whether the particular circuit meets the required bias stability.

TABLE 4-1   TRANSISTOR CIRCUIT CONFIGURATIONS

Input signal to	Output signal from	Circuit name	Acronym
Base	Collector	Common emitter	CE
Base	Emitter	Common collector or emitter follower	CC EF
Emitter	Collector	Common base	CB

BIASING STABILITY   133

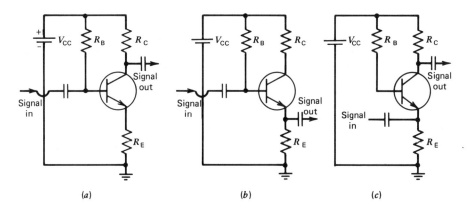

FIGURE 4-12 Fixed bias for various configurations. (a) Common emitter. (b) Common collector. (c) Common base.

A general definition of bias stability with respect to temperature is given by

$$S(T) = \frac{\Delta I_C/I_{CQ}}{\Delta T/T} \qquad (4\text{-}12)$$

The bias stability factor, $S(T)$, is the ratio of *relative* (or %) changes in I_{CQ} (drifting away from the desired I_{CQ}) to relative (or %) changes in temperature. (Many texts define the stability factor in terms of absolute, rather than relative, changes in I_C. Thus, a circuit with a stability factor of $S = 1$ is the most stable, and $S > 1$ indicates bias instability. This implies that, for example, a 1 mA change in I_C from 1 to 2 mA is as significant as a 1 mA change in I_C from 10 to 11 mA. Clearly, the latter represents a much smaller instability than the former.)

It should be noted that temperature changes affect a number of transistor parameters. Transistor leakage, V_{BE} and β are all affected by temperature variations. $S(T)$ represents the *total* effect of temperature variations.

EXAMPLE 4-2

A biasing circuit was designed to operate at a 20°C junction temperature with $I_{CQ} = 5$ mA. At a 40°C junction temperature it was found that $I_{CQ} = 6$ mA. Find $S(T)$.

SOLUTION

$$\Delta T/T = \frac{40 - 20}{20} = 1 \qquad (100\%)$$

$$\Delta I_C/I_{CQ} = \frac{6 - 5}{5} = 1/5 \qquad (20\%)$$

$$S(T) = \frac{1/5}{1} = 0.2$$

134 THE BIPOLAR TRANSISTOR

It is sometimes possible to calculate $S(T)$ from circuit values. Consideration must then be given to the three major factors that determine I_C that are affected by temperature that is, leakage current, V_{BE}, and β, and the way they combine to produce $S(T)$. (These factors are not necessarily additive.) The discussion of temperature stability is quite complex, and only qualitative analysis will be used when considering $S(T)$ for the various biasing circuits.

Of the three factors mentioned above, variations in β is the most crucial (in most applications). As noted earlier, β varies drastically from transistor to transistor of the same type and also varies with temperature (as well as with I_{CQ} itself). The β stability factor is defined as

$$S(\beta) = \frac{\Delta I_C / I_{CQ}}{\Delta \beta / \beta} \qquad (4\text{-}13)$$

The significance of $S(\beta)$ can be demonstrated by considering a transistor for which $\Delta\beta/\beta = 3$ (this 3 to 1 range of β is quite common). A stability factor of 1, $S(\beta) = 1$, will yield a 3 to 1 change in I_{CQ} corresponding to the 3 to 1 β change. This may easily shift the Q-point out of the active region, rendering the circuit inoperative. The most stable bias circuit is one for which $S(\beta) = 0$ (and $S(T) = 0$). This implies that I_{CQ} is *independent* of β. (See voltage divider biasing circuit, Sec. 4-7.4.)

It is important to note that bias stability is more a function of circuit design than of transistor characteristics. Some circuits are inherently unstable while others exhibit a very high degree of stability. It is the function of the circuit designer to select a suitable circuit and then calculate the appropriate resistor values.

4-7 BIASING CIRCUITS

A number of biasing circuits will be analyzed. Some of these are so unstable that they *cannot* be used in amplifier applications. They are, however, useful for digital or switching applications when bias stability is not very important.

As noted earlier, the circuits presented are all applicable to all transistor configurations (CE, CB, CC). In Chapter 5, when discussing transistor amplifiers, we will examine the effect of biasing circuits on the ac performance of the overall circuit.

4-7.1 Fixed Bias

Figure 4-13a shows a fixed bias circuit. The desired I_{CQ} is obtained by selecting a suitable base current I_{BQ} since $I_C = \beta \times I_B$. (The subscript Q indicates that the quantity involved is for the Q-point.) The circuit design is then concerned with selecting R_B to obtain the required I_B. In this circuit I_C is strongly dependent on β. (This will become clearer from the discussion of $S(\beta)$ for this circuit; see Ex. 4-4.)

In the computation of I_C or I_B, both circuit equations, KVL and KCL, and device equations, $I_C = \beta I_B$, will be used.

BIASING CIRCUITS   135

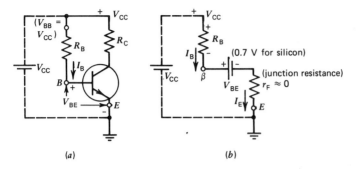

FIGURE 4-13   Fixed bias. (V_{CC} is used here as both the collector and base supply. The base supply may be separate from V_{CC}.) (a) Actual circuit (silicon transistor). (b) Base portion equivalent circuit (V_{BE} = 0.7 V for silicon).

I_B is computed by applying KVL to the base circuit, after replacing the base-emitter portion with an equivalent circuit.

The base-emitter junction is essentially a forward biased diode that can be replaced by the equivalent circuit as shown in Fig. 4-13b (see Sec. 2-5.3).

r_F (the diode resistance) in Fig. 4-13b may be neglected since it is very small relative to other resistors. Applying KVL to the base emitter loop yields two voltage drops: the $I_B \times R_B$ drop and V_{BE}.

$$V_{CC} - I_B \times R_B - V_{BE} = 0 \tag{4-14}$$

Solving Eq. (4-14) for R_B and I_B, respectively, we get

$$R_B = \frac{V_{CC} - V_{BE}}{I_B} \tag{4-14a}$$

$$I_B = \frac{V_{CC} - V_{BE}}{R_B} \tag{4-14b}$$

To find I_C, the device relation $I_C = \beta I_B$ is used.

EXAMPLE 4-3

The circuit of Fig. 4-13 that used a silicon transistor with a β = 100 has to be biased so that I_{CQ} = 5 mA. V_{CC} is given as 20 V. Find R_B. (Note that for silicon V_{BE} = 0.7 V.)

SOLUTION

First find I_{BQ}

$$I_{BQ} = \frac{I_{CQ}}{\beta} = (5 \times 10^{-3})/100 = 50 \ \mu A$$

136 THE BIPOLAR TRANSISTOR

Using Eq. (4-14b), we get

$$R_B = (V_{CC} - V_{BE})/I_B = (20 - 0.7)/(50 \times 10^{-6})$$
$$= 19.3/(50 \times 10^{-6}) = 386 \text{ k}\Omega$$

As noted, the stability of the fixed bias circuit is very poor. This is demonstrated by the following example:

EXAMPLE 4-4

In Ex. 4-3 assume that the β changes to 200. (This often happens when the transistor is replaced.) Find:
 (a) I_{CQ}.
 (b) $S(\beta)$.

SOLUTION

(a) I_B will be the same as in Ex. 4-3 since none of the elements in the loop have been changed. Hence, $I_B = 50$ µA, and $I_C = \beta \times I_B = 200 \times 50 \times 10^{-6} = 10$ mA.

(b) To find $S(\beta)$ note that $\Delta\beta = 100$ and $\Delta\beta/\beta = 100/100 = 1$ (or 100% change). $\Delta I_C = 5$ mA and $\Delta I_C/I_{CQ} = (5 \text{ mA})/(5 \text{ mA}) = 1$; hence,

$$S(\beta) = \frac{\Delta I_C/I_{CQ}}{\Delta\beta/\beta} = 1/1 = 1 \tag{4-13}$$

4-7.2 Bias with Feedback–Emitter Resistance

The fixed bias circuit, as was noted, is very unstable. Changes in β or temperature strongly affect I_{CQ}. A circuit that was expected to have a particular I_{CQ} may wind up with double or triple the expected current because of changes in β or the temperature, or both. To combat this instability, the concept of feedback is used. The idea is to use the anticipated change in I_C to control I_C. The fact that I_C, for example, has increased is fed back to the base circuit in such a manner as to cause a reduction in I_C. The net effect of this feedback arrangement is a much smaller change in I_C. The feedback connection tends to *negate* or oppose changes in I_C. This type of feedback is called negative feedback. A very simple way of introducing this feedback into the fixed bias circuit is by the addition of an emitter resistance R_E, sometimes called the emitter feedback resistor (Fig. 4-14a). Here, an increase in I_C results in an increase in V_E, the voltage across R_E, and a decrease in I_B. Note that R_E and the voltage V_E are part of the base current loop. The decrease in I_B tends to minimize the increase in I_C.

In Ex. 4-4 it was found that for fixed bias $S(\beta) = 1$. This means that any change in β is directly reflected in a change in I_{CQ} on a one to one basis (doubling of I_{CQ}, as in Ex. 4-4, can easily render the circuit useless).

BIASING CIRCUITS   137

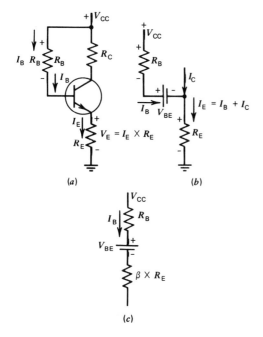

FIGURE 4-14 Fixed bias with R_E. (a) Circuit. (b) Equivalent circuit (base part). (c) Base input equivalent.

Note that the fixed bias circuit shown in Fig. 4-13 can be used *only* in the CE configuration. This is because the signal can be fed in at the base only and the output taken from the collector only. Because this circuit suffers from bias instability, it is used practically only in switching application.

Figure 4-14a shows a modified fixed bias circuit with an emitter resistance. Figure 4-14b shows the equivalent base loop. We again apply KVL to the base loop, recognizing, however, that the current through R_E is I_E and the voltage across R_E is $I_E \times R_E$.

Applying KVL to Fig. 4-14b, we get

$$V_{CC} - I_B \times R_B - V_{BE} - I_E \times R_E = 0 \qquad (4\text{-}15)$$

In Eq. (4-15) there are two unknowns: I_B and I_E (if we assume that R_B and R_E are given). The device equation $I_E \approx I_C = \beta \times I_B$ is used as the second simultaneous equation permitting a solution for I_B and I_C (the assumption that β is large results in $I_E \approx I_C$, otherwise, $I_E = (\beta+1)I_B$). Substituting the device equation $I_C = \beta \times I_B$ in Eq. (4-15) yields

$$V_{CC} - I_B \times R_B - V_{BE} - \beta I_B \times R_E = 0$$
$$V_{CC} - V_{BE} = I_B(R_B + \beta R_E)$$
$$I_B = \frac{V_{CC} - V_{BE}}{R_B + \beta R_E} \qquad (4\text{-}15a)$$

138 THE BIPOLAR TRANSISTOR

Equation (4-15a) implies that as far as I_B is concerned, the circuit looks like the one shown in Fig. 4-14c. KVL applied to Fig. 4-14c yields

$$V_{CC} - I_B R_B - V_{BE} - I_B(\beta \times R_E) = 0 \qquad (4\text{-}16)$$

Note that the resistor R_E is replaced by a resistor $\beta \times R_E$ (β times larger). Equation (4-16) yields the same expression obtained in Eq. (4-15a) from the original circuit. It is worth repeating that, in terms of the base current, the effect of R_E is multiplied by β!

The stability of this circuit is somewhat better than that of the simple fixed bias circuit. It depends on the relative magnitudes of R_E and R_B/β. It can be shown that for very large R_E ($R_E \gg R_B/\beta$) the operating point is nearly independent of β, that is, $S(\beta) \approx 0$. From Eq. (4-15a)

$$I_C = \beta \times I_B = \beta \left[\frac{(V_{CC} - V_{BE})}{R_B + \beta R_E} \right] = \frac{V_{CC} - V_{BE}}{R_B/\beta + R_E} \qquad (4\text{-}17)$$

If $R_E \gg R_B/\beta$, Eq. (4-17) becomes

$$I_C = (V_{CC} - V_{BE})/R_E \quad \text{(neglecting } R_B/\beta) \qquad (4\text{-}17a)$$

Equation (4-17a) shows that I_C is completely independent of β. Unfortunately, practical circuits rarely fulfill the condition $R_B/\beta \ll R_E$, so some dependence on β still exists in this biasing circuit.

EXAMPLE 4-5

In Fig. 4-14a, $R_E = 1.5$ kΩ, $R_B = 100$ kΩ, $V_{CC} = 20$ V, $\beta = 200$, and the transistor is silicon. Find $S(\beta)$.

SOLUTION

First find I_{CQ}.

$$I_{CQ} = \frac{V_{CC} - V_{BE}}{R_B/\beta + R_E} = \frac{20 - 0.7}{\dfrac{100 \text{ k}\Omega}{200} + 1.5 \text{ k}\Omega}$$

$$= \frac{19.3}{2 \text{ k}\Omega} = 9.65 \text{ mA} \qquad (4\text{-}17)$$

$$I_{CQ} = 9.65 \text{ mA}$$

Assume $\Delta\beta$ of 20, so that

$$\Delta\beta/\beta = 20/200 = 1/10 \quad (10\%)$$

The new β is equal to $\beta + 20 = 220$. ΔI_C can be calculated by finding the new I_{CQ} (for $\beta = 220$).

BIASING CIRCUITS

$$I_{CQnew} = \frac{20 - 0.7}{\frac{100\ k\Omega}{220} \times 1.5\ k\Omega} = 9.8744\ mA$$

$$\Delta I_C = I_{CQnew} - I_{CQ} = 9.8744 - 9.65 = 0.2244\ mA$$

$$I_C/I_{CQ} = 0.2244/9.65 = 0.0232$$

and

$$S(\beta) = \Delta I_C/I_{CQ}/\Delta\beta/\beta = \frac{0.0232}{1/10} = 0.232$$

Note that the reasonably good stability in Ex. 4-5 has been obtained with only a 3 to 1 ratio of R_E to R_B/β (1.5 kΩ)/(0.5 kΩ).

$S(\beta)$ can be found directly, for this circuit, from the expression

$$S(\beta) = R_B/(R_B + \beta R_E)$$

See Appendix B.

$$S(\beta) = \frac{100\ k\Omega}{(100\ k\Omega + 200 \times 1.5\ k\Omega)} = 0.25$$

The difference in $S(\beta)$ is due to the very small β increment involved in the differential method used to get the latter formula for $S(\beta)$. The formula is also more accurate.

The circuit shown in Fig. 4-14a can be used in all three transistor configurations. This is shown in Fig. 4-15. Note that Fig. 4-15b has been drawn to conform to the accepted convention in which signal flow is shown from left to right. It is, of course, the same bias circuit as in Fig. 4-14. While the basic bias circuit may be the same for all three configurations, the selected resistance values may differ substantially. In the CC circuit of Fig. 4-15c, for example, $R_C = 0$ (we need no collector resistor) and R_E is large (compared to R_E in the other circuit configurations).

In the fixed bias with emitter resistance circuit (Fig. 4-14), I_B depends on $R_B + \beta R_E$. In transistor circuit design, a particular biasing point I_{CQ} is chosen first, thus establishing I_{BQ}. It is then necessary to select both R_B and R_E to fit that requirement. R_E is usually selected so that the voltage across R_E, V_E is larger than V_{BE}. ($V_E \simeq$ 3 to five times V_{BE}.) This means that for silicon where $V_{BE} = 0.7\ V$, $V_E = I_E \times R_E = 3 \times V_{BE} = 2.1\ V$.

This assumption simplifies the design and

$$R_E \approx \frac{3 \times V_{BE}}{I_E} \simeq \frac{3 \times V_{BE}}{I_{CQ}}\ \Omega \tag{4-18}$$

Equation (4-18) permits the selection of R_E directly from the desired I_{CQ}.

140 THE BIPOLAR TRANSISTOR

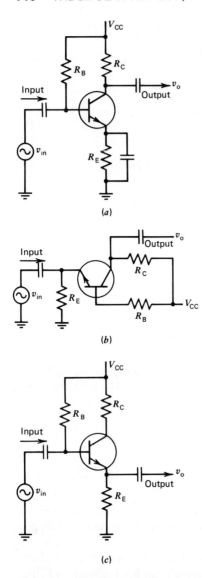

FIGURE 4-15 Three transistor amplifier configurations using fixed bias with R_E. (All capacitors are open circuits for dc). (a) Common emitter (CE). (b) Common base (CB). (c) Common collector (CC or EF).

EXAMPLE 4-6

In Fig. 4-15b, given $V_{CC} = 30$ V, $I_{CQ} = 4$ mA, and a silicon transistor, $\beta = 50$, find
 (a) R_E. (Use $V_E = 3\, V_{BE}$)
 (b) R_B.

SOLUTION

(a) $R_E = 3 \times V_{BE}/I_{CQ} = (3 \times 0.7)/(4 \times 10^{-3}) = 2.1/(4 \times 10^{-3}) = 525\ \Omega$

(b) $I_B = (V_{CC} - V_{BE})/(R_B + \beta R_E)$

Rearranging, we get

$$R_B + \beta R_E = (V_{CC} - V_{BE})/I_B$$

and

$$R_B = (V_{CC} - V_{BE})/(I_C/\beta - \beta R_E)$$

where $I_B = I_C/\beta$. Substituting given values yields

$$R_B = (30 - 0.7)/(4\ \text{mA}/50 - 50 \times 525) = 366\ \text{k}\Omega - 26\ \text{k}\Omega$$

$R_B \simeq 340\ \text{k}\Omega$

The introduction of R_E makes the operating point less dependent on β. It also reduces the effect of changes in V_{BE} (which are usually temperature induced). This is the basis for selecting $V_E > V_{BE}$. The variations in V_{BE} will be swamped by the changes in V_E. V_E tends to change in the opposite direction of V_{BE}, which neutralizes the temperature induced changes in V_{BE}.

4-7.3 Self-Bias Circuit

Figure 4-16 shows a typical self-bias circuit. The stability of this biasing circuit is somewhat better than the fixed bias with R_E circuit. The reasons for the improved stability will become clear from the analysis that follows.

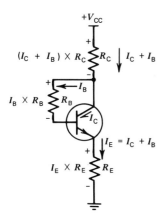

FIGURE 4-16 Self-bias circuit.

142 THE BIPOLAR TRANSISTOR

The base loop includes all three resistors: R_C, R_B, R_E. The base current is supplied from the collector (not from V_{CC}); hence, it passes through R_C. Applying KVL (from V_{CC} toward GND), we get

$$V_{CC} - (I_C + I_B) \times R_C - I_B R_B - V_{BE} - I_E R_E = 0 \qquad (4\text{-}19)$$

$$V_{CC} - V_{BE} - I_E(R_C + R_E) - I_B R_B = 0 \qquad (4\text{-}19a)$$

Note that the substitution $I_E = I_C + I_B$ was used in Eq. (4-19a). Equation (4-19a) is identical in form to Eq. (4-15). R_E, in Eq. (4-15), is replaced by $R_E + R_C$. This makes the effective R_E resistance larger and further stabilizes the circuit. The solutions for I_B and I_C are obtained directly from Eqs. (4-15a) and (4-17), respectively

$$I_B = (V_{CC} - V_{BE})/[R_B + \beta(R_C + R_E)] \qquad (4\text{-}20)$$

$$I_C = \beta I_B = (V_{CC} - V_{BE})/(R_B/\beta + R_E + R_C) \qquad (4\text{-}21)$$

EXAMPLE 4-7

In Fig. 4-16, given $I_{CQ} = 5$ mA $V_{CC} = 20$ V, $V_{CQ} = 10$ V, $\beta = 100$, silicon transistor ($V_{BE} = 0.7$), find
  (a) R_E. (Use $V_E = 3\, V_{BE}$)
  (b) R_C.
  (c) R_B.

SOLUTION

(a) We assume $V_E = 3 \times V_{BE} = 3 \times 0.7 = 2.1$ V. Since the voltage V_E is produced by the current I_E in R_E ($I_E \approx I_{CQ}$)

$$R_E = V_E/I_{CQ} = 2.1/(5 \times 10^{-3}) = 420\ \Omega$$

(b) $V_C = V_{CC} - (I_B + I_C)R_C$
(KVL is applied to the loop containing V_{CC}, $(I_B + I_C)R_C$, and V_C.) Rearranging, we get

$$R_C = (V_{CC} - V_C)/(I_B + I_C) = (V_{CC} - V_c)/(I_C/\beta + I_C)$$

Substituting the given values yields

$$R_C = (20 - 10)/(5\ \text{mA}/100 + 5\ \text{mA}) \approx 10/(5\ \text{mA}) = 2\ \text{k}\Omega$$

(c) $I_B = \dfrac{V_{CC} - V_{BE}}{R_B + \beta(R_E + R_C)}$

Solving for R_B yields

$$R_B = \frac{V_{CC} - V_{BE}}{I_B} - \beta(R_E + R_C) = \frac{V_{CC} - V_{BE}}{I_C/\beta} - \beta(R_E + R_C)$$

BIASING CIRCUITS

Substituting given values and those found in (a) and (b) above, we get

$$R_B = \frac{20 - 0.7}{5 \text{ mA}/100} - 100(420 + 2000) = 144 \text{ k}\Omega$$

Note that the self-bias circuit can be designed with $R_E = 0$. The solution will follow the exact procedure outlined above with the value of R_E set to zero. Equations (4-20) and (4-21) then become

$$I_B = \frac{V_{CC} - V_{BE}}{R_B + \beta R_C} \qquad (4\text{-}20a)$$

$$I_C = \frac{V_{CC} - V_{BE}}{\frac{R_B}{\beta} + R_C} \qquad (4\text{-}21a)$$

The self-bias connection is another example of the use of negative feedback to stabilize the biasing operation—more precisely, to hold I_C constant at the quiescent point. Here, a change in I_C affects a change in V_C (collector voltage) that changes I_B in such a way as to minimize the change in I_C. For example, if I_C should decrease, V_C would increase ($V_C = V_{CC} - I_C R_C$, the term $I_C R_C$ decreases), tending to increase I_B that works to negate the original decrease in I_C.

4-7.4 Voltage-Divider Biasing

In the discussion of the fixed bias with emitter resistance circuit (Sec. 4-7.2), it was noted that in order to obtain β independence (and improve stability), it is necessary that $R_B/\beta \ll R_E$, or $R_B \ll \beta R_E$. While it was nearly impossible to satisfy this condition in the previous biasing circuits, in the voltage-divider circuit this condition is easily met. Figure 4-17 shows the voltage-divider biasing circuit. (We show two supplies V_{BB} and V_{CC}. Usually, $V_{BB} = V_{CC}$ and the battery V_{BB} is replaced by the dashed line, allowing V_{CC} to be used as the base bias supply voltage.)

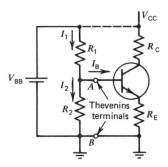

FIGURE 4-17 Voltage divider bias circuit. (If $V_{BB} = V_{CC}$, then V_{BB} may be replaced by the dashed line.)

144   THE BIPOLAR TRANSISTOR

The analysis of this circuit follows, almost exactly, the procedure used in analyzing the circuit of Fig. 4-14. Indeed, the two circuits can be made to look alike by the use of Thevenin's theorem. (In the analysis it is assumed that $V_{BB} = V_{CC}$. If this is not true, replace all references to V_{CC} by V_{BB}. Fig. 4-18a shows the portion of the circuit for which we now obtain a Thevenin's equivalent (points A and B). Figure 4-18b shows the Thevenin's equivalent for terminals A-B. From Thevenin's theorem

$$V_{Th} = \frac{V_{CC} \times R_2}{R_1 \times R_2} = V_{ABOC} \qquad (4\text{-}22)$$

$$R_{Th} = \frac{R_1 \times R_2}{R_1 + R_2} = R_{AB} \qquad (4\text{-}23)$$

Figure 4-18c shows the complete circuit with the Thevenin's equivalent substituted for the original base circuit. Figure 4-18c is identical in form to Fig. 4-14. In Fig. 4-18c V_{Th} replaces the V_{CC} supply used for the base circuit and R_{Th} replaces R_B. (The collector supply voltage V_{CC} is the same in both figures.) With Eq. (4-15a) applied, I_B becomes

$$I_B = \frac{V_{Th} - V_{BE}}{R_{Th} + \beta R_E} \qquad (4\text{-}24)$$

or

$$I_B = \frac{V_{Th} - V_{BE}}{\beta R_E} \quad (\text{if } R_{Th} \ll \beta R_E) \qquad (4\text{-}24a)$$

From Eq. (4-17a)

$$I_C = \frac{V_{Th} - V_{BE}}{\frac{R_{Th}}{\beta} + R_E} \qquad (4\text{-}25)$$

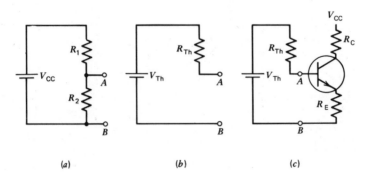

FIGURE 4-18   The use of Thevenin's equivalent. (a) Original portion of circuit. (b) Thevenin's equivalent. (c) Complete circuit with Thevenin's equivalent.

Again, if

$$\frac{R_{Th}}{\beta} \ll R_E \text{ (or } R_{Th} \ll \beta R_E\text{),}$$

then

$$I_C \simeq \frac{V_{Th} - V_{BE}}{R_E} \quad (4\text{-}25a)$$

Equation (4-25a) shows that the operating point I_{CQ} is β independent. A change in β will have no effect on I_C since β does not appear in Eq. (4-25a). $\frac{R_{Th}}{\beta} \ll R_E$ or $R_{Th} \ll \beta R_E$ require that $R_1 \| R_2 \ll \beta R_E$ (that is, either $R_1 \ll \beta R_E$ or $R_2 \ll \beta R_E$).

The steps involved in the analysis are:

1. Find V_{Th}. This voltage is the result of the voltage divider rule:

$$\frac{V_{CC} \times R_2}{R_1 + R_2} = V_{Th}$$

Note that V_{Th} does not consider the effects of I_B (no load or open circuit voltage at the base: $V_{Th} = V_{BOC}$).
2. Find R_{Th}. This is the parallel combination of R_1 and R_2.
3. Use Eqs. (4-24) and (4-25) to find I_B and I_C.

EXAMPLE 4-8

In Fig. 4-17, $R_1 = 12$ kΩ, $R_2 = 3$ kΩ, $R_E = 0.47$ kΩ, $R_C = 1.2$ kΩ, $V_{CC} = 15$ V, $\beta = 100$, silicon transistor. Find I_B, I_C using the exact method. (Do not assume that $R_1 \| R_2 \ll \beta R_E$.)

SOLUTION

Step 1 $\quad V_{Th} = \dfrac{15 \times 3 \text{ k}\Omega}{12 \text{ k}\Omega + 3 \text{ k}\Omega} = 3$ V $\quad\quad\quad\quad\quad\quad\quad\quad\quad$ (4-22)

Step 2 $\quad R_{Th} = 12$ k$\|3$ k $= 2.4$ kΩ

Step 3 $\quad I_B = \dfrac{3 - 0.7}{2.4 \text{ k}\Omega + 100 \times 0.47 \text{ k}\Omega} = 46.56$ μA $\quad\quad$ (4-23)

$\quad\quad\quad\quad I_C = \beta I_B = 4.656$ mA $\quad\quad\quad\quad\quad\quad\quad\quad\quad\quad\quad\quad$ (4-24)

or

$I_C = \dfrac{3 - 0.7}{2.4 \text{ k}/\beta + 0.47 \text{ k}} = 4.656$ mA $\quad\quad\quad\quad\quad\quad\quad$ (4-25)

EXAMPLE 4-9

(a) Solve the problem of Ex. 4-8, this time, use the approximate approach, assuming $R_1\|R_2 \ll \beta R_E$.
(b) Show that the latter condition holds.
(c) Find the error introduced by the approximation.

SOLUTION

(a) V_{Th} and R_{Th} have been found in Ex. 4-8. $V_{Th} = 3$ V, $R_{Th} = 2.4$ kΩ.

$$I_B \simeq \frac{V_{Th} - 0.7}{\beta R_E} = \frac{3 - 0.7}{100 \times 0.47 \text{ k}} = 48.9 \text{ μA} \qquad (4\text{-}24a)$$

$$I_C = \beta I_B = 4.89 \text{ mA}$$

or

$$I_C \simeq \frac{V_{Th} - 0.7}{R_E} = 4.89 \text{ mA} \qquad (4\text{-}25a)$$

(b) $\beta R_E = 100 \times 0.47 = 47$ kΩ

$$R_1\|R_2 = 2.4 \text{ k}\Omega$$
$$2.4 \text{ k}\Omega \ll 47 \text{ k}\Omega$$

(c) The error introduced by the approximation used in Ex. 4-9 is

$$\frac{4.89 - 4.656}{4.656} \times 100 = 5\%$$

This error can be roughly given by the ratio of R_{Th} to βR_E

$$\frac{R_{Th}}{\beta R_E} = \frac{2.4}{47} = 0.051 = 5.1\%$$

The voltage divider biasing circuit can be used with $R_E = 0$ (emitter returned directly to ground) and it can be used for all three basic amplifier configurations.

The design of this biasing circuit requires some seemingly arbitrary choices. We must select R_E (see Sec. 4-7.2) and either R_1 or R_2. The choice of R_C does not affect the bias current I_{CQ}. (It is very important, however, when V_{CEQ} is considered.) It is common to choose $R_2 \ll \beta R_E$ (use minimum β). A good choice, somewhat arbitrary, is:

$$R_2 = \beta R_E/10 \qquad (4\text{-}26)$$

The condition $R_1\|R_2 \ll \beta R_E$ will now hold. Needless to say, I_{CQ} must be known. It is usually obtained from the basic choice of an operating point. The steps involved in the design of the voltage divider biasing circuit are:

BIASING CIRCUITS   147

1. Select the operating point (I_{CQ}, V_{CEQ}).
2. Select $V_E \approx 3 V_{BE}$.
3. Find R_E from Eq. (4-18).
4. Get R_2 from Eq. (4-26).
5. Find V_B from $V_B = V_E + V_{BE}$.
6. Solve for R_1 from the voltage divider relation

$$V_B \approx \frac{V_{CC} \times R_2}{R_1 + R_2}$$

$$R_1 = \frac{V_{CC} \times R_2}{V_B} - R_2$$

7. Find R_C from KVL applied to the collector loop; assume V_{CEQ} is given.

$$V_{CC} - V_{CEQ} - I_{CQ}(R_C + R_E) = 0 \quad (I_C \approx I_E)$$

$$R_C = \frac{V_{CC} - V_{CEQ}}{I_{CQ}} - R_E$$

EXAMPLE 4-10

Find R_E, R_1, R_2, and R_C for a voltage divider biasing circuit with the following parameters:

$\beta = 100$, Silicon transistor
$I_{CQ} = 5$ mA
$V_{CC} = 20$ V, $V_{CEQ} = 10$ V
(use V_{CC} to bias the transistor).
Use standard 10% resistors.

SOLUTION

The circuit is shown in Fig. 4-17 (where $V_{BB} = V_{CC}$). To find R_E use Eq. (4-18)

$$R_E = \frac{3 \times V_{BE}}{I_C} = \frac{3 \times 0.7}{5 \times 10^{-3}} = 420 \ \Omega \tag{4-18}$$

Since 420 Ω is a standard 10% value, $R_E = 420 \ \Omega$

$$R_2 = \frac{\beta \times R_E}{10} = \frac{100 \times 420}{10} = \underline{4.2 \ k\Omega} \tag{4-26}$$

(4.2 kΩ is a standard 10% resistance). V_B (base voltage) can now be found

$$V_B = V_E + 0.7 \ V$$

and

$$V_E = I_C \times R_E = 5 \times 10^{-3} \times 420 = 2.1 \ V$$
$$V_B = 2.1 + 0.7 = 2.8 \ V$$

148 THE BIPOLAR TRANSISTOR

Note that we use the approximate method $V_B \simeq V_{Th}$.

$$V_B = \frac{V_{CC} \times R_2}{R_1 + R_2}$$

Solving for R_1, we get

$$R_1 = \frac{V_{CC} \times R_2}{V_B} - R_2$$

Substituting value yields

$$R_1 = \frac{20 \times 4200}{2.8} - 4200 = 25800 \; \Omega$$

We use $R_1 = 27 \; k\Omega$. Since this is the closest 10% value, R_C can be found as outlined in Step 7 above

$$R_C = \frac{V_{CC} - V_{CEQ}}{I_{CQ}} - R_E = \frac{20 - 10}{5 \; mA} - 420 = 1580 \; \Omega$$

Use $R_C = 1.5 \; k\Omega$.

It is expected that in the last example, the calculated value of I_{CQ}, based on the chosen design values, will not agree with the $I_{CQ} = 5 \; mA$ required. This is because of the compromises made in selecting the resistors to conform to the restriction of using 10% resistors only. (Clearly, 5% tolerance would have permitted better choices at a higher cost.) $S(\beta)$ for the voltage divider circuit is

$$S(\beta) = \frac{R_{Th}}{R_{Th} + \beta R_E} \quad \text{(Appendix C)}$$

In the voltage divider circuit R_{Th} is usually selected $R_{Th} \ll \beta R_E$ [see Eq. (4-26)] $S(\beta)$ then can be estimated by

$$S(\beta) \approx \frac{R_{Th}}{\beta R_E}$$

Since $R_{Th} \ll \beta R_E$ hence, $S(\beta) \ll 1$. In Ex. 4-10 $R_1 = 27 \; k\Omega$, $R_2 = 4.2 \; k\Omega$, and $R_E = 420 \; \Omega$. R_{Th} becomes

$$R_{Th} = R_1 \| R_2 = 27 \; k\Omega \| 4.2 \; k\Omega \simeq 3.6 \; k\Omega$$

$$S(\beta) = \frac{R_{Th}}{R_{Th} + \beta R_E} = \frac{3.6 \; k\Omega}{3.6 \; k\Omega + 100 \times 420}$$

$$\approx \frac{3.6}{100 \times 420} = 0.08$$

This is an excellent stability factor, indicating that I_C is very nearly independent of β.

4-7.5 Bias Circuits with Two or Three Supplies

An examination of the various biasing circuits discussed (as well as many other biasing circuits) will show that Fig. 4-14 is representative of most biasing circuits. It was already noted that the voltage divider circuit can be reduced to Fig. 4-14. Clearly, the fixed bias circuit is identical to Fig. 4-14 with $R_E = 0$. It is not always clear what voltages should be used in the analysis. Let us take an example in which all three transistor terminals are returned (through resistors) to three different supply voltages and the biasing circuit is of the type of Fig. 4-14. Figure 4-19 shows such a circuit. To obtain a general solution, it is useful to assume that V_{BB}, V_{CC}, and V_{EE} are all positive. (V_{EE} is more likely to be negative in NPN transistor circuits. The general solution, however, will include the case in which *any* voltage may be negative.)

Applying KVL to the input loop in Fig. 4-19, we get

$$V_{BB} - I_B R_B - V_{BE} - I_E \times R_E - V_{EE} = 0$$

If we substitute $I_E \approx I_C = \beta I_B$ and regroup,

$$V_{BB} - V_{EE} - V_{BE} - I_B(R_B + \beta R_E) = 0 \qquad (4\text{-}27)$$

$$I_B = \frac{V_{BB} - V_{EE} - V_{BE}}{R_B + \beta R_E}$$

Equation (4-27) is the same as Eq. (4-15a) where V_{CC} in Eq. (4-15a) has been replaced by $(V_{BB} - V_{EE})$, or Eq. (4-24) where V_{Th} is replaced by $(V_{BB} - V_{EE})$.

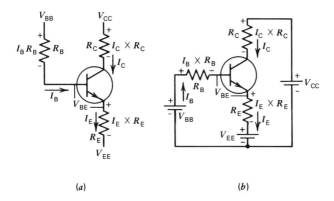

FIGURE 4-19 Representative biasing circuit. (a) Circuit showing polarities of currents and voltage drops. (b) Batteries shown for all voltage sources.

150 THE BIPOLAR TRANSISTOR

To find I_C we use $I_C = \beta I_B$ or simply rewrite Eq. (4-27) as

$$I_C = \frac{V_{BB} - V_{EE} - V_{BE}}{\dfrac{R_B}{\beta} + R_E} \qquad (4\text{-}28)$$

EXAMPLE 4-11

For the circuit shown in Fig. 4-20a find:
  (a) I_B (I_{BQ}).
  (b) I_C (I_{CQ}).

These are the quiescent operating currents. The condition $R_2 \ll \beta R_E$ holds (1.8 kΩ ≪ 56 kΩ). This circuit, due to the fact that $R_C = 0$, is suitable only for the common collector configuration (see Ch. 5).

SOLUTION

Note that

$$V_{EE} = -5 \text{ V}$$

To solve the problem, we find V_{Th} (V_{AB}) and R_{Th}

$$V_{Th} = \frac{5 \times R_2}{R_1 + R_2} = \frac{5 \times 1800}{8200 + 1800} = 0.9 \text{ V}$$

$$R_{Th} = R_1 \| R_2 = \frac{8200 \times 1800}{8200 + 1800} = 1.476 \text{ k}\Omega$$

Fig. 4-20b shows the transformed circuit with the Thevinin's equivalent. Comparing Fig. 4.20a to 4-19a, we note that the only difference is the substitution in Fig. 4-20a of V_{Th} for V_{BB} and R_{Th} for R_B. Equation (4-28) may now be used.

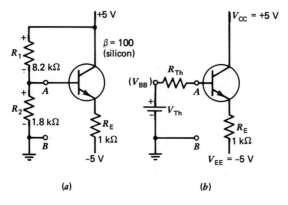

FIGURE 4-20 Circuit for Ex. 4-11. (a) Actual circuit. (b) Circuit with Thevenin's equivalent.

(a) $I_B \simeq \dfrac{V_{Th} - V_{EE} - 0.7}{R_{Th} + \beta R_E}$

Since $R_{Th} \ll \beta R_E$ (due to the fact that $R_2 \ll \beta R_E$),

$$I_B \simeq \dfrac{V_{Th} - V_{EE} - 0.7}{\beta R_E} = \dfrac{0.9 - (-5) - 0.7}{100 \times 1000}$$

$$= \dfrac{5.9 - 0.7}{100 \text{ k}\Omega} = 52 \times 10^{-6} \text{ A}$$

$$I_B = 52\ \mu A$$

(b) $I_C = \beta I_B = 100 \times 52 \times 10^{-6} = \underline{5.2 \text{ mA}} = I_{CQ}$ or,

$$I_C = \dfrac{V_{Th} - V_{EE} - V_{BE}}{R_E} \quad \text{(Eq. (4-25) modified)}$$

$$I_C = \dfrac{0.9 - (-5) - 0.7}{1000} = 5.2 \text{ mA}$$

EXAMPLE 4-12

Find I_C and V_C for the circuit in Fig. 4-21a.

SOLUTION

While this circuit may be converted to the general circuit of Fig. 4-19, a slightly different Thevenin's equivalent is shown in Fig. 4-21b. Note that the Thevenin's equivalent was computed from point A (the base) to the -5 V terminal, point B,

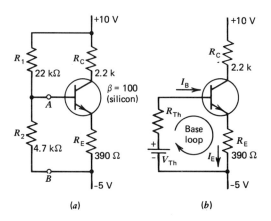

(a)   (b)

FIGURE 4-21 Circuits for Ex. 4-12. (a) Circuit. (b) Circuit with Thevenin's equivalent.

152 THE BIPOLAR TRANSISTOR

and *not* to ground. Here,

$$V_{Th} = \frac{10 - (-5) \times 4.7 \text{ k}\Omega}{22 \text{ k}\Omega + 4.7 \text{ k}\Omega} = 2.64 \text{ V}$$

$10 - (-5) = 15$ V is the voltage across $R_1 + R_2$.

$$R_{Th} = R_1 \| R_2 = 22 \text{ k}\Omega \| 4.7 \text{ k}\Omega = 3.87 \text{ k}\Omega$$

Applying KVL to the base loop in Fig. 4-21b, we get

$$V_{Th} - I_B \times R_{Th} - 0.7 - I_E \times R_E = 0$$

By substituting $I_E \approx I_C = \beta I_B$ and regrouping, we get

$$I_B = \frac{V_{Th} - 0.7}{R_{Th} + \beta R_E} = \frac{2.64 - 0.7}{3.87 \text{ k} + 100 \times 0.39 \text{ k}} = 45 \text{ }\mu\text{A}$$

$$I_C = \beta I_B = 100 \times 45 \times 10^{-6} = \underline{4.5 \text{ mA}}$$

$$V_C = V_{CC} - I_C R_C$$
$$= 10 - 4.5 \times 10^{-3} \times 2.2 \times 10^3 = \underline{0.1 \text{ V}}$$

In Ex. 4-12, the only source in the base loop is V_{Th}. The -5V (V_{EE}) is outside the loop and does not affect the current in the loop. This is a result of the particular way in which the Thevenin's equivalent was developed, from base to the -5V terminal. The two supply circuits, Figs. 4-20 and 4-21, are quite common in many amplifier applications. They permit one, for example, to obtain a collector quiescent voltage of 0V (see Ex. 4-12) that is often very convenient. (There is no need for a capacitor to block the dc voltage when connecting this stage to another circuit.)

4-8 DC LOAD LINE—CLASSES OF OPERATION

The selection of an operating point (Q-point) involves both I_C and V_{CE}. So far, our attention was focused on the choice of I_C (establishing I_{CQ}) and the circuits used in setting up the desired I_C. The value of V_{CE}, once I_C has been established, depends on the value of the collector resistance R_C. Figure 4-22 shows three Q-points. The initial Q-point (the one the circuit was designed for) is at the center of the V_{CE} versus I_C plot. Q' results from *too high* a value for R_C, while Q'' stems from a very low R_C. Note that I_{CQ} is about the *same* for *all three* operating points.

The evaluation of V_{CE} can be done by applying KVL to the collector-emitter loop (Fig. 4-23). In Fig. 4-23a all sources are assumed to have positive polarities (V_{EE} and V_{CC}). This permits us to obtain a general solution in which the numerical values, positive or negative, may be directly inserted into the general solution. (The base circuit is not shown in Fig. 4-23 since it is not relevant to the present computations.)

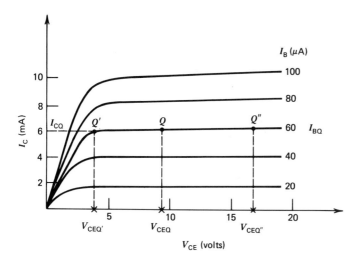

FIGURE 4-22 Shift in Q-point.

Figure 4-23b shows the battery connections consistent with Fig. 4-23a, clearly outlining the collector-emitter loop. The application of KVL to this loop (with the assumption $I_C \simeq I_E$) yields

$$V_{CC} - I_C \times R_C - V_{CE} - I_C R_E - V_{EE} = 0 \qquad (4\text{-}29)$$

Regrouping, we get

$$(V_C - V_{EE}) - I_C(R_C + R_E) = V_{CE} \qquad (4\text{-}29a)$$

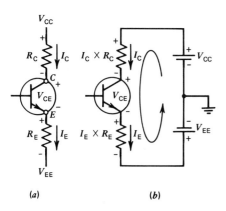

(a) (b)

FIGURE 4-23 V_{CE} calculations. (a) Circuit. (b) Showing batteries, collector-emitter loop.

154 THE BIPOLAR TRANSISTOR

Equation 4-29 is a linear relation between I_C and V_{CE}, where the total supply voltage $V_{CC} - V_{EE}$ and the total resistance in the loop $R_C + R_E$ are the parameters. Clearly, a change in either parameter will alter the relation between I_C and V_{CE}. Eq. (4-29) is the dc load line equation. A plot of Eq. (4-29a), which is a straight line, gives the value of V_{CE} for any I_C and vice versa. The plot is called the dc load line.

EXAMPLE 4-13

Figure 4-24a gives a biasing circuit and Fig. 4-24b gives the characteristic curves (I_C, versus V_{CE}) for the transistor used in Fig. 4-24a.
   (a) Obtain the load line equation.
   (b) Draw the load line.
   (c) Select a Q-point in the center of the active region and find I_{CQ}, V_{CEQ}, and I_{BQ}.

SOLUTION

(a) The load line equation is

$$20 - (-5) - I_C \times (1000 + 250) = V_{CE} \qquad (4\text{-}29a)$$

$$25 - 1250 \times I_C = V_{CE}$$

(b) The plot of the load line requires the establishment of *two* $I_C - V_{CE}$ points. In the table below we chose (conveniently) points for $I_C = 0$ and $V_{CE} = 0$

	I_C	V_{CE}	Notes
Point A	0	25 V	V_{CE} = total supply voltage
Point B	25/1250 = 20 mA	0	$I_C = I_{Cmax} = I_{Csat}$ (collector saturation current)

(This table has been obtained by inserting $I_C = 0$ in the equation above and solving for V_{CE}, and similarly solving for I_C when $V_{CE} = 0$. These two points are selected arbitrarily. Any two points would do.) The line connecting point A to point B is the dc load line (Fig 4-24b).

(c) The operating point *must* lie on this line (Fig. 4-24b). For Q to be centered, $I_{CQ} = 1/2\, I_{Csat}$

$$I_{CQ} = 10 \text{ mA}, \qquad V_{CEQ} = 12.5 \text{ V } (Q \text{ in Fig. 4-24}b)$$

$I_{BQ} = 100 \text{ μA}$. (R_B can be found as shown in Secs. 4-6 and 4-7.)

EXAMPLE 4-14

In the problem of Ex. 4-13 it was found that $I_{CQ} = 10$ mA and $I_{BQ} = 100$ μA ($\beta = 100$). Assume these two values, and plot the load line and find Q for the

DC LOAD LINE—CLASSES OF OPERATION 155

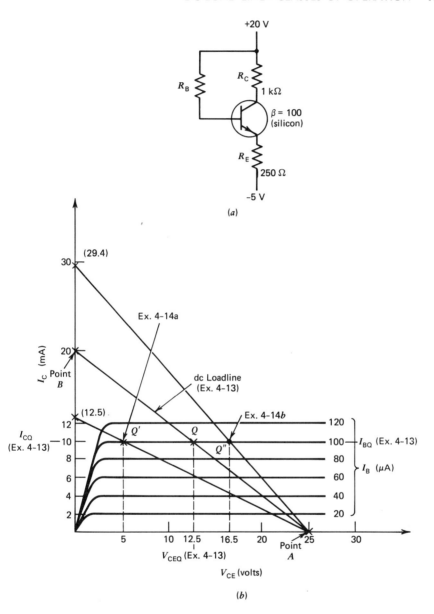

FIGURE 4-24 Circuit for Ex. 4-13 and 4-14. (a) Circuit diagram. (b) I_C versus V_{CE} curves.

circuit shown, with

(a) $R_C = 1.75 \text{ k}\Omega$
(b) $R_C = 600 \text{ }\Omega$

156 THE BIPOLAR TRANSISTOR

SOLUTION

(a) The new load line equation is

$$25 - I_C \times (1750 + 250) = V_{CE}$$

The two points are

I_C	V_{CE}
0	25 V (equals total supply voltage)
$I_{Csat} = \dfrac{25}{2000} = 12.5 \text{ mA}$	0

Since I_B is assumed to be that of Ex. 4-13, $I_B = 100$ μA; the Q-point is at the intersection of the load line with the curve $I_B = 100$ μA, shown as Q' in Fig. 4-24b. $\underline{I_{CQ} = 10 \text{ mA}} \quad \underline{V_{CEQ} = 5 \text{ V.}}$

(b) The load line equation is

$$25 - I_C \times (600 + 250) = V_{CE}$$

The two points are

I_C	V_{CE}
0	25 V
$I_{Csat}\dfrac{25}{850} = 29.4 \text{ mA}$	0

The Q-point shown as Q'' in Fig. 4-24b yields

$$I_{CQ} = \underline{10 \text{ mA}}, \; V_{CEQ} = \underline{16.5 \text{ V.}}$$

Two separate supplies were used in Exs. 4-13 and 4-14. It is often preferable to use a single supply. The solution, however, would follow exactly the procedure used in Ex. 4-13. If there is no emitter supply (emitter "returned" to ground), we simply substitute $V_{EE} = 0$ in Eq. (4-29a). Similarly, $V_{CC} = 0$ when only an emitter supply is used (collector "returned" to ground). It is always important to guarantee proper voltage polarities to the transistor (collector base junction reverse bias, base-emitter forward bias).

4-9 BIASING FOR SWITCHING APPLICATIONS

It was previously assumed that the operating point, the Q-point, is to be selected near the center of the active region. This is usually true for amplifiers in analog applications. The design of switching circuits, which are extensively used in digital applications, requires biasing at one extreme of the active region (actually outside

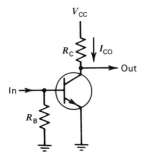

FIGURE 4-25 Biasing at the cutoff.

the active region). Most switching circuits are biased at cut off, that is, $I_{CQ} = 0$ $V_{CE} = $ supply voltage. $I_{CQ} = 0$ implies $I_{BQ} = 0$. To bias at cut off, we need only guarantee zero base current. Figure 4-25 shows a simple circuit where $I_{BQ} = 0$ (no source voltage in the base-emitter loop). This biasing circuit functions properly only if I_{CO}, the collector emitter leakage current, is extremely small. Fortunately, in most modern low-power silicon transistors, I_{CO} is less than one nA (at 25°C). It should be noted that the circuit of Fig. 4-25 is not suitable for power applications in which I_{CO} can be appreciable. (See Ch. 8.) The operation of the circuit, shown in Fig. 4-25, with digital signals is discussed in Sec. 5-9.

4-10 NPN VERSUS PNP

For convenience, all examples in the preceding sections used NPN transistors. The circuits in these examples could have used PNP transistors instead. This would entail providing negative dc supply voltages for the collector and base (instead of the positive supply voltages used for the NPN circuits). The analysis of these PNP circuits follows the same principles used for the NPN analysis.

A simple way of handling the various *negative* supply voltages involved in the PNP circuits is to *temporarily* (for analysis purposes only) *reverse* the polarity of all supply voltages and use the equations used in the analysis of the corresponding NPN circuits. Once a numerical solution has been obtained, its polarity must again be reversed.

EXAMPLE 4-15

For the circuit in Fig. 4-26 find:
(a) I_{BQ} and I_{CQ}.
(b) V_{CEQ}.
(c) V_{CQ} (collector to ground).

SOLUTION

Here, $V_{CC} = -20$ V, $V_{EE} = +10$ V. To use Eqs. (4-24) and (4-25), all polarities are reversed, voltages $V'_{CC} = 20$ V, $V'_{EE} = -10$ V. Note that $V_{BB} = V_{CC}$ and $V_{BE} = 0.35$ V (germanium).

158 THE BIPOLAR TRANSISTOR

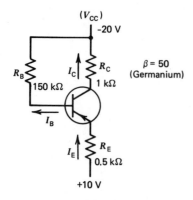

FIGURE 4-26 Circuit for Ex. 4-15.

(a) $I'_B = \dfrac{20 - (-10) - 0.35}{150 \text{ k}\Omega + 50 \times 0.5 \text{ k}\Omega}$ (4-27)

$I'_B = \dfrac{30 - 0.35}{175 \text{ k}\Omega} = 169 \text{ }\mu\text{A}$

$I'_C = \beta \times I_B = 50 \times 169 \times 10^{-6} = 8.45 \text{ mA}$

Reversing polarities again, we get

$$I_B = -169 \text{ }\mu\text{A} \qquad I_C = -8.45 \text{ mA}$$

(b) Apply KVL to the collector-emitter loop (with reversed polarities for the current as well as voltages).

$$20 - I'_C \times R_C - V'_{CE} - I'_C \times R_E - (-10) = 0$$

$V'_{CE} = 30 - I'_C(R_C + R_E)$ (4-26)

$V'_{CE} = 30 - 8.45 \times 10^{-3} \times (1000 + 500) = 17.3 \text{ V}$

To get the actual voltage, reverse the polarity again.

$$V_{CE} = -17.3 \text{ V}$$

(c) $V'_C = 20 - I'_C \times R_C = 20 - 8.45 \times 10^{-3} \times 1000$

$= 11.55 \text{ V}$

$V_C = -11.55 \text{ V}$

4-11 MISCELLANEOUS BIASING CIRCUITS

In many circuit applications, only a single voltage supply, either negative *or* positive, is available to the designer. It is, however, often desirable to use both NPN and PNP transistors. The normal biasing circuits discussed so far require a positive

MISCELLANEOUS BIASING CIRCUITS

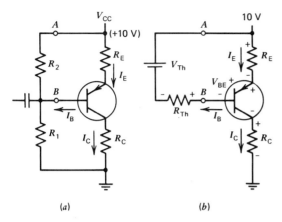

FIGURE 4-27 PNP transistor with positive supply voltages-voltage div. biasing. (a) Circuit. (b) Using Thevenin's equivalent.

supply for NPN transistors and a negative supply for PNP circuits. It is nevertheless possible to use the NPN with negative supplies and the PNP with positive supply voltages. This is accomplished by interchanging the collector emitter connections. Figures 4-27 and 4-28 show two such "inverted" circuits. To obtain I_B and I_C use can again be made of KVL and Thevenin's equivalent where applicable.

Figure 4-27b shows the use of a Thevenin's equivalent. (A solution could also be obtained by the use of a Thevenin's equivalent for point B to GND.) Here,

$$V_{Th} = \frac{V_{CC} \times R_2}{R_1 + R_2}$$

$$R_{Th} = R_1 \| R_2$$

Applying KVL to the base loop, we get

$$-I_E R_E - V_{BE} - I_B R_{Th} + V_{Th} = 0$$

$$I_B = \frac{V_{Th} - V_{BE}}{R_{Th} + \beta R_E} \tag{4-30}$$

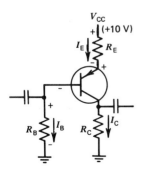

FIGURE 4-28 PNP with positive supply—fixed biasing.

160 THE BIPOLAR TRANSISTOR

Equation (4-30) is identical to Eq. (4-24). But then the corresponding circuits (Fig. 4-18c and Fig. 4-27b) are also identical. Note that V_{CC} here is outside the base loop because of the way the Thevenin's equivalent was developed (points B-A rather than B-GND).

Applying KVL to the base circuit in Fig. 4-28 (the loop contains the V_{CC} supply, R_E, V_{BE}, and R_B), we get

$$V_{CC} - I_E R_E - V_{BE} - I_B R_B = 0$$

$$I_B = \frac{V_{CC} - V_{BE}}{R_B + \beta R_E} \tag{4-31}$$

Compare Eq. (4-31) to Eq. (4-15a) and Fig. 4-14 to Fig. 4-28.

SUMMARY OF IMPORTANT TERMS

Base current—I_B The current in the base terminal of the transistor. It controls the collector current.

Bias The setting up of dc currents and voltages in the transistor as a preliminary to its use in signal processing.

Bias-feedback A class of bias circuits that incorporate feedback methods to stabilize bias currents. Feedback, here it's negative feedback, is a method by which the *controlled* parameter (I_C, for example, as controlled by I_B) is used to affect (partially control) the controlling parameter (here, allowing I_C or I_E to control I_B).

Bias-feedback with R_E The incorporation of an emitter resistor R_E through which feedback between I_E and I_B is established.

Bias—fixed A biasing circuit with no feedback to stabilize the bias currents.

Bias—self A feedback bias circuit in which the base current is obtained from the collector voltage.

Bias—stability A measure of how constant the operating point is (I_{CQ} and V_{CEQ}) in the face of changes in β, $S(\beta)$, or the temperature $S(T)$.

Bipolar Junction Transistor (BJT) A three-terminal device, with the terminals emitter, base, collector, constructed by a molecular (metallurgical) junction of two back to back p-n junctions: Thus, two types of BJTs exist.

N	Collector
P	Base
N	Emitter

and

P	Collector
N	Base
P	Emitter

referred to as NPN and PNP, respectively.

Collector characteristics A plot of I_C (collector current) versus V_{CE} (collector to emitter voltage) for a number of different values of base current (I_B).

Collector current (I_C) The current through the collector terminal of the transistor.

Common base (CB) A circuit used to process signals in which the input is connected (directly or indirectly) to the emitter and output taken from the collector.

Common collector (CC or EF) Input to base output from emitter.

Common emitter (CE) Input to base output from emitter.

Current amplification The phenomenon in which the current in one part of the circuit or device is controlled by (proportional to) a smaller current in another part of the circuit or device. For example, I_C and I_B, respectively. (See β_{ac}, β_{dc}.)

Cut off The state of a transistor in which $I_C = 0$.

"Early" effect A transistor phenomenon which causes I_C to vary (increase) with changes (increases) in V_{CE} even though I_B is held constant.

Emitter current (I_E) The current in the emitter terminal of the transistor.

Load line A graphical method used in representing the various ac signals and dc levels in a common emitter transistor circuit.

Operating point (the Q-point) This describes the base current, collector current, and collector emitter voltage associated with the particular bias circuit (I_{BQ}, I_{CQ}, V_{CEQ}).

Saturation An operating mode for which I_C is the maximum possible current for the particular circuit.

Transistor See Bipolar Junction Transistor.

h_{FE} ($= \beta_{dc}$) Defined as I_C/I_B (dc current ratio).

h_{fe} ($= \beta_{ac}$) Defined as $\Delta I_C/\Delta I_B$ or i_c/i_b (ac current ratio). (Used in the hybrid equivalent circuit, Ch. 5.)

r_e The dynamic resistance of the B-E junction of the forward-biased transistor. Given approximately by $0.026/I_E$ for 300°K. (Used in the dynamic resistance equivalent circuit, Ch. 5.)

α Defined as i_c/i_e or $\Delta I_C/\Delta I_E$.

β_{ac} See h_{fe}.

β_{dc} See h_{FE}.

PROBLEMS

4-1. Show the various internal current carriers for a PNP transistor. Redraw Fig. 4-1, 4-2, and 4-3 for PNP transistor.

4-2. For the transistor voltages shown in Fig. 4-29 state whether the transistor is

(a) saturated (or close to it);
(b) cut off (or close to it);
(c) in the active region.

4-3. What is the main cause for variations in base width? Explain.

4-4. Figure 4-30 shows the results of measurements done on "actively" biased transistors. For each case determine

(a) whether it is a PNP or NPN transistor;
(b) whether it is a germanium or silicon transistor;
(c) β.

4-5. Classify the circuits of Fig. 4-31 in terms of their configuration: CE, CB, or CC.

4-6. Find $S(\beta)$ for the circuit in Fig. 4-31b. (Assume that the dc resistance of the transformer windings is zero.)

Hint: Assume a β and a $\Delta\beta$ and compute I_C for both cases (β and $\beta + \Delta\beta$).

162 THE BIPOLAR TRANSISTOR

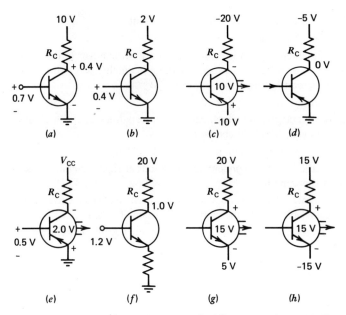

FIGURE 4-29 (a) Silicon transistor. (b) Silicon transistor. (c) Germanium transistor. (d) Silicon or germanium transistor. (e) Germanium transistor. (f) Germanium transistor. (g) Silicon transistor. (h) Silicon transistor.

4-7. Which of the circuits in Fig. 4-31 has the best overall stability? Give your reasons.

4-8. The silicon transistor circuit in Fig. 4-17 has the following parameters: $V_{CC} = 20$ V, $R_C = 1.5$ kΩ, $R_1 = 16$ kΩ, $R_2 = 4$ kΩ, $R_E = 500$ Ω, $\beta = 100$. Find:

(a) I_{BQ}.
(b) I_{CQ}.
(c) V_{CEQ}.

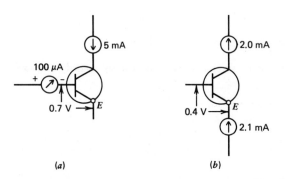

FIGURE 4-30

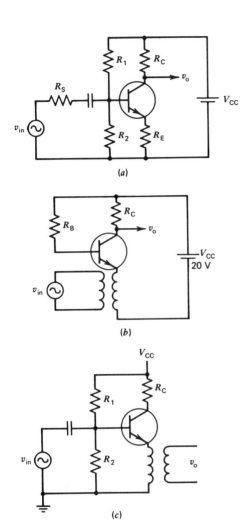

FIGURE 4-31

4-9. The transistor of Fig. 4-31c is a silicon transistor. The values are

$$R_1 = 90 \text{ k}\Omega, \ R_2 = 10 \text{ k}\Omega, \ R_C = 3 \text{ k}\Omega, \ \beta = 50, \ V_{CC} = 10 \text{ V}$$

Find:

(a) I_{BQ}.
(b) I_{CQ}.
(c) V_C.

(Assume resistance of windings = 0.)
Hint: Use Thevenin's equivalent for voltage divider.

164 THE BIPOLAR TRANSISTOR

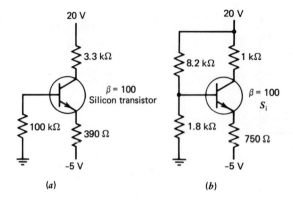

FIGURE 4-32

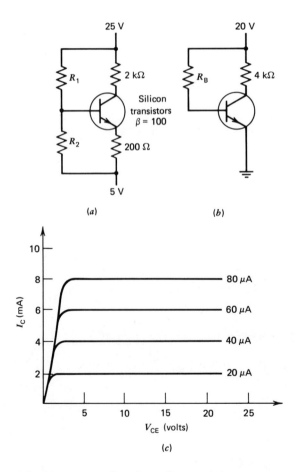

FIGURE 4-33   (c) Transistor characteristics.

4-10. In Fig. 4-31b (silicon transistor) it is desired to operate the transistor exactly at the center of the active region. The parameters are

$$V_{CC} = 30 \text{ V}, R_C = 3 \text{ k}\Omega, \beta = 100.$$

Find: I_{CQ}, I_{BQ}, and R_B.
(Assume resistance of windings = 0.)
Hint: $V_{CEQ} = 1/2 V_{CC}$. $I_{CQ} = 1/2 I_{Csat}$.

4-11. Find the operating point (I_{CQ}, V_{CEQ}) for the circuits shown in Fig. 4-32a.

4-12. Find the operating point for the circuit shown in Fig. 4-32b.

4-13. Draw the dc load line for the circuits shown in Fig. 4-33. In each case select an operating point as close to the center of the active region as possible and find the appropriate biasing resistor(s). (Use $R_2 = \beta R_E/10$.)

4-14. Find I_{BQ}, I_{CQ}, and V_{CEQ} for the circuit given in Fig. 4-34.

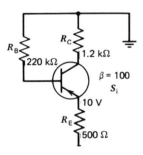

FIGURE 4-34

4-15. Find I_{BQ}, I_{CQ}, and V_{CQ} for the circuit shown in Fig. 4-35, given $R_C = 1.5 \text{ k}\Omega$, $R_B = 100 \text{ k}\Omega$, $\beta = 50$, silicon transistor, $V_{CC} = 15 \text{ V}$.

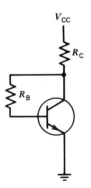

FIGURE 4-35

166   THE BIPOLAR TRANSISTOR

4-16. (a) Find the minimum value of R_C required to produce saturation in the circuit of Fig. 4-34 (keep $R_B = 220$ kΩ).
  (b) find what the largest value of R_B ($R_C = 1.2$ kΩ) that will produce saturation in Fig 4-34.

4-17. Repeat Problem 4-16 for the case in which $R_E = 0$.

4-18. (a) Find the lowest β that will cause saturation of the circuit in Fig. 4-34.
  (b) Repeat (a) above with $R_E = 0$.
  (c) Which of these two circuits has a better stability, $S(\beta)$? Explain in detail.

4-19. Design a voltage divider biasing circuit with R_E, to be used in a CE configuration. Given a 30 V battery and a PNP germanium transistor with $\beta = 75$, find all resistor values required to obtain

$$V_{CEQ} = 1/2 V_{CC}, \qquad I_{CQ} = 1/2 I_{Csat}.$$

Use $R_C = 2$ kΩ.
Sketch the circuit.

4-20. In Fig. 4-27a $R_1 = 12$ kΩ, $R_2 = 2.2$ kΩ, $R_E = 680$ Ω, $R_C = 1.8$ kΩ, $V_{CC} = 12$ V, $\beta = 100$, germanium transistor.
Find:

  (a) I_{BQ}.
  (b) I_{CQ}.
  (c) V_{CQ} (collector to GND voltage).
  (d) V_{CEQ}.

4-21. In Fig. 4-28, $R_B = 220$ kΩ, $R_E = 1$ kΩ, $R_C = 2.2$ kΩ, $\beta = 50$, $V_{CC} = 20$ V, germanium transistor.
Find:

  (a) I_{BQ}.
  (b) I_{CQ}.
  (c) V_{CEQ}.
  (d) V_{CQ}.

CHAPTER

5
THE TRANSISTOR AS AN AMPLIFIER (SMALL SIGNAL)

5-1 CHAPTER OBJECTIVES

This chapter is concerned with transistor circuits that are used to amplify or process signals. There are many different such circuits, and the student will learn to analyze some of the common classes of amplifier circuits. Important characteristics, such as voltage gain (the measure of voltage amplification) and current gain, will be described for the various classes of circuits. Specific equivalent circuits will be developed, giving the student an understanding of circuit performance and the ability to evaluate the most important circuit characteristics. In addition to voltage and current gain, the meanings of the output resistance (Thevenin's resistance of the equivalent output circuit) and input resistance (the resistance the circuit presents to the signal source) are clearly analyzed.

It is expected that the student, after completing this chapter, will have all the tools and analytical knowledge to analyze any single transistor amplifier circuit.

5-2 INTRODUCTION

Chapter 4 discussed various methods of "setting up" the transistor so that it could amplify signals. The word amplify, and hence the use of the word amplifier, in this context does not necessarily imply that the signal output is larger (amplified) than the input. A less common term but more accurate is the term process. The transistor, after it is properly biased, is ready to process signals. It is worth noting that "signals" refers to any voltage or current, ac or dc, that the circuit is designed

168 THE TRANSISTOR AS AN AMPLIFIER (SMALL SIGNAL)

to process. It is clear, for example, that the supply voltages used in the biasing circuits are *not* signals. These are *not* processed by the circuits; they do *not* serve as input and do *not* produce an output.

In contrast to Chapter 4, this chapter is concerned with the performance of a transistorized circuit (circuit and transistor) with signal voltages and currents, primarily ac signals. (See Sec. 7-4 and Chs. 10 and 11 for dc amplifiers.) A number of assumptions are made in the discussion that follows. First, it is assumed that the amplitude of the signal driving the circuit is small enough so that transistor behavior may be considered linear. This assumption is actually a consequence of the assumption that β_{ac} is a fixed quantity for a given transistor. As long as the variations in the collector current are small (small signal), β_{ac} is indeed constant. Throughout the presentation we develop equivalent circuits (replacing the transistor with an equivalent circuit) that, in many cases, are approximate only. The equivalence is always based on the *ac* behavior *only*. As a result, the calculated voltages and currents do not include any dc (biasing) values. (Separate dc calculations yield the dc values that have been discussed in Ch. 4 in connection with biasing.)

To simplify matters somewhat, it is assumed that capacitors are ideal, that is, large capacitors that are intended to serve as shorts (by pass) or connecting capacitors (coupling capacitors) are indeed shorts. These are shorts only as far as *ac* voltages and currents are concerned. They are open circuits for dc voltages and currents. (This assumption is reexamined in the discussion of the frequency response of the circuits in Ch. 8.)

A note regarding the use of notation is in order here. Lowercase letters will denote ac quantities. Resistors, however, since they are both ac and dc components will be denoted by capital R. Thus, the *ac* collector current will be referred to as i_c while I_C will refer to the dc collector current. Similarly, v_c is the ac collector voltage and i_B the ac base current, etc.

5-3 IMPORTANT AMPLIFIER PARAMETERS

It is often convenient and very meaningful to describe the amplifier in terms of its function, rather than in terms of the circuitry used. There are four basic parameters that are commonly used to describe the behavior of the amplifier. These are defined with the aid of the amplifier block diagram of Fig. 5-1.

A_v, *the voltage gain*, is defined as the ratio of voltage output (v_o) to voltage input (v_{in})

$$A_v = \frac{v_o}{v_{in}} \qquad (5\text{-}1)$$

v_{in} is the voltage applied to the amplifier input and v_o is the output voltage, the ac voltage across the load.

A_i, *the current gain*, is the ratio of output current i_o to input current i_{in}, where the output current is the current in the load.

$$A_i = \frac{i_o}{i_{in}} \qquad (5\text{-}2)$$

IMPORTANT AMPLIFIER PARAMETERS

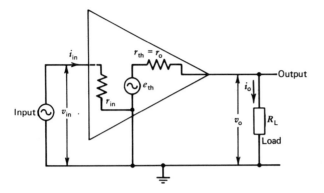

FIGURE 5-1 Amplifier block diagram.

The word "load" refers to the particular component, or circuit, that the amplifier is driving. It is the user who, in the final analysis, decides what the load is—whether it is a relay connected to the collector, the collector resistor itself, or a speaker connected to the emitter through the 50-foot cable. To evaluate A_v and A_i we must first define the load.

The two parameters A_v and A_i describe the voltage and current response of the amplifier. We compare how much larger, or smaller, the output is than the input.

The fact that i_{in} is produced by v_{in} leads to the conclusion that the amplifier input section behaves like a resistance, as far as voltages and currents are concerned. That resistance, termed the amplifier *input resistance* (or impedence), is defined by Ohms law.

$$r_{in}^{(1)} = \frac{v_{in}}{i_{in}} \qquad (5\text{-}3)$$

It is usually *not* a physical resistance. It is that resistance that requires v_{in} to provide the current i_{in}. r_{in} may be considered the load resistance of the driving source v_{in} or of a previous amplifier stage. Low values of r_{in} would require high currents to be provided by the driving source, which as demonstrated later, usually presents serious problems.

The output terminals of the amplifier can be viewed as the terminals of a voltage source, which, as expected, has some internal resistance (or impedance). This voltage source is then connected to drive the load. The internal resistance of the source is a very important parameter since it limits the current that can be supplied to the load. This internal resistance is the Thevenin's (or Norton's) resistance (or impedance) of the source. It is assumed that it is possible to obtain the Thevenin's equivalent for the output terminals. When applied to amplifier circuits, this internal resistance or Thevenin's resistance is called the *output resistance* of the amplifier.

---

[1] To preserve generality, the term z_{in} should be used. Since the circuits are assumed to be purely resistive, the r_{in} is used instead.

170 THE TRANSISTOR AS AN AMPLIFIER (SMALL SIGNAL)

More precisely, it is the Thevenin's resistance of the output terminals of the amplifier.

The output resistance r_o (or z_o) is defined in Thevenin's terms (see Ch. 1) as

$$r_o = \frac{v_{oc}}{i_{sc}} = \frac{\text{open circuit voltage}}{\text{short circuit current}} \quad (5\text{-}4)$$

$$= \frac{v_{TH}}{i_N} = r_{TH}$$

where v_{oc} is the open circuit voltage at the output terminals (load removed) and i_{sc} is the short circuit current between these terminals (effectively, i_o for $R_L = 0$).

r_o cannot be measured directly with an ohmmeter no more than the resistance of a dc circuit can be measured while connected to the battery. However, r_o does not exist other than when all biasing voltages are applied. r_o can be measured by measuring v_{oc} directly (remove R_L and measure v_o) and instead of measuring i_{sc} directly (short circuiting the output may damage the circuit), a measurement of v_o is made with a known R_L in the circuit. The output resistance r_o is then given by

$$r_o = \frac{v_{oc} - v_o}{v_o/R_L} = \frac{v_{oc} - v_o}{v_o} \times R_L \quad (5\text{-}4)$$

The last expression can be derived by realizing that r_o (Fig. 5-1) is given by

$$r_o = \frac{v_{oc} - v_o}{i_o} = \frac{v_{TH} - v_o}{i_o} \quad (5\text{-}4a)$$

($v_{oc} = v_{Th}$) is the Thevenin's voltage. Consequently, $v_{oc} - v_o$, the voltage across r_o, divided by i_o, the current through r_o, equals r_o. By definition

$$i_o = \frac{v_o}{R_L} \quad (5\text{-}4b)$$

The substitution of Eq. (5-4b) in Eq. (5-4a) yields Eq. (5-4).

The significance of r_{in} and r_o can best be illustrated by a cascade connection of two amplifiers (Fig. 5-2). The output voltage of amplifier 1, v_o, is applied to the input of amplifier 2. v_o can be given in terms of v_{Th}, r_o, and r_{in}.

$$v_o = \frac{v_{Th} \times r_{in}}{r_o + r_{in}} \quad \text{(voltage division)} \quad (5\text{-}6)$$

(Here, r_{in} is the load of amplifier 1.)

From Eq. (5-6), it is clear that no matter what v_{Th} is, v_o will be small compared to v_{Th} (only a small portion of the *potentially* available output voltage v_{Th} will *actually* be available) if either r_o is large, or r_{in} is small ($r_o \gg r_{in}$). It is usually desirable to keep r_o small and r_{in} large.

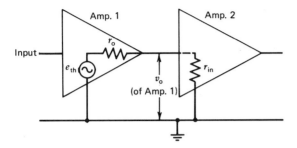

FIGURE 5-2 Cascaded amplifiers.

An amplifier parameter that is of interest for power amplifiers is the power gain A_p. It is defined as

$$A_p = P_o/P_{in} \qquad (5\text{-}7)$$

where P_o is the power delivered to the load (R_L) and P_{in} is the signal power input to the amplifier.

Equation (5-7) can be written in terms of voltage and current gains, A_v and A_i. The power output P_o can be expressed as

$$P_o = v_o \times i_o \qquad (5\text{-}8)$$

(assuming rms values) and P_{in} as

$$P_{in} = v_{in} \times i_{in} \qquad (5\text{-}9)$$

From Eqs. (5-8) and (5-9) by substitution into Eq. (5-7), we get

$$A_p = \frac{v_o \times i_o}{v_{in} \times i_{in}} = A_v \times A_i \qquad (5\text{-}10)$$

The specific values of these five parameters (A_v, A_i, r_o, r_{in}, A_p) depend on the specific circuit used. No wonder then, that ac circuit analysis (for active circuits) largely concentrates on evaluating these parameters. Since the circuits were assumed to be linear, it is clear that the evaluation of v_o and i_o can be done by first obtaining A_v and A_i. Equations (5-1) and (5-2) can be rewritten as

$$v_o = A_v \times v_{in} \qquad (5\text{-}1a)$$
$$i_o = A_i \times i_{in} \qquad (5\text{-}2a)$$

5-4 TRANSISTOR EQUIVALENT CIRCUITS

Chapter 1 of this text discusses various equivalent circuits and their significance. It is now possible to apply these equivalent circuit techniques to the transistor. Our interest at the moment is the transistor itself, rather than the circuit configured

172 THE TRANSISTOR AS AN AMPLIFIER (SMALL SIGNAL)

around the transistor. In order to be able to apply the various circuit laws, it is essential that the device be replaced by a *linear* circuit. (Recall that the transistor was assumed to be linear and hence we expect to be able to find a *linear* equivalent circuit.)

5-4.1 The Hybrid Parameters

One of the most popular equivalent circuits for the transistor, up until a few years ago, was the hybrid equivalent circuit, utilizing what are commonly called the hybrid parameters. In recent years, other types of equivalent circuits have been developed, in particular the dynamic resistance technique. While we expect to rely on the latter approach for most, if not all, of our analysis, we nevertheless present a brief description of the hybrid approach. This is necessary because only after we have made appropriate approximations in this equivalent circuit, can the dynamic resistance technique be developed.

The hybrid approach views the transistor (in a particular configuration) as a four-terminal circuit: two input and two output terminals. If, for example, the common emitter (CE) configuration (see Sec. 4-5.1) is used as a reference, the two input terminals are the base and the emitter; the two output terminals are the collector and the emitter (Fig. 5-3). As expected, the emitter terminal is common to both input and output. Bear in mind that this discussion relates to the ac behavior of the transistor and that the dc setup has been accomplished, even though the dc biasing components are not shown at this point.

In the hybrid approach, two sets of equivalent circuits are developed, one for the input terminals *B-E* and one for the output terminals *C-E*.

The input circuit and the output circuit are replaced, respectively, by their Thevenin's and Norton's equivalents (Fig. 5-3b). The parameters are all expressed in terms of *h* symbols to denote that the *h*ybrid approach is being used. A second subscript *e* indicates that the equivalent was developed for a common emitter configuration (a common *base* equivalent, for example, would have a *b* as a second subscript, for example, h_{ib}).

An explanation of the various *h* parameters follows:

h_{ie} This is the Thevenin's resistance of the input circuit, hence the subscript *i*. h_{ie} is usually less than a few kΩ.

h_{re} The input Thevenin's voltage is found to depend on the collector voltage v_{ce}. Note that there are no *real* sources in the input circuit. The term $h_{re} \times v_{ce}$ is a "reflected," or "fed back" voltage, caused by the imperfection of the transistor (the subscript *r* stands for *r*eflected or *r*everse coupling). The magnitude of h_{re} is very small, about 10^{-7}; as a result, the Thevenin's source, $h_{re} \times v_{ce}$, is very small. If, for example, $v_c = 10$ V and $h_{re} = 10^{-7}$, then $h_{re} \times v_{ce} = 10 \times 10^{-7} = 1$ μV. It is reasonable to ignore this voltage source and represent the input circuit by a degenerated Thevenin's equivalent, where the $h_{re} \times v_c$ source is replaced by a short (Fig. 5-3c). Note that h_{re} is *unitless*.

h_{fe} The current source in the output is given as $h_{fe} \times i_B$. We recognize this as the collector current i_c and h_{fe} as β_{ac}. h_{fe} is the *f*orward current transfer ratio,

TRANSISTOR EQUIVALENT CIRCUITS   173

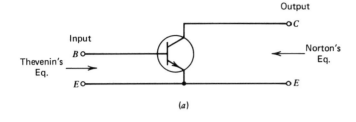

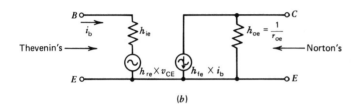

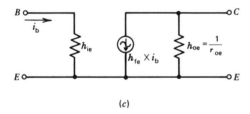

FIGURE 5-3 The hybrid equivalent. (a) CE configuration (showing the two pairs of terminals). (b) The Thevenin's and Norton's equivalent. (c) Degenerate input equivalent ($h_{re} = 0$).

relating input current i_B to output current i_c. In most small signal transistors h_{fe} is 100 or higher ($h_{fe} = \beta_{ac} = \frac{i_c}{i_B}$).

h_{oe} This is the typical Norton's *conductance* in the output circuit. The term r_{oe} is often used instead of h_{oe}. r_{oe} is the Norton's *resistance* and thus $r_{oe} = 1/h_{oe}$. (The use of r_{oe} is for convenience only since most of us are accustomed to dealing with resistance in ohms rather than conductance in mhos.) r_{oe} is usually in the order of 100 kΩ or more and may often be neglected in circuit analysis.

One of the major drawbacks of the h-parameters approach is the fact that all parameters must be given by the manufacturer. A measurement of most of them would involve the use of special instruments or measuring circuits. In addition, the h-parameters are assumed to be constant over the whole active region, an assumption that is very approximate at best. We are also reminded that, at least theoretically, every configuration (CE, CB, CC) has its own set of h-parameters, which complicates the analysis. (Conversion between the different parameters is

174 THE TRANSISTOR AS AN AMPLIFIER (SMALL SIGNAL)

possible.) Since, as noted earlier, this approach for the most part will not be used in the text, we demonstrate its use by a single, simple example.

EXAMPLE 5-1

In the circuit of Fig. 5-4, $h_{ie} = 1\text{ k}\Omega$, h_{re} – negligible, $h_{fe} = 100$, $r_{oe} = 100\text{ k}\Omega$. Find the voltage gain

$$A_v = \frac{v_o}{v_{in}}$$

SOLUTION

The hybrid equivalent circuit is shown in Fig. 5-4b. The capacitor is considered a short for ac and V_{CC} is effectively 0 V, or ground, for ac. (These points will be clarified later.) v_o is the voltage across $R_C \| r_{oe}$. The current through this parallel combination is $i_c = h_{fe} \times i_B$; hence,

$$v_o = -i_c \times (r_{oe} \| R_C) = -h_{fe} \times i_B \times (r_{oe} \| R_C).$$

The negative sign is due to the direction of the current i_c (from ground up). To find v_o, i_B must first be found.

$$i_B = v_{in}/h_{ie} = v_{in}/1\text{ k}\Omega.$$

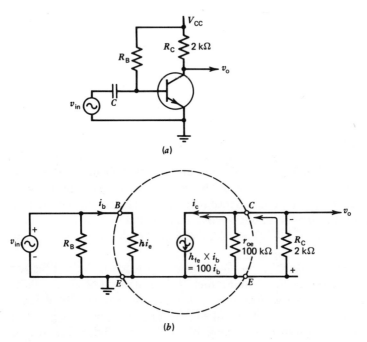

FIGURE 5-4 Circuits for Ex. 5-1. (a) Actual circuit. (b) Hybrid equivalent circuit (neglecting $h_{re} \times V_c$).

(v_{in} is the voltage input across the base-emitter terminals.) Substituting for i_B, we get

$$v_o = -h_{fe} \times i_B \times (r_{oe} \| R_C)$$

$$= -h_{fe} \times \frac{v_{in}}{h_{ie}} \times (r_{oe} \| R_C)$$

$$v_o = -100 \frac{v_{in}}{1000} \times (100 \text{ k}\Omega \| 2 \text{ k}\Omega)$$

$$\approx -100 \times \frac{v_{in}}{1000} \times 2000 \quad (100 \text{ k}\Omega \| 2 \text{ k}\Omega \approx 2000 \text{ }\Omega)$$

$$\frac{v_o}{v_{in}} \approx -\frac{100 \times 2000}{1000} = -200$$

Example 5-1 was solved by the use of the equivalent circuit, Ohm's law *and nothing more*. Simple *circuit* laws were applied after the *device* was replaced by an equivalent circuit.

5-4.2 The Dynamic Resistance r_e and the Equivalent Circuit

The main objectives of the dynamic approach are, first, to obtain an equivalent circuit for the transistor that is independent of the configuration. The same equivalent will be used for all configurations, CE, CB, and CC. Second, to make transistor performance somewhat dependent upon the dc operating point, in particular I_C (dc), more closely representing the real transistor.

In the "active" transistor the base-emitter junction is forward biased. It can then be represented as a source V_{BE} (0.7 V for silicon) in series with the forward resistance, r_f (Fig. 5-5b). The circled transistor symbol in Fig. 5-5b represents an ideal transistor. The B-E junction of the *ideal* transistor is forward biased and thus, a *perfect short*. This *ideal* transistor is included in the circuit to remind us that we are dealing with a transistor and not just a battery and resistor. Since we are interested in an ac

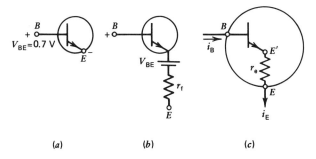

FIGURE 5-5 Base-emitter equivalent. (a) Forward-biased junction (Silicon). (b) Eq. circuit. (c) B-E ac equivalent.

176 THE TRANSISTOR AS AN AMPLIFIER (SMALL SIGNAL)

equivalent, only ac voltages and currents are of interest; the resistance r_f must be the dynamic (ac) resistance of the junction from Chapter 2 (at 300°K)

$$r_f = \frac{0.026 \text{ V}}{I_D \text{ A}} = \frac{26 \text{ mV}}{I_D \text{ mA}} \qquad (2\text{-}5a)$$

Since the dynamic resistance is now applied to a transistor, r_f is renamed r_e. The subscript e denotes the fact that this dynamic (equivalent) resistance is placed in the emitter circuit. In the transistor, the current I_E, the emitter dc current, is used instead of I_D and

$$r_e \, (\Omega) = \frac{0.026 \text{ V}}{I_E \text{ A}} = \frac{26 \text{ mV}}{I_E \text{ mA}} \qquad (5\text{-}11)$$

The dc battery in Fig. 5-5b is replaced by a short in conformance with the superposition theorem, since we are solving the circuit for ac only. The final B-E equivalent circuit is shown in Fig. 5-5c. The B-E junction has been replaced by a resistance r_e, on the emitter side, in series with an ideal B-E junction where

$$v_B = v_E' \qquad (5\text{-}12)$$

The ac voltage at the base and at the *ideal* emitter are the same. This does *not* mean that v_B, the base voltage, equals v_E the emitter terminal voltage (of the real transistor). It is extremely important to note *that the current i_B (the current entering the base) is not equal to i_e, the emitter current*, even though there is no ac voltage drop $v_{BE'}$.

To complete the equivalent circuit, our attention turns to the collector circuit. As noticed in the hybrid approach, the current in the collector is i_c where $i_c = \beta_{ac} i_B$ and there is a resistance r_{oe} from collector to emitter. We will "borrow" these two parameters. To account for the current i_c, a current source i_c is introduced into the collector leg. (See Fig. 5-6.) The resistor r_{oe} that has been renamed r_{ce} to

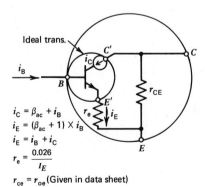

FIGURE 5-6 Complete transistor equivalent circuit.

indicate that it represents a collector-emitter equivalent resistance is shown connected between the collector and emitter (of the *real* transistor). The inner circle in Fig. 5-6 is the ideal transistor where

$$v_B = v_{E'} \qquad (5\text{-}12)$$

The current i_c is flowing to the emitter only (collector to emitter) and not to the base, so that $i_E = i_B + i_C$. The outer circle in Fig. 5-6 contains the complete equivalent of the transistor. *The equations shown in Fig. 5-6 are an essential part of the equivalent circuit.*

In the foregoing discussions β_{ac} represented the ac current gain. For the sake of simplicity it is assumed, from here on, that the symbol β refers to both ac and dc current gain, since the two are usually quite close in value. β_{ac}, which is often denoted h_{fe}, and r_{oe} are usually provided by the manufacturer. These will be referred to as β and r_{ce}, respectively.

The dynamic equivalent circuit is not dependent on the particular configuration involved. That is not to say that the ac performance of these configurations is the same. It is an equivalent circuit of the *three-terminal* device, namely the transistor. It also makes the transistor performance dependent on I_E (dc) since r_e depends on I_E. (This dependence upon I_E has indeed been borne out by measurements.)

The relation between the h-parameters and the parameters in the dynamic equivalent approach can be readily demonstrated. For example, the ac resistance that would be measured between base and emitter in the h-parameter equivalent is clearly h_{ie} (excluding circuit components).

$$r_{BE} = h_{ie} \qquad \text{(See Fig. 5-4b)} \qquad (5\text{-}13)$$

An expression for r_{BE} using the dynamic equivalent circuit can be obtained by the use of KVL. The source v_{in} and i_{in} in Fig. 5-7 are the equivalent of an ac ohmeter. Clearly, in Fig. 5-7,

$$r_{BE} = \frac{v_{in}}{i_{in}} \qquad (5\text{-}14)$$

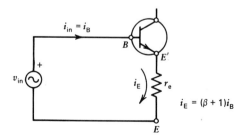

FIGURE 5-7 B-E resistance.

By KVL

$$v_{in} - i_E r_e = 0$$
$$v_{in} = i_E \times r_e = (\beta+1) \times i_B \times r_e$$
(recall that $i_E = (\beta+1)i_B$)

Since $i_B = i_{in}$

$$v_{in} = (\beta+1)i_{in} \times r_e$$

and

$$\frac{v_{in}}{i_{in}} = (\beta+1)r_e = r_{BE}$$

From Eq. (5-13) we get

$$r_{BE} = h_{ie} = (\beta+1)r_e \qquad (5\text{-}14a)$$

In Eq. (5-14a), h_{ie} is constant while r_e varies with I_E (dc). Since h_{ie} is given by the manufacturer as a fixed value, Eq. (5-14a) really holds only for a particular I_E.

5-5 TRANSISTOR CIRCUITS—ac ANALYSIS

As might be expected, ac analysis concentrates on ac aspects of circuit performance. All dc sources are "eliminated" to conform to the superposition theorem, permitting us to deal separately with ac (or dc) behavior. (The word "eliminate" is taken to mean: open all ideal current sources, replace by a short all ideal voltage sources.) As a result, only ac sources remain in the circuit. All dc supply batteries are replaced by shorts. (This is based on the assumption that the dc supply voltages are all ideal.) Figure 5-8 shows the changes that take place when all dc sources are eliminated. (The student is again reminded that the dc biasing supplies and components are essential to the operation of the circuit. The analysis, however, is done separately for ac and dc.) The batteries shown in Fig. 5-8a represent the various dc sources, commonly shown as V_{CC}, V_{EE}, etc. Replacing these batteries by shorts results in the ac circuit shown in Fig. 5-8b. (It is also assumed that C_c is large enough to be considered a short circuit.) The next step in the process of obtaining the ac equivalent circuit involves replacing the transistor by *its* ac equivalent circuit (Fig. 5-8c). Recall that r_e must be obtained from the dc analysis [Eq. (5-11)]. The circuit in Fig. 5-8c can be analyzed using standard circuit theory.

The process of obtaining the full ac equivalent of transistor circuits consists of two steps. First, redraw the *circuit* components (resistors, capacitors, inductors, voltage and current sources) in terms of their àc significance; second, replace the three-terminal device, the transistor, with its functional equivalent circuit. This procedure is applicable to all circuit configurations. It is understandable, however,

TRANSISTOR CIRCUITS—ac ANALYSIS

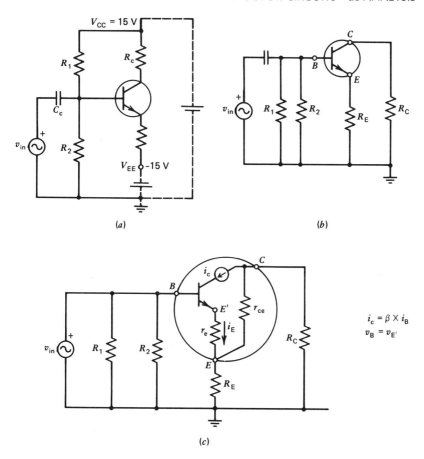

FIGURE 5-8 The ac equivalent circuit. (a) Circuit. (b) Eliminating dc sources. (c) The complete equivalent circuit.

that the operation of the circuits, as reflected by the results of the analysis, depends upon the exact circuit, namely the configuration used, and the component values.

The general description of the CE, CB, and CC configurations was discussed in Sec. 4-5.1. Many different circuits, differing in component values and circuit details, fall into these categories. Consequently, it is foolhardy to attempt to obtain a single formula applicable to all circuits of a single type. *Warning to the wise: formulae should not be used unless they are clearly understood, and when no formulae are given, develop the solution by the application of basic circuit theory.*

The distinction between the three categories is made here because it is possible to establish an *approximate* general behavior for *each type*. That is, the parameters A_v, A_i, r_{in}, r_o can be given a range of values for each of the three configurations. The steps involved in the analysis itself are much the same for *all* configurations. (This classification of transistor circuits permeates the literature, making it a convenient approach.) The ac analysis of all transistor circuits involves four basic steps.

1. Obtain I_E (dc emitter current) by applying dc methods (Sec. 4-7.4).
2. Find $r_e = \dfrac{0.026}{I_E} = \dfrac{26 \text{ mV}}{I_E \text{ mA}}$
3. Draw the equivalent ac circuit with all appropriate simplification, for example, short all capacitors, show all dc supplies as GND, etc.
4. Apply ac circuit theory to solve the circuit, that is, find the various parameters, A_v, A_i, r_{in}, r_o.

5-5.1 Common Emitter (CE)

A CE circuit is shown in Fig. 5-9a. Figure 5-9b shows the complete equivalent circuit. The resistance r_{ce} is usually a large resistance (compared to R_E and r_e). This permits us to somewhat simplify the equivalent circuit, as shown in Fig. 5-9c. Note that r_{ce} is now connected between ground and the collector, placing it in parallel with R_C.

VOLTAGE GAIN, A_v

The ac current i_c produces a voltage drop across $r_{ce} \| R_C$, which is v_o (since i_c is flowing through this parallel combination) with the polarities shown (Fig. 5-9c). Note that i_c is flowing from ground (0 V) "UP" toward v_o. This produces a phase reversal (negative polarity) for v_o. In other words, v_o and i_c have a 180° phase difference. v_o is given as

$$v_o = -i_c \times (R_c \| r_{ce}) \qquad (5\text{-}15)$$

To find v_o, i_c must be found, and since $i_c = \beta i_B$, i_B must be found first. This can be done by applying KVL to the base loop (Fig. 5-10).

$$v_{in} - i_e \times r_e - i_e \times R_E = 0 \qquad (5\text{-}16)$$
$$v_{in} - i_e(r_e + R_E) = 0 \qquad (5\text{-}16a)$$

Equation (5-16) contains all the voltage drops in the loop. (Recall that the base emitter ac voltage in an ideal transistor is zero, $v_{BE'} = 0$.) It is important to realize that in the given circuit, $v_B = v_{in}$, which may not always be the case. Substituting $i_e = (\beta + 1) \times i_B$ in Eq. (5-16a) and solving for i_B, we get

$$v_{in} - (\beta + 1) \times i_B \times (r_e + R_E) = 0$$
$$v_{in} = i_B \times (\beta + 1) \times (r_e + R_E)$$

$$i_B = \dfrac{v_{in}}{(\beta + 1)(r_e + R_E)} \qquad (5\text{-}17)$$

$$i_B \approx \dfrac{v_{in}}{\beta(r_e + R_E)} \qquad (5\text{-}17a)$$

In Eq. (5-17a) we assumed that $\beta \gg 1$ (neglect the 1 compared to β). This is true for a vast majority of transistors and will be assumed throughout this text. It is clear

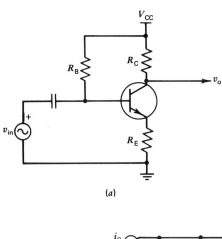

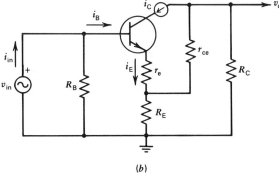

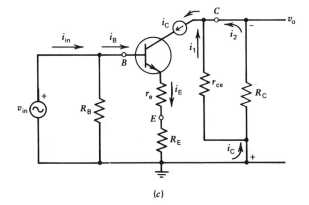

FIGURE 5-9 CE amplifier stage. (a) The circuit. (b) Equivalent circuit. (c) Simplified equivalent circuit.

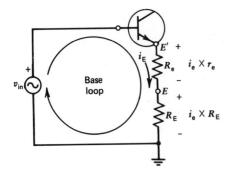

FIGURE 5-10 KVL in base loop.

that this assumption is equivalent to assuming that $i_e \approx i_c$ or $i_B \ll i_c$. In much of the analysis that follows we will, as a matter of course, *substitute* i_c *for* i_e. Multiplying Eq. (5-17a) by β yields i_c.

$$i_c = \beta i_B = \beta \times \frac{v_{in}}{\beta \times (r_e + R_E)} = \frac{v_{in}}{r_e + R_E} \qquad (5\text{-}18)$$

This last result could have been obtained by simply realizing that $v_{E'} = v_{in}$, since there is no ac voltage drop between the base and the ideal emitter. (E' is the emitter of the ideal transistor.) Since $v_{E'} = v_{in}$, it follows that

$$i_e = \frac{v_{E'}}{r_e + R_E} = \frac{v_{in}}{r_e + R_E} \approx i_c$$

Combining Eq. (5-15) and (5-18), we get

$$v_o = -i_c \times R_C \| r_{ce} = -\frac{v_{in}}{r_e + R_E} \times R_C \| r_{ce}$$

and finally,

$$A_v = \frac{v_o}{v_{in}} = -\frac{R_C \| r_{ce}}{r_e + R_E} \qquad (5\text{-}19)$$

Equation (5-19) is the *typical* CE voltage gain expression. The negative sign indicates that there is a phase reversal between v_{in} and v_o (180° phase shift).

It is common to "couple" (connect) the output voltage v_o to a load, be it a resistor or earphones, etc. This load, referred to as R_L (or Z_L when it is not a pure resistance), is connected in parallel with r_{ce} and R_C. Equation (5-19) is then modified to account for R_L

$$A_v = \frac{v_o}{v_{in}} = -\frac{R_{CN}}{r_e + R_E} \qquad (5\text{-}19a)$$

where

$$R_{CN} = R_L \| R_C \| r_{ce}$$

R_{CN} may, in fact, be thought of as the *total* resistance from collector to ground (as obtained from the ac equivalent circuit). Similarly, the denominator contains the total resistance emitter to GND. An important reminder: A_v is unitless; it is the ratio of two voltages.

EXAMPLE 5-2

Find the voltage gain A_v for the circuit in Fig. 5-11, without C_E, the emitter capacitor.

SOLUTION

Step 1 Find I_E using dc analysis (Sec. 4-7.4). We find $I_E = 5.3$ mA. (The student is invited to verify this value.)

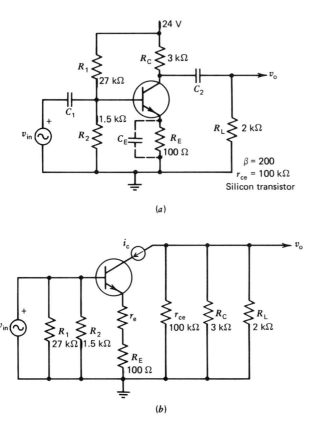

FIGURE 5-11 Typical CE circuit. (a) Actual circuit. (b) Equivalent circuit (simplified).

184 THE TRANSISTOR AS AN AMPLIFIER (SMALL SIGNAL)

Step 2 $r_e = \dfrac{26}{I_E} = \dfrac{26}{5.3} = 4.9\ \Omega \quad \left(\dfrac{26\ \text{mV}}{I_E\ \text{mA}}\right)$

Step 3 Obtain the equivalent circuit, shown in Fig. 5-11b. (All capacitors are represented by shorts, V_{CC} replaced by GND.)

Step 4 Follow the procedure of finding i_B, i_c, and v_o (in terms of v_{in}) or use Eq. (5-19a). Note that

$$R_{CN} = r_{ce}\|R_C\|R_L = 100\ \text{k}\Omega\|3\ \text{k}\Omega\|2\ \text{k}\Omega$$
$$R_{CN} = 100\ \text{k}\Omega\|(3\times2)/(3+2)\ \text{k}\Omega = 100\ \text{k}\Omega\|1.2\ \text{k}\Omega$$
$$R_{CN} \approx 1.2\ \text{k}\Omega \quad (\text{Since } 100\ \text{k}\Omega \gg 1.2\ \text{k}\Omega).$$

$$A_v = -\dfrac{R_{CN}}{r_e + R_E} = -\dfrac{1200}{4.9 + 100} = -11.44 \qquad (5\text{-}19a)$$

Note that here $r_e \ll R_E$; hence, the sum $R_E + r_e \simeq R_E$. This leads to

$$A_v \approx -\dfrac{1200}{100} = -12 \quad (\text{less than } 5.0\% \text{ error})$$

The approximations made in Ex. 5-2, namely neglecting r_{ce} because $r_{ce} \gg R_L\|R_C$ and ignoring r_e ($r_e \ll R_E$) are usually good, in particular the first one. r_e may not always be neglected, since R_E is often zero or shorted by a capacitor.

EXAMPLE 5-3

In Fig. 5-11 connect the capacitor C_E from emitter to ground (shown in dashed lines in Fig. 5-11) and find A_v.

SOLUTION

Connecting C_E as shown places an *ac* short across R_E. $R_E = 0$. The dc circuit is unaltered. We then have

$$A_v = -\dfrac{1200}{r_e + R_E} = -\dfrac{1200}{4.9 + 0} = -245$$

(Neglecting r_e here results in a zero denominator.)

The voltage gain calculated in Ex. 5-3 is large and somewhat typical of the CE configuration, particularly when R_E is bypassed.

INPUT RESISTANCE, r_{in}

r_{in}, for the CE configuration (Fig. 5-9), can be found as follows. r_{in} is defined as v_{in}/i_{in}, the resistance (or impedance) the source v_{in} "sees." It may be considered to consist of two parts, the resistor R_B in parallel with the resistance presented by the base to ground transistor portion r_{inB} (Fig. 5-12). R_B is a circuit component while r_{inB} is dependent on the transistor (and R_E, as we see shortly). It is clear that

$$r_{inB} = v_B/i_B$$

Applying KVL to the base loop, we get (what we already got in Eq. (5-17a) for $v_{in} = v_B$)

$$i_B = \frac{v_B}{\beta(r_e + R_E)}$$

and

$$\frac{v_B}{i_B} = \beta(r_e + R_E) = r_{inB} \qquad (5\text{-}20)$$

Equation (5-20) gives a general representation for r_{inB}, the input resistance "looking" directly into the base. No matter what the circuit configuration is, the resistance presented by or the equivalent resistance of base to ground is given by Eq. (5-20). When "looking" into the base, we "see" the *emitter (ideal) to ground* resistance ($r_e + R_E$) as if it were a much larger resistance, multiplied by β. When transferring emitter resistance to the base side, it must be multiplied by β.

The total resistance r_{in} is now given (for Fig. 5-9) by

$$r_{in} = R_B \| r_{inB} = R_B \| \beta(r_e + R_E)$$

For the voltage divider bias circuit, there are two parallel resistors, R_1 and R_2, taking the place of R_B above. In general, R_{BN} can be used to denote all ac resistance base to ground and express r_{in} as

$$r_{in} = R_{BN} \| \beta(r_e + R_E) \qquad (5\text{-}20a)$$

EXAMPLE 5-4

Find r_{in} for the circuit of Fig. 5-11a:
(a) Without the capacitor C_E.
(b) With the capacitor C_E connected.

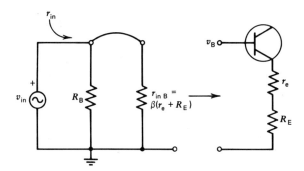

FIGURE 5-12 r_{in} in CE circuit.

186   THE TRANSISTOR AS AN AMPLIFIER (SMALL SIGNAL)

SOLUTION

From the equivalent circuit (Fig. 5-11b), it can be seen that r_{in} consists of $R_1 \| R_2 \| r_{inB}$. From Ex. 5-2 $r_e = 4.9\ \Omega$. Hence,

(a) $r_{in} = 27\ k\Omega \| 1.5\ k\Omega \| 100(4.9+100)\ \Omega = 1.42\ k\Omega \| 10.49\ k\Omega = \underline{1.25\ k\Omega}$

(b) The capacitor C_E effectively shorts R_E (makes $R_E = 0$). Thus, $r_{in} = 27\ k\Omega \| 1.5\ k\Omega \| 100(4.9+0) = 1420 \| 490\ \Omega = \underline{364\ \Omega}$

Note that shorting R_E in Ex. 5-4 reduces the input resistance substantially. Furthermore, neglecting r_e in part (a) would introduce a small error ($\approx 5\%$), while in part (b) we simply *cannot* neglect r_e (it is the only ac emitter resistance). r_{in} for the CE configuration is typically a few $k\Omega$ or less.

CURRENT GAIN, A_i

The current gain A_i may be calculated from the voltage gain A_v, the input resistance r_{in}, and the load resistance R_L. The definition of A_i was given as

$$A_i = \frac{i_o}{i_{in}} \qquad (5\text{-}2)$$

where i_o is the ac current in the load R_L, and i_{in} is the current from the ac source v_{in} into r_{in} (Fig. 5-13). Since v_o is the voltage across R_L, i_o can be written as

$$i_o = \frac{v_o}{R_L}$$

and similarly,

$$i_{in} = \frac{v_{in}}{r_{in}}$$

The ratio of these two expressions is

$$\frac{i_o}{i_{in}} = \frac{v_o/R_L}{v_{in}/r_{in}} = \frac{v_o}{v_{in}} \times \frac{r_{in}}{R_L} \qquad (5\text{-}21)$$

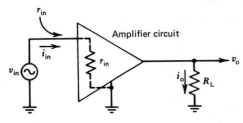

FIGURE 5-13 A_i calculation.

Note that $i_o/i_{in} = A_i$ and $v_o/v_{in} = A_v$; hence, from Eq. (5-21)

$$A_i = A_v \times \frac{r_{in}}{R_L} \qquad (5\text{-}21a)$$

Equation (5-21a) does not depend on the circuit configuration at all; it is a result of the basic definitions of the various parameters. (This equation is sometimes referred to as the transistor–gain–impedance relation, TGIR.) Caution must be exercised in our definition of r_{in} and our specification of R_L. Again, a reminder— A_i is unitless, it is a ratio of two currents.

EXAMPLE 5-5

Find the current gain for Fig. 5-11:
(a) Without C_E.
(b) With C_E.

SOLUTION

(a) In the previous examples we found that in Fig. 5-11 without C_E, $A_v = -11.44$ and $r_{in} = 1.25$ kΩ. Since $R_L = 2$ kΩ, using Eq. (4-21a), we get

$$A_i = -11.44 \times \frac{1.25 \text{ k}\Omega}{2 \text{ k}\Omega} = -7.15$$

(b) With C_E connected, we found in previous examples $A_v = -245$ and $r_{in} = 364$ Ω. Hence, from Eq. (4-21a)

$$A_i = -245 \times \frac{364 \text{ }\Omega}{2 \text{ k}\Omega} = -44.6$$

Two observations are in order in connection with Ex. 5-5. First, the minus sign for A_i indicates that while i_{in} flows "toward" ground, i_o flows "away" from ground; these currents are of opposite polarity, 180° out of phase. Second, A_i in a CE circuit can *never* exceed β (in magnitude).

OUTPUT RESISTANCE, r_o

To obtain r_o the method used in obtaining the Thevenin's resistance is applied— namely, *remove the load and "look" into the circuit at the two nodes the load was connected to.* (Use the "eyes" of the load to do the "looking.") r_o (which is r_{Th} as defined in Sec. 5-3) can be found by eliminating all sources (independent sources) and calculating the resistance between the collector and GND (Fig. 5-14). The current source $i_c = \beta i_B$ is a dependent current source; it depends on i_B. However, the elimination of v_{in} as part of the Thevenin's process effectively eliminates, opens,

188 THE TRANSISTOR AS AN AMPLIFIER (SMALL SIGNAL)

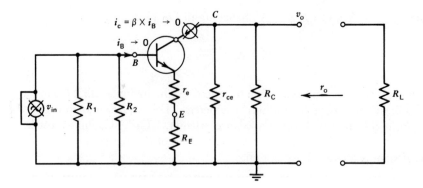

FIGURE 5-14 r_o computation. (Eliminate all ideal sources—short v_{in} open i_c).

the i_c source. The output resistance then becomes

$$r_o = r_{ce} \| R_C \qquad (5\text{-}22)$$

r_o can usually be approximated by R_C since $r_{ce} \gg R_C$.

EXAMPLE 5-6

Find r_o for Fig. 5-11.

SOLUTION

Here, the presence (or absence of C_E) has no significance.

$$r_o = r_{ce} \| R_C \qquad (5\text{-}22)$$
$$r_o = 100 \text{ k}\Omega \| 2 \text{ k}\Omega \approx 2 \text{ k}\Omega \qquad (2\% \text{ error})$$

POWER GAIN, A_p

The power gain A_p is usually only considered when dealing with power circuits. Nevertheless, it is computed for Ex. 5-2, which deals with a low power circuit for the purposes of demonstration only. Based on the definition $A_p = A_v \times A_i$, for Ex. 5-2 with C_E in the circuit,

$$A_p = (-245)(-44.6) = 10927$$

The following two examples are comprehensive in nature and may serve as a review for the CE configuration.

EXAMPLE 5-7

For the circuit shown in Fig. 5-15, find A_v, r_{in}, A_i, and r_o:
  (a) Without the capacitor C_E.
  (b) With the capacitor C_E.
Approximations introducing errors of less than 10% are acceptable but must be justified.

SOLUTION

dc calculations to find r_e (ignore ac sources; open all capacitors).

$$V_{Th} = \frac{15 \times 2.4 \text{ k}\Omega}{(24 + 2.4) \text{ k}\Omega} = 1.36 \text{ V}$$

$$R_{Th} = 24 \text{ k}\Omega \| 2.4 \text{ k}\Omega = 2.18 \text{ k}\Omega$$

KVL

$$V_{Th} - I_B R_{Th} - 0.7 - I_E \times R_{ET} = 0$$

where $R_{ET} = R_{E1} + R_{E2}$.

$$I_E = \frac{V_{Th} - 0.7}{R_{Th}/\beta + R_{ET}}$$

Since $R_{Th}/\beta \ll R_{ET}$ (more than 10 to 1), we may neglect R_{Th} and

$$I_E = \frac{V_{Th} - 0.7}{R_{ET}} = \frac{1.36 - 0.7}{490} = 1.35 \text{ mA}$$

$$r_e = \frac{26}{1.35} = \underline{19.3 \; \Omega}$$

ac calculation (Fig. 5-15c)
  (a) A_v

$$A_v = -\frac{R_{CN}}{r_e + R_{E1} + R_{E2}}$$

$$= -\frac{\text{(all resistance collector to GND)}}{\text{(all resistance emitter to GND)}}$$

$$A_v = -\frac{r_{ce} \| R_C \| R_L}{r_e + R_{E1} + R_{E2}} \quad (5\text{-}19)$$

$$= -\frac{200 \times 10^3 \| 2.4 \times 10^3 \| 3.3 \times 10^3}{19.3 \; \Omega + 100 \; \Omega + 390 \; \Omega}$$

$$\approx -\frac{2.4 \times 10^3 \| 3.3 \times 10^3}{490}$$

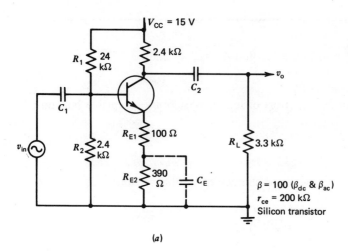

(a)

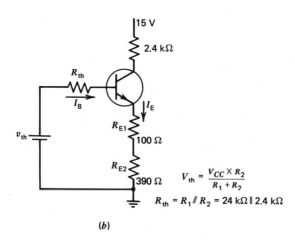

(b)

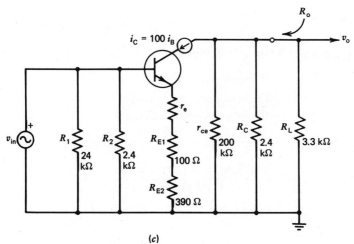

(c)

The approximations made above are
1. Neglect r_{ce} since $r_{ce} \gg R_C \| R_L$ (better than 10 to 1).
2. Neglect r_e since $r_e \ll (R_{E1} + R_{E2})$ (better than 10 to 1).

Each approximation introduces less than a 10% error.

$$A_v \approx -\frac{1390}{490} = -2.8.$$

r_{in}

$$r_{in} = R_1 \| R_2 \| r_{inB}$$
$$r_{inB} = \beta(r_e + R_{E1} + R_{E2}) \tag{5-20}$$
$$r_{inB} \approx \beta(R_{E1} + R_{E2}) \quad \text{(since } r_e \ll (R_{E1} + R_{E2})\text{)}$$
$$r_{inB} = 100(490) = 49 \text{ k}\Omega$$
$$r_{in} = 2.4 \text{ k}\Omega \| 24 \text{ k}\Omega \| 49 \text{ k}\Omega = \underline{2.1 \text{ k}\Omega}$$

A_i

$$A_i = A_v \times \frac{r_{in}}{R_L} = -2.8 \times \frac{2100}{3300} = \underline{-1.78} \tag{5-21}$$

r_o

$$r_o = r_{ce} \| R_C \approx R_C \quad (r_{ce} \gg R_C) \tag{5-22}$$
$$r_o \approx \underline{2.4 \text{ k}\Omega}$$

(b) The introduction of C_E is equivalent, for ac consideration, to replacing R_{E2} with a short—that is, $R_{E2} = 0$. The dc calculations remain unaffected; hence, $r_e = 19.3 \ \Omega$ as before.

A_v

$$A_v = -\frac{R_{CN}}{r_e + R_{E1} + R_{E2}}$$
$$= -\frac{200 \times 10^3 \| 2.4 \times 10^3 \| 24 \times 10^3}{19.3 + 100} = \frac{-2.4 \times 10^3 \| 24 \times 10^3}{119.3}$$

Again, r_{ce} may be neglected but *not* r_e since r_e is now about 20% of R_E (not negligible).

$$A_v = -\frac{1390}{119.3} = \underline{-11.65}$$

FIGURE 5-15 (*opposite*) CE with voltage divider bias. (a) Actual circuit. (b) Biasing equivalent. (c) ac equivalent circuit (no C_E).

192 THE TRANSISTOR AS AN AMPLIFIER (SMALL SIGNAL)

r_{in}

$r_{in} = R_1 \| R_2 \| r_{inB}$

$r_{inB} = \beta(r_e + R_{E1}) = 100(19.3 + 100) = 11.93 \text{ k}\Omega$ (4-20)

$r_{in} = 2.4 \text{ k}\Omega \| 24 \text{ k}\Omega \| 11.93 \text{ k}\Omega = \underline{1.84 \text{ k}\Omega}$

A_i

$A_i = A_v \dfrac{r_{in}}{R_L} = -11.65 \times \dfrac{1.840 \text{ k}\Omega}{3.3 \text{ k}\Omega} = \underline{-6.5}$ (4-21)

r_o

$r_o \approx R_C = \underline{2.4 \text{ k}\Omega}$

EXAMPLE 5-8

For the circuit of Fig. 5-16a (self bias CE configuration), find:
(a) A_v.
(b) r_{in}.
(c) r_o.

SOLUTION

dc calculations to obtain r_e
I_{CQ} is found from Eq. (4-21)

$$I_{CQ} = \dfrac{V_{CC} - 0.7}{\dfrac{R_B}{\beta} + (R_E + R_C)} = \dfrac{20 - 0.7}{\dfrac{300 \times 10^3}{100} + 200 + 2000}$$

Here, $R_B = R_{B1} + R_{B2} = 150 \text{ k}\Omega + 150 \text{ k}\Omega = 300 \text{ k}\Omega$.

$$I_{CQ} = \dfrac{19.3}{5.2 \times 10^3} = 3.7 \times 10^{-3} = 3.7 \text{ mA}$$

To find r_e (since $I_C \approx I_E$),

$$r_e = \dfrac{26}{3.7} = 7 \text{ }\Omega$$

The ac equivalent circuit is shown in Fig. 5-16b (neglecting r_{CE}). Note that, for ac analysis, R_{B1} is connected from collector to ground (in parallel with R_C), and R_{B2} is connected from base to ground.

$A_v = \dfrac{-R_{CN}}{R_E + r_e} = \dfrac{-R_C \| R_{B1}}{R_E + r_e} = \dfrac{2 \text{ k}\Omega \| 150 \text{ k}\Omega}{200 + 7} = \underline{-9.5}$

$r_{in} = R_{B2} \| r_{inB} = R_{B2} \| \beta(R_E + r_e)$
$= 150 \text{ k}\Omega \| 100 \times 207 = \underline{18.2 \text{ k}\Omega}$

$r_o = R_C \| R_{B1} = 2 \text{ k}\Omega \| 150 \text{ k}\Omega = 1970 \text{ }\Omega \approx \underline{2 \text{ k}\Omega}$

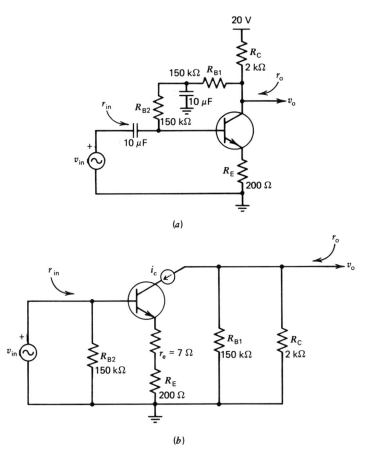

FIGURE 5-16 Self bias CE circuit for Ex. 5-8. (a) Circuit. (b) ac equivalent circuit (neglecting r_{ce}).

5-5.2 Common Base (CB)

In the common base circuit the input signal is fed to the emitter, directly or indirectly, and the output is obtained from the collector (Fig. 5-17). Figures 5-17a and 5-17b represent the identical circuit. The circuit was drawn once with the transistor on its side and once with the transistor upright. Many texts prefer the "lying down" version (nobody knows why). This text, however, will draw all transistors upright since it is consistent with the circuits already presented and reduces confusion. Figure 5-17 shows a capacitor C_B (dashed lines). The analysis that follows considers the circuit where C_B is *not* connected. Subsequently, the effects of connecting C_B will be investigated.

The first step in the analysis is to obtain the *full* equivalent circuit for the circuit and the device (Fig. 5-17c). As in the case of the CE circuit, the device equivalent circuit is modified by connecting r_{ce} to GND rather than to the emitter E. (The equivalent circuit is based on the understanding that for ac, V_{CC} becomes ground and all capacitors represent shorts.)

194 THE TRANSISTOR AS AN AMPLIFIER (SMALL SIGNAL)

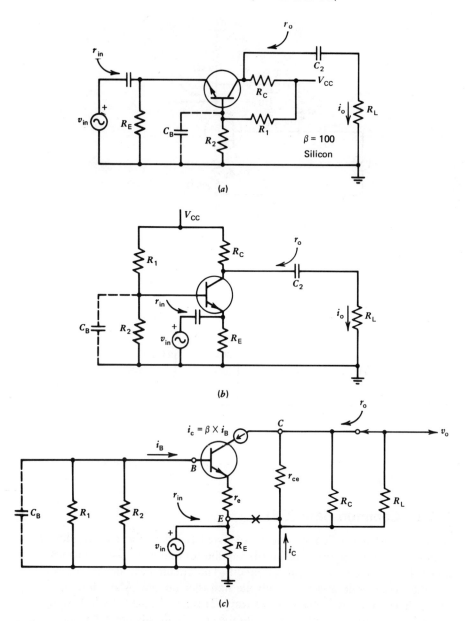

FIGURE 5-17 Common base (CB) configuration. (a) Actual circuit-transistor shown on its side. (b) Same circuit transistor shown upright. cc) Equivalent circuit (ac).

VOLTAGE GAIN, A_v

The computation of v_o, in the process of obtaining A_v, does *not* involve R_E in this particular circuit. This is a result of the fact that R_E is connected in parallel with the ac source v_{in}. (The Thevenin's equivalent of a voltage source with a non-zero resistor across it is the voltage source alone. The resistor is completely disregarded.) The KVL applied to the base-emitter loop does not contain an R_E term, but rather the voltage across R_E, namely v_{in}.

v_o is given by

$$v_o = -i_c \times R_{CN} \qquad (5\text{-}15)$$

where $R_{CN} = r_{ce}\|R_C\|R_L$.

As before, assume $i_e \simeq i_c$ and calculate i_c by applying KVL to the input loop (base-emitter). The polarities are shown in Fig. 5-17c.

$$-i_B \times (R_1\|R_2) - i_e \times r_e - v_{in} = 0$$

(KVL applied in the CW direction)

$$v_{in} = -\frac{i_c}{\beta}(R_1\|R_2) - i_c \times r_e \quad (\text{here, } i_b = i_c/\beta \text{ and } i_e \simeq i_c)$$

$$v_{in} = -i_c\left(\frac{R_1\|R_2}{\beta} + r_e\right) \qquad (5\text{-}23)$$

$$i_c = -v_{in}\bigg/\left(\frac{R_1\|R_2}{\beta} + r_e\right) \qquad (5\text{-}23a)$$

The term $R_1\|R_2$ can be generalized to mean all resistance between base and ground, R_{BN}.

$$i_c = -v_{in}\bigg/\left(\frac{R_{BN}}{\beta} + r_e\right) \qquad (5\text{-}23b)$$

Substituting Eq. (5-23b) in the expression for v_o [Eq. (5-15)], we get

$$v_o = -i_c R_{CN} = -\left(-v_{in}\bigg/\frac{R_{BN}}{\beta} + r_e\right) \times R_{CN}$$

$$v_o = v_{in} \times \frac{R_{CN}}{\frac{R_{BN}}{\beta} + r_e}$$

and

$$A_v = \frac{v_o}{v_{in}} = \frac{R_{CN}}{\frac{R_{BN}}{\beta} + r_e} \qquad (5\text{-}24)$$

In the CB circuit, v_o and v_{in} are *in phase* because, as Eq. (5-24) indicates, A_v is always positive.

The effect of connecting C_B would be to place a short across the combination $R_1\|R_2$ resulting in the elimination of this term, $R_{BN} \to 0$. Under these conditions the voltage gain A_v becomes

$$A_v = \frac{R_{CN}}{r_e} \qquad (5\text{-}24a)$$

In the circuit of Fig. 5-17, the input signal v_{in} was connected directly to the emitter. In Fig. 5-18 the input signal is connected in series with R_E. The transformer (assumed to have a 1:1 turns ratio for convenience only) couples the signal to the transistor. The ac equivalent circuit simply shows v'_{in} (the prime mark is used to denote that it is the transformed v_{in}, for the 1:1 turns ratio $v'_{in} = v_{in}$) connected in series with R_E. (Note that if the dc resistance of the transformer is assumed to be zero, the dc equivalent circuit, for bias computations, is identical to Fig. 5-17.)

The voltage gain of Fig. 5-18 can be obtained by a careful comparison with Fig.

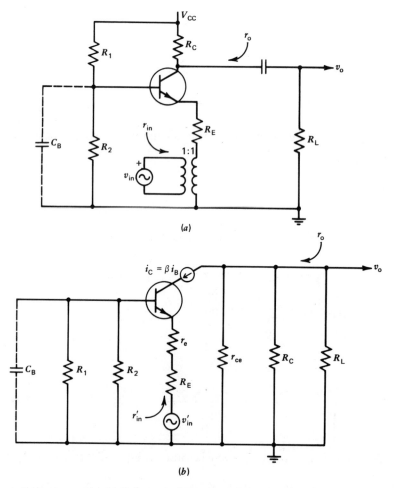

FIGURE 5-18 CB with input in series with R_E. (a) Actual circuit. (b) ac equivalent (for transformer ratio of 1:1 $r_{in} = r'_{in}$ and $v'_{in} = v_{in}$).

5-17. The only difference, as far as the input signal is concerned, is that in Fig. 5-18 r_e is in series with R_E, both being part of the base-emitter loop. We may then substitute $(R_E + r_e)$ for r_e in Eq. (5-24).

$$A_v = \frac{R_{CN}}{\dfrac{R_{BN}}{\beta} + R_E + r_e} \qquad (5\text{-}25)$$

As before, the introduction of C_B reduces R_{BN} to zero and Eq. (5-25) becomes

$$A_v = \frac{R_{CN}}{R_E + r_e} \qquad (5\text{-}25a)$$

The student is urged to develop Eq. (5-25) by applying the steps used in obtaining Eq. (5-24) to the circuit of Fig. 5-18.

EXAMPLE 5-9

Find the voltage gain for the circuit in Fig. 5-19:
(a) Without C_B.
(b) With C_B in the circuit.

SOLUTION

The circuit of Fig. 5-19 is a combination of Figs. 5-17 and 5-18 as far as the input circuit is concerned. Based on our discussion in connection with Fig. 5-17, it is expected here that R_{E2} will have no effect on the voltage gain A_v. (The voltage across R_{E2} is v_{in} regardless of the value of R_{E2}.) $R_{E1} + r_e$, which are a part of the base-emitter loop, *will* figure in the computation of A_v.
(a) Using Eq. (5-25) and realizing that $R_{E1} = R_E$, we get

$$A_v = \frac{R_{CN}}{\dfrac{R_{BN}}{\beta} + R_{E1} + r_e}$$

$$R_{CN} = r_{ce}\|R_C\|R_L = 200 \text{ k}\Omega\|2 \text{ k}\Omega\|3 \text{ k}\Omega = 1.2 \text{ k}\Omega$$

$$R_{BN} = R_1\|R_2 = 16 \text{ k}\Omega\|4 \text{ k}\Omega = 3.2 \text{ k}\Omega$$

$$A_v = \frac{1200}{\dfrac{3200}{200} + 100 + 4.65} = \frac{1200}{120.65} \approx 10$$

($r_e = 4.65 \text{ }\Omega$ was obtained from dc analysis not shown.) Note that neglecting r_{ce} and r_e introduces very little error (less than 3%).
(b) The introduction of C_B reduces R_{BN} to zero and hence,

$$A_v = \frac{R_{CN}}{R_{E1} + r_e} = \frac{1200}{104.65} = 11.5$$

198 THE TRANSISTOR AS AN AMPLIFIER (SMALL SIGNAL)

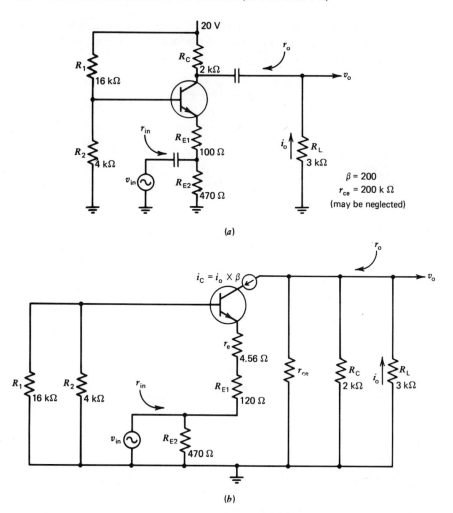

FIGURE 5-19 CB circuit. (a) Circuit. (b) Equivalent circuit. (r_e = 4.56 Ω from dc analysis).

INPUT RESISTANCE, r_{in}

The input resistance r_{in} is the resistance seen by the ac source, as indicated in Figs. 5-17, 5-18, and 5-19. While all three figures represent common base configurations, they differ in circuit detail and hence, yield slightly different expressions for r_{in}. As in the common emitter case, r_{in} is going to be calculated in two steps. First, find r_{inE}, namely the input resistance "looking" into the emitter directly, disregarding, temporarily, all other components. Figure 5-20 shows the portion of the circuit that is involved in the computation of r_{inE}. The measurement of resistance involves the use of a voltage source and the measurement or calculation of the current it

TRANSISTOR CIRCUITS—ac ANALYSIS

produces. Precisely such an approach is taken when calculating r_{inE}—namely, $v_m/i_m = r_{inE}$, where v_m and i_m are introduced for the sake of the computation alone (Fig. 5-20). Although this method is not used in *meaursing* r_{inE}, it may serve us well in our *calculations* of r_{inE}. By the application of KVL to the input loop, we get (in the CW direction)

$$v_m - i_B \times R_{BN} - i_e r_e = 0 \tag{5-26}$$

Since $i_e = i_m$ and $i_B \approx i_e/\beta = i_m/\beta$, Eq. (5-26) can be rewritten as

$$v_m - \frac{i_m}{\beta} \times R_{BN} - i_m \times r_e = 0$$

$$v_m - i_m\left(\frac{R_{BN}}{\beta} + r_e\right) = 0; \qquad v_m = i_m \times \left(\frac{R_{BN}}{\beta} + r_e\right)$$

and

$$\frac{v_m}{i_m} = r_{inE} = \frac{R_{BN}}{\beta} + r_e \tag{5-26a}$$

R_{BN} represents the total base to ground resistance $R_1 \| R_2$ in this case. We are now in a position to combine r_{inE} with the other circuit components to obtain r_{in} (Fig. 5-21). In Fig. 5-21a that represents the circuit of Fig. 5-17

$$r_{in} = R_E \| r_{inE} \tag{5-27}$$

Both R_E and r_{inE} are connected across the input voltage; they are effectively in parallel. Combining Eqs. (5-26a) and (5-27) yields

$$r_{in} = R_E \left\| \left(\frac{R_{BN}}{\beta} + r_e\right) \right. \tag{5-27a}$$

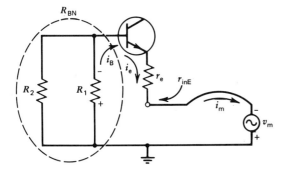

FIGURE 5-20 Calculating r_{inE}.

200 THE TRANSISTOR AS AN AMPLIFIER (SMALL SIGNAL)

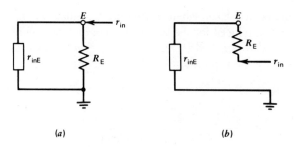

FIGURE 5-21 (a) r_{in} for circuit of Fig. 5-17 ($r_{in} = r_{inE} \| R_E$). (b) r_{in} for circuit of Fig. 5-18 ($r_{in} = R_E + r_{inE}$).

In Fig. 5-18, R_E is in series with r_{inE} as represented again in Fig. 5-21b. Here,

$$r_{in} = R_E + r_{inE} = R_E + \frac{R_{BN}}{\beta} + r_e \qquad (5\text{-}27b)$$

Both Eqs. (5-27a) and (5-27b) are applicable to the CB configuration. The difference between these two expressions result from differences in circuit details (another example of the importance of *understanding* the formulae). It should be noted that R_{BN} could be shorted ($R_{BN} = 0$) by the addition to the circuit of a base to ground capacitor C_B.

EXAMPLE 5-10

Find r_{in} for the circuit of Fig. 5-19:
  (a) Without C_B.
  (b) With C_B.

SOLUTION

As already noted in the previous example, Fig. 5-19 is a combination of the circuits of Figs. 5-17 and 5-18. Figure 5-22 shows clearly how r_{in} can be viewed. It is clear that in this case r_{in} consists of

$$R_{E2} \| (R_{E1} + r_{inE})$$

R_{E1} is the R_E of Fig. 5-21b and R_{E2} is the R_E of Fig. 5-21a.

(a) $r_{inE} = \dfrac{4 \text{ k}\Omega \| 16 \text{ k}\Omega}{200} + 4.65 \qquad (5\text{-}26)$

$\qquad = \dfrac{3200}{200} + 4.65 = 20.65 \ \Omega$

$\qquad r_{in} = 470 \| (100 + 20.65) = \underline{96 \ \Omega}$

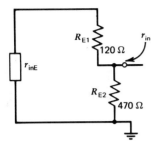

FIGURE 5-22 r_{in} for Ex. 5-10.

(b) Placing C_B in the circuit reduces r_{inE} to
$$r_{inE} = r_e = 4.65 \ \Omega \ (R_{BN} = 0)$$
and
$$r_{in} = 470 \| (100 + 4.65) = 86 \ \Omega$$

r_e could be neglected in both parts (a) and (b). r_{in} for the CB configuration is usually below about 500 Ω (often substantially less than 500 Ω).

CURRENT GAIN A_i

To obtain the current gain, we may simply apply the general formula given by Eq. (5-21) (the TGIR).

$$A_i = A_v \times \frac{r_{in}}{R_L} \tag{5-21}$$

EXAMPLE 5-11

Find A_i for the circuit of Fig. 5-19:
(a) Without C_B.
(b) With C_B.

SOLUTION

(a) Since we found in previous examples that $A_v = 10$, $r_{in} = 96 \ \Omega$ and $R_L = 3 \ k\Omega$ is given

$$A_i = 10 \times \frac{96}{3000} = \underline{0.32} \tag{5-21}$$

(b) Since $A_v = 11.5$, $r_{in} = 86 \ \Omega$, $R_L = 3 \ k\Omega$

$$A_i = 11.5 \times \frac{86}{3000} = \underline{0.33} \tag{5-21}$$

It is clear, from a simple inspection of the typical CB circuit, that $A_i < 1$. The source provides the current i_e and the absolute maximum that may flow in the load is i_c—clearly, $i_c/i_e < 1$, ($i_c < i_e$).

OUTPUT RESISTANCE, r_o

It is from the "point of view" of the load R_L that r_o is evaluated (the equivalent Thevenin's resistance of the source that drives R_L). Consequently, R_L is excluded, by definition, from being part of r_o. An inspection of the CB equivalent circuit (Figs. 5-17, 5-18, and 5-19) clearly shows that for CB, as for the CE, configurations

$$r_o = r_{ce} \| R_C \qquad (5\text{-}28)$$

(We eliminate all ideal sources and then obtain r_o.) Note that while A_v, r_{in}, A_i are all different for the three figures (5-17, 5-18, 5-19), r_o is identical, since the output circuit is the same in all cases.

EXAMPLE 5-12

Find r_o for Fig. 5-19.

SOLUTION

By inspection (opening current source i_c)

$$r_o = r_{ce} \| R_C = 200 \text{ k}\Omega \| 2 \text{ k}\Omega \approx 2 \text{ k}\Omega$$

r_{ce} may be neglected in this example.

5-5.3 Common Collector (CC) or Emitter Follower (EF)[1]

As will become apparent from the detailed analysis, the most useful function of the emitter follower is impedance (or resistance) transformation, that is, it has a high input resistance (r_{in}) and a low output resistance r_o. The importance of impedance transformation is illustrated by Fig. 5-23. We are attempting here to couple a source, say a microphone, with a 10 kΩ internal resistance R_s to a load R_L of 100 Ω. Fig. 5-23a shows that if the source is directly coupled to the load, the voltage across R_L is very small, $v_L \approx \frac{1}{100} \times v_s$. By interjecting a circuit such as the EF, we find that the input to the EF, v_{in}, is nearly equal to v_s. By voltage division

$$v_{in} = \frac{v_s \times r_{in}}{R_s + r_{in}} = \frac{v_s \times 100 \text{ k}\Omega}{110 \text{ k}\Omega} = 0.9 \, v_s$$

---

[1] The more common name of this circuit configuration is emitter following (EF), while the term common collector (CC) is consistent with the terminology used for the other two configurations.

TRANSISTOR CIRCUITS—ac ANALYSIS   203

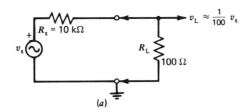

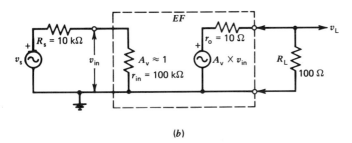

FIGURE 5-23   EF as impedance transformer. (a) Source coupled directly to load. (b) EF interjected between source and load.

If it can be assumed that the voltage gain of the EF is unity, $A_v = 1$ (this will be confirmed later), then

$$v_o = v_{in} = 0.9\, v_s$$

v_o here is the output voltage with R_L removed (no load output). By voltage division

$$v_L = \frac{v_o \times R_L}{r_o + R_L} = \frac{v_o \times 100}{110} = 0.9\, v_o = 0.81\, v_s \qquad (v_o = 0.9\, v_s)$$

The last expression gives the output voltage across R_L (output with the load connected). The high r_{in} and low r_o of the interjected circuit permit us to transfer the source voltage v_s to the load, in spite of the very large difference in the source and load resistance. As far as the source is concerned, it "sees" a 100 kΩ load, namely r_{in}. As for the load R_L, it is driven from a low impedance source r_o. (See also Fig. 5-2 and the related discussion.)

VOLTAGE GAIN, A_v

A typical EF circuit and its equivalent is shown in Fig. 5-24.
 Voltage gain (actually a voltage loss, since it will be found that $A_v < 1$) of the EF circuit can be obtained by simple application of voltage division. In Fig. 5-24c the total emitter to ground resistance is represented by R_{EN}, which by inspection of Fig. 5-24b equals

$$R_{EN} = r_{ce} \| R_E \| R_L$$

204 THE TRANSISTOR AS AN AMPLIFIER (SMALL SIGNAL)

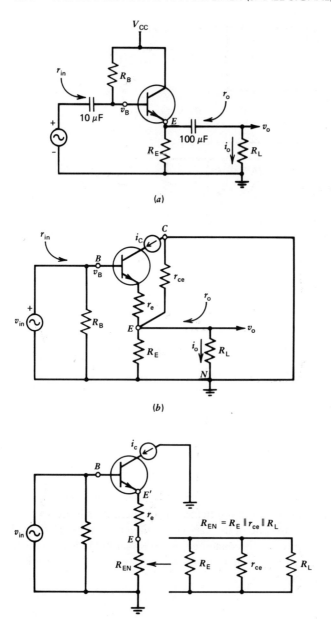

FIGURE 5-24 EF circuit. (a) Actual circuit. (b) Equivalent circuit. (c) Simplified equivalent circuit.

r_{ce} is directly connected from emitter to collector, but since the collector is grounded, r_{ce} is directly between the emitter and ground. Since, for the ideal transistor, $v_B = v_{E'}$ (E' denoting the emitter of an ideal transistor) and since $v_{in} = v_B$, we find that $v_{E'} = v_{in}$. By voltage division between r_e and R_{EN}

$$v_o = \frac{v_{E'} \times R_{EN}}{r_e + R_{EN}} = \frac{v_{in} \times R_{EN}}{r_e + R_{EN}}$$

and

$$A_v = \frac{v_o}{v_{in}} = \frac{R_{EN}}{r_e + R_{EN}} \quad (5\text{-}29)$$

Equation (5-29) clearly shows that $A_v < 1$. In the most ideal case, in which r_e may be assumed negligible ($r_e \ll R_{EN}$), we find

$$A_v \approx \frac{R_{EN}}{R_{EN}} = 1 \quad (5\text{-}29a)$$

INPUT RESISTANCE, r_{in}

The input portion of the EF circuit is much like that of the CE configuration. In both cases the input voltage is applied to the base. As a result, the expressions for r_{in} may be expected to be very much alike. The equivalent circuit (Fig. 5-24c) is almost a duplicate of Fig. 5-9c, as far as the base-emitter loop is concerned. The R_E of Fig. 5-9c becomes R_{EN} in Fig. 5-24c. All other components are identical. It stands to reason, then, that r_{in} for the EF can be directly obtained from r_{in} for the CE [Eq. (5-20a)]

$$r_{in} = R_{BN} \| r_{inB} = R_{BN} \| \beta(r_e + R_{EN}) \quad (5\text{-}30)$$

As in the common emitter case, Eq. (5-30) consists of two parallel parts: R_{BN}, all resistance base to ground, and r_{inB}, the input resistance of the base terminal. It is important to note that unlike the CE expression, r_{in} in Eq. (5-30) depends on the load resistance R_L, since R_{EN} includes the effect of R_L. If, for a moment, the effects of r_{ce} and R_E may be neglected (which in many cases is quite reasonable), it is then apparent that R_L is transformed into $\beta \times R_L$ when viewed *through* the base terminal. Under this assumption $R_{EN} = R_L$ and $r_{inB} = \beta \times R_{EN} = \beta \times R_L$.

CURRENT GAIN, A_i

As in all previous cases Eq. (5-21) may be used to obtain the current gain.

$$A_i = A_v \times \frac{r_{in}}{R_L} \quad (5\text{-}21)$$

Since the voltage gain of the EF circuit is often unity, $A_v \approx 1$, A_i may be approximated by

$$A_i \approx r_{in}/R_L$$

206 THE TRANSISTOR AS AN AMPLIFIER (SMALL SIGNAL)

This approximation is valid only if $r_e \ll R_{EN}$.

OUTPUT RESISTANCE, r_o

To find r_o, first remove R_L, since r_o is the Thevenin's resistance r_{Th} of the source driving R_L. (r_o is "seen" by R_L.) r_o may be broken down into two distinct parts: $R_E \| r_{ce}$, shown by dashed lines in Fig. 5-25, and r_{oE} that is the Thevenin's resistance measured "looking" directly into the emitter terminal (Fig. 5-25). Clearly, $R_E \| r_{ce}$ is in parallel with r_{oE} and

$$r_o = r_{oE} \| (R_E \| r_{ce}) \tag{5-31}$$

To find r_{oE}, we temporarily ignore $R_E \| r_{ce}$ and proceed to use the method for finding the Thevenin's resistance. Eliminating all sources, in particular shorting out v_{in}, leaves the base to ground resistance as $R_{BN} \| R_s$. The internal resistance of the source R_s has been included here (while not shown in previous circuits) because it is an important part of r_o. r_{oE} may be found by the technique used in Sec. 5-5.2 as shown in Fig. 5-25 where

$$r_{oE} = \frac{v_m}{i_m} \quad \begin{array}{l}(r_{ce} \text{ and } R_E \text{ are ignored} \\ \text{when calculating } r_{oE})\end{array}$$

It goes without saying that $i_e = i_m$. Applying KVL to the base-emitter loop, we get

$$-i_B \times R_s \| R_{BN} - i_e \times r_e + v_m = 0$$

Since $i_b \approx i_e/\beta$ and $i_e = i_m$

$$v_m = \frac{i_e}{\beta} \times R_s \| R_{BN} + i_e \times r_e$$

$$= i_e \times \left(\frac{R_s \| R_{BN}}{\beta} + r_e\right) = i_m \left(\frac{R_s \| R_{BN}}{\beta} + r_e\right)$$

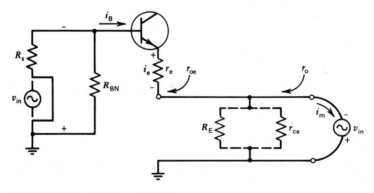

FIGURE 5-25 Finding r_o for EF.

and

$$r_{oe} = \frac{v_m}{i_m} = \frac{R_s \| R_{BN}}{\beta} + r_e \qquad (5\text{-}31a)$$

The total output resistance r_o is

$$r_o = R_E \| r_{ce} \left(\frac{R_s \| R_{BN}}{\beta} + r_e \right) \qquad (5\text{-}31b)$$

In many EF circuits a small R_C is introduced. This is done to limit the dc current through the transistor in case of a short across the output terminals, emitter to ground. (The full V_{CC} would otherwise be directly across C-E in case of a short.) This small R_C affects the dc operation only and has no bearing on any of the ac parameters.

EXAMPLE 5-13

Fig. 5-26 shows an EF circuit to which a 100 Ω load is to be connected. $r_e = 13$ Ω (found from dc calculations not shown).
(a) Find v_o, r_{in}, r_o for the case where R_L is not connected.
(b) Find v_o and r_{in} after R_L has been connected.
(c) Find A_i when R_L is connected.

SOLUTION

In parts (a) and (b) we will neglect the effect of R_s on A_v and assume that $v_s = v_{in} = 0.1$ V. (v_s is, essentially, the no load voltage at the terminals of the generator.) The equivalent circuit is shown in Fig. 5-26b.
(a) To find v_o, A_v is found first.

$$A_v = \frac{R_{EN}}{r_e + R_{EN}}$$

$$= \frac{4.7 \text{ k}\Omega \| 100 \text{ k}\Omega}{13 + 4.7 \text{ k}\Omega \| 100 \text{ k}\Omega} \approx 1 \quad (\text{since } r_e \ll R_{EN})$$

Here, $R_{EN} = R_E \| r_{ce} = 4.7 \text{ k}\Omega \| 100 \text{ k}\Omega = 4.5 \text{ k}\Omega$ and $r_e = 13$ Ω. v_o becomes

$$v_o = A_v \times v_{in} = \underline{0.1 \text{ V}}$$

To find r_{in},

$$r_{in} = R_{BN} \| r_{inB} = R_1 \| R_2 \| \beta(r_e + R_{EN})$$
$$r_{in} = (100 \text{ k}\Omega \| 100 \text{ k}\Omega) \| 200(13 + 4500) = \underline{47.4 \text{ k}\Omega}$$

208 THE TRANSISTOR AS AN AMPLIFIER (SMALL SIGNAL)

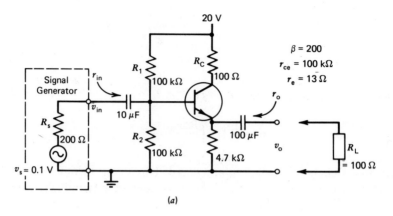

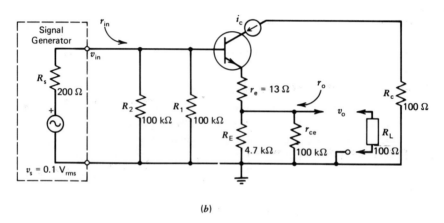

FIGURE 5-26 EF circuit for Ex. 5-13. (a) Circuit. (b) Equivalent circuit (no load).

r_o is given by

$$r_o = \frac{R_s \| R_{BN}}{\beta} + r_e \approx \frac{R_s}{\beta} + r_e$$

$$= \frac{200}{200} + 13 = \underline{14 \; \Omega}$$

(Note that $R_{BN} \gg R_s$.)

(b) To find v_o and r_{in} with R_L connected, the previous computations are repeated with R_L made a part of R_{EN}

$$R_{EN} = R_E \| r_{ce} \| R_L$$

$$R_{EN} = 47 \text{ k}\Omega \| 100 \text{ k}\Omega \| 100 \; \Omega \approx 100 \; \Omega$$

$$(R_{EN} \approx R_L)$$

Hence,

$$A_v = \frac{R_{EN}}{r_e + R_{EN}} \approx \frac{R_L}{r_e + R_L} = \frac{100}{13 + 100} = \underline{0.88}$$

and

$$v_o = v_{in} \times A_v = 0.1 \times 0.88 = \underline{0.088 \text{ V}}$$
$$r_{in} = R_{BN}\|\beta(r_e + R_{EN})$$
$$r_{in} = R_{BN}\|\beta(r_e + R_L) = 50\text{ k}\Omega\|200(13 + 100)$$
$$r_{in} = \underline{15.6 \text{ k}\Omega}$$
$$(R_{BN} = 100\text{ k}\Omega\|100\text{ k}\Omega = 50\text{ k}\Omega)$$

A_v and v_o for this case may also be found by using the parameters obtained in part (a). In part (a), $r_o = 14\ \Omega$ and $v_o = 0.1$ V. We now replace the emitter output portion with its Thevenin's equivalent (Fig. 5-27) and calculate v_o (with R_L) by voltage division

$$v_o = \frac{0.1\ R_L}{r_o + R_L} = \frac{0.1 \times 100}{14 + 100} = 0.88$$

r_o need not be recomputed in part (b), since the connection of R_L does *not* affect r_o. $r_o = 14\ \Omega$ as in part (a).

(c) To find A_i, the data obtained in part (b) and Eq. (5-21) may be used: $A_v = 0.88$, $r_{in} = 15.6\text{ k}\Omega$, $R_L = 100\ \Omega$

$$A_i = 0.88 \times \frac{15.6 \times 10^3}{100} = 137.$$

Note that $A_i < \beta$. A_i cannot exceed β.

5-6 SUMMARY OF SINGLE STAGE TRANSISTOR AMPLIFIERS

The following discussion and table is intended to give only a general view of the three transistor configurations. The formulae presented are general only and may be applied to particular circuits only after clearly understanding the meaning of the different symbols and how they relate to the specific circuit. The terms high, low, etc. used in Table 5-1 are relative only, and even then only for the run-of-the-mill circuits. The methods used to obtain the various formulae relied on a uniform device equivalent circuit. The performance of the different circuits, however, varies substantially.

210 THE TRANSISTOR AS AN AMPLIFIER (SMALL SIGNAL)

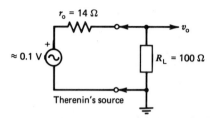

FIGURE 5-27 Thevenin's equivalent of output section from Fig. 5-26.

The following general comment with regard to the impedance transformation between the base and the emitter may give the student some added insight into the analysis. In general, when "looking" into the base, we "see" the emitter to ground resistance *multiplied* by β. When "looking" into the emitter, we "see" the base to ground resistance *divided* by β. These two observations are schematically illustrated in Fig. 5-28.

TABLE 5-1 SUMMARY OF FORMULAE

	CE Circuit	CB Circuit		CC (EF) Circuit
Voltage gain A_v	$-\dfrac{R_{CN}}{r_e + R_E}$ (high)	$\dfrac{R_{CN}}{\dfrac{R_{BN}}{\beta} + r_e}$ (Fig. 4-17)	$\dfrac{R_{CN}}{\dfrac{R_{BN}}{\beta} + R_E + r_e}$ (Fig. 4-18)	$\dfrac{R_{EN}}{r_e + R_{EN}} < 1$
Input resistance r_{in}	$R_{BN} \| \beta(r_e + R_E)$ (medium)	$\left(r_e + \dfrac{R_{BN}}{\beta}\right) \| R_E$ (Fig. 4-17) (low)	$r_e + \dfrac{R_{BN}}{\beta} + R_E$ (Fig. 4-18) (low)	$R_{BN} \| \beta(r_e + R_{EN})$ (medium to high)
Current gain A_i	$A_v \times \dfrac{r_{in}}{R_L} < \beta$	$A_v \times \dfrac{r_{in}}{R_L} < 1$		$A_v \times \dfrac{r_{in}}{R_L} < \beta$
Output resistance r_o	$R_C \| r_{ce}$ (medium)	$R_C \| r_{ce}$ (medium)		$R_E \| \left(r_e + \dfrac{R_s \| R_{BN}}{\beta}\right)$ (low)

Notes: The following explanation refers to the ac equivalent circuits:
  R_{CN} total resistance between collector to ground ($R_C \| r_{ce} \| R_L$).
  R_{BN} total resistance between base to ground (R_B or $R_1 \| R_2$, etc.).
  R_{EN} total resistance from emitter to ground ($R_E \| r_{ce} \| R_L$).

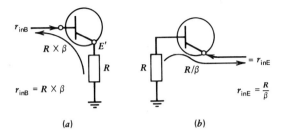

FIGURE 5-28 Base-emitter impedance transformations. (a) "Looking" into the base. (b) "Looking" into the emitter (ideal emitter).

5-7 GRAPHICAL ANALYSIS AND DISTORTION

One of the basic assumptions underlying the foregoing analysis is that the transistor is a linear device or can be considered linear in the operating range, namely, the active region. Linearity means that the gain of the transistor circuit does not depend on the amplitude of the input signal. A 0.1 V signal must be amplified precisely as much as a 1.0 V signal is. This implies that β is constant in this region and does not depend on I_c. In previous discussions it was noted, however, that β does vary somewhat with changes in the operating point. β does depend somewhat on I_c.

In addition, the dynamic resistance, which also affects the gain of the circuit, also depends upon I_C. The analysis, so far, assumed small ac signals and hence, a small ΔI_C or small collector current variations. This made the assumption that β and r_e are constant a reasonable one.

When large signals are applied to the transistor, we might expect different amplification factors to be applied to different portions of the ac waveform. Since r_e *decreases* with increases in I_C, the upper (higher base current and collector current) portion of the input waveform would be amplified more than the lower or negative portions. This behavior is demonstrated graphically by use of the load line (see Sec. 4-8).

The dc load line for the circuit in Fig. 5-29a is shown in Fig. 5-29b, superimposed on the collector characteristics of the transistor used. The dc load line represents the KVL equation

$$V_{CC} - I_C R_C - V_{CE} = 0$$
$$20 - I_C \times 2000 - V_{CE} = 0$$
$$20 - V_{CE} = I_C \times 2000$$

This is a linear (straight line) relation between I_C and V_{CE}. Two points are required for the plot of this equation. (Note the two points $V_{CE} = 20$ V, $I_C = 0$ and $V_{CE} = 0$, $I_C = I_{C(max)} = 10$ mA.) With dc analysis I_{BQ} is found as $I_{BQ} = 50$ μA. The

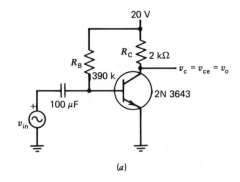

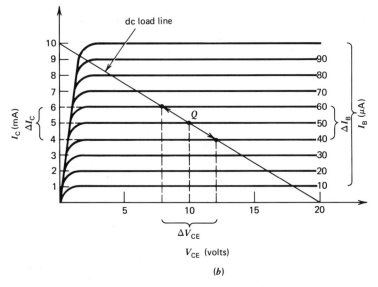

FIGURE 5-29 The dc load line (in a CE configuration). (a) The circuit. (b) Collector characteristics with dc load line.

quiescent operating point, the Q-point, is found at the intersection of the load line and the line $I_B = 50$ μA. (See Fig. 5-29b.)

The ac signal, applied to the base, causes a change in the base current ΔI_B, which in turn produces a change in I_C, ΔI_C, and ΔV_{CE}. In Fig. 5-29b we have used, as an example, $\Delta I_B = \pm 10$ μA (from 50 to 40 μA and up to 60 μA) and obtained

$$\Delta I_C = \pm 1 \text{ ma}$$
$$\Delta V_{CE} = \pm 2 \text{ V}$$

This value of ΔV_{CE} is consistent with Eq. (5-15) with r_{ce} neglected. From Eq. (5-15)

$$v_o = -i_c \times R_C \quad \text{(neglecting } r_{ce}\text{)}$$

In Fig. 5-29 $v_o = v_{CE}$. Since $v_o (= v_{CE})$ and i_c are ac quantities that can be represented by ΔV_{CE} and ΔI_C, respectively, Eq. (5-15) may be rewritten as

$$\Delta V_{CE} = -\Delta I_C \times R_C$$

In the example used in Fig. 5-29 $\Delta I_C = 1$ mA, $R_C = 2$ kΩ, and $\Delta V_{CE} = -\Delta I_C \times R_C = -10^{-3} \times 2 \times 10^{+3} = -2$ V. For $\Delta I_C = -1$ mA, a decrease in I_C, ΔV_{CE} becomes $+2$ V. (Note that the ∓ 2 V change in V_{CE} was caused by a ± 10 µA change in I_B.)

It is worth noting that the slope of the load line that is defined as $\Delta I_C/\Delta V_{CE}$ can be found from Eq. (5-32) as

$$\text{slope} = \frac{\Delta I_C}{\Delta V_{CE}} = -\frac{1}{R_C} \quad (5\text{-}32a)$$

For $R_C = 2$ kΩ, as in Fig. 5-29, the slope becomes

$$-\frac{1}{2 \text{ k}\Omega} = -0.5 \text{ m}\mho \quad \text{(or } -0.5 \text{ mA per volt)}$$

For a one-volt change in V_{CE}, I_C must change by 0.5 mA. The load line equation must always be satisfied. As a result, when I_B changes, it appears as if the operating point slides up and down the load line.

A *time* axis may be added to the plot of Fig. 5-29b as shown in Fig. 5-30. The ac base signal consists of a current sine wave with an amplitude of 10 µA. The resultant ΔI_C and ΔV_{CE} are current and voltage sine waves with 1 mA and 2 V amplitudes, respectively.

From Fig. 5-30 it is clear that the maximum, undistorted ΔI_C is ± 5 mA (0 to 10 mA), the maximum ΔV_{CE} is ± 10 V (0 to 20 V), and the maximum, undistorted ΔI_B is ± 50 µA (0 to 100 µA). Should ΔI_B exceed this maximum, the result is distortion. A portion of the wave will be "clipped" off (Fig. 5-30), since the ac input base current causes both saturation and cut off.

In Fig. 5-30, the constant base current lines (for example, $I_B = 10$ µA, 20 µA, etc.) were evenly spaced throughout. This implies a constant β. The β for ΔI_B of 10 µA, from $I_B = 80$ µA to $I_B = 90$ µA, equals

$$\beta = \frac{\Delta I_C}{\Delta I_B} = \frac{1 \text{ mA}}{10 \text{ µA}} = 100$$

For the same $\Delta I_B = 10$ µA, evaluated this time in the range of $I_B = 10$ to 20 µA;

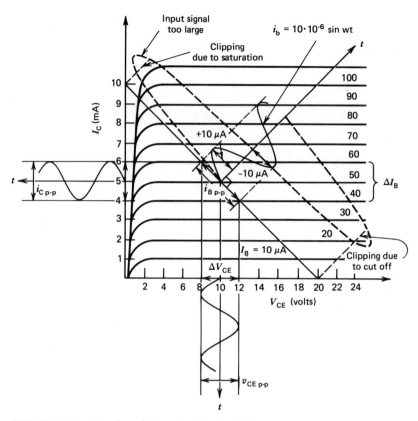

FIGURE 5-30 Signal analysis using the load line.

we again obtain

$$\beta = \frac{\Delta I_C}{\Delta I_B} = \frac{1 \text{ mA}}{10 \text{ μA}} = 100$$

β is indeed constant for the characteristic curves of Fig. 5-30.

The fact that β was assumed constant (and the collector characteristics plotted accordingly) results in distortion-free amplification, other than saturation and cut off as noted before.

The collector characteristics obtained by actual measurements are more likely to be somewhat nonlinear. Figure 5-31 shows, somewhat exaggerated, nonlinear characteristics. Note that β varies throughout. Around $I_B = 20$ μA

$$\beta \approx \frac{\Delta I_C}{\Delta I_B} = \frac{1.5 \text{ mA} - 0.75 \text{ mA}}{20 \text{ μA} - 10 \text{ μA}} = 75$$

while around $I_B = 60 \ \mu A$,

$$\beta = \frac{8.4 \ mA - 6 \ mA}{60 \ \mu A - 50 \ \mu A} = 240$$

A ΔI_B signal of $\pm 10 \ \mu A$, as shown in Fig. 5-31, produces an output that is asymmetrical. ΔI_C changes by 2.4 mA in one direction (6 to 8.4 mA) and by -1.5 mA (6 to 4.5 mA) in the other direction. (Similar asymmetry is exhibited by ΔV_{CE}.) If ΔI_B, the input signal is a sine wave with peak to peak of 20 μA ($\pm 10 \ \mu A$), the output is not a sine wave anymore (or a very distorted sine wave). The ΔI_C and ΔV_{CE} shown in Fig. 5-31 are a result of a sinusoidal base signal. It is important to note that had ΔI_B been kept very small, in the order of 1 μA, the distortion would have been almost nonexistant, since β over that small a range can safely be assumed constant. In other words, the transistor behaves quite linearly over the small signal $\pm 1 \ \mu A$.

The distortion associated with nonlinearities in transistor characteristics within the active region is by far more pervasive and difficult to eliminate than the one caused by saturation and cut off. The fidelity of the hi-fi amplifier, for example, is largely dependent on the minimization of these nonlinearities. Transistors and

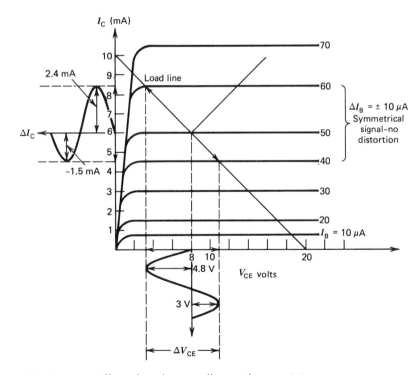

FIGURE 5-31 Effect of nonlinear collector characteristics.

5-8 THE AC LOAD LINE

In the previous discussion, the ac collector voltage ΔV_{CE} ($= v_{CE}$) was obtained graphically from a plot of the dc load line relating V_{CE} to I_c. In this plot only R_C plays a role. In practice, the output voltage v_{CE} is dependent on the total ac resistance R_{CN} collector to ground, including the effects of a load resistor R_L that may be ac (capacitively) coupled to the collector. Consequently, Eq. (5-32) must be rewritten

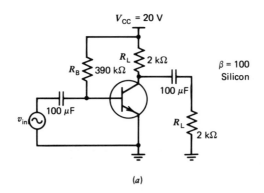

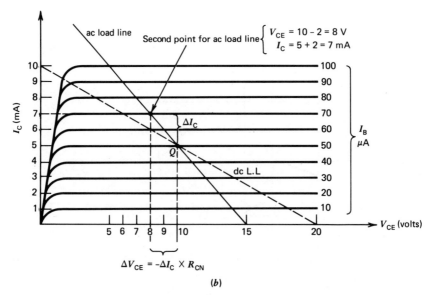

FIGURE 5-32 ac Load Line. (a) Circuit. (b) Load lines.

to include all resistance or equivalent resistance from collector to ground

$$v_o = -i_c \times R_{CN} \tag{5-32b}$$

Equation (5-32a) indicates that the ac load line must have a slope of

$$\text{Slope} = \frac{\Delta I_c}{\Delta V_{CE}} = -\frac{1}{R_{CN}} \tag{5-32c}$$

R_{CN} is typically $R_C \| R_L$

To obtain the ac load line, it is first necessary to obtain the Q-point since the ac load line must pass through this point. It is then possible to draw a line through the Q-point with a slope of $-1/R_{CN}$. Since the load line is a straight line, only two points are required for its plot. One point is the Q-point. The other can be obtained by selecting an arbitrary value of ΔI_c and finding the corresponding ΔV_{CE} from Eq. (5-32b). The ac load line, which represents Eq. (5-32b), gives a relation between the ac quantities ΔI_c and ΔV_{CE}, while the dc load line [Eq. (5-32)] relates the dc values I_C to V_{CE}. Once the ac load line has been obtained, the graphical signal analysis is the same as that for the dc load line.

Figure 5-32b shows both the dc and ac load lines for the circuit shown in Fig. 5-32a. Note that here

$$R_C = 2\ k\Omega, \qquad R_L = 2\ k\Omega$$

and

$$R_{CN} = R_C \| R_L = 1\ k\Omega$$

The ac load line passes through the Q-point, $V_{CQ} = 10$ V, $I_{CQ} = 5$ mA. A second point on the line is obtained by choosing a $\Delta I_C = 2$ mA (quite arbitrarily) and finding ΔV_{CE}

$$\Delta V_{CE} = -\Delta I_C \times R_{CN} = -2 \times 10^{-3} \times 10^3 = -2\ \text{V}$$

EXAMPLE 5-14

Find the ac load line and v_o, graphically, for the circuit and collector characteristics given in Figs. 5-33a and 5-33b. Assume the ac signal produces a peak-to-peak base current of 100 μA (±50 μA), and that the dc resistance of the transformer windings is zero.

SOLUTION

To obtain the ac load line, the operating point, Q, must first be found based upon dc considerations. The dc equivalent circuit is shown in Fig. 5-33c, and the dc load line is shown superimposed on the collector characteristics. Since it was

218 THE TRANSISTOR AS AN AMPLIFIER (SMALL SIGNAL)

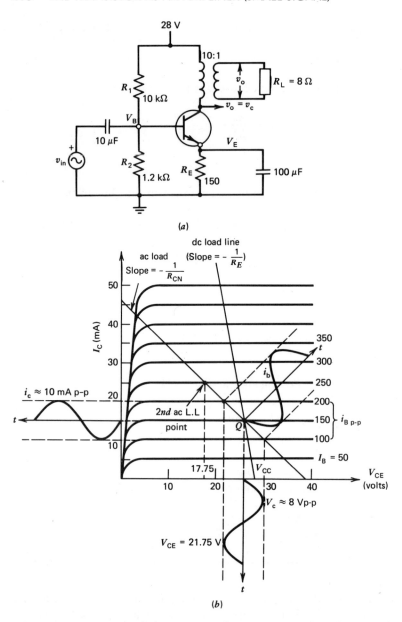

FIGURE 5-33 Circuit and characteristics for Ex. 5-14. (a) Actual circuit. (b) Collector characteristics with superimposed graphical analysis. (c) dc equivalent circuit. (d) ac equivalent circuit ('actual' transistor not its ac dynamic equivalent).

assumed that the dc resistance of the primary of the transformer is zero, the dc load line equation becomes

$$V_{CC} - I_C R_E = V_{CE} \quad (I_E \approx I_C)$$

THE AC LOAD LINE

FIGURE 5-33 (Continued)

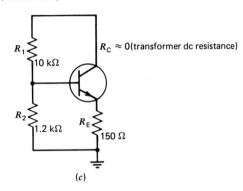

(c)

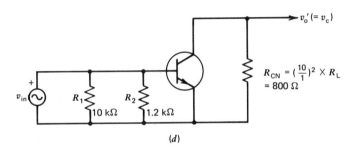

(d)

The plot of this equation gives the dc load line. To find the Q-point, we compute I_{CQ}, assuming a high enough β so that V_B can be obtained by simple voltage division.

$$V_B = \frac{28 \times 1.2 \text{ k}\Omega}{10 \text{ k}\Omega + 1.2 \text{ k}\Omega} = 3 \text{ V}$$

$$V_E = V_B - 0.7 = 3.0 - 0.7 = 2.3 \text{ V}$$

$$I_E = \frac{V_E}{R_E} = \frac{2.3}{150} = 15.3 \text{ mA}$$

$$I_{CQ} \approx I_E = 15.3 \text{ mA} \approx 15 \text{ mA}$$

$$V_{CEQ} = V_{CC} - I_{CQ} \times R_E = 28 - 15 \times 10^{-3} \times 150 = 25.7 \text{ V}$$

The Q-point is the intersection of $I_C = 15$ mA with the dc load line. To find the ac load line, we find R_{CN}, the ac resistance transformed through a transformer with a 10:1 turns ration.

$$R_{CN} = \left(\frac{10}{1}\right)^2 \times 8 = 800 \text{ }\Omega$$

Selecting $\Delta I_C = 10$ mA (quite arbitrarily) and calculating ΔV_{CE}, we get

$$\Delta V_{CE} = -10 \times 10^{-3} \times 800 = -8.0 \text{ V}$$

220 THE TRANSISTOR AS AN AMPLIFIER (SMALL SIGNAL)

The second point for the ac load line is

$$I_C = I_{CQ} + \Delta I_C = 15 + 10 = 25 \text{ mA}$$
$$V_{CE} = V_{CEQ} + \Delta V_{CE} = 25.7 + (-8) = 17.7 \text{ V}$$

The base input signal i_B is shown swinging between 100 and 200 µA (± 50 µA), producing corresponding v_C and i_C. Note that $v_C = v'_o$ where v'_o is the voltage across the primary of the transformer (v_C is not v_o). To get v_o, we transform v'_o to the secondary of the transformer.

$$v_o = \frac{1}{10} v'_o = \frac{1}{10} \times 8 = \underline{0.8 \text{ V}_{p-p}}$$

5-9 THE TRANSISTOR AS A SWITCH

The preceding discussion of transistor operation was restricted to transistors operating in the active region (never reaching saturation or cut off). Some very popular and widespread applications, however, involve both saturation and cut off. We refer, of course, to digital applications. For these applications circuits are operated between two, and only two, voltage levels; we will call them V_{High} and V_{Low} (V_H, V_L). Signals are either "high" or "low." Intermediate voltage levels are not permitted. ("High" may be defined as a voltage *above* a certain value and "low" as a voltage *below* a certain value.)

The transistor may be switched between saturation and cut off to produce the "low" and "high" levels in response to a digital input signal. For the purposes of this discussion we choose to define V_H as 2.4 V, or higher, and V_L as less than 0.4 V. (In most digital integrated circuits the definitions are V_H = 2.4 to 5 V and V_L = 0 to 0.4 V.) The analysis concerns itself with only these two voltage levels. Figure 5-34 shows a simple transistor circuit (of the common emitter type) with an input signal alternating between V_H and V_L. To obtain v_o the circuit must be analyzed for each of the input levels separately. (Note that the analysis considers the dc operation rather than ac.)

For $v_{in} = V_H$, that is for the time period during which v_{in} = 2.4 V or more, the circuit looks like Fig. 5-35a. Under these conditions the transistor is expected to be "on," saturated, resulting in $v_o \approx 0$ V or $v_o = V_L$ (collector-emitter are effectively shorted). To obtain the required value of R_B, note that

$$I_{C(sat)} = V_{CC}/R_C \tag{5-33}$$

The *minimum* base current required to produce this current is

$$I_{B(sat)} = \frac{I_{C(sat)}}{\beta} = \frac{V_{CC}}{R_C}/\beta = \frac{V_{CC}}{R_C \times \beta} \tag{5-34}$$

The term *minimum* base current is used because any base current larger than the

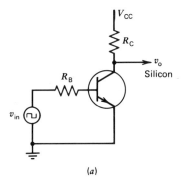

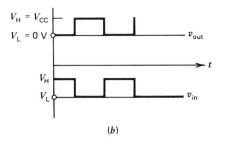

FIGURE 5-34 Transistor as a switch. (a) Circuit. (b) Input and output waveforms.

one obtained from Eq. (5-34) will result in saturation. Since V_H must cause saturation (produce a base current equal to or larger than $I_{B(sat)}$), we may write (by KVL)

$$V_H - I_{B(sat)} \times R_B - V_{BE} = 0$$

$$R_B = \frac{V_H - V_{BE}}{I_{B(sat)}} \quad (5\text{-}35)$$

(V_{BE} is the forward base-emitter voltage drop, about 0.7 V for silicon transistors and 0.35 V for germanium transistors.) Equation (5-35) gives the *largest* R_B that will *still* result in saturation when V_H is applied. Selecting R_B less than the value indicated by Eq. (5-35) will "over" saturate or overdrive the transistor. In most digital applications R_B is selected to overdrive the transistor, thus guaranteeing saturation for worst case conditions.

When $v_{in} = V_L$ (in our case $V_L = 0.4$ V or less), the transistor should be cut off, yielding $v_o = V_{CC}$ or $v_o = V_H$. Disregarding transistor leakage currents (which are extremely small), it is clear that in Fig. 5-35b the transistor is indeed cut off. $V_L = 0.4$ V is not sufficient to forward bias the base-emitter junction of the silicon transistor and hence, $I_B = 0$ results in $I_C = 0$ and $v_o = V_{CC}$. The output and input waveforms are shown in Fig. 5-34b.

222 THE TRANSISTOR AS AN AMPLIFIER (SMALL SIGNAL)

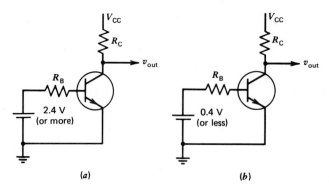

FIGURE 5-35 "High" and "Low" equivalent circuits. (a) $V_{in} = V_H$. (b) $V_{in} = V_L$.

EXAMPLE 5-15

The logic levels of a system (digital voltage levels) are defined as

$$V_H = 2.4 \text{ to } 5 \text{ V}$$
$$V_L = 0 \text{ to } 0.6 \text{ V}$$

(a) Select a suitable (maximum) R_B from the circuit in Fig. 5-36a.
(b) Can a germanium transistor be used, and if not, how can the circuit be changed to allow the use of a germanium transistor?

SOLUTION

To account for "worst" case situations, we use $V_H = 2.4$ V and $V_L = 0.6$ V. (In contrast, $V_H = 5$ V and $V_L = 0$ V represent a "best" case condition.)
(a) From Eq. (5-33)

$$I_{C(sat)} = \frac{V_{CC}}{R_C} = \frac{10}{1 \text{ k}\Omega} = 10 \text{ mA}$$

and

$$I_{B(sat)} = \frac{I_{C(sat)}}{\beta} = \frac{10 \times 10^{-3}}{100} = 100 \times 10^{-6} \text{ A} = 100 \text{ μA} \quad (5\text{-}34)$$

and

$$R_B = \frac{V_H - 0.7}{I_{B(sat)}} = \frac{2.4 - 0.7}{100 \times 10^{-6}} = 17 \text{ k}\Omega \quad (5\text{-}35)$$

R_B must be 17 kΩ or less to produce saturation for the minimum V_H.
(b) Should a germanium transistor be used, with a $V_{BE} = 0.35$ V (forward bias base-emitter voltage) and $R_B = 17$ kΩ. The transistor will *not* be cut off with v_{in}

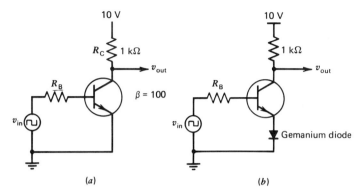

FIGURE 5-36 Circuits for Ex. 5-15. (a) Silicon transistor. (v) Germanium transistor.

$= V_L = 0.6$ V. Applying KVL (for the interval when $V_{in} = 0.6$ V), we get

$$0.6 - I_B \times 17 \text{ k}\Omega - 0.35 = 0$$

$$I_B = \frac{0.6 - 0.35}{17 \text{ k}\Omega} = 14.7 \times 10^{-6} = 14.7 \text{ }\mu\text{A}$$

and

$$I_C = \beta \times I_B = 100 \times 14.7 \times 10^{-6} = 1.47 \text{ mA}$$

A possible solution is shown in Fig. 5-36b. The germanium diode in the emitter increases the voltage required to produce a non-zero I_B to $0.35 + 0.35 = 0.7$ V. (Both the B-E junction and the series diode must be forward biased.) No base current (and no collector current) can flow with only 0.6 V (worst case V_L) applied to the base. The transistor stays cut off with V_L applied to the base, resulting in $v_o = V_H$ as desired.

Transistors can also be used as switches in analog applications, effectively serving as substitutes for mechanical switches that control the flow of signals (ac or dc). In recent years, however, this application has been almost completely taken over by field effect transistors. A discussion of analog switching applications will be presented as it applies to field effect transistors (Ch. 6).

SUMMARY OF IMPORTANT TERMS

Amplifier A circuit used to process signals. Usually, the current output, voltage output, or both are larger than the input current or input voltage, respectively.

Current gain (A_i) The ratio of the output signal current (signal current through the load) to the input signal current in an amplifier circuit.

Dynamic resistance (r_e) The dynamic resistance of the B-E junction, forward biased. Used in the dynamic equivalent circuit of the transistor.

Hybrid equivalent circuit A transistor equivalent circuit representing the transistor in a

224   THE TRANSISTOR AS AN AMPLIFIER (SMALL SIGNAL)

CE configuration. It utilizes a Thevenin's equivalent for the input (base) circuit and a Norton's equivalent for the output (collector) circuit.

Hybrid parameters   The various parameters (h_{fe}, h_{re}, h_{oe}, h_{ie}) used in the hybrid equivalent circuit.

Input resistance (r_{in})   The resistance "seen" when "looking" into the input terminals of an amplifier.

Load—line—ac   A graphic method used to represent the various ac voltages and currents in a transistor.

Output resistance (r_o)   The Thevenin's (or Norton's) resistance of the output section of an amplifier. (The equivalent of the output section is a Thevenin's or Norton's circuit.)

Power gain   The ratio of signal power output to signal power input P_o/P_{in}.

Voltage gain (A_v)   The ratio of output signal voltage to input signal voltage in an amplifier circuit.

h_{FE} ($= \beta_{dc}$)   Defined as $\dfrac{I_C}{I_B}$ (dc current ratio).

h_{fe} ($= \beta_{ac}$)   Defined as $\dfrac{\Delta I_c}{\Delta I_B}$ or $\dfrac{i_c}{i_b}$ (ac current ratio).

h_{ie} The input resistance of the transistor in a CE configuration (used in the hybrid equivalent circuit).

h_{oe} $\left(= \dfrac{1}{r_{oe}}\right)$ The collector-emitter ac conductance of a transistor in a CE configuration. Used in the hybrid equivalent circuit.

h_{re} The reverse voltage feedback factor in a CE configuration. Used in the hybrid equivalent circuit.

$I_{B(sat)}$ The base current necessary to produce $I_{C(sat)}$.

$I_{C(sat)}$ The saturation collector current. Maximum collector current limited by V_{CC} and $R_C + R_E$.

r_{ce} See h_{oe}.

PROBLEMS

5-1.   Given an amplifier with the following characteristics (stand alone):
Amplifier 1

$$r_{in} = 3 \text{ k}\Omega$$
$$A_v = 25$$
$$r_o = 1 \text{ k}\Omega$$

The amplifier is used to drive earphones with $r_{phones} = 100\ \Omega$ from a magnetic tape recording head with a source resistance of $R_s = 8\text{ k}\Omega$. (See Fig. 5-37.)

(a) Find the total voltage gain A_v. (From unloaded source to output across the load.)
(b) Find the total current gain (A_i).
(c) Find the total power gain (A_p).

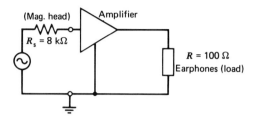

FIGURE 5-37

5.2. A preamplifier and postamplifier are inserted in the system of Fig. 5-37 as shown in Fig. 5-38. The characteristics of the two added amplifiers are identical

$$r_{in} = 200 \text{ k}\Omega$$
$$A_v = 1$$
$$r_o = 10 \text{ }\Omega$$

(a) Find the total voltage gain (A_v).
(b) Find the total current gain (A_i).
(c) Find the total power gain (A_p).

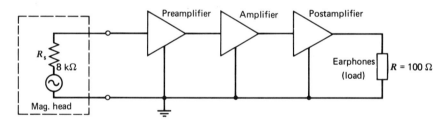

FIGURE 5-38

5-3. Assume that in problems 5-1 and 5-2 the source produces 100 µV rms under no load conditions.

(a) Find the output voltage (v_o) and output power (P_o) for the circuit in Problem 5-1.
(b) Find v_o and P_o for the circuit in Problem 5-2.

5-4. Find A_v, r_{in}, and r_o for the circuit of Fig. 5-39. (Compute r_e first.) Draw the equivalent circuit.

5-5. Find v_o, i_o, r_{in}, and r_o for Fig. 5-40.

5-6. Find A_v, A_i, r_{in}, and r_o for Fig. 5-41.

5-7. For the circuit of Fig. 5-42:

(a) Draw the ac equivalent circuit.
(b) Calculate A_v, A_i, r_{in}, and r_o.

226 THE TRANSISTOR AS AN AMPLIFIER (SMALL SIGNAL)

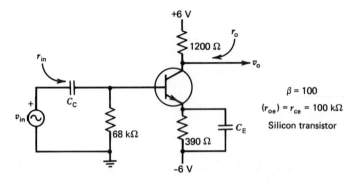

FIGURE 5-39

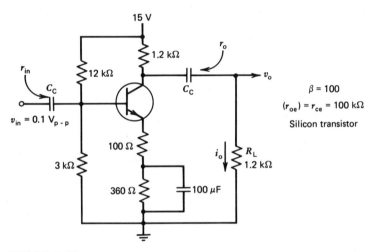

FIGURE 5-40

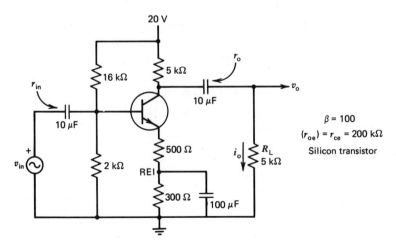

FIGURE 5-41

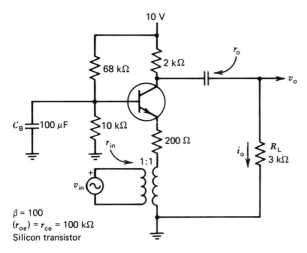

FIGURE 5-42

5-8. For the circuit of Fig. 5-43:

    (a) Draw the ac equivalent circuit.
    (b) Find v_o, i_o, r_{in}, and r_o.

5-9. For Fig. 5-44, Calculate A_v, r_{in}, r_o, and v_o.

5-10. In Fig. 5-44 connect a 100 Ω load across the output ($R_L = 100\ \Omega$) and assume that the source has an internal resistance $R_s = 500\ \Omega$. (Neglect the effect of R_s on the voltage gain.)

    (a) Calculate r_o, r_{in}.
    (b) Calculate A_v, A_i, A_p.
    (c) Calculate v_o, i_o, P_o.

5-11. In Fig. 5-45 the operating point is $V_{CQ} = 1/2 V_{CC}$, $I_{CQ} = 5$ mA.

$$V_E = 2\ \text{V}.\ (\approx 3 \times V_{BE})$$

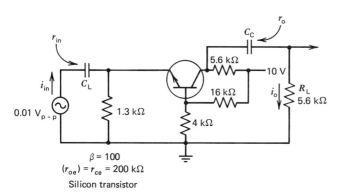

FIGURE 5-43

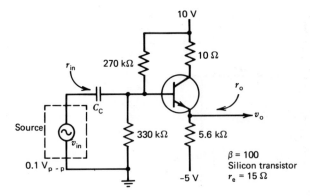

FIGURE 5-44

The parameters of the stage are

$$A_v = -10$$

$$R_2 = \beta(R_{E1} + R_{E2})/10 \quad \begin{pmatrix} \text{Explain why } R_2 \text{ is selected so} \\ \text{that } R_2 \ll \beta(R_{E1} + R_{E2}) \end{pmatrix}$$

$$R_1 = 6 \text{ k}\Omega$$

(a) Find R_C, R_{E1}, R_{E2}, R_1, R_2.
(b) Find A_i, r_{in}, and r_o for your circuit.
(c) Find C_E so that $X_{CE} = \dfrac{R_{E1} + r_e}{10}$ at 100 Hz.

Hint: You must use both ac and dc criteria. (You may use reasonable approximations.)

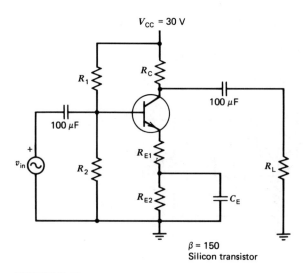

FIGURE 5-45

5-12. (a) For Problem 5-11, replace all values for R_C, R_{E1}, R_{E2}, R_1, R_2 with standard 20% resistors and calculate Av, r_{in} and r_o.
(b) Compare the results of Problem 5-11 with those of Problem 5-12.

5-13. Find A_v, A_i, r_{in}, and r_o for the circuit of Fig. 5-46.

5-14. (a) Draw the dc and ac load lines for the circuit of Fig. 5-46 (draw approximate collector characteristics so that $\beta = 100$ for all values of I_B.)
(b) What is the maximum v_{in} and v_o before clipping sets in. Draw the signal input and output on the same plot as part (a), given that $i_B = 20 \times 10^{-6} \sin \omega t$ and $I_{BQ} = 45 \mu A$. (Use Av = -9.5)

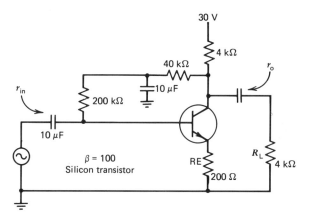

FIGURE 5-46

5-15. For the circuit of Fig. 5-47 calculate r_{in}, r_o, A_v, and A_i.

5-16. The input voltage for Problem 5-15 is 0.05 V_{rms}. Find v_o, i_o.

5-17. An emitter follower (EF) is used to drive an 16 Ω speaker. The source has a high internal resistance requiring $r_{in} \geq 5$ kΩ. It is required that the voltage gain (base to emitter) be 0.8. Use $R_E = 470$ Ω with a fixed bias, and $V_{CC} = 20$ V.

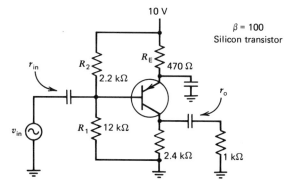

FIGURE 5-47

230   THE TRANSISTOR AS AN AMPLIFIER (SMALL SIGNAL)

(a) Sketch the circuit.
(b) Determine the required minimum β.
(c) Find R_B (the biasing resistor) assuming the β is that found in part (b).

5-18. In Fig. 5-48 $|v_{o1}| = |v_{o2}|$, $I_{CQ} = 6.5$ mA

(a) Find R_E and R_C.
(b) Plot v_{o1} and v_{o2} showing the relative phase. (Assume a sinusoidal input signal.)
Hint: To get $v_{o1} = v_{o2}$ it is necessary to make $R_C = R_E$ (explain why).

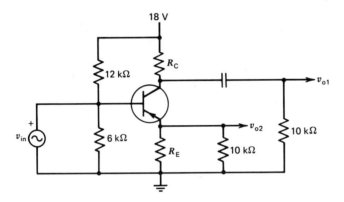

FIGURE 5-48

5-19. For the circuit of Fig. 5-49 find:

(a) A_v.
(b) r_{in}.
(c) A_i.
(d) r_o.

5-20. For the circuit of Fig. 5-49 find:

(a) v_o.
(b) i_o.

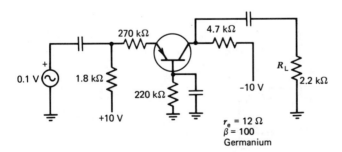

FIGURE 5-49

5-21. In the circuit of Fig. 5-50 the input voltage switches between 0.4 to 2.4 V.

(a) With $R_L = 5.6$ kΩ and $\beta = 100$, determine R_B so that saturation is just reached (when $v_{in} = 2.4$ V).
(b) With $R_L = 5.6$ kΩ and $R_B = 250$ kΩ, find the minimum β that will produce saturation.

5-22. In Fig. 5-50 v_{in} varies between 0.4 to 2.4 V. $R_L = 5.6$ kΩ, $R_B = 47$ kΩ, $\beta = 100$.

(a) Find v_o corresponding to the two input levels.
(b) If v_{in} is changed to 1.0 to 2.4 V, will the transistor cut off? Find v_o.
Hint: To find v_o for the case where $v_{in} = 1.0$ V, replace the voltage divider $R_C - R_L$ and V_{CC} by its Thevenin's equivalent. $V_{Th} = 2.5$ V, $R_{Th} = 2.8$ kΩ.

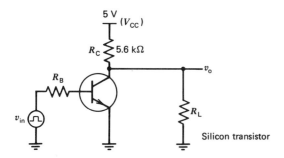

FIGURE 5-50

CHAPTER

6

FIELD EFFECT TRANSISTORS (FET)

6-1 CHAPTER OBJECTIVES

In this chapter the student will learn about the physics of the various types of Field Effect Transistors (FETs) and their circuit applications. The FETs covered include the Junction FET (JFET), the Metal Oxide FET (MOSFET), and the Vertical MOSFET (VMOSFET). Both enhancement and depletion modes of operation are presented.

The student will gain the ability to analyze and design different types of biasing circuits, and to find the operating point using Shockley's equation and its graphic representation. Both single and double supply biasing problems are explained.

For ac (small signal) operation, the chapter presents methods of analysis for the typical JFET amplifier configurations, such as common source and common drain (source follower). The calculations of various parameters, voltage gain, input resistance and output resistance, are demonstrated, showing the student how to compute these parameters for a variety of circuits, and giving him or her an understanding of their importance. The student is taught to find g_{mQ} (the operating g_m) after the Q-point has been determined.

The last part of this chapter presents a simple introduction to switching applications, both digital and analog.

6-2 INTRODUCTION

The field effect transistor (FET), like the bipolar transistor (Chap. 4), is a semiconductor device. The principles of operation of the FET, however, differ substantially from those of the bipolar transistors. One of the major functional differences is that, in the FET, the current through the device is controlled by a *voltage* (which sets up an electric field) rather than by a current as in the bipolar transistor.

In fact, the control terminal of the FET, called the *gate*, requires *no current*. Any current that might flow in the gate circuit is a result of leakage, and does not affect the controlled (output) current through the device. As a result, the input resistance (into the gate) is inherently very high: 10^7–10^{12} Ω for the gate of the FET, compared to 10^3–10^4 Ω for the bipolar transistor base.

Since we are dealing with a semiconductor, one may expect to encounter both N type and P type. In addition, there are two basic categories of FETs: the *junction field effect transistor* (JFET) and the *metal oxide semiconductor field effect transistor* (MOSFET). These differ in their physical structure as well as in their electrical performance.

6-3 THE FUNDAMENTALS OF THE JFET

The operation of the JFET can best be explained in terms of the variations in the depletion region of a reverse-biased diode. It is useful, therefore, to review briefly the depletion theory applied to diodes.

As noted in Chapter 2, the diode is constructed by metalurgically joining two oppositely doped semiconductor elements, P material and N material. As the P-N junction is formed, a depletion region results near the junction (Fig. 6-1a). In this region, there are very few current carriers (mobile electrons or holes). By application of a *reverse bias* to the junction (Fig. 6-1b), the width of the depletion region is increased. This phenomenon is the essence of the operation of the JFET.

The JFET consists of a doped semiconductor core called a channel, surrounded by an oppositely doped semiconductor ring called the gate. Figure 6-2 shows an N channel surrounded by a P gate. This structure is referred to as an N channel JFET. (Interchanging the N and P materials produces a P channel JFET.) The actual structure of the JFETs that are available commercially may differ somewhat from that shown in Fig. 6-2. This figure is illustrative only and permits a simple analysis of the operation of the JFET.

The P-N junction shown in Fig. 6-2 forms a diode and behaves like a diode. The depletion region depends on the bias applied to the junction. Figure 6-3a

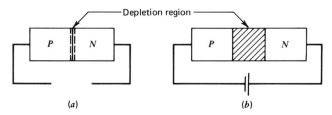

FIGURE 6-1 Diode depletion region. (a) No bias. (b) Reverse-biased junction showing enlarged depletion region.

234 FIELD EFFECT TRANSISTORS (FET)

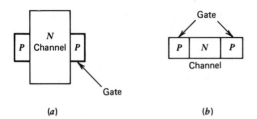

FIGURE 6-2 Basic JFET structure. (a) Side view. (b) Top view.

shows an unbiased P-N junction, while Fig. 6-3b shows a P-N junction with a large reverse bias. Note that the depletion region in the first case is minimal, leaving the channel *open* for conduction; while in the second case, the reverse bias is large enough to extend the depletion region until the channel is completely cut off. The channel is effectively blocked, not allowing conduction through the channel.

So far, bias has been applied between the channel (both ends of the channel) and the gate (G). The operation of the JFET involves the current flow through the channel from the top, which is called the *drain* (D), to the bottom, which is called the *source* (S). This requires the application of a voltage *across* the channel, between the drain and the source (V_{DS}). The operation is somewhat complicated, therefore, since the voltage that is applied to the gate to reverse bias the junction combines with V_{DS}. This produces biasing conditions that vary appreciably from the drain-gate region to the source-gate region (Fig. 6-4). The voltage between the gate and drain is $|V_{DS}| + |V_{GS}|$. (Note the polarities of the two voltages.)

If we use the values shown in Fig. 6-4a (these values were chosen for illustrative purposes only), the reverse bias *drain* to *gate* is 4.0 V. The voltage between the *source* and *gate* is 1 V (the source is grounded), yielding only a 1 volt reverse bias in this region. As a result, the depletion region is substantially larger in the drain-gate region than in the source-gate region. As Fig. 6-4a illustrates, the channel,

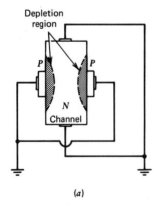
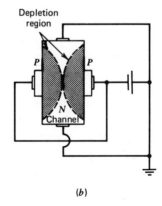

FIGURE 6-3 The depletion region in the JFET. (a) No bias. (b) Substantial reverse bias.

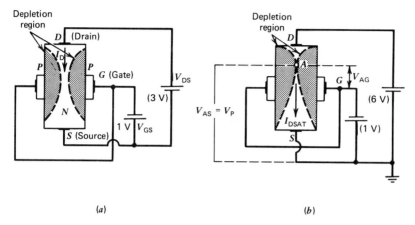

FIGURE 6-4 Bias voltages in the JFET. (a) Low V_{DS}. (b) Pinch off.

under the particular bias shown, is conducting. Current flows from drain to source, since the depletion region is not large enough to block conduction. This channel current is essentially ohmic. That is, it conforms to Ohm's law, in which the applied voltage is V_{DS}, the current is I_D, and the resistance is the channel resistance.

As V_{DS} is increased, the depletion region deepens. The depletion regions from the two sides of the channel eventually meet at point A (Fig. 6-4b). At first glance, one is tempted to conclude, erroneously, that conduction has been blocked since the channel has been "pinched off." However, the current I_D that flows under this "pinch off" condition is similar to the injected current in a bipolar transistor. It consists of carriers injected by the applied V_{DS}. To be more precise, the carriers are injected by the voltage V_{AS}, the voltage between point A and S. Note that the "pinch off" current is almost entirely dependent on this internal voltage V_{AS}. As V_{DS} is increased, beyond the voltage required to produce "pinch off," the voltage V_{AS} does not change. This can be explained by the fact that the channel resistance between the drain terminal and point A increases with any increase in V_{DS} because of the further enlargement of the depletion region. The increase in channel resistance increases the voltage drop between the drain and point A, tending to keep the voltage V_{AS} constant. As V_{DS} increases, a larger portion of this voltage is dropped across the channel between the drain and point A. Since V_{AS} remains constant, so does I_D, even beyond "pinch off." The drain-gate reverse-bias voltage (V_{DG}) that is required to produce "pinch off" (the condition in which the two depletion regions just touch) is called V_P. The drain current under "pinch off" is a saturation current, $I_{D(sat)}$. (It does not increase with increasing V_{DS}.)

In the foregoing discussion, V_{GS} was held constant and the device behavior was investigated as V_{DS} increased. Now, assume that $V_{DS} \gg V_P$. ($V_{DS} \gg V_P$ and $V_{GS} = 0$ results in $V_{DG} \gg V_P$ and "pinch off.") For $V_{GS} = 0$, the depletion region in the gate-source region is minimal. The channel resistance in this region is relatively low. As a result, $I_{D(sat)}$ (with $V_{GS} = 0$) is large. As will be seen later, this is the maximum $I_{D(sat)}$ that can flow in the transistor. It is called I_{DSS}. As the reverse bias, gate to source, V_{GS} increases, the channel resistance in this region increases, reducing the $I_{D(sat)}$. Thus, for example, if $I_{DSS} = 6$ mA (for $V_{GS} = 0$), it may be expected that for $V_{GS} = -1$ V, $I_{D(sat)} = 4$ mA, and so on. If V_{GS} is increased further (larger reverse bias), a point is reached where $I_{D(sat)} = 0$. The transistor is then cut off. This happens when $V_{GS} = V_P$—that is, when the *gate-source* region is "pinched off." (The gate-drain region was assumed "pinched off" since it was assumed that $|V_{DS}| \gg |V_P|$.) Conduction can be cut off completely by applying a $|V_{GS}| \geq |V_P|$.

To summarize, $I_{D(sat)}$ is the drain current (approximately constant) when $V_{DS} \gg V_P$. I_{DSS} is the $I_{D(sat)}$ for $V_{GS} = 0$. $I_{D(sat)}$ decreases as $|V_{GS}|$ is increased (increasing the reverse bias in the gate-source region). $I_{D(sat)}$ reaches zero as V_{GS} approaches V_P.

A reexamination of Fig. 6-4 leads to the conclusion that since the drain and the source are the same type of semiconductor (the N-type in the N channel JFET and the P-type in the P channel JFET), it should be possible to operate the transistor with drain and source interchanged. Indeed, such reversible transistors are available (2N4220 through 2N4222, Motorola MPF102 through MPF107, MPF151 through MPF156, and more). This interchangeability is usually a function of the physical structure. (Recall a similar discussion for bipolar transistors.)

6-4 JFET dc PARAMETERS AND CHARACTERISTICS

The plot of I_D versus V_{DS} (for $V_{GS} = 0$) is shown in Fig. 6-5. The zero gate to source voltage implies a minimum depletion in the gate-source region. That means the gate-source biasing (0 V) causes no increase in channel resistance. Figure 6-

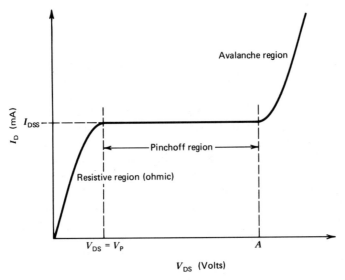

FIGURE 6-5 I_D versus V_{DS} ($V_{GS} = 0$).

JFET DC PARAMETERS AND CHARACTERISTICS   237

5, then, shows the I_D versus V_{DS} plot for the case in which the channel has the lowest possible resistance, so the drain current is the maximum possible (for the particular transistor). The value of this saturation current, I_{DSS}, is characteristic of the transistor and is usually supplied by the manufacturer. I_{DSS} *can be defined* as the drain current for the conditions: $|V_{DS}| > |V_P|$ and $V_{GS} = 0$.

The region (Fig. 6-5) for which $|V_{DS}| < |V_P|$ is referred to as the resistive or ohmic region. It is also called the linear region. In this range, the FET behaves, approximately, like a resistor and Ohm's law serves as an approximation of the I_D, V_D relation.

The region of $|V_D| > |V_P|$ is called the "pinch off" or saturation region. In this region I_D is approximately constant. As already noted (Chap. 2), the diode (P-N junction) eventually enters the avalance region when it is subject to very large reverse voltage (that is, V_{DS}). This is shown by the rapid increase in I_D as V_{DS} goes beyond point A in Fig. 6-5.

As noted before, the drain saturation current, $I_{D(sat)}$, for $V_{DS} > V_P$ depends on V_{GS}, the gate to source bias. For $V_{GS} = 0$, this current is at a maximum. (We refer to this current as I_{DSS}.) As V_{GS} increases (more reverse bias), $I_{D(sat)}$ decreases until cut off is reached. The family of curves (drain characteristics), with V_{GS} as the parameter (Fig. 6-6), shows the variations in $I_{D(sat)}$ as V_{GS} varies. For $V_{GS} = 0$ (top curve), $I_{D(sat)} = 4.9$ mA (point a in Fig. 6-6a). As V_{GS} increases (becoming more negative), $I_{D(sat)}$ decreases. For $V_{GS} = -1.0$ V, $I_{D(sat)} = 2.6$ mA (point b), and $I_{D(sat)} = 0$ for $V_{GS} \approx -3.4$ V. This means that -3.4 V applied between gate and source cuts off conduction ($I_D = 0$). Note that a larger V_{GS} (say, -5 V) is

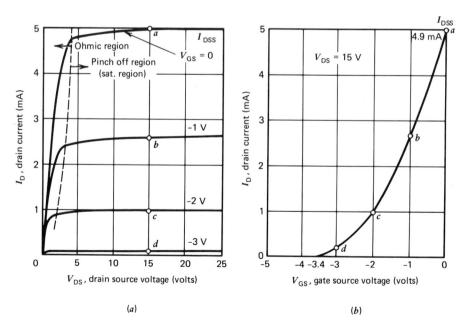

FIGURE 6-6 Characteristics of the 2N4221 JFET (courtesy of Motorola Corp).

238 FIELD EFFECT TRANSISTORS (FET)

more than enough to cut off conduction. The minimum V_{GS} that produces cut off (-3.4 V here) is the "pinch off" voltage, V_P (sometimes referred to as $V_{GS(OFF)}$).

Two important transistor parameters can be obtained from Fig. 6-6a. $I_{DSS} = 4.9$ mA and $V_P = -3.4$ V. These parameters are usually given by the transistor manufacturer and they *describe* the behavior of the particular transistor type.

6-4.1 Shockley's Equation

It is sometimes convenient to get a relation between V_{GS} and I_D directly, for the range where $|V_{DS}| > |V_P|$. This relation is referred to as the common source transfer characteristic. This plot can be obtained from the drain characteristic of Fig. 6-6a by choosing a constant V_{DS} (say, $V_{DS} = 15$ V) and reading off the I_D-V_{GS} pairs (points *a*, *b*, *c*, and *d* in Fig. 6-6a). These points are plotted on a graph of I_D versus V_{GS} (Fig. 6-6b).

The graph of Fig. 6-6b can be represented algebraically by Shockley's equation

$$I_D = I_{DSS} (1 - V_{GS}/V_P)^2 \tag{6-1}$$

This equation is a valid approximation of the behavior of all JFETs. Using this mathematical approximation, we can derive the plot of I_D versus V_{GS} for *any* JFET, provided the two parameters I_{DSS} and V_P are given. In Fig. 6-6b, for example, we can *calculate* I_D for points *a*, *b*, *c*, and *d* from Shockley's equation and plot I_D versus V_{GS}. The values of I_{DSS} and V_P must, of course, be known.

EXAMPLE 6-1

Given $V_P = -3.4$ V, $I_{DSS} = 4.9$ mA (typical parameters for the 2N4221), calculate I_D using Shockley's equation for
 (a) $V_{GS} = -1$ V (Fig. 6-6b, point *b*).
 (b) $V_{GS} = -2$ V (Fig. 6-6b, point *c*).
 (c) $V_{GS} = -3$ V (Fig. 6-6b, point *d*).

SOLUTION

Write Shockley's equation with the given I_{DSS} and V_P. (All currents are in mA.)

$$I_D = 4.9(1 - V_{GS}/-3.5)^2$$

(a) Substitute $V_{GS} = -1$ in the equation above

$$I_D = 4.9(1 - (-1)/(-3.5))^2 = 2.5 \text{ mA}$$

(b) Substitute $V_{GS} = -2$ V

$$I_D = 4.9(1 - (-2/-3.5))^2 = 0.9 \text{ mA}$$

(c) Substitute $V_{GS} = -3$ V

$$I_D = 4.9(1 - (-3/-3.5))^2 = 0.1 \text{ mA}$$

The values calculated in Ex. 6-1 agree closely with the respective values obtained from the graph given in Fig. 6-6a. (This graph is supplied by the manufacturer.)

In many instances, the manufacturer of the JFET provides only the values of I_{DSS} and V_P, with the understanding that the transfer characteristics (such as those shown in Fig. 6-7) follow Shockley's equation. The user, if he so desires, can plot these characteristics by getting at least three points on that curve. Convenient choices for these three points are:

(1) $V_{GS} = 0$, $I_D = I_{DSS}$;
(2) $V_{GS} = V_P$, $I_D = 0$;
(3) $V_{GS} = V_P/2$, $I_D = I_{DSS}/4$.

The first two points represent the extremities of the graph, while the third point is a midpoint (of sorts). Point (3) results from substituting $V_{GS} = V_P/2$ in Shockley's

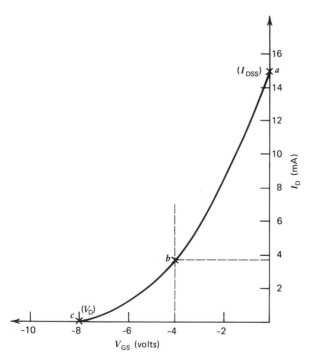

FIGURE 6-7 Transfer characteristics of the 2N4722 (Ex. 6-2).

240   FIELD EFFECT TRANSISTORS (FET)

equation. It is the point at which V_{GS} is *one-half* the pinch off voltage, producing an I_D that is a *quarter* of the I_{DSS}.

EXAMPLE 6-2

Obtain a three-point plot of the transfer characteristic of the 2N4222, given $V_P = -8.0$ V and $I_{DSS} = 15$ mA.

SOLUTION

The three-point plot is shown in Fig. 6-7. Points *a* and *c*, the extremities of the plot, are given by I_{DSS} and V_P, respectively. Point *b* is obtained for $V_{GS} = V_P/2 = -8/2 = -4.0$ V.

$$I_D = I_{DSS}(1 - V_{GS}/V_P)^2 \quad (6\text{-}1)$$
$$I_D = I_{DSS}(1 - (V_P/2)/V_P)^2$$
$$= I_{DSS}(1 - 1/2)^2 = (1/4)I_{DSS}$$

Since $I_{DSS} = 15$ mA, the coordinates of point *b* are:

$$I_D = (1/4)15 = 3.75 \text{ mA}$$

and

$$V_{GS} = (1/2)V_P = -4.0 \text{ V}$$

The three-point plot of Ex. 6-2 may be somewhat inaccurate, since only three points were used to plot this nonlinear curve. It is, however, quite suitable for most JFET applications.

To simplify the analysis of JFET circuits, the curve of Fig. 6-7 is used as a universal curve for all JFETs. The coordinates will be rescaled to account for the specific values of I_{DSS} and V_P.

The drain and transfer characteristics of the N-channel and P-channel transistors differ only in the *polarities* of the various quantities. All the P-channel parameters have opposite polarities than those of the N-channel (Fig. 6-8). For convenience both P and N-channel FETs will be represented by unipolar plots (the original N-channel plots) and the axis will be simply labeled appropriately. Fig. 6-8, *d* and *f*, are redrawn in Fig. 6-9. The various parameters have the polarities appropriate for the P-channel JFET, in disagreement with the conventional positive and negative directions of the axis.

6-5   JFET BIASING

The JFET, like the bipolar transistor, requires the establishment of a dc operating point before it can be used in signal applications. The analysis and design of the JFET biasing circuits are simpler than their counterparts for bipolar transistors.

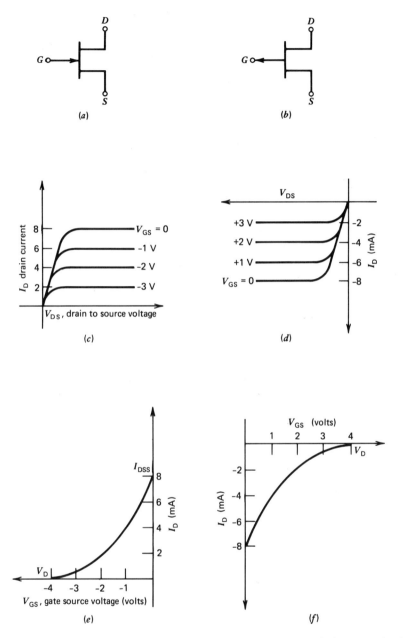

FIGURE 6-8 Characteristics of N and P channel JFETs (used $|V_p| = 4$ V, $|I_{Dss}| = 8$ mA). (a) N-channel JFET symbol. (b) P-Channel JFET symbol. (c) N-channel drain characteristics. (d) P-channel drain characteristics. (e) N-channel transfer characteristics. (f) P-channel transfer characteristics.

242 FIELD EFFECT TRANSISTORS (FET)

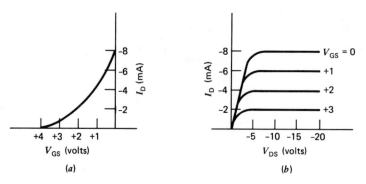

FIGURE 6-9 Relabeled P-channel characteristics. (a) Transfer characteristics. (b) Drain characteristics.

This is because *no current flows in the gate*. The gate current, usually given in the JFET data sheets, is the gate leakage current, I_{GSS}, and is of the order of a few nA, insignificant in most applications. In addition, the JFET is used, for the most part, in two configurations: common source (CS) and common drain (CD). The definitions of the JFET configurations follow the corresponding definitions of the bipolar configurations and will not be repeated here. A common gate (CG) configuration is so rarely used that it will not be covered in this text.

As noted before, our graphic analysis will utilize the transfer characteristic (V_{GS} versus I_D) that are the graphic representation of Shockley's equation [Eq. (6-1)]. However, in some cases, strictly mathematical analysis will be used relying on Eq. (6-1) (and the appropriate circuit equations).

The operating point (the Q-point) of the JFET is established by V_{GSQ} (the gate to source voltage associated with the Q-point). The selection of a particular value for V_{GSQ} is usually based on some performance consideration such as the need to select a linear operating range or a high ac gain operating point, etc. Since V_{GS} and I_D are related (Shockley's equation), the operating V_{GS}, (V_{GSQ}) produces the quiescent I_D, (I_{DQ}).

A number of different biasing circuits will be discussed. The analysis of these circuits involves device relationships, Shockley's equation, as well as the circuit laws, KVL and KCL. The purpose of the biasing circuits is to establish a desired V_{GSQ}, which results in a corresponding I_{DQ}. The analysis will then be concerned with the calculation of V_{GSQ} and I_{DQ} and is based on:

a. The gate will be considered an open circuit. There is no gate current.
b. The drain-source voltage V_{DS} is assumed large enough for the transistor to be operating in the "pinch off" region, where I_D is approximately constant for a given V_{GS}.
c. The transfer characteristic (and/or the drain characteristic) will be used for graphic analysis, in conjunction with KVL.
d. Shockley's equation (Eq. 6-1) and KVL will be used in the purely mathematical approach.

6-5.1 Fixed Bias

A fixed bias circuit is shown in Fig. 6-10. Since no current is flowing in R_G, V_G (gate to ground voltage) is equal to the battery voltage. $V_G = -2.0$ V and $V_{GS} = -2.0$ V. This is the quiescent gate to source voltage, V_{GSQ}. To obtain the corresponding I_{DQ}, either the drain or transfer characteristic (Fig. 6-11) or Eq. (6-1) may be used. The Q-point is shown in Figs. 6-11a and 6-11b. From the two graphs, $I_{DQ} = 1$ mA. To use Eq. (6-1), we simply substitute $V_{GS} = -2.0$ V. (Note that $I_{DSS} = 5$ mA and $V_P = -3.5$ V for the 2N4221.)

$$I_D = I_{DSS}(1 - V_{GS}/V_P)^2 = 5(1 - -2/-3.5)^2 \tag{6-1}$$

$$I_D = .92 \text{ mA}$$

The difference between the I_D obtained graphically and that obtained from Eq. (6-1) is due to the inaccuracies in the given V_P and I_{DSS}. Also, recall Eq. (6-1) is only an approximate representation of the transistor behavior.

The fixed bias circuit (Fig. 6-10) requires a separate biasing battery V_{GG} in addition to the drain supply voltage V_{DD}. As a result, it is rarely used.

6-5.2 Self Bias

By introducing a source resistor, R_S (Fig. 6-12), V_{GG} may be eliminated. The gate source bias is established by the voltage drop across R_S, $I_D \times R_S$. Here, the drain current I_D, which is itself dependent on V_{GS}, is used to establish the desired V_{GS}. This is typical in feedback biasing, in which the current in the output portion of the device (drain-source) is used to provide biasing for the input portion (gate source).

To find the operating point, Q, KVL is applied to the gate-source loop, and use is made of the device relationships, I_D versus V_{GS} [whether graphic or Eq. (6-1)]. KVL applied to the input loop yields the general equation

$$V_G - V_{GS} - V_S = 0 \tag{6-2}$$

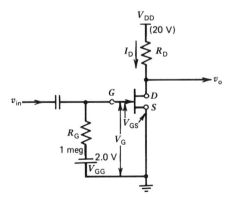

FIGURE 6-10 Fixed bias (N-channel), transistor (2N4221).

244 FIELD EFFECT TRANSISTORS (FET)

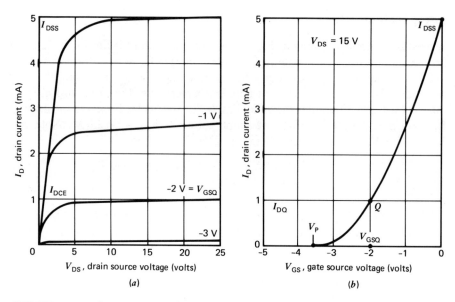

FIGURE 6-11 Characteristics of the 2N4221 JFET (courtesy of Motorola Corp). (a) Q-point on drain characteristics. (b) Q-point on transfer characteristics.

Since $V_S = I_D R_S$ (the drain and source current are identical, as there is no current entering at the gate), Eq. (6-2) becomes

$$V_G - V_{GS} - I_D R_S = 0 \tag{6-2a}$$

or

$$V_{GS} = V_G - I_D R_S$$

This is the equation of the input load line representing a relation between I_D and V_{GS}. For the circuit of Fig. 6-12, $V_G = 0$. (There is no current through R_G; hence, no voltage across it.) We get

$$0 - V_{GS} - I_D R_S = 0$$

$$V_{GS} = -I_D R_S \tag{6-3}$$

Combining Eq. (6-3) with Eq. (6-1), solving two simultaneous equations, yields

$$I_D = I_{DSS}(1 - (-I_D R_s / V_P)^2 \tag{6-4}$$

Equation (6-4) has a single unknown, I_D. (I_{DSS}, V_P, and R_S are all given values.) It is then possible to solve Eq. (6-4) for I_D, and find V_{GS} from Eq. (6-3) or Eq. (6-1). But this path to a solution is somewhat complex. Instead, a simpler graphic solution will be used.

As noted, two simultaneous equations, (6-1) and (6-3), must be solved. Both of

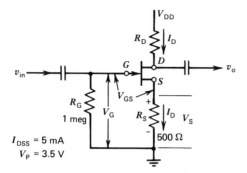

FIGURE 6-12 Self bias (2N4221).

these equations relate I_D to V_{GS} and may be plotted on a I_D versus V_{GS} coordinate system. Note that the plot of Eq. (6-1) is given by the transfer characteristic (Fig. 6-11b). The graphic solution, then, requires a plot of Eq. (6-3) superimposed on that of Eq. (6-1). The intersection of the two graphs gives the Q-point (I_{DQ} and V_{GSQ}).

EXAMPLE 6-3

Find the Q-point, graphically, for the circuit shown in Fig. 6-12. $I_{DSS} = 5$ mA; $V_P = -3.5$ V. Use the universal transfer characteristics.

SOLUTION

Figure 6-13 is the rescaled universal transfer characteristic. The scales of the coordinates are usually given as shown (0 to 1.0); for example, 0.5 on the I_D scale means that $I_D = 0.5 I_{DSS}$. Similarly, 0.6 on the V_{GS} scale means $V_{GS} = 0.6 V_P$. For the rescaled plot, these points represent $I_D = 0.5 \times 5 = 2.5$ mA, and $V_{GS} = 0.6 \times (-3.5) = -2.1$ V, respectively. Equation (6-3) can be plotted on the same set of axes.

$$V_{GS} = -I_D R_S \quad (6-3)$$
$$V_{GS} = -I_D \times 500$$

This is a straight line, so that only two points are required for the plot.

Point 1: $I_D = 0$ $\quad V_{GS} = 0$
Point 2: $I_D = 4$ mA $\quad V_{GS} = -4 \times 10^{-3} \times 500 = -2$ V.

[These two points are quite arbitrary; select I_D, then find V_{GS} from Eq. (6-3).] The plot of Eq. (6-3) is shown in Fig. 6-13. The intersection of Eq. (6-1) and Eq. (6-3) is the Q-point. (This is the graphic solution for the two simultaneous equations.)

$$V_{GSQ} = -1.15 \text{ V} \quad\quad I_{DQ} = 2.3 \text{ mA}$$

246 FIELD EFFECT TRANSISTORS (FET)

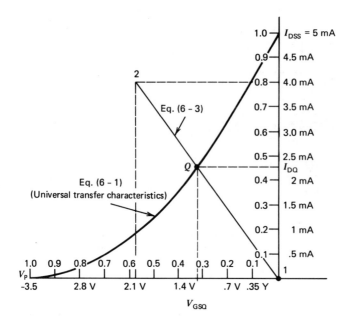

FIGURE 6-13 Finding the Q-point (Ex. 6-3).

The general steps involved in the graphic solution are:

1. Apply KVL to the input (gate-source) circuit to obtain the equation for the input loadline. [Eq. (6-2)].
2. Relable and rescale the universal JFET transfer characteristic. (This makes the curve representative of the specific transistor with the given I_{DSS} and V_P.)
3. Plot the input loadline using the same set of coordinates as the JFET characteristic.
4. The intersection of the characteristic and the load line is the Q-point.
5. Read off V_{GSQ} and I_{DQ}.

These four steps are applicable to all common JFET biasing circuits.

6-5.3 Voltage Divider Biasing

A variation of the self-bias circuit (Fig. 6-12) is shown in Fig. 6-14. The analysis of this circuit follows the four steps outlined above and is essentially the same as that for the self-bias circuit. (Sec. 6-5.2).

Applying KVL to the gate-source loop, we get

$$V_G - V_{GS} - I_D R_S = 0 \tag{6-2a}$$

For this circuit, $V_G \neq 0$. Hence, we calculate V_G (using voltage division)

$$V_G = \frac{V_{DD} \times R_2}{R_1 + R_2} \tag{6-5}$$

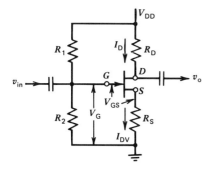

FIGURE 6-14 Voltage divider bias circuit.

Eq. (6-2) and (6-5) combine to yield the load line equation.

$$\frac{V_{DD} \times R_2}{R_1 + R_2} - V_{GS} - I_D R_S = 0 \qquad (6\text{-}6)$$

To get the Q-point, steps 2, 3, and 4 above are used.

Note One convenient point for plotting the load line is

$$I_D = 0, \qquad V_{GS} = V_G.$$

[Substitute $I_D = 0$ in Eq. (6-2a).]

EXAMPLE 6-4

In Fig. 6-14, $V_{DD} = 12$ V, $R_1 = 1.8$ MΩ, $R_2 = 200$ kΩ, $R_S = 500$ Ω, $R_D = 1$ kΩ, $V_P = -4$ V, $I_{DSS} = 6$ mA.
  (a) Find the operating point (V_{GSQ}, I_{DQ}).
  (b) Find V_{DS}.

SOLUTION

(a) A universal transfer curve is relabeled with $V_P = -4$ V, and $I_{DSS} = 6$ mA and the axes rescaled accordingly (Fig. 6-15). Note that the scale is continued in the positive direction of V_{GS}. To apply KVL [Eq. (6-2a) and (6-6)], we first calculate V_G [Eq. (6-5)].

$$V_G = \frac{V_{DD} \times R_2}{R_1 + R_2} = \frac{12 \times 200 \text{ k}\Omega}{1800 \text{ k}\Omega + 200 \text{ k}\Omega} = 1.2 \text{ V}$$

Eq. (6-2a) [and (6-6)] becomes $V_{GS} = 1.2 - I_D \times 500$ (the gate-source load line equation). To plot this load line, two points are needed. One convenient point is $I_D = 0$, $V_{GS} = V_G = 1.2$ V. To find a second point, select an arbitrary value for I_D and find V_{GS} from the load line equation. (Make sure this point is within

248 FIELD EFFECT TRANSISTORS (FET)

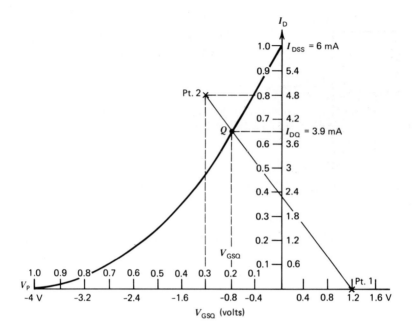

FIGURE 6-15 Voltage divider biasing—graphic solution.

the confines of your coordinate system.) Choosing $I_D = 4.8$ mA and substituting into Eq. (6-2a), we get

$$V_{GS} = 1.2 - 4.8 \times 10^3 \times 500 = -1.2 \text{ V} \quad \text{(pt. 2)}$$

The second point is $I_D = 4.8$ mA, $V_{GS} = -1.2$ V. Steps 3 and 4 yield

$$\underline{V_{GSQ} = -0.8 \text{ V}} \qquad \underline{I_{DQ} = 3.9 \text{ mA}}$$

Note, again, that the input load line is given by Eq. (6-2a) and that a convenient point used in plotting the load line is always $I_D = 0$, $V_{GS} = V_G$. It is useful to calculate V_G (recall that $V_G = 0$ for the circuit of Fig. 6-12) as the first step in obtaining the plot of the input load line.

(b) To find V_{DS}, KVL is applied to the output loop, the drain-source circuit.

$$V_{DD} - I_D \times R_D - V_{DS} - I_D \times R_S = 0$$
$$V_{DD} - I_D(R_D + R_S) = V_{DS}$$
$$12 - 3.9 \times 10^3(1000 + 500) = \underline{6.15 \text{ V}}$$

EXAMPLE 6-5

Obtain V_{GSQ} and I_{DQ} for the circuit in Fig. 6-16a.

JFET BIASING 249

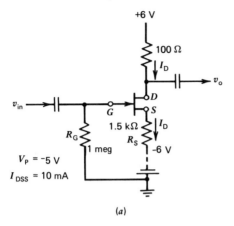

(a)

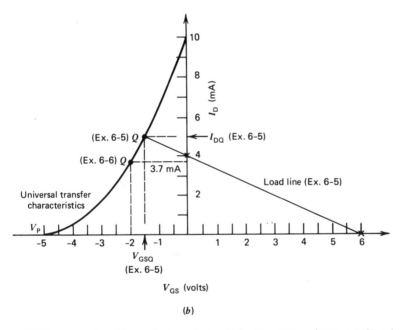

(b)

FIGURE 6-16 Graphic analysis and circuit for Exs. 6-5 and 6-6. (a) Circuit. (Dashed lines show Source battery connections.) (b) Transfer characteristics with load line.

SOLUTION

Applying KVL to the gate-source circuit (including the -6 V source supply voltage) yields

$$V_G - V_{GS} - I_D R_S + 6 = 0$$

Since $V_G = 0$ (gate connected to GND through R_G)

$$V_{GS} = 6 - I_D R_S = 6 - I_D \times 1500$$

The two points for the load line are

Point 1: $I_D = 0$, $V_{GS} = +6$ V
Point 2: $I_D = 4$ mA, $V_{GS} = 0$ V

(The second point is selected arbitrarily.) The load line and the Q-point are shown in Fig. 6-16b. I_{DQ} and V_{GSQ} are

$$V_{GSQ} = -1.5 \text{ V}, \qquad I_{DQ} = 5 \text{ mA}$$

Note that the -6 V source supply, in Ex. 6-5, results in $+6$ V voltage being applied to the gate (so that for $I_D = 0$, $V_{GS} = +6$ V).

EXAMPLE 6-6

(a) Find R_S for a self bias circuit so that $V_{GSQ} = -2$ V. (N channel JFET.)
(b) Find R_D so that $V_{DS} = \frac{1}{2}V_{DD}$, a typical operating point for amplifier circuits, given $V_{DD} = 24$ V, $I_{DSS} = 10$ mA, $V_p = -5$ V (use the curve used in Ex. 6-5. It has the same V_p and I_{DSS}).

SOLUTION

(a) The circuit is shown in Fig. 6-17. From the universal curve (Fig. 6-16b) we find that for $V_{GS} = -2$ V, $I_D = 3.7$ mA. Since for this circuit

$$V_{GS} = -I_D R_S \qquad (6\text{-}3)$$

R_S is obtained from Eq. (6-3),

$$R_S = \frac{V_{GS}}{-I_D} = \frac{-2}{-3.7 \times 10^{-3}} = 540 \ \Omega$$

$$V_{DD} - I_D \times R_D - V_{DS} - I_D \times R_S = 0$$

$$R_D + R_S = \frac{V_{DD} - V_{DS}}{I_D}$$

(b) R_D is obtained by the application of KVL to the drain source circuit.

Since $V_{DS} = \frac{1}{2}V_{DD} = 12$ V

$$R_D + R_S = \frac{24 - 12}{3.7 \times 10^{-3}} = 3240 \ \Omega$$

$$R_D = 3240 - R_S = 3240 - 540 = 2700 \ \Omega$$

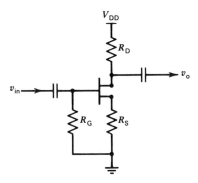

FIGURE 6-17 Self-bias circuit for Ex. 6-6.

6-6 JFET ac ANALYSIS (SMALL SIGNAL)

The preliminaries of establishing a dc operating point out of the way, we may now investigate the operation of the JFET when ac signals are involved. As noted before, only two configurations will be analyzed, common source and common drain (or source follower).

The basic operation of the JFET centers around the gate *voltage* controlling the drain *current*. Shockley's equation, which is a fundamental representation of the JFET, relates dc gate-source *voltage* V_{GS} to drain *current* I_D. The ac analysis must then be concerned with changes in V_{GS}, ΔV_{GS}, and the corresponding changes in I_D, ΔI_D; in other words, the analysis is concerned with the effects of *ac* voltages applied between gate and source v_{GS} (lowercase symbols representing ac quantities). A graphic demonstration of this ac voltage-current relation is shown in Fig. 6-18b. For the sake of clarity, it is assumed that the JFET has an $I_{DSS} = 10$ mA, $V_P = -4$ V, and that the Q-point was selected as $V_{GSQ} = -2$ V (Fig. 6-18a). As shown, v_{in} is the same as v_{GS}. Indeed, v_{in} *in this circuit* is effectively connected between gate and source. (The capacitor is assumed to serve as a short circuit for ac signals.) The 0.8 V amplitude of v_{in} causes V_{GS} to vary between -2.8 V $(-2 - 0.8)$ and 1.2 V $(-2 + 0.8)$, $\Delta V_{GS} = \pm 0.8$ V (1.6 V p-p). These variations in V_{GS} produce corresponding changes in I_D (1 to 4.8 mA) $\Delta I_D = 3.8$ mA p-p. Because of the nonlinearity of the V_{GS} versus I_D curve, the ac drain current lost its symmetry; it is no longer a pure sinusoid. The JFET has introduced a substantial distortion. While the input signal is a pure sinusoid, the output current i_D is not. If we used a substantially smaller v_{in}, the distortion would be much less pronounced. If distortion is of importance (and it usually is in most analog applications), the input signal must be limited to a rather very small amplitude. (Recall similar considerations for the BJT, bipolar junction transistor.)

6-6.1 g_m—Transconductance

We now proceed to formulate, mathematically and graphically, the relation between *changes* in V_{GS} and corresponding *variations* in I_D. Our interest is focused on the

252 FIELD EFFECT TRANSISTORS (FET)

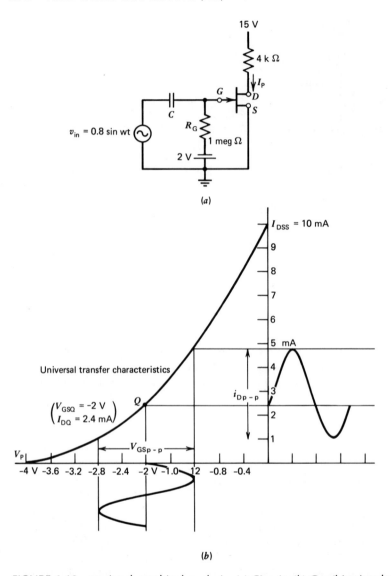

FIGURE 6-18 ac signal graphical analysis. (a) Circuit. (b) Graphic signal analysis.

relation $\Delta I_D/\Delta V_{GS}$ or i_D/v_{GS} that relates the *output* current i_D to an input voltage v_{GS}. This ratio is not a constant. Its value depends not only on the characteristics of the transistor (V_p and I_{DSS}) but also on the operating point.

The ratio $\Delta I_D/\Delta V_{GS}$ or i_D/v_{GS} is defined as the transconductance of the FET denoted by g_m.

$$g_m = \frac{\Delta I_D}{\Delta V_{GS}} = \frac{i_D}{v_{GS}}$$

The term "transconductance" points to the fact that g_m is a "transfer" parameter between input and output and has the units of conductance, mhos.[1]

The significance of g_m is that with g_m known, i_D can be found for a given v_{GS} from

$$i_D = g_m \times v_{GS} \tag{6-7a}$$

This, in turn, permits calculation of the output voltage (at the drain or source) and finally the voltage gain, v_o/v_{in}.

The detailed evaluation, using differential calculus, is given in Appendix D. It relies on the fact that for very *small values of* ΔV_{GS}, ($\Delta V_{GS} \to 0$),

$$\frac{\Delta I_D}{\Delta V_{GS}} = \frac{dI_D}{dV_{GS}}$$

that is, the derivative of I_D with respect to V_{GS}. The expression for g_m is given by

$$g_m = \frac{-2 I_{DSS}}{V_P}\left(1 - \frac{V_{GS}}{V_P}\right) \tag{6-8}$$

where I_{DSS} and V_P are device constants given by the manufacturer.

Equation (6-8) shows that g_m is dependent, among other things, on V_{GS}, the dc gate-source voltage. Recall that the process of biasing the transistor entails the establishment of V_{GSQ} (V_{GS} for the Q-point). It is now apparent that the selection of the operating point V_{GSQ} affects the ac operation directly, by establishing a specific value for g_m, that is, g_{mQ}. As before, the subscript Q indicates the association with the Q-point.

To simplify matters somewhat, Eq. (6-8) is rewritten in the form

$$g_m = g_{mo} \times \left(1 - \frac{V_{GS}}{V_P}\right) \tag{6-9}$$

where

$$g_{mo} = -\frac{2 I_{DSS}}{V_P} \tag{6-10}$$

The numerical value of g_{mo} is characteristic of the transistor. An examination of Eq. (6-8) shows that g_{mo} is the value of g_m when

$$V_{GS} = 0 \quad \text{and} \quad I_D = I_{DSS}$$

g_m (and g_{mo}) are always positive in spite of the negative sign in Eqs. (6-8) and (6-10). This is so because I_{DSS}/V_P is always negative ($-I_{DSS}/V_P$ always positive).

---

[1] The unit of conductance has been recently renamed the Siemen in honor of a noted scientist by this name.

254   FIELD EFFECT TRANSISTORS (FET)

For N channel devices, V_P is negative and I_{DSS} positive, while the reverse holds true for P channel devices. Similarly, V_{GS}/V_P is always positive since V_{GS} and V_P always have the same sign.

EXAMPLE 6-7

Obtain the expression for g_m, given $I_{DSS} = 6$ mA, $V_P = -3$ V.

SOLUTION

First evaluate g_{mo}.

$$g_{mo} = -\frac{2I_{DSS}}{V_P} = -\frac{2 \times 6 \times 10^{-3}}{-3} \qquad (6\text{-}10)$$

$$= 4 \times 10^{-3}\ \mho$$

$$g_{mo} = 4\ \text{m}\mho = 4000\ \mu\mho$$

$$g_m = 4 \times 10^{-3} \times \left(1 - \frac{V_{GS}}{-3}\right) = 4 \times 10^{-3} \times \left(1 - \left|\frac{V_{GS}}{V_P}\right|\right)$$

$$\left(\left|\frac{V_{GS}}{V_P}\right|\ \text{the magnitude of the quotient}\right)$$

6-6.2 g_m and the Q-point—g_{mQ}

The value of g_m is essential to the evaluation of the *ac* performance of the circuit. Since g_m depends on V_{GS} [Eq. (6-9)], the calculation of g_{mQ} requires that V_{GSQ} be found first. The evaluation of g_{mQ}, once V_{GSQ} is known, may take two forms: algebraic using Eq. (6-9), and graphic.

The algebraic method requires the evaluation of Eq. (6-9) for a given V_{GS}. Clearly, g_{mo} must be known, which requires that V_P and I_{DSS} be given.

EXAMPLE 6-8

Find g_{mQ} for the circuit of Fig. 6-19a.

SOLUTION

V_{GSQ} is found with the aid of the universal curve rescaled for the given transistor parameters (Fig. 6-19b). The input load line is drawn from the KVL equation

$$V_G - V_{GS} - I_D R_S = 0$$

Since $V_G = 0$

$$V_{GS} = -I_D R_S = -I_D \times 500$$

JFET AC ANALYSXIS (SMALL SIGNAL)

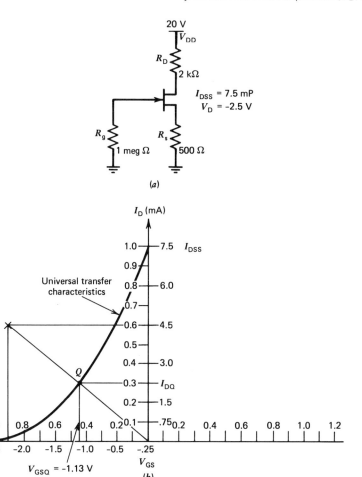

FIGURE 6-19 Finding a and g_m (algebraically). (a) The circuit. (b) Finding V_{GSQ} and g_{mQ}.

The two points of the load line are

$$I_D = 0 \qquad V_{GS} = 0$$
$$I_D = 4.5 \text{ mA} \qquad V_{GS} = -2.25 \text{ V}$$

The Q-point is found at

$$V_{GSQ} = -1.13 \text{ V} \qquad I_{DQ} = 2.25 \text{ mA}$$

To get g_m, we find g_{mo}

$$g_{mo} = -\frac{2I_{DSS}}{V_P} = -\frac{2 \times 7.5 \times 10^{-3}}{-2.5} = 6 \times 10^{-3} \, \Omega \qquad (6\text{-}10)$$

256 FIELD EFFECT TRANSISTORS (FET)

and

$$g_m = g_{mo}\left(1 - \frac{V_{GSQ}}{V_P}\right) = 6 \times 10^{-3}\left(1 - \frac{-1.13}{-2.5}\right) \quad (6-9)$$
$$= 3.29 \times 10^{-3} \, \Omega = \underline{3290 \, \mu\mho}$$

The graphic solution for g_{mQ} utilizes the plot of Eq. (6-9). Equation (6-9) is a *linear* relation between g_m and V_{GS} (g_{mo} and V_P are given values; g_m and V_{GS} are the variables). The plot is then a straight line on a g_m versus V_{GS} set of coordinates. We will use the V_{GS} abscissa of the universal curve and give a g_m scale in place of, or in addition to, the I_D ordinate (Fig. 6-20).

Two convenient points for the g_m versus V_{GS} plot are

$$V_{GS} = 0 \quad g_m = g_{mo} \quad \text{(point 1)}$$
$$V_{GS} = V_P \quad g_m = 0 \quad \text{(point 2)}$$

These points are obtained by substituting the chosen V_{GS} values into Eq. (6-9). These two points and the g_m plot are shown in Fig. 6-20. Note that for $V_{GS} = 0$, $g_m = g_{mo}$. For the same point $I_D = I_{DSS}$. With reference to the I_D scale, the g_{mo} point on the g_m scale will coincide with the I_{DSS} point on the I_D scale. It is useful to plot the I_D versus V_{GS} and the g_m versus V_{GS} graphs using V_{GS} as a *common* abscissa, as shown in Fig. 6-20. To find g_{mQ} graphically, we first find V_{GSQ} (see Ex. 6-8) on the V_{GS} versus I_D plot and then find g_{mQ} for V_{GSQ} on the g_m plot.

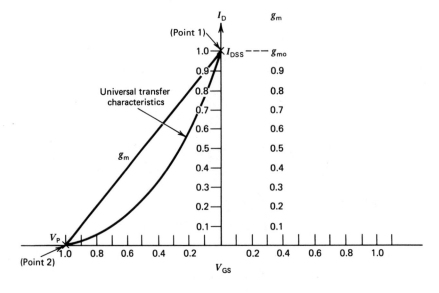

FIGURE 6-20 g_m Plot.

JFET AC ANALYSXIS (SMALL SIGNAL)

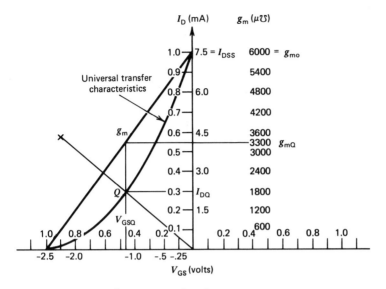

FIGURE 6-21 Finding g_{mQ} graphically.

EXAMPLE 6-9

Find g_{mQ} graphically for Ex. 6-8.

SOLUTION

To obtain V_{GSQ}, proceed as in Ex. 6-8 (Fig. 6-19). To find g_m, we first establish the g_m scale.

$$g_{mo} = \frac{-2I_{DSS}}{V_p} = \frac{-2 \times 7.5 \times 10^{-3}}{-2.5}$$
$$= 6 \times 10^{-3} \, \Omega = 6000 \, \mu\mho$$

A separate g_m scale is added, as shown in Fig. 6-21 and the g_m graph (a straight line) is then plotted. g_{mQ} is found by extending a line vertically at V_{GSQ} to intersect the g_m graph. The value of g_m at this intersection is g_{mQ}. Here, $g_{mQ} = 3300 \, \mu\mho$. The slight inaccuracy in g_{mQ} (3300 $\mu\mho$ compared to 3290 $\mu\mho$ found algebraically) is due to plotting inaccuracies.

The graphic method of finding g_{mQ} may be summarized as follows:

1. Draw the input load line to find Q.
2. Plot g_m as a straight line connecting $V_{GS} = 0$ on the abscissa to $I_D = I_{DSS}$, or $g_m = g_{mo}$ on the ordinate.
3. Find g_{mo} [Eq. (6-10)] and provide the g_m scale on the ordinate.
4. Proceed *vertically* from the Q-point to intersect with the g_m graph. Read g_m (on the g_m scale) at the intersection.

EXAMPLE 6-10

Find g_{mQ} for the circuit of Fig. 6-22a.

SOLUTION

The load line is plotted as shown in Fig. 6-22b. The two points used for the load line plot are $I_D = 0$, $V_{GS} = 0$ and $I_D = 3$ mA, $V_{GS} = -I_D R_S = -3 \times 10^{-3} \times 600 = -1.8$ V. (The load line equation is $V_{GS} = -I_D \times R_S$.) Q is found as shown in Fig. 6-21b.

$$V_{GSQ} = -1.2 \text{ V}$$

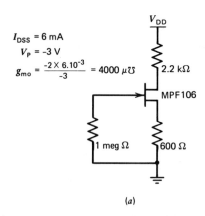

$I_{DSS} = 6$ mA
$V_P = -3$ V
$g_{mo} = \dfrac{-2 \times 6.10^{-3}}{-3} = 4000 \ \mu\mho$

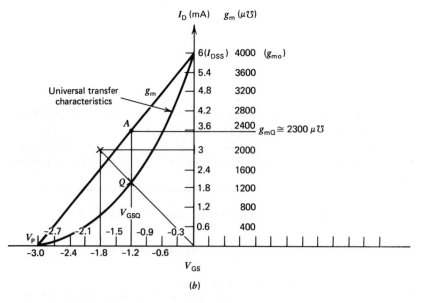

FIGURE 6-22 (a) Circuit and (b) graphic solution for Ex. 6-10.

The ordinate is given a g_m scale, and g_{mQ} is read off corresponding to point A, the intersection of V_{GSQ} with the g_m plot.

$$g_{mQ} = 2300 \ \mu \mho.$$

The student is urged to proceed solving Ex. 6-10 algebraically. Notice how time-consuming the algebraic method is.

6-6.3 The JFET Equivalent Circuit

Similar to the approach used for the BJT, it is again necessary to replace the device with some sort of a *circuit*, functionally equivalent to the JFET. This is necessary to permit the use of conventional circuit analysis tools such as KVL, KCL, and superposition. Again, the assumption is made that the various voltage-current relationships in the device are linear, (over a very small operating range). It is then permissible to apply other circuit rules to the device, for example, Thevenin's and Norton's theorems.

The operation of the JFET centers around two important parameters, the *control voltage* v_{GS} and the *controlled current* i_D. It is natural, then, to consider the gate-source terminals as the input section of the device and the drain, or drain-source, as the output terminals. As a result, the equivalent circuit for the JFET contains two separate parts, an input equivalent and an output equivalent.

Since the gate is essentially one terminal of a reverse biased diode, the current into the gate (or out of it in the P channel device) consists of a diode leakage current. This current is so small that we may consider the gate an open circuit. The equivalent of the input section, gate-source, is reduced to an open circuit (Fig. 6-23).

The *current* i_n, the output portion of the device or drain-source, is determined by v_{GS} [Eq. (6-7)]. Since it is the output *current* that is related to v_{GS}, not some output voltage, it is logical to select a Norton's (current source) equivalent for this portion of the device. As shown in Fig. 6-23c, the drain-source equivalent is given by an ideal current source, $i_D = g_m v_{GS}$ [Eq. (6-7a)]. In general, the Norton's equivalent is expected to include a resistance (in parallel with the current source) such as r_d (Fig. 6-23d). r_d represents the drain to source ac resistance. It is defined as $r_d = \Delta V_{DS}/\Delta I_D$, the ac drain-source voltage divided by the ac drain current for a particular V_{GS}. A typical value for r_d may be obtained by referring to the drain characteristic (Fig. 6-24). A rather large $\Delta V_{DS} = 15$ V produces only a small $\Delta I_D = 0.2$ mA (the computation is done for $V_{GS} = -1$ V).

$$r_d = \frac{\Delta V_{DS}}{\Delta I_D} = \frac{15}{0.2 \times 10^{-3}} = 75 \ k\Omega$$

In many applications, this large value of r_d may turn out to have little effect on the circuit and will be neglected.

The complete equivalent circuit for the JFET is shown in Fig. 6-23e. It must

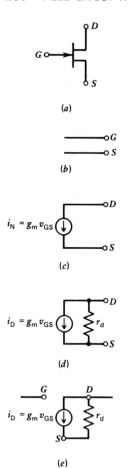

FIGURE 6-23 Development of the equivalent circuit. (a) The three-terminal device. (b) Input equivalent (open circuit). (c) Output (drain-source) equivalent, idealized (Norton current source). (d) Drain source equivalent with r_d. (e) The complete equivalent circuit.

be emphasized that it is the *ac* equivalent circuit and all the parameters shown are ac quantities.

To obtain the equivalent circuit of the FET in a particular circuit, the operating g_m (g_{mQ}) is calculated as outlined in Sec. 6-6.2. r_d is usually given or may be calculated from the drain characteristics at the given V_{GSQ}.

Since an increase in V_{GS} (ΔV_{GS} positive) results in a decrease in V_D (ΔV_D negative), the *direction* of the drain current source was selected to reflect this relative polarity.

In the following discussion and examples we adopt the basic ac circuit assumptions used in the various BJT configurations. First, all capacitors are considered short circuits; second, all dc supply voltages are eliminated, replaced by a short (V_{DD}

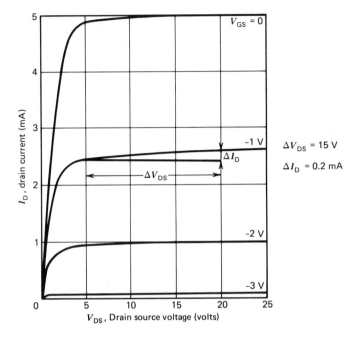

FIGURE 6-24 r_d on the drain characteristics—2N4221 (courtesy of Motorola Corp).

becomes GND), in conformance with the superposition theorem. Recall that we are considering ac behavior only. Our analysis, again, concerns the basic amplifier parameters A_v, A_i, r_{in}, r_o, and A_p.

6-6.4 Common Source Circuit (CS)

A typical common source circuit (input to gate, output from drain) is shown in Fig. 6-25a. The ac equivalent of the device and the circuit is shown in Fig. 6-25b. Note that the capacitor C_S shorts the resistor R_S connecting the source terminal to ground.

The computation of the voltage gain requires that g_m be known. It is assumed here that g_m has been computed for the circuit as described in Sec. 6-6.2.

A_v, VOLTAGE GAIN

In the equivalent circuit (Fig. 6-25b), r_d and R_D are in parallel and v_o is the voltage across this parallel combination. Hence,

$$v_o = -i_D \times r_d \| R_D \tag{6-11}$$

The minus sign is necessary because v_o is negative with respect to ground. (The

FIELD EFFECT TRANSISTORS (FET)

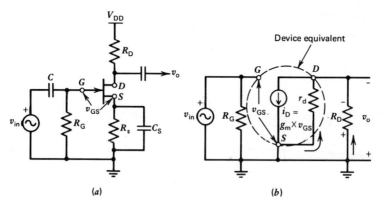

FIGURE 6-25 Common source circuit with fixed bias. (a) The circuit. (b) The equivalent circuit.

current is flowing from ground to v_o.) Since,

$$i_D = g_m v_{GS} \tag{6-7b}$$

v_o becomes

$$v_o = -g_m v_{GS} \times r_d \| R_D \tag{6-11a}$$

In Fig. 6-25 $v_{in} = v_{GS}$; as a result, Eq. (6-11a) may be written as

$$v_o = -g_m v_{in} \times r_d \| R_D$$

and

$$\frac{v_o}{v_{in}} = -g_m \times r_d \| R_D = A_v \tag{6-12}$$

In many applications $r_d \gg R_D$, so that, $r_d \| R_D \approx R_D$ and Eq. (6-12) becomes

$$A_v \approx -g_m R_D \tag{6-12a}$$

(g_m here is the operating g_m—namely, g_{mQ}.)

The circuit in Fig. 6-25a has been used because it affords a simple solution. Many common source circuits, however, do not include the capacitor C_S as shown in Fig. 6-26a with the equivalent circuit shown in Fig. 6-26b. Two important observations can be made. First, $v_{in} \neq v_{GS}$ and r_d is *not* parallel to R_D. As a result, the calculation of v_o and hence, A_v becomes somewhat more complex. We first make the assumptions $r_d \gg R_D$ and $r_d \gg R_S$. Both are reasonable since r_d is usually 100 kΩ or more, while R_D and R_S are a few kΩ. This permits us to neglect (open) r_d in the equivalent circuit, resulting in the simplified equivalent circuit shown in

JFET AC ANALYSXIS (SMALL SIGNAL) 263

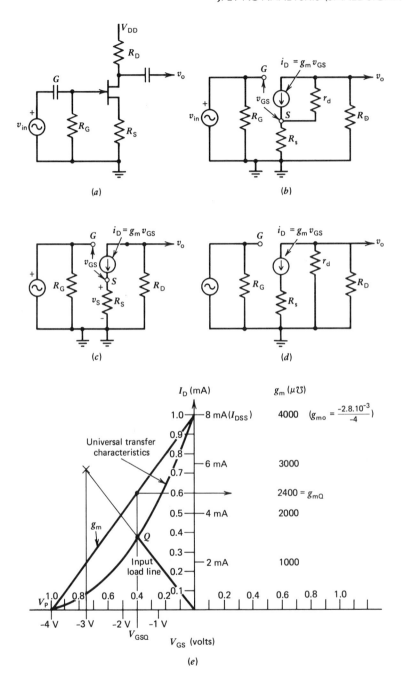

FIGURE 6-26 Common source with self bias—no C_s. (a) The circuit. (b) Eq. circuit. (c) Simplified eq. circuit. (d) Eq. circuit with $r_d \parallel R_D$. (e) V_{GSQ} and g_{mQ} for Ex. 6-11.

Fig. 6-26c where

$$v_o = -i_D R_D = -g_m v_{GS} R_D \qquad (6\text{-}13)$$

To obtain v_o/v_{in}, v_{GS} is expressed in terms of v_{in}.
Applying KVL to the input loop

$$v_{in} - v_{GS} - i_D R_S = 0 \qquad (6\text{-}14)$$

by substituting $i_D = g_m \times v_{GS}$ this becomes

$$v_{in} - v_{GS} - g_m v_{GS} R_S = 0$$
$$v_{in} = v_{GS}(1 + g_m R_S)$$
$$v_{GS} = v_{in}/(1 + g_m R_S) \qquad (6\text{-}15)$$

Substituting Eq. (6-15) in Eq. (6-13), we get

$$v_o = [-g_m v_{in}/(1 + g_m R_S)] R_D$$

and

$$\frac{v_o}{v_{in}} = -g_m R_D/(1 + g_m R_S) = A_v \qquad (6\text{-}16)$$

Equation (6-16) is reasonably accurate (less than 10% error) for $r_d \simeq 5R_S$ and $r_d \simeq 5R_D$.

A better approximation can be obtained by assuming that r_d is connected from drain to GND, not to the source (Fig. 6-26d). A_v then becomes

$$A_v = -\frac{g_m R_D \| r_d}{1 + g_m R_S} \qquad (6\text{-}17)$$

A precise analysis of this circuit is shown in Appendix E, yielding

$$A_v = \frac{v_o}{v_{in}} = -\frac{g_m R_D}{1 + g_m R_S + \dfrac{R_S + R_D}{r_d}} \qquad (\text{E-}5)$$

EXAMPLE 6-11

Find A_v for the circuit of Fig. 6-26a where

$$V_P = -4\text{ V}, \quad I_{DSS} = 8\text{ mA}, \quad V_{DD} = 25\text{ V},$$

$$R_S = 500\ \Omega, \quad r_d = 20\text{ k}\Omega, \quad \text{and}$$

(a) $R_D = 6\text{ k}\Omega$.

(b) $R_D = 3\ k\Omega$.
(c) $R_D = 1\ k\Omega$.
Use Eq. (6-16) that neglects r_d and Eq. (6-17).

SOLUTION

First, find V_{GSQ} and g_{mQ}. (See Exs. 6-8 and 6-9.) The graphic solution is shown in Fig. 6-26e. R_D does not affect V_{GSQ} or g_{mQ}; consequently, $g_{mQ} = 2400\ \mu\mho$ for all three parts of this example.
(a) $R_D = 6\ k\Omega$
From Eq. (6-16)

$$A_v = -\frac{g_m R_D}{1 + g_m R_S} = \frac{-2400 \times 10^{-6} \times 6 \times 10^3}{1 + 2400 \times 10^{-6} \times 500} \quad (6\text{-}16)$$

$$= \frac{-14.4}{2.2} = -6.5$$

From Eq. (6-17)

$$A_v = -\frac{g_m \times R_D \| r_d}{1 + g_m R_S} = -\frac{2400 \times 10^{-6} \times 6 \| 20}{2.2} = -5.0$$

If we use the more elaborate Eq. (E-5),

$$A_v = -5.7$$

Equation (6-16) results in a 14% error relative to Eq. (E-5). Equation (6-17) results in a 12% error relative to Eq. (E-5).
(b) $R_D = 3\ k\Omega$

$$A_v = -\frac{2400 \times 10^{-6} \times 3 \times 10^3}{2.2} = \frac{-7.2}{2.2} = -3.3 \quad (6\text{-}16)$$

$$A_v = \frac{2400 \times 10^{-6} \times 3\ k\|20\ k}{2.2} = \frac{-6.26}{2.2} = -2.85 \quad (6\text{-}17)$$

Using Eq. (E-5), we find that the gain is $A_v = -3.0$. Errors of 10% and 5%, respectively (relative to $A_v = -3.0$).
(c) $R_D = 1\ k\Omega$

$$A_v = -\frac{2400 \times 10^{-6} \times 10^3}{2.2} = -\frac{24}{2.2} = -1.1 \quad (6\text{-}16)$$

$$A_v = \frac{2400 \times 10^{-6} \times 1\ k\|20\ k}{2.2} = -\frac{2.28}{2.2} = -1.04 \quad (6\text{-}17)$$

If we use Eq. (E-5), the gain is $A_v = -1.05$. Errors of 4.7% and 1%, respectively (relative to $A_v = -1.05$).

Note in Ex. 6-4 how small the error becomes as R_D decreases relative to r_d. Equation (6-17) is the better approximation yet simple when compared to Eq. (E-5). Only Eq. (6-16) or (6-17) will be used in this text. Equation (E-5) is given for reference purposes only.

EXAMPLE 6-12

Modify the circuit used in Ex. 6-11 by connecting a capacitor across R_S. (Fig. 6-27) Solve for A_v with
(a) $R_D = 6\ k\Omega$.
(b) $R_D = 3\ k\Omega$.
(c) $R_D = 1\ k\Omega$.

SOLUTION

The dc parameters are not affected by the introduction of the capacitor. Hence, $g_{mQ} = 2400\ \mu\mho$. The capacitor placed across R_S effectively shorts R_S, that is, $R_S \to 0$. The ac equivalent circuit becomes that shown in Fig. 6-27b. Two solutions will be given, one neglecting r_d [Eq. (6-12a)], the other including the effect of r_d [Eq. (6-12)].
(a) $R_D = 6\ k\Omega$

$$A_v = -g_m R_D = -2400 \times 10^{-6} \times 6 \times 10^3 \qquad (6\text{-}12a)$$
$$= -14.4 \quad \text{(neglecting } r_d\text{)}$$

$$A_v = -g_m R_D \| r_d$$
$$= 2400 \times 10^{-6} \times \frac{6 \times 10^3 \times 20 \times 10^3}{2 \times 10^3} \qquad (6\text{-}12)$$
$$= -11.1 \quad \text{(the effect of } r_d \text{ included)}$$

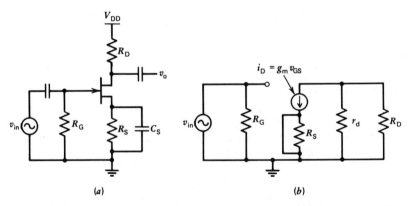

FIGURE 6-27 The CS circuit with C_S. (a) The circuit. (b) The ac eq. circuit.

The error in the approximate answer, relative to the latter solution, is about 30%. Clearly, in this case the exact expression Eq. (6-12) ought to be used.

(b) $R_D = 3\ k\Omega$

$$A_v = -2400 \times 10^{-6} \times 3 \times 10^3 = -7.2 \qquad (6\text{-}12a)$$

$$A_v = 2400 \times 10^{-3} \times \frac{3 \times 10^3 \times 20 \times 10^3}{23 \times 10^3} \qquad (6\text{-}12)$$

$$= -6.3$$

Error $\approx 14\%$

(c) $R_D = 1\ k\Omega$

$$A_v = -2400 \times 10^{-6} \times 10^{-3} = -2.4 \qquad (6\text{-}12a)$$

$$A_v = -2400 \times 10^{-6} \times \frac{10^3 \times 20 \times 10^3}{21 \times 10^3} \qquad (6\text{-}12)$$

$$= -2.3$$

Error $\approx 4\%$

In part (c) the error is small, making the approximate method quite acceptable.

EXAMPLE 6-13

Find v_o for the circuit of Fig. 6-28a.

SOLUTION

I_{DSS}, V_P, and R_S are given in Ex. 6-11. Again, $g_{mQ} = 2400\ \mu\mho$. The ac equivalent circuit is shown in Fig. 6-28b. Note that R_S is effectively zero (due to C_S). v_o is the voltage across the parallel combination of $R_L \| R_D \| r_d$

$$v_o = -\overbrace{g_m v_{GS}}^{i_D} \times (R_L \| R_D \| r_d)$$

Since $v_{GS} = v_{in} = 0.5\ V$

$$v_o = -2400 \times 10^{-6} \times 0.5 \times 6\ k\Omega \| 3\ k\Omega \| 50\ k\Omega$$

$$\approx \underline{-2.4\ V}$$

It is reasonable here to neglect $r_d = 50\ k\Omega$ compared to $R_L \| R_D = 2\ k\Omega$.

Example 6-13 points to the need to modify Eqs. (6-12), (6-12a), (6-16), (6-17), and (6-18) in cases in which an additional resistance R_L is connected from drain to ground. R_D is replaced in all these equations by $R_D \| R_L$, since in all circumstances these two resistors are, indeed, in parallel.

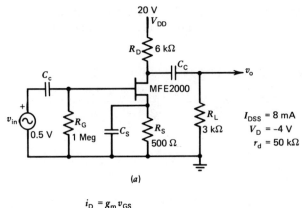

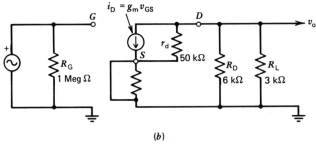

FIGURE 6-28 CS circuit with R_L. (a) Circuit. (b) Equivalent circuit.

Equation (6-12) becomes

$$A_v = -g_m R_D \| R_L \| r_d \qquad (6\text{-}12c)$$

Equation (6-12a) becomes

$$A_v = -g_m R_D \| R_L \qquad (6\text{-}12d)$$

Equation (6-16) becomes

$$A_v = \frac{-g_m R_D \| R_L}{1 + g_m R_S} \qquad (6\text{-}16a)$$

Equation (6-17) becomes

$$A_v = \frac{-g_m R_D \| R_L \| r_d}{1 + g_m R_S}$$

r_d in the above expressions may be neglected if $R_D \| R_L \ll r_d$—that is, when the combination $R_D \| R_L$ is much smaller than r_d. As a rule of thumb, neglect r_d if it is about 10 times (or more) larger than $R_D \| R_L$ ($r_d \geq 10 \times R_D \| R_L$). This means that Eqs. (6-12d) and (6-16a) are to be used.

EXAMPLE 6-14

Remove C_S in Fig. 6-28 and find A_v and v_o. $g_{mQ} = 2400\ \mu\mho$ (see Ex. 6-11)

SOLUTION

The equivalent circuit is shown in Fig. 6-29. Since $R_D \| R_L \ll r_d$, it is reasonable to neglect the effects of r_d. Hence,

$$A_v = \frac{-g_m R_D \| R_L}{1 + g_m R_S} \tag{6-16a}$$

$$= \frac{-2400 \times 10^{-6} \times 6\ \mathrm{k\Omega} \| 3\ \mathrm{k\Omega}}{1 + 2400 \times 10^{-6} \times 500} = \underline{-2.18}$$

Note that the exact solution [Eq. (E-5)] leads to $A_v = \underline{-2.13}$, a 2% difference. From the definition of A_v,

$$v_o = A_v \times v_{in}$$
$$= -2.18 \times 0.5 = \underline{-1.09\ \mathrm{V}}$$

It is worth repeating that the minus sign in A_v and v_o simply indicates a 180° phase shift (phase reversal) between v_{in} and v_o.

A_i, CURRENT GAIN AND r_{in}, INPUT RESISTANCE

Our initial assumption, as the equivalent of the input section indicates, is that there is no current into the gate. This means that i_{in} is the current into R_G only. Since this i_{in} can be made arbitrarily small, by increasing R_G (up to 10 MΩ), it is not meaningful to evaluate A_i. As far as the value of i_o is concerned, it can always be found once v_o is known

$$i_o = \frac{v_o}{R_L}$$

The fact that the only input current is that flowing through R_G indicates that

$$r_{in} = R_G \tag{6-18}$$

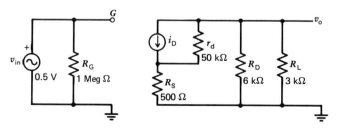

FIGURE 6-29 Equivalent circuit for Ex. 6-14.

R_G is effectively the only resistance that v_{in} has to drive.

r_o, OUTPUT RESISTANCE

By definition, r_o is the Thevenin's resistance of the source as viewed by the load. In other words, the *load* "looks" into the source to determine r_o.

In Fig. 6-30a the equivalent circuit of Fig. 6-29 is redrawn to show clearly the two terminals for which r_o is computed. To obtain r_o we follow the procedures for obtaining R_{Th}. First, remove the load as shown in Fig. 6-30a; then "eliminate" all independent sources (see Ch. 1)—that means, open current sources and replace voltage sources by a short (Fig. 6-30b). v_{in}, an *independent* voltage source, is replaced by a short. Since i_D is a *dependent* current source (it depends on v_{GS}), we find that the "opening" of i_D is consistent with the elimination of v_{in}. Clearly, $v_{in} = 0$ yields $v_{GS} = 0$, which in turn causes $i_D = 0$, effectively opening i_D.

The measurement of r_o, that is, connecting a v_m and measuring i_m, yields

$$r_o = R_D \| (r_d + R_S) \tag{6-19}$$

The series combination $r_d + R_S$ is in parallel with R_D.

As we have noted before, r_d is usually in the order of many kΩ; R_S is usually about 1 kΩ or less. As a result, since $R_S \ll r_d$, Eq. (6-19) may be approximated by

$$r_o = R_D \| r_d \tag{6-19a}$$

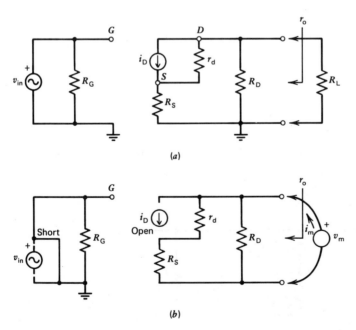

FIGURE 6-30 r_o for Fig. 6-29. (a) The equivalent circuit. (b) "Eliminating" all sources.

JFET AC ANALYSXIS (SMALL SIGNAL)

This is consistent with the modified equivalent circuit of Fig. 6-26d. In many cases $R_D \ll r_d$ and the expression for r_o may be simplified further

$$r_o \approx R_D \qquad (6\text{-}19b)$$

EXAMPLE 6-15

For the circuit in Fig. 6-28 find
(a) r_{in}.
(b) r_o.

SOLUTION

(a) $r_{in} = R_G = 1 \text{ M}\Omega$.
(b) The capacitor C_S, connected across R_S, effectively shorts R_S; hence, $R_S = 0$. Eq. (6-19) becomes

$$r_o = R_D \| (r_d + 0) = R_D \| r_d$$

which is Eq. (6-19a) given for the case in which R_S was neglected. Hence,

$$r_o = R_D \| r_d = 6 \text{ k}\Omega \| 50 \text{ k}\Omega = \underline{5.36 \text{ k}\Omega}$$

The approximation $r_o = R_D$ would produce a 12% error.

EXAMPLE 6-16

Find r_o for Fig. 6-28 with C_S removed.

SOLUTION

$$r_o = R_D \| (r_d + R_S) \qquad (6\text{-}19)$$
$$= 6 \text{ k}\Omega \| (50 \text{ k}\Omega + 5 \text{ k}\Omega)$$
$$\approx 6 \text{ k}\Omega \| 50 \text{ k}\Omega$$

Note that here $R_S \ll r_d$ and

$$r_o \approx 5.36 \text{ k}\Omega$$

The error introduced by the last approximation is about 1%.

6-6.5 Common Drain Circuit—Source Follower

The common drain circuit is usually called the source follower (SF). It is primarily used to provide a relatively low output resistance. The output of the SF, with its low r_o, can be used to drive higher loads (lower values of R_L) when compared to the drive capabilities of the CS circuit.

272 FIELD EFFECT TRANSISTORS (FET)

A typical SF circuit is shown in Fig. 6-31a. The equivalent circuit is shown in Fig. 6-31b and redrawn in Fig. 6-31c. The student is urged to verify that Fig. 6-31b and Fig. 6-31c are indeed *identical*. Note that v_o is taken *from the source*.

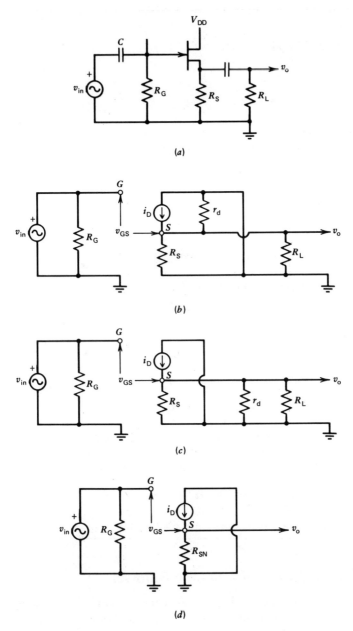

FIGURE 6-31 The SF circuit. (a) The circuit. (b) Equivalent circuit. (c) Equivalent circuit redrawn. (d) Replacing $R_d \parallel r_d \parallel R_L$ with R_{SN}.

Since, as shown in Fig. 6-31c, R_S, r_d, and R_L are all in parallel, these resistors have been replaced by $R_{SN} = R_S \| r_d \| R_L$ in Fig. 6-31d.

A_v, VOLTAGE GAIN

From Fig. 6-31d, v_o is

$$v_o = i_D \times R_{SN}$$

Here, v_o is positive (compared to Fig. 6-11), since the polarity of i_D is such to produce a positive v_o, as shown. Substituting $i_D = g_m v_{GS}$, we get

$$v_o = g_m \times v_{GS} \times R_{SN} \tag{6-20a}$$

To obtain v_{GS} in terms of v_{in}, we follow the analysis used for the CS circuit [Eqs. (6-14) and (6-15)]. Note that the only difference between the CS and SF, as far as the input section is concerned, is the substitution of R_{SN} for R_S. The result is, from Eq. (6-15),

$$v_{GS} = \frac{v_{in}}{1 + g_m R_{SN}} \tag{6-21}$$

If we combine Eqs. (6-20a) and (6-21)

$$v_o = \frac{g_m v_{in}}{1 + g_m R_{SN}} \times R_{SN}$$

and

$$\frac{v_o}{v_{in}} = \frac{g_m R_{SN}}{1 + g_m R_{SN}} = A_v \tag{6-22}$$

It is clear from Eq. (6-22) that $A_v < 1$ (compare with A_v for the EF circuit).

A comparison of Eq. (6-22) with (6-16a) shows that they are indeed similar. The term in the denominator is modified to account for the total resistance, source to ground. In the CS circuit it was R_S, while here it is $R_{SN} = R_S \| r_d \| R_L$. In the numerator R_{SN} replaces $R_D \| R_L$, since here the voltage v_o is across R_{SN}, while for the CS circuit, v_o is taken across $R_D \| R_L$. The sign reversal has been discussed above.

For the circuits where $r_d \gg R_L \| R_S$, the parallel combination $R_S \| r_d \| R_L$ may be approximated by $R_{SN} \approx R_S \| R_L$.

r_{in}, INPUT RESISTANCE

Here, as in the CS circuit,

$$r_{in} = R_G \tag{6-23}$$

(Gate-source is an open-circuit.)

r_o, OUTPUT RESISTANCE

To simplify the calculation of r_o, we perform the analysis in two parts. First, we find r_{os}—that is, the output resistance when "looking" directly into the source (source to ground). Second, we combine this r_{os} with the *resistance* seen by R_L (Fig. 6-32) to yield r_o.

In order to evaluate r_{os}, we again "eliminate" all independent sources. In this case $v_{in} \rightarrow 0$ (Fig. 6-32c). i_D, however, can *not* be eliminated; this is so because in this circuit eliminating v_{GS}, by setting $v_{in} = 0$, does *not* result in $i_D = 0$. We can see that the measuring voltage v_m is essentially connected between gate and source; hence, even though $v_{in} = 0$, v_{GS} is non-zero, $v_{GS} = v_m$. It is apparent from the circuit that $i_m = i_D$. By definition $r_{os} = v_m/i_m$ since, as noted above, $v_{GS} = v_m$ and $i_m = i_D$.

$$r_{os} = \frac{v_{GS}}{i_D}$$

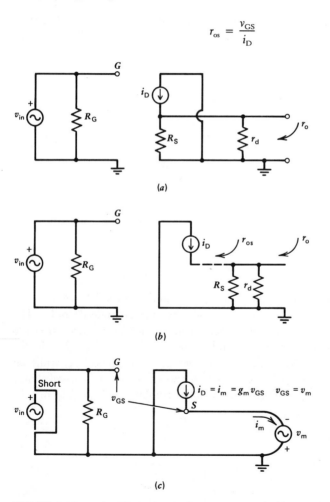

FIGURE 6-32 r_o for SF. (a) Equivalent circuit excluding R_L. (b) r_{os} and $R_s \parallel r_d$ shown separately. (c) Applying Thevenin's principles to find r_{os}.

JFET AC ANALYSXIS (SMALL SIGNAL)

From Eq. (6-7b) ($i_D = g_m v_{GS}$), we get

$$\frac{v_{GS}}{i_D} = \frac{1}{g_m}$$

As a result,

$$r_{os} = \frac{v_{GS}}{i_D} = \frac{1}{g_m} \quad (6\text{-}24)$$

Since r_{os} is in parallel with $R_S \| r_d$, r_o is given by

$$r_o = r_{os} \| R_S \| r_d = \frac{1}{g_m} \| R_S \| r_d \quad (6\text{-}25)$$

As in many earlier circuits, a simplification is possible for the case in which $r_d \gg R_S$

$$r_o \cong \frac{1}{g_m} \| R_S \quad (6\text{-}25a)$$

neglecting r_d. Equation (6-25a), given in detail, becomes

$$r_o = \frac{\frac{1}{g_m} \times R_S}{\frac{1}{g_m} + R_S} = \frac{R_S}{1 + g_m R_S} \quad (6\text{-}25b)$$

r_o is relatively small and is usually close to $1/g_m$.

EXAMPLE 6-17

For the circuit shown in Fig. 6-31a, $R_G = 0.5$ MΩ, $R_S = 6$ kΩ, $R_L = 3$ kΩ, $r_d = 50$ kΩ, $g_{mQ} = 3000$ μ℧. Find:
(a) A_v.
(b) r_{in}.
(c) r_o.

SOLUTION

(a)
$$A_v = \frac{g_m R_{SN}}{1 + g_m R_{SN}} \quad (6\text{-}22)$$

$R_{SN} = r_d \| R_S \| R_L = 50 \text{ kΩ} \| 6 \text{ kΩ} \| 3 \text{ kΩ}$
$= 50 \text{ kΩ} \| 2 \text{ kΩ} = 1.9 \text{ kΩ}$

$$A_v = \frac{3000 \times 10^{-6} \times 1.9 \times 10^3}{1 + 3000 \times 10^{-6} \times 1.9 \times 10^3} = 0.85$$

276 FIELD EFFECT TRANSISTORS (FET)

(b) $r_{in} = R_G = 0.5\ M\Omega$

(c) $$r_o = \frac{1}{g_m}\|R_S\|r_d \qquad (6\text{-}25)$$

$$r_o = \frac{1}{3000 \times 10^{-6}}\|6\ k\Omega\|50\ k\Omega$$

$$= 333\|6000\|50000 = 311\ \Omega$$

Neglecting r_d, we have $r_o \cong 1/g_m\|R_S = 313\ \Omega$ (a very small error).

Note that r_o is very close to $1/g_m = 333\ \Omega$. This is particularly true when $R_S \gg 1/g_m$ and $r_d \gg 1/g_m$. [Eq. (6-25b) yields $r_o = 313\ \Omega$. Since it is identical to Eq. (6-25), $r_o = 1/g_m\|R_S$.]

6-7 CS AND SF APPLICATIONS

The following examples are intended to demonstrate the analysis of a variety of CS and SF circuits. The reader must realize, however, that only a few selected circuits will be discussed.

EXAMPLE 6-18

Obtain the ac equivalent circuit for Fig. 6-33a. Make reasonable approximations ($\pm 10\%$ error) and find:

(a) A_v.
(b) r_{in}.
(c) r_o.

SOLUTION

In order to obtain the ac equivalent circuit, we must first find g_{mQ}. This involves dc analysis to find V_{GSQ}.

$$V_G = \frac{20 \times 1.0}{9.0 + 1.0} = 2\ V$$

The input load line equation is

$$V_G - V_{GS} - I_D R_{ST} \quad 0 \quad (R_{ST} = R_{S1} + R_{S2})$$

$$2 - I_D \times 540 = V_{GS}$$

V_{GSQ} is found at the intersection of the load line and the transfer characteristic (Shockley's equation).

$$V_{GSQ} = -0.32\ V \quad \text{(Fig. 6-33b)}$$

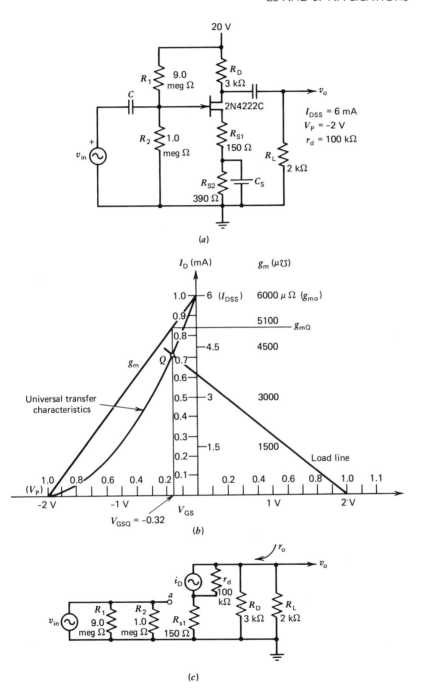

FIGURE 6-33 Circuits for Ex. 6-18. (a) The circuit. (b) Graphic analysis. (c) Equivalent circuit.

278   FIELD EFFECT TRANSISTORS (FET)

The g_m plot is obtained from the two points:

1. $g_m = 0$ (at $V_{GS} = V_P$)

2. $g_m = g_{mo} = \dfrac{-2I_{DSS}}{V_P}$

   $= \dfrac{-12 \times 10^{-3}}{-2} = 6000 \ \mu\mho$ (at $V_{GS} = 0$)

The graphic solution for g_{mQ} is shown in Fig. 6-33b.

$$g_{mQ} = 5100$$

(A much more complex algebraic solution, which is less subject to error, yields $g_{mQ} = 5070 \ \mu\mho$. Note how close the simple graphic solution is to the more elaborate algebraic one.) The ac equivalent circuit is shown in Fig. 6-33c.

Since $r_d \gg R_L \| R_D$ and $r_d > R_S$, r_d may be neglected. Note that for the dc analysis, $R_{ST} = R_{S1} + R_{S2} = 540 \ \Omega$, while for ac purposes $R_S = R_{S1} = 150 \ \Omega$ since R_{S2} is shorted by C_S.

(a) $A_v = \dfrac{-g_m R_D \| R_L}{1 + g_m R_S}$

$R_D \| R_L = 3 \text{ k}\Omega \| 2 \text{ k}\Omega = 1.2 \text{ k}\Omega$

$A_v = \dfrac{-5100 \times 10^{-6} \times 1.2 \times 10^3}{1 + 5100 \times 10^{-6} \times 150} = \underline{-3.47}$

(b) Here, $r_{in} = R_1 \| R_2 = 9.0 \text{ M}\Omega \| 1.0 \text{ M}\Omega$
$= \underline{900 \text{ k}\Omega}$

(c) $r_o \approx R_D = \underline{3 \text{ k}\Omega}$ (The error introduced by neglecting r_d is about 3%.)

EXAMPLE 6-19

The circuit shown in Fig. 6-34 drives a 8 kΩ load; it must be biased so that $V_{GSQ} = -1$ V, $V_{DQ} = 16$ V ($\frac{1}{2} V_{DD}$). The required voltage gain is -10. Find:
(a) R_D.
(b) R_{S1} and R_{S2}.

SOLUTION

The solution involves both dc and ac considerations. At $V_{GSQ} = -1$ V ($\frac{1}{2} V_P$), $I_D = \frac{1}{4} I_{DSS} = 2$ mA
   (a) R_D is found by applying the dc KVL to the output loop.

$$V_{DD} - I_D R_D - V_D = 0, \quad R_D = \dfrac{V_{DD} - V_D}{I_D}$$

CS AND SF APPLICATIONS 279

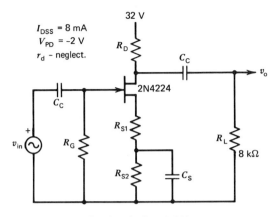

FIGURE 6-34 SF circuit (Ex. 6-19).

Since $V_D = 16$ V and $I_D = 2.0$ mA (the operating point),

$$R_D = \frac{32 - 16}{2 \times 10^{-3}} = 8 \text{ k}\Omega$$

(b) We first find $R_{ST} = R_{S1} + R_{S2}$, based on dc considerations. The dc KVL applied to the gate-source loop yields

$$V_{GS} = -I_D R_S, \qquad R_{ST} = \frac{-V_{GS}}{I_D}$$

$$R_{ST} = -\frac{-1}{2 \times 10^{-3}} = 500 \ \Omega$$

($V_{GS} = -1$ V is the Q-point.) To obtain R_{S1}, ac considerations are used.

$$|A_v| = \frac{g_{mQ} R_D \| R_L}{1 + g_{mQ} R_{S1}}, \qquad R_{S1} = \frac{R_D \| R_L}{|A_v|} - \frac{1}{g_{mQ}}$$

here, g_{mQ} is ½ g_{mo}, since $V_{GSQ} = ½ V_P$. Hence,

$$g_{mo} = \frac{-2 \times 8 \times 10^{-3}}{-2} = 8000 \ \mu\mho$$

$$g_{mQ} = 4000 \ \mu\mho$$

$$R_{S1} = \frac{8 \text{ k}\Omega \| 8 \text{ k}\Omega}{10} - \frac{1}{4000 \times 10^{-6}} = 150 \ \Omega$$

$$R_{S2} = R_{ST} - R_{S1} = 500 - 150 = 350 \ \Omega$$

280 FIELD EFFECT TRANSISTORS (FET)

EXAMPLE 6-20

The CS and SF circuits are sometimes combined, resulting in a circuit with two outputs, one from the drain v_{oD} and one from the source v_{oS} (Fig. 6-35). As shown, each drives a 1 kΩ load. Neglect r_d. Find:
(a) v_{oD}.
(b) v_{oS}.
(c) r_{o1}.
(d) r_{o2}.

SOLUTION

From dc graphic analysis (not shown), we get $g_{mQ} = 2750$ μΩ. (The student is requested to verify this result.)
(a) To get v_{oD}, find the appropriate voltage gain.

$$A_{vD} = \frac{v_{oD}}{v_{in}} = \frac{-g_m R_D \| R_{LD}}{1 + g_m R_S \| R_{LS}} \qquad (6\text{-}16a)$$

Note that as far as ac signals are concerned, the effective resistance source to GND is $R_S \| R_{LS}$.

$$A_{vD} = \frac{-2750 \times 10^{-6} \times 6\,\text{k}\Omega \| 1\,\text{k}\Omega}{1 + 2750 \times 10^{-6} \times 6\,\text{k}\Omega \| 1\,\text{k}\Omega} = -0.7$$

$$v_{oD} = A_{vD} \times v_{in} = -0.7 \times 1.5 = \underline{-1.05\ \text{V}}$$

(b) $A_{vS} = \dfrac{g_m R_S \| R_{LS}}{1 + g_m R_S \| R_{LS}}$

$\qquad = \dfrac{2750 \times 10^{-6} \times 6\,\text{k}\Omega \| 1\,\text{k}\Omega}{1 + 2750 \times 10^{-6} \times 6\,\text{k}\Omega \| 1\,\text{k}\Omega}$

$\qquad A_{vS} = \underline{0.7}$

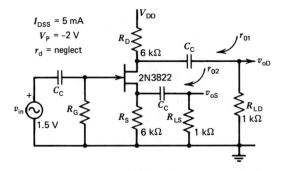

FIGURE 6-35   CS and SF combination (Ex. 6-20).

$(|A_{vs}| = |A_{vD}|$ since $R_S = R_D$ and $R_{LS} = R_{LD}$.)
$|v_{oD}| = |v_{os}| = 0.7 \times 1.5 = 1.05$ V

Note that although both outputs have the same amplitude, 1.05 V, they are 180° apart. This circuit is called a phase splitter. It is used in conjunction with push-pull type power amplifiers. (See Ch. 8.)

(c) $r_{o1} = 6$ kΩ.

(d) $r_{o2} = \dfrac{1}{g_{mQ}} \| R_S = \dfrac{1}{2750 \times 10^{-6}} \| 6$ k$\Omega = 343$ Ω

EXAMPLE 6-21

For the circuit shown in Fig. 6-36a find:
(a) The Q-point (V_{GSQ}, I_{DQ}, and V_{DSQ}).
(b) A_v.
(c) v_o.
(d) r_{in}.
(e) r_o.

SOLUTION

The transistor is a P channel, with a transfer characteristic as shown in Fig. 6-36b. Note that I_D is negative and V_{GS} positive. It is convenient to temporarily reverse all dc voltage and current polarities and assume the transistor to be an N channel type. This leads to the transfer characteristics of Fig. 6-36c. The ac performance of the P and N channel does not depend on the dc polarities of these transistors. No polarity reversal is needed in ac analysis. The dc load line is obtained from the equation $V_G - V_{GS} - I_D R_S = 0$ (Fig. 6-36c) where

$$V_G = \dfrac{28 \times 1 \text{ M}\Omega}{(22 + 1) \text{ M}\Omega} = 1.2 \text{ V} \quad \text{and} \quad R_S = 1.2 \text{ k}\Omega$$

(a) From the load line

$$V_{GSQ} = -1.05 \text{ V} \qquad I_{DQ} = 1.8 \text{ mA}$$

To get the actual V_{GSQ} and I_{DQ} for the P channel transistors, we reverse the polarities

$$V_{GSQ} = +1.05 \text{ V}$$
$$I_{DQ} = -1.8 \text{ mA}$$

$$\begin{aligned} V_{DSQ} &= V_{DD} - I_D \times R_D - I_D \times R_S \\ &= 32 - 1.8 \times 10^{-3} \times 10 \times 10^3 - 1.8 \times 10^{-3} \times 1.2 \times 10^3 \\ &= 11.8 \text{ V} \end{aligned}$$

$V_{DSQ} = -11.8$ V   (again reverse polarity)

282 FIELD EFFECT TRANSISTORS (FET)

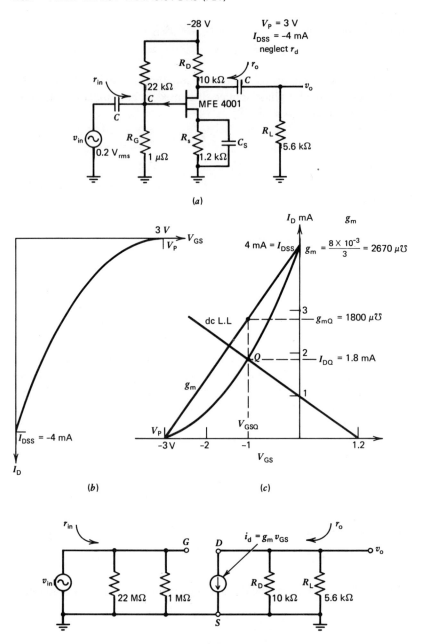

FIGURE 6-36 *P*-channel CS circuit (Ex. 6-21). (a) The circuit. (b) *P*-channel transfer characteristics. (c) Inverted characteristics. (d) The equivalent circuit.

(b) Find g_{mQ} on the g_m graph (Fig. 6-36c) $g_{mQ} = 1800\ \mu\mho$. The equivalent circuit is shown in Fig. 6-36d. It is identical to that of a similar circuit using an N channel transistor. Clearly, *no polarity reversal* is called for in ac computations.

$$A_v = -g_m \times R_D \| R_L$$
$$= -1800 \times 10^{-6} \times 10\ k \| 5.6\ k = \underline{-6.5}$$

(c) $v_o = A_v \times v_{in} = -6.5 \times 0.2 = 1.3\ V_{rms}.$
(d) $r_{in} = 22\ M \| 1\ M = 0.96\ M\Omega.$
(e) $r_o = R_D = 10\ k\Omega.$

6-7.1 FET Circuits with Varying V_P and I_{DSS}

So far, all FETs were assumed to have specific values for V_P and I_{DSS}. The specification sheet for these devices, however, gives a range of V_P and I_{DSS}. For example, the data sheet for the MPF107, N channel FET (made by Motorola, Inc.) gives the range for V_P (V_{GSOFF} in Motorola's notation) as -2.0 to -6.0 V and for I_{DSS} from 8.0 to 20 mA. The minimum and maximum transfer characteristics corresponding to this range of values are shown in Fig. 6-37. A typical load line drawn for descriptive purposes only is shown to intersect the two graphs at Q_1 and Q_2. The actual operating point may lie anywhere between Q_1 and Q_2

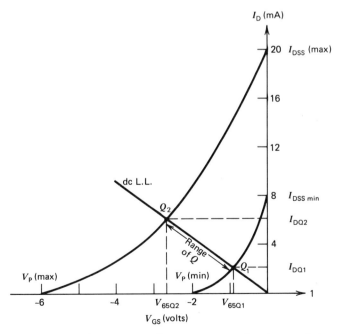

FIGURE 6-37 Minimum and maximum transfer characteristics.

on the load line. g_m for any Q in this range can only be estimated if specific V_P and I_{DSS} are not known. Note that g_{mo} ranges between

$$\text{Max. } g_{mo} = \frac{-2I_{DSS(max)}}{V_{P(min)}}$$

$$= \frac{40 \times 10^{-3}}{2} = 20{,}000 \ \mu\mho$$

$$\text{Min. } g_{mo} = \frac{-2I_{DSS(min)}}{V_{P(max)}}$$

$$= \frac{16 \times 10^{-3}}{6} = 2670 \ \mu\mho$$

The range of g_{mo} is rather extreme. The actual range of g_{mo} is substantially less and is often specified in the data sheet. For the M_{PF} 107 the range of g_{mo} is given as 4000 to 8000 $\mu\mho$.

6-8 MOSFET (METAL-OXIDE SEMICONDUCTOR FET)[1]

In the MOSFET (unlike the JFET), the gate is completely insulated from the channel by a thin (about 1 μm) layer of silicon oxide. This permits operation with gate-source or gate-channel voltages *above* and *below* zero. Both positive and negative voltages are acceptable. Recall that the P-N (or N-P) gate to channel structure of the JFET requires that this junction always be reverse biased. The insulated gate of the MOSFET further reduces substantially the gate current. In the JFET, the gate current, which consists of a reverse P-N junction current, is of the order of nanoamps (10^{-9} A). In the MOSFET, the gate current is less than one picoamp (10^{-12} A). This translates into an input-gate resistance of about 10^8 to 10^{10} Ω for the JFET and 10^{12} Ω or more for the MOSFET.

The operation of the JFET relies on the depletion principle. The gate to source voltage tends to reduce (deplete) carrier concentrations in the channel, thereby increasing channel resistance and reducing channel current. The MOSFET, on the other hand, may operate in both the depletion as well as the enhancement mode. In the enhancement mode, channel conduction is *enhanced*, carrier concentration increased by the gate to source voltage, V_{GS}.

The MOSFETs are available as N-channel or P-channel structures, operating in either depletion or enhancement modes or both.

6-8.1 MOSFET Structure

A typical N-channel MOSFET structure capable of depletion and enhancement operation is shown in Fig. 6-38. A silicon oxide is used to insulate electrically the gate from the channel. Channel conduction is controlled by the electrical field

---

[1] Also called IGFET, insulated gate FET.

MOSFET (METAL-OXIDE SEMICONDUCTOR FET)

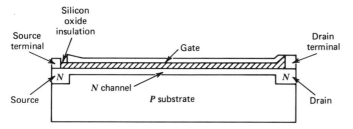

FIGURE 6-38 N-channel MOSFET structure—cross section. (Illustrative only.)

induced in the channel by the gate voltage. The process of induction is similar to that taking place in a capacitor. A positive voltage (positive charges) on the gate induces negative charges in the channel. The induced negative charges increase the already relatively high concentration of the free electrons in the channel (assuming an N channel MOS), and hence, increase channel conductivity. This is the enhancement mode. Channel conduction is increased by gate voltage.

A negative voltage applied to the gate induces positive charges in the channel, *reducing* electron concentration and with it the channel conduction. This depletion mode of operation is similar to the JFET operation. With sufficient negative voltage applied to the gate the channel may be completely cut off. The gate voltage producing channel cut off is called V_P (as in the JFET) or $V_{GS(OFF)}$.

The pure enhancement mode MOSFET does not contain a physical channel and operates with an induced channel, sometimes referred to as a "virtual" channel. The structure of the N-channel MOSFET (called the N-channel even though there is no physical channel) is the same as that of the depletion or enhancement transistor with the N-channel eliminated. The drain and source are separated by a P-type material (Fig. 6-39a).

With no voltage applied to the gate, the drain-source structure is essentially a back-to-back connection of two N-P junctions (NP-PN). There is no channel conduction other than leakage current. A positive voltage applied to the gate induces a negative charge on the surface immediately under the gate. This has the effect of "inverting" the P material in that region—that is, producing a virtual or induced N-channel. Conduction may then take place through the channel. It is worth noting that the gate voltage must exceed a certain threshold voltage V_T before conduction can take place. (This V_T usually ranges between 1 to 2 V for the silicon gate structure, which is most frequently used in digital applications.) Fig. 6-39b shows the virtual channel induced by the gate voltage. Because of the close proximity of gate to "channel," they are separated by a 1 μm oxide insulator; relatively small gate voltages produce large charge distributions in the "channel."

6-8.2 MOSFET Characteristics

MOSFETs have two basic types of drain and transfer characteristics: one type for the enhancement mode transistor and one for the depletion/enhancement mode. We also note that the N-channel and P-channel transistors differ only in the polarity

286 FIELD EFFECT TRANSISTORS (FET)

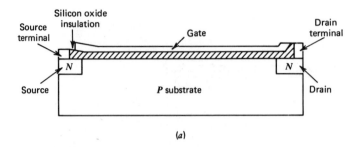

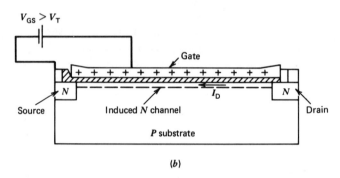

FIGURE 6-39 Enhancement mode MOSFET. (a) Basic structure—no physical channel. (b) Showing induced N channel.

of the various currents and voltages. The general shape of the curves is, essentially, identical. In the following discussion typical N-channel transistors will be referred to. The equivalent (or matched) P-channel characteristics may be obtained by simply reversing all polarities.

A set of characteristics, drain and transfer, for the depletion/enhancement transistor are shown in Fig. 6-40. V_{GS} may range from about -3 V, at which point the channel is cut off, to about $+5$ V for which $I_D = 20$ mA. Because the particular transistor has a maximum drain current rating of 20 mA, the plots are restricted to I_D below 20 mA.

The general shape of the transfer characteristics are similar to those for the JFET. The MOSFET, however, may operate with $V_{GS} > 0$. As a result, I_{DSS} is *not* the maximum drain current as it is for the JFET. For the MOSFET, I_{DSS} is simply another point on the plot. In particular, it is I_D when $V_{GS} = 0$, as defined in the discussion of the JFET.

An approximate mathematical representation of the transfer characteristics may again be given by Shockley's equation

$$I_D = I_{DSS}\left(1 - \frac{V_{GS}}{V_P}\right)^2 \tag{6-1}$$

Fig. 6-40c shows how well Eq. (6-1) agrees with the transfer characteristics given by the manufacturer. When evaluating Eq. (6-1) for $V_{GS} > 0$, V_{GS}/V_P is negative and the term in parenthesis is larger than 1 indicating that for $V_{GS} > 0$, $I_D > I_{DSS}$.

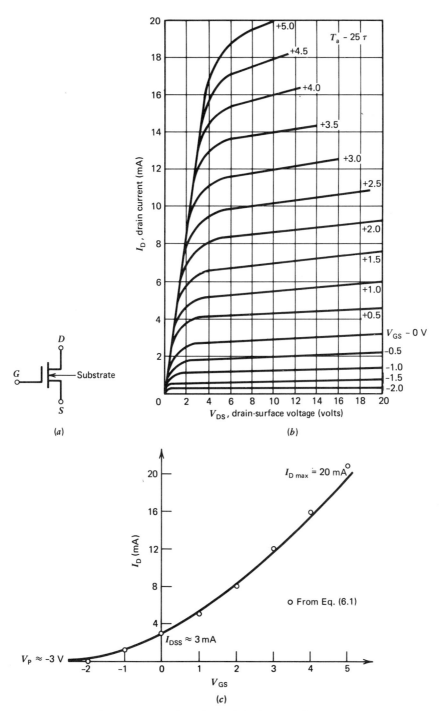

FIGURE 6-40 Characteristics of N-channel depletion-enhancement MOSFET-2N3797 (courtesy of Motorola Corp). (a) Symbol. (b) Drain characteristics. (c) Transfer characteristics. V_{GS} redrawn on linear scale.

288 FIELD EFFECT TRANSISTORS (FET)

Figure 6-41 shows the typical enhancement characteristics for an N-channel transistor. The threshold voltage V_T is about 2 V. That means that for $V_{GS} < 2$ V, the drain current I_D is very small and consists of a leakage current.

The approximate mathematical description of the transfer characteristics of the

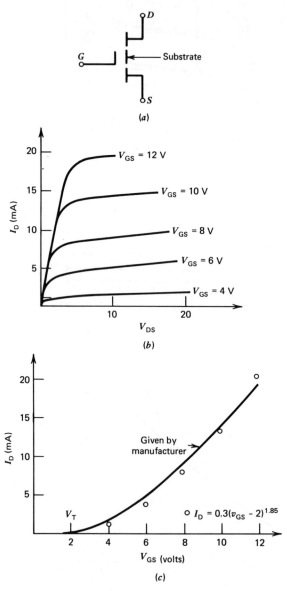

FIGURE 6-41 Characteristics of enhancement MOSFET (2N4351). (a) Symbol. (b) Drain characteristics (typical). (c) Transfer characteristics.

typical enhancement MOSFET is given by

$$I_D = K(V_{GS} - V_T)^a \qquad (6\text{-}26)$$

I_D – mA, V_{GS} and V_T – volts

Here, both K and a are selected to approximate the characteristics for a specific transistor. For convenience we use $a = 2$, and K is typically 0.3 mA/V². (As we see in Fig. 6-41c, a very good fit is accomplished with $a = 1.85$.) If we use these typical values for a and K, the transfer characteristics for the enhancement MOSFET are approximated, rather crudely, by

$$I_D = 0.3(V_{GS} - V_T)^2 \qquad (6\text{-}27)$$

Note that Eq. (6-27) is valid only for $V_{GS} > V_T$, since for $V_{GS} \leq V_T$ the transistor is cut off, $I_D = 0$. For the P-channel MOSFET, where V_T and V_{GS} are negative, the same expression may be used as in Eq. (6-27). This time, however, the expression holds only for $V_{GS} < V_T$; V_{GS} more negative than V_T. (For $V_{GS} \geq V_T$, $I_D = 0$.)

For the sake of completeness, Fig. 6-42 gives the transfer characteristics of depletion/enhancement and enhancement MOSFET of both N-channel and P-channel types. The symbols shown are used by most manufacturers. A variety of other symbols are, and have been, used by a number of manufacturers. We will use the symbols given here exclusively.

The value $I_{D(max)}$ refers to the maximum permissible drain current as specified by the manufacturer. It varies substantially from one transistor type to another.

Note that the only difference between the characteristics of the P and N channel of the same type are the polarities of the voltage and current. Many manufacturers represent both the P and N channel in the first two quadrants, assigning appropriate polarities to V_{GS} and I_D (Fig. 6-43).

6-9 MOSFET BIASING

All the biasing circuits used for the JFET are applicable to the MOSFET. The method of finding the Q-point is essentially that used for the JFET. The differences between the JFET and the MOSFET are a function of the *range* of operation of the devices. V_{GS} for the depletion/enhancement MOSFET can be positive *or* negative. A Q-point such that $V_{GS} = 0$ is then somewhere in the *middle* of the operating range and may be used in amplifier applications (Fig. 6-44). Similarly, the enhancement MOSFET may be biased directly from V_{DD}, with no emitter resistor (Fig. 6-45). These circuits *cannot* function with a JFET.

EXAMPLE 6-22

Find Q for the circuit of Fig. 6-46a. Use the dc load line method. The transfer characteristics are given in Fig. 6-46b.

290 FIELD EFFECT TRANSISTORS (FET)

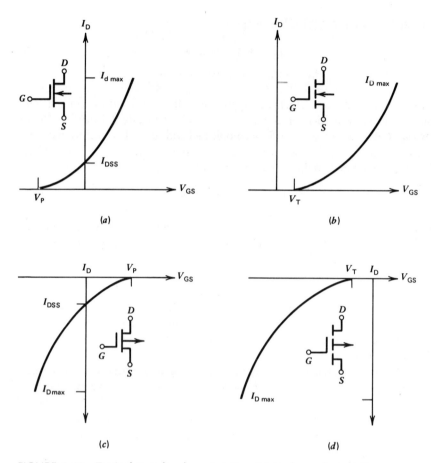

FIGURE 6-42 Typical transfer characteristics. (a) N channel, depletion-enhancement. (b) N channel, enhancement. (c) P channel, depletion-enhancement. (d) P channel, enhancement.

SOLUTION

The load line is shown in Fig. 6-46b. The load line equation is

$$V_G - V_{GS} - I_D R_S = 0$$

where

$$V_G = \frac{V_{DD} \times R_2}{R_1 + R_2} = \frac{20 \times 5 \text{ M}\Omega}{5 \text{ M}\Omega + 5 \text{ M}\Omega} = 10 \text{ V}$$

Rewriting the load line equation, we get

$$V_{GS} = 10 - I_D \times 500$$

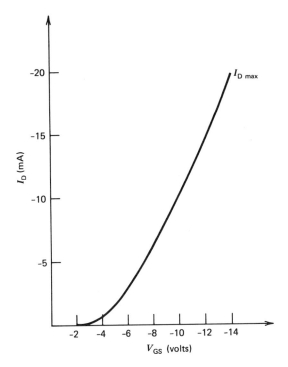

FIGURE 6-43 P-channel MOSFET transfer characteristics (2N4352) with reversed polarities.

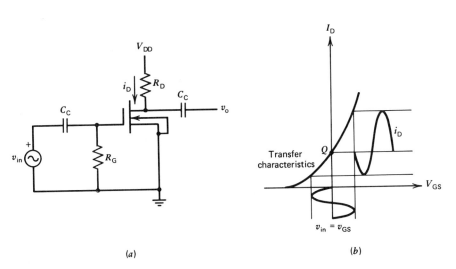

FIGURE 6-44 Depletion-enhancement FET biased at $V_{GS} = 0$. (a) Circuit. (b) Q-point and signal waveforms.

292 FIELD EFFECT TRANSISTORS (FET)

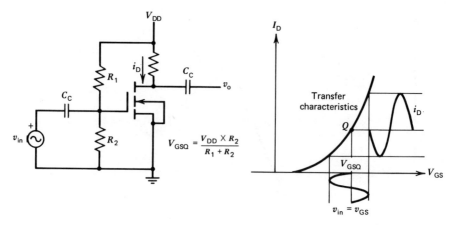

FIGURE 6-45 Enhancement MOSFET biasing. (a) Circuit. (b) Q-point and waveforms.

The plot of the load line is obtained from:

I_D	V_{GS}
0	10 V
10 mA	5 V

The Q-point is found at the intersection of the load line and the characteristic curve

$$I_{DQ} = \underline{6.4 \text{ mA}}$$
$$V_{GSQ} = \underline{6.8 \text{ V}}$$

Another circuit that is commonly used with enhancement MOSFETs and could not be used with the JFETs is the self bias with no source resistor (Fig. 6-47). To obtain the Q-point, KVL is applied to the gate-source section.

$$V_D - V_{GS} = 0$$

(There is no voltage across R_G.)
Since $V_D = V_{DD} - I_D R_D$, the KVL can be written as

$$V_{DD} - I_D R_D - V_{GS} = 0 \qquad (6\text{-}28)$$

$$V_{GS} = V_{DD} - I_D R_D \qquad (6\text{-}28a)$$

Here, R_D plays a role similar to that of R_S. (Compare this KVL equation with those for the voltage divider JFET bias.) The load line, Eq. (6-28a), may be plotted on

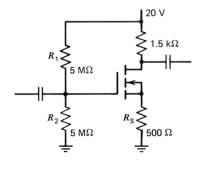

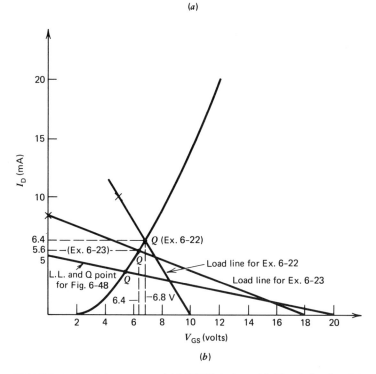

FIGURE 6-46 Enhancement MOSFET biasing. (a) Circuit for Ex. 6-22. (b) Load line solution for Exs. 6-22 and 6-23 and Fig. 6-48.

the transfer characteristics and a Q-point obtained. Two convenient points for the load line plot are

$$I_D = 0, \qquad V_{GS} = V_{DD}$$

and

$$I_D = \frac{V_{DD}}{R_D}, \qquad V_{GS} = 0$$

294   FIELD EFFECT TRANSISTORS (FET)

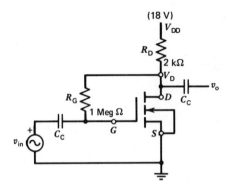

FIGURE 6-47   Self bias MOSFET ($V_D = V_G$).

EXAMPLE 6-23

In Fig. 6-47 let $R_D = 2 \text{ k}\Omega$, $V_{DD} = 18$ V. Obtain the Q-point using the transfer characteristics given in Fig. 6-46.

SOLUTION

The two points for the load line plot are $I_D = 0$, $V_{GS} = V_{DD} = 18$ V (substitute $I_D = 0$ in Eq. (6-28a) and solve for V_{GS}) and

$$I_D = \frac{V_{DD}}{R_D} = \frac{18 \text{ V}}{2 \text{ k}\Omega} = 9 \text{ mA}, \qquad V_{GS} = 0$$

[substitute $V_{GS} = 0$ in Eq. (6-28a) and find I_D]. The plot of the load line and the Q-point are shown in Fig. 6-46b.

$$I_{DQ} = \underline{5.6 \text{ mA}} \qquad V_{GSQ} = \underline{6.4 \text{ V}}$$

A bias circuit combining the self bias technique with a source resistor is shown in Fig. 6-48. To obtain the operating point, KVL is used

$$V_{DD} - I_D R_D - V_{GS} - I_D R_S = 0 \tag{6-29}$$

$$V_{GS} = V_{DD} - I_D(R_D + R_S) \tag{6-29a}$$

Equation (6-29a) is the same as Eq. (6-28a) with R_D in Eq. (6-28a) replaced by $R_D + R_S$. The solution for the Q-point follows the steps shown in Ex. 6-23. The load line and Q-point have been plotted in Fig. 6-46b for the values shown in Fig. 6-48. The load line equation is

$$V_{GS} = 20 - I_D(3.3 \text{ k}\Omega + 1.0 \text{ k}\Omega) \tag{6-29a}$$
$$= 20 - I_D(4.3 \text{ k}\Omega)$$

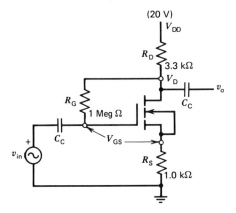

FIGURE 6-48 Self bias with R_s.

6-10 MOSFET ac ANALYSIS

The ac analysis of MOSFETs follows the methods used for the JFET. In particular, the depletion/enhancement FET, yields the same expression for g_m

$$g_m = \frac{-2I_{DSS}}{V_P}\left(1 - \frac{V_{GS}}{V_P}\right) \qquad (6\text{-}9)$$

As a result, the g_m plot is again a straight line as shown in Fig. 6-49. (The specific V_P and I_{DSS} shown are used in Ex. 6-24.) The value $g_{mo} = -2I_{DSS}/V_P$ is nothing more than a point on the g_m line for which $V_{GS} = 0$ and $I_D = I_{DSS}$. g_{mo} for the MOSFET is *not* the maximum g_m, unlike g_{mo} for the JFET.

EXAMPLE 6-24

Find g_{mQ} for the circuit in Fig. 6-50. Use the characteristics given in Fig. 6-49.

SOLUTION

The g_m plot is shown in Fig. 6-49. The load line is plotted from the KVL

$$V_G - V_{GS} - I_D R_S = 0$$

where

$$V_G = \frac{V_{DD} \times R_2}{R_1 + R_2} = \frac{18 \times 2 \text{ M}\Omega}{4 \text{ M}\Omega} = 9 \text{ V}$$

The KVL becomes

$$V_{GS} = 9 - I_D \times (560)$$

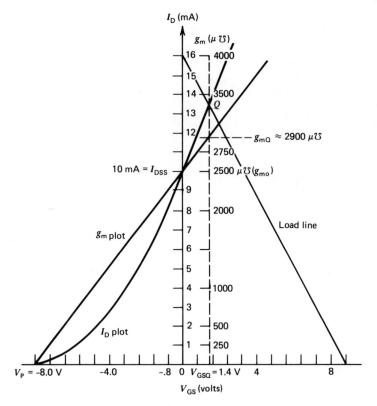

FIGURE 6-49 g_m plot for depletion-enhancement MOSFET.

The points used for the load line plot are

$$I_D = 0, \qquad V_{GS} = 9 \text{ V}$$

$$I_D = \frac{9}{560} = 16 \text{ mA}, \qquad V_{GS} = 0$$

V_{GS} at the Q-point is

$$V_{GSQ} = 1.4 \text{ V}$$

The g_m scale is based upon Eq. (6-9). Note that here

$$g_{mo} = \frac{-2I_{DSS}}{V_P} = \frac{-2 \times 10 \times 10^{-3}}{-8}$$

$$= 2500 \text{ μ}\Omega$$

g_{mQ} is the g_m associated with the Q-point. With $V_{GS} = 1.4$ V

$$g_{mQ} = 2900 \text{ μ}\mho$$

MOSFET AC ANALYSIS 297

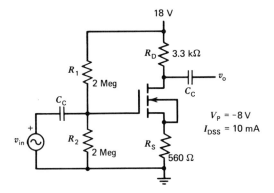

FIGURE 6-50 CS circuit using depletion-enhancement MOSFET. (Ex. 6-23).

Note that $g_{mQ} > g_{mo}$, which is impossible for the JFET since V_{GS} cannot exceed 0 V.

The amplifier parameters A_v, r_o, r_{in} can be found for any MOSFET circuit by using the identical equations and equivalent circuits used for the corresponding JFET circuit, and using the g_m as obtained in Ex. 6-24.

The g_m plot for the enhancement MOSFET is shown in Fig. 6-51. It is a plot of

$$\frac{dI_D}{dV_{GS}} = g_m = 2 \times K(V_{GS} - V_T) \tag{6-30}$$

(g_m in m℧, V_{GS} and V_T in V,   K in mA/V²)

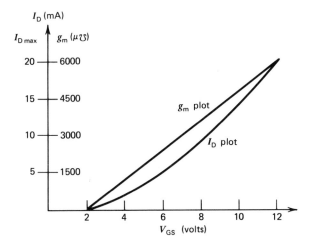

FIGURE 6-51 g_m plot for enhancement MOSFET.

298 FIELD EFFECT TRANSISTORS (FET)

Equation (6-30) is obtained from Eq. (6-26) with $a = 2$. The I_D and g_m plots of the enhancement MOSFET are not easily generalized to allow a single standard plot to be used in all enhancement MOSFET calculations. This requires specific graphs for each transistor. The values of V_T and K determine the shapes of these graphs. The plots in Fig. 6-49 were done for $V_T = 2$ V and $K = 0.3$ mA/V². The two points used for the g_m graph were: $V_{GS} = V_T = 2$ V; $g_m = 0$, and $V_{GS} = 12$ V (an arbitrary point for which $I_D = 20$ mA) and, by Eq. (6-30), for this V_{GS}

$$g_m = 2 \times 0.3(12 - 2) = 6 \text{ m}\mho = 6000 \text{ }\mu\mho \qquad (6\text{-}30)$$

6-11 FETS IN SWITCHING APPLICATIONS

The computer of today contains literally thousands of FETs operating in a digital mode. These digital transistors are either saturated (or near saturation) or cut off. The voltage, drain to source, V_{DS} is either near zero, when in saturation, or high when in cut off. Besides these digital applications, the FET is often used to switch analog voltages, disconnect an analog signal from one load and connect it to another (replacing the mechanical or electromechanical switches). The dominance of FETs in digital applications is largely due to the ease with which those transistors can be manufactured in integrated circuit form. Many hundreds of MOSFETs, with suitable circuitry, packaged in a small (1" × ½") package. The importance of FETs in analog switching applications is due to the resistive nature of the FET. A BJT, when saturated, *does* have a collector to emitter voltage. V_{CE} is made up of V_{CB} and V_{BE} when both these junctions are forward biased. These two forward diode voltage drops are effectively subtracted due to the junction polarities, to yield V_{CE}. Nevertheless, V_{CE} is rarely zero. This introduces spurious voltages when the transistor is used to switch analog voltages.

The FET behaves essentially like a resistor: very high resistance when the FET is open and low resistance when the FET is saturated. There is no voltage drop V_{DS} other than that due to the signal current.

6-11.1 Analog Switching

A simple analog switching application is shown in Fig. 6-52a. Two signal sources are remotely selected by control signals 1 and 2. By grounding control 1, Q_1 is "on" (saturated). If at the same time, Q_2 is cut off by control 2, only source 1 will be fed through to the load R_L. The state of the control lines can be interchanged, turning Q_2 "on" and Q_1 "off," disconnecting source 1, v_1, from the load and connecting v_2 to R_L. Q_1 and Q_2 (and any number of additional FETs) operate as switches.

There are two parameters of utmost importance in analog switching. First, the "on resistance" of the FET, $r_{ds(on)}$—that is, the resistance of the drain-source path when the transistor is fully "*on.*" Second, the "off resistance," which is essentially the "off" channel leakage resistance. This parameter is usually given in terms of

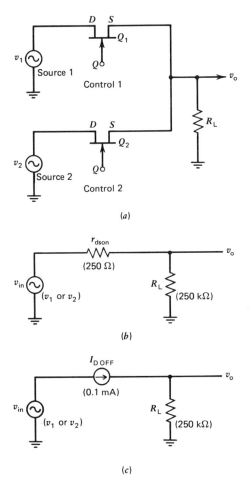

FIGURE 6-52 Analog signal switching (multiplexing). (a) Simplified circuit. (b) "On" state equivalent circuit. (c) "Off" state equivalent circuit.

the drain cut-off current, $I_{D(off)}$. For example, the 2N3824 JFET is specified to have $r_{D(on)} = 250\ \Omega$ maximum and $I_{D(off)} = 0.1$ nA maximum. In the "on" state, $r_{DS(on)}$ forms a voltage divider with R_L, resulting in a loss in the switched signal voltage. To minimize this loss, we require that $r_{DS(on)} \ll R_L$. On the other hand, $I_{D(off)}$ allows a *disconnected* signal to affect v_o, introducing coupling between the disconnected source and the output. This phenomenon is referred to as cross coupling or "cross talk." In general, it is relatively easy to select FETs that provide an almost perfect switching function. That is, no voltage drop for the "on" state and no feed through (perfectly open) in the "off" state. Over all accuracy of 0.1% or better is easily attainable. Clearly, as the number of sources to be switched, multiplexed as it is called, is increased, the accuracy deteriorates.

If we select, for example, $R_L = 250\ k\Omega$ and $Q_1 = Q_2 = 2N3824$ in Fig. 6-

300　FIELD EFFECT TRANSISTORS (FET)

52a, we find that the maximum voltage drop across $r_{DS(on)}$ in the "on" state is (see Fig. 6-52b)

$$v_{drop} = \frac{v_{in} \times r_{DS(on)}}{r_{DS(on)} + R_L} = \frac{v_{in} \times 250}{250 + 250 \text{ k}\Omega}$$

$$\approx 0.001 \, v_{in} \quad \text{(about 0.1\% of } v_{in}\text{)}$$

$$v_o = 0.999 \, v_{in}$$

At the same time, the contribution of the leakage current to v_o (the cross talk from one "open switch") is, at the most,

$$I_{D(off)} \times R_L = 0.1 \times 10^{-9} \times 250 \times 10^3 = 2.5 \times 10^{-5} = 25 \, \mu V$$

Recall that $r_{DS(on)} = 250 \, \Omega$, $I_{D(off)} = 0.1$ nA (see Fig. 6-52c).

6-11.2　Digital Switching

A JFET can be turned "on" by setting $V_{GS} = 0$ and turned "off" by applying an input to the gate so that $|V_{GS}| > |V_P|$. That is, the magnitude of V_{GS} is larger than the magnitude of V_P. More specifically, V_{GS} is more negative than V_P for the N-channel FET and more positive than V_P for the P-channel type. In both cases the result is cut off. For the enhancement MOSFET, cut off is obtained when V_{GS} is less than V_T; the transistor is operating below the threshold voltage. The switching is done by an input square-wave (or bilevel) signal. Figure 6-53 gives a typical switched JFET with the specific input amplitude required to turn the transistor "on" and "off." Note that the JFET is a P channel with $V_P = 3$ V. This requires that the input amplitude be larger than 3 V (it is 5 V) to cut the transistor "off." For the interval when v_{in} is zero, that transistor saturates, $I_D = I_{DSS}$. R_L must be selected so that

$$|I_{DSS} \times R_L| \geq V_{DD}$$

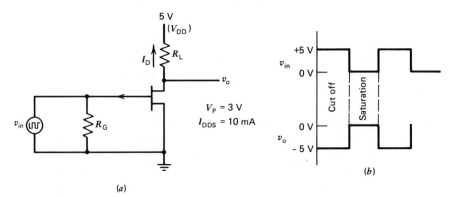

FIGURE 6-53　Digital JFET circuit (P channel). (a) The circuit. (b) Input and output waveforms.

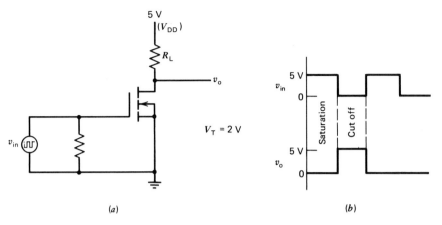

FIGURE 6-54 Enhancement MOSFET in a digital circuit. (a) The circuit. (b) Input and output waveforms.

This simply means that the full V_{DD} is dropped across R_L, so that for saturation $v_o = 0$ V. For the specific parameters shown in Fig. 6-53, $R_L = 1$ kΩ is appropriate since

$$|I_{DSS} R_L| = 10 \times 10^{-3} \times 10^3 = 10 \text{ V} \geq |V_{DD}|$$

so that v_o under saturation is zero. Note that I_D under saturation is less than I_{DSS} since

$$|I_{DSAT}| = \frac{|V_{DD}|}{R_L} = \frac{|-5|}{1 \text{ k}\Omega} = 5 \text{ mA}$$

A digital circuit using an enhancement MOSFET (N channel) is shown in Fig. 6-54. To obtain "cut off," v_{in} must be below V_T; this takes place for the interval when $v_{in} = 0$. For the interval in which $v_{in} = 5$ V, the transistor is turned on and the particular I_D can be found from the appropriate transfer characteristics at the point $V_{GS} = 5$ V. R_L must again be selected so that

$$I_D \times R_L \geq V_{DD}.$$

6-12 VMOS FET

The MOSFET has been, until recently, limited to low power applications (less than 1 W). The main reason for this restriction is the fact that the current in the FET I_D flows horizontally along the top surface, just below the gate. (See Fig. 6-39.) This thin and relatively long layer has very severe current limitations. To increase current carrying capacity, the physical size of the channel must be increased, providing a wider channel. This increases the gate to channel area and,

302 FIELD EFFECT TRANSISTORS (FET)

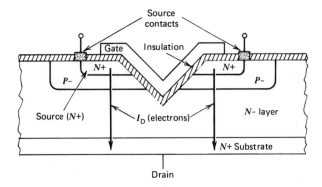

FIGURE 6-55 VMOS FET Structure.

hence, the gate to channel capacity. This, in turn, makes the transistor useless for all but very low frequency applications.

The "vertical" MOS, VMOS, overcomes these basic disadvantages inherent in the MOSFET structure by constructing the transistor so that the current I_D flows vertically. Figure 6-55 shows a typical cross-section of a VMOS transistor. The current (electrons) flows downward through the P^- [1] channel and the N^- [1] intermediate layer to the N^+ drain. The current path can be made very short and relatively very wide, allowing for a high current flow.

A positive gate to source voltage, above the threshold voltage, enhances current flow. For example, the VN66AF VMOS (Siliconix Inc.) has an $I_D = 0.65$ A at $V_{GS} = 5$ V and an $I_D = 2$ A at $V_{GS} = 10$ V.

Figure 6-56 gives the drain and transfer characteristics, as well as the g_m versus I_D plot for a typical VMOS, the VN66AF. Note the high I_D and V_{DS}. This is a power device capable of dissipating a maximum of 15 W, substantially more than the power capabilities of the standard MOSFET or JFET.

The VMOS has a number of important advantages. Its input resistance is very high, like the MOSFET, while its power capabilities are also high, like that of the BJT. In addition, as we see from Fig. 6-56b, the relation between V_{GS} and I_D is nearly linear (above threshold), not proportional to V_{GS}^2 as for the MOSFET. This results in a constant g_m and undistorted amplification, particularly of large signals (see Chapter 9). Note also that the g_m for the typical VMOS is very high, compared to the MOSFET, making very high gains possible.

In the BJT an increase in temperature causes an increase in I_C. This, in turn, causes an increase in device power dissipation resulting in an increase in device temperature that again increases I_C and so on, producing what is called "thermal runaway." A large I_C (specific values depend on the particular transistor) may cause

---

[1] The minus (−) or plus (+) signs next to the material type N^-, N^+, etc. indicates the level of concentration of the particular dopant. P^+ means P material with *high* dopant concentrations. N^- means N material with *low* concentrations of donor atoms.

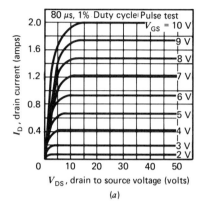
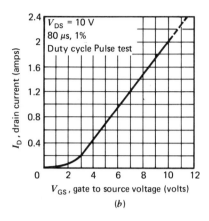
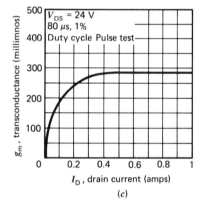

FIGURE 6-56 (a) Drain characteristics. (b) Transfer characteristics. (c) g_m vs I_D. (Courtesy Silicomix Inc.)

"thermal runaway" to take place and eventually this self-sustained process will destroy the transistor because of the excessive power dissipation. On the other hand, the current in the VMOS *decreases* with increases in temperature, thus completely eliminating the thermal runaway problem. A large I_D may cause an increase in the device temperature that will tend to *decrease* I_D, reducing the power dissipated in the VMOS and thus lowering the temperature, etc.

This reduced current with an increase in temperature permits paralleling of VMOS, for higher current applications, with no precaution against "current hogging."[2] There is no "current hogging" as it occurs in BJTs. The load current automatically divides between the two parallel VMOS FETs, while for BJTs one

---

[2] Current "hogging," where one of the paralleled BJTs conducts most of the current, is a result of initial imbalances between the transistors and is aggravated by temperature induced increases in I_C, as noted in connection with the thermal runaway.

of the paralleled transistors may "hog" mostly the load current and consequently overheat and be destroyed.

There also are speed advantages. The VMOS can turn "on" and turn "off" high currents, at rates many times faster than the BJT. For example, the 2N6657 VMOS can switch 1 A "on" or "off" in about 4 ns, compared to a rate of 100 ns or more for a suitable BJT.

The various analysis methods described in connection with the MOSFET are fully applicable to the VMOS, when used with the appropriate VMOS characteristic curves.

SUMMARY OF IMPORTANT TERMS

"Boot-strap" circuit A FET circuit used to increase r_{in}. It consists of a SF with R_G connected from gate to source.

Channel The bulk portion of the FET structure, through which the current (controlled by the gate-to-source voltage) passes.

Common drain A FET amplifier configuration in which the input signal is connected to the gate and the output is taken from the source. Also called a source follower, it has, characteristically, a relatively low output impedance and voltage gain less than 1.

Common source A FET configuration with the input at the gate and the output from the drain. Has a relatively high output impedance and gain.

Depletion A mode of FET operation in which the appropriate gate-source voltage keeps I_D below I_{DSS}.

Drain One of the three terminals of the FET. It is at one end of the FET channel and is structured to *receive* the current passing through the channel.

Enhancement A mode of operation in which I_D, the drain current, can be enhanced to exceed I_{DSS} by appropriate gate-source voltage. This is applicable to the MOSFET structure, not to a JFET.

Equivalent circuit Related to the FET, it is a representation of the FET that simulates FET behavior. It consists of linear circuit elements (such as current sources, resistors, etc.) and allows FET circuits to be analyzed by the use of circuit theorems.

FET Field Effect Transistor. A semiconductor structure in which the current through a semiconductor channel is controlled by a voltage applied to an adjacent semiconductor material of opposite polarity. The operation involves the depletion region phenomenon.

Fixed bias A biasing circuit that depends solely upon external voltages (and resistors). The FET current I_D does *not* affect the operating point V_{GSQ}.

Gate The "control" terminal of a FET. The *voltage* applied to this terminal affects the current I_D. There is practically no gate current. (Any gate current is the result of leakage.)

JFET A FET in which the depletion phenomenon induced into the channel by the gate is achieved by a structure similar to that of a junction diode (P and N materials brought together on the molecular level).

MOSFET A FET in which the gate is isolated from the channel by a metal oxide (silicon oxide). Both depletion and enhancement can take place. Gate current is practically zero (well below that for a JFET).

Pinch-off region The area in the drain characteristic (a plot of I_D versus V_{DS} for different values of V_{GS}) in which $V_{DS} > V_P$, where V_{DS} is the drain-source voltage and V_P is a fixed value, the pinch-off voltage. (See below.)

SUMMARY OF IMPORTANT TERMS 305

Pinch-off voltage The smallest voltage (in magnitude) specified by the device manufacturer, which, when applied between gate and source, results in $I_D = 0$. (For example, if $V_P = -3$ V, then for $V_{GS} = -3$ V or any voltage more negative than -3 V, $I_D = 0$.)

Self bias A biasing circuit (usually with a source resistor R_S) in which the bias voltage V_{GSQ} depends on I_D. (There is a dc feedback effect.)

Shockley's equation An equation relating V_{GS} to I_D in a JFET, first introduced by Dr. Shockley.

$$I_D = I_{DSS}\left(1 - \frac{V_{GS}}{V_P}\right)^2$$

Source follower (See Common drain.)

Thermal runway A condition in which an increase in I_C increases device temperature, in turn increasing I_C until the maximum power dissipation limits are exceeded, thereby destroying the transistor.

Transconductance (g_m) The ratio of $\Delta I_D/\Delta V_{GS}$ (change in drain current divided by change in gate-source voltage). Has the units of mhos $\mho$.

VMOS A MOSFET constructed in such a way that the current I_D flows vertically down (or up) the structure. It has a higher power capability than the standard MOS and relatively wider frequency response.

Voltage divider biasing A biasing circuit that uses a voltage divider from the supply voltage to produce V_{GSQ}.

g_m (Transconductance) For the JFET,

$$g_m = \frac{-2I_{DSS}}{V_P}\left(1 - \frac{V_{GS}}{V_P}\right)$$

For enhancement MOSFETs,

$$g_m = aK(V_{GS} - V_T)$$

where K and a are approximation constants and V_T is the threshold voltage. Usually, $a = 2$ and K depends on the specific MOSFET.

g_{mo} In a JFET, this depends on I_{DSS} and V_p.

$$g_{mo} = -2I_{DSS}/V_p$$

This is the first term in the expression for g_m. It is g_m for the case in which

$$V_{GS} = 0, \qquad I_D = I_{DSS}.$$

g_{mQ} The g_m at a particular operating point.

I_D The drain current, flowing from source to drain. In a JFET, I_D is approximated by Shockley's equation

$$I_D = I_{DSS}\left(1 - \frac{V_{GS}}{V_P}\right)^2$$

This equation may also be used for the enhancement/depletion MOSFET. For the enhancement MOSFET, the approximation is

$$I_D = K(V_{GS} - V_T)^2$$

306 FIELD EFFECT TRANSISTORS (FET)

where V_T is the threshold voltage, and K and a are constants. Usually, $a = 2$ and K is given by the manufacturer for the specific device.

I_{DSS} The drain current under the conditions $V_{GS} = 0$ and $V_{DS} > V_P$.

r_d The internal resistance of the current source used to represent the drain-source behavior (in the device equivalent circuit). It is defined as $\Delta V_{DS}/\Delta I_D$ for very small increments.

$r_{DS(on)}$ The drain to source resistance for an "on" FET.

r_{in} Input resistance. $r_{in} = v_{in}/i_{in}$.

r_o The output resistance of a circuit. The Thevenin resistance of the circuit when represented as a Thevenin source.

$V_{GS(off)}$ Same as V_P

V_P See Pinch-off voltage.

V_T The threshold voltage for the enhancement MOSFET. It is only a voltage greater in magnitude than V_T when applied between gate and source, it will produce drain current. Applicable to the enhancement MOSFET.

PROBLEMS

A number of universal transfer characteristics are provided at the end of this section.

6-1. Define (or explain) the following terms:

    (a) "Pinch-off" region.
    (b) "Ohmic" region.
    (c) Pinch-off voltage, V_P.
    (d) I_{DSS}.
    (e) Transfer characteristics.
    (f) Drain characteristics.

6-2. Plot Shockley's equation for $V_P = -5$ V, $I_{DSS} = 8$ mA (use V_{GS} increments of 0.5 V).

6-3. Repeat Problem 6-2 using the 3-point plot and compare the two.

6-4. For the circuit of Fig. 6-57, find

    (a) I_D.
    (b) V_D.

Use Shockley's equation.

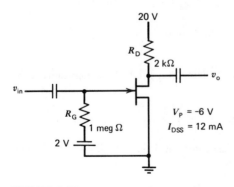

FIGURE 6-57

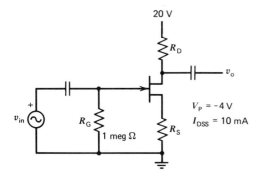

FIGURE 6-58

6-5. Repeat Problem 6-4 using the transfer characteristics.

6-6. For the circuit of Fig. 6-58 $R_D = 3.3\ \text{k}\Omega$, $R_S = 1\ \text{k}\Omega$. Find:

(a) V_{GSQ}.
(b) I_{DQ}.
(c) V_{DS}.
(d) V_S.

Use transfer characteristics.

6-7. Repeat Problem 6-6 for $R_D = 3.3\ \text{k}\Omega$, $R_S = 500\ \Omega$.

6-8. For the circuit of Fig. 6-59 $R_D = 2.7\ \text{k}\Omega$, $R_S = 1.5\ \text{k}\Omega$. Find:

(a) V_{GSQ}.
(b) I_{DQ}.
(c) V_{DQ}.

6-9. Repeat Problem 6-8 for $R_D = 2.7\ \text{k}\Omega$, $R_S = 750\ \Omega$.

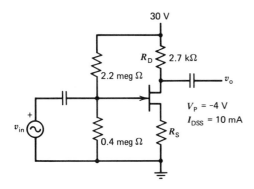

FIGURE 6-59

308 FIELD EFFECT TRANSISTORS (FET)

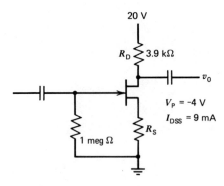

FIGURE 6-60

6-10. For the circuit in Fig. 6-60 $V_{GSQ} = -1.5$ V. Find:

(a) I_{DQ}.
(b) R_S.
(c) V_D.
(d) Show the operating point (the Q-point) on the transfer characteristic.

6-11. Find the operating point for the circuit of Fig. 6-61 (V_{GSQ}, I_{DQ}, V_{DQ}).

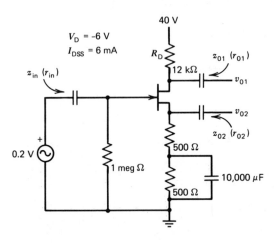

FIGURE 6-61

6-12. For the circuit of Fig. 6-62 find:

(a) V_{GSQ}.
(b) I_{DQ}.
(c) V_{DQ}.

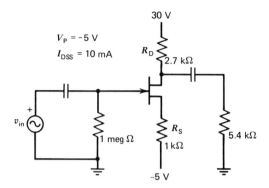

FIGURE 6-62

6-13. Find the operating point for the circuit of Fig. 6-63 (V_{GSQ}, I_{DQ}, V_{DQ}).

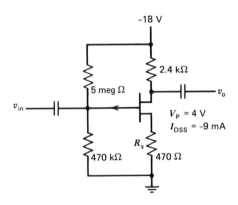

FIGURE 6-63

6-14. Find the g_{mQ} for Problems

(a) 6-5.
(b) 6-7.
(c) 6-8.
(d) 6-10.
(e) 6-12.
(f) 6-13.

6-15. For the circuit of Fig. 6-58 with $R_D = 3.9$ kΩ, $R_S = 1$ kΩ, $g_m = 1000$ $\mu\mho$, find:

(a) A_v.
(b) r_{in}.
(c) r_o.

310 FIELD EFFECT TRANSISTORS (FET)

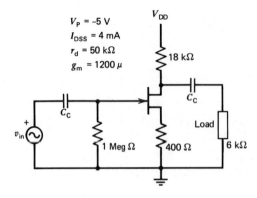

FIGURE 6-64

6-16. For the circuit in Fig. 6-64 find:

(a) A_v.
(b) r_{in}.
(c) r_o.

6-17. For the circuit in Fig. 6-65 find:

(a) v_o.
(b) r_{in}.
(c) r_o.

6-18. Repeat Problem 6-17 with C_S removed.

6-19. For the circuit of Fig. 6-61 with $g_m = 1200\ \mu\mho$, find:

(a) v_{o1}.
(b) v_{o2}.
(c) r_{o1}.
(d) r_{o2}.

Neglect r_d.

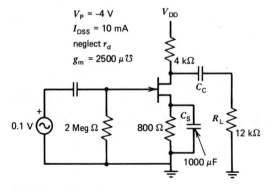

FIGURE 6-65

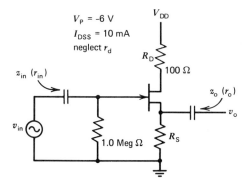

FIGURE 6-66

6-20. For the circuit of Fig. 6-66

(a) Give the equivalent circuit. Find A_v and r_o (z_o) for the following circuit values:
(b) $R_S = 0.5$ kΩ.
(c) $R_S = 1.0$ kΩ.
(d) $R_S = 2$ kΩ.
(e) $R_S = 5$ kΩ.

(Note that g_{mQ} must be recalculated for each value of R_S.)

6-21. For the circuit of Fig 6-67 calculate A_v and r_o for

(a) $R_{S1} = 1$ kΩ, $R_{S2} = 4$ kΩ
(b) $R_{S1} = 2$ kΩ, $R_{S2} = 3$ kΩ
(c) $R_{S1} = 5$ kΩ, $R_{S2} = 0$

(Note that $R_{S1} + R_{S2} = 5$ kΩ; hence, g_{mQ} is invariant.)

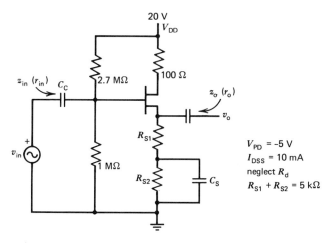

FIGURE 6-67

312 FIELD EFFECT TRANSISTORS (FET)

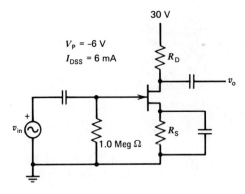

FIGURE 6-68

6-22. For the circuit of Fig. 6-68 find:

(a) R_S.
(b) R_D.
(c) r_o (z_o).
(d) r_{in} (z_{in}).

The circuit operates with $V_{GSQ} = -2$ V and $A_v = -10$.

6-23. A FET amplifier stage is to be designed so that it has the following characteristics:

$$A_v = -8$$
$$r_{in} = 1 \text{ M}\Omega$$

It is to be biased so that $V_{GSQ} = \frac{1}{2} V_P$ and $V_D = \frac{1}{2} V_{DD}$. The circuit used with the FET specifications is given in Fig. 6-69. Find R_{S1}, R_{S2}, R_D, R_G. Hint: $R_{S1} + R_{S2}$, and R_D are found from dc considerations.

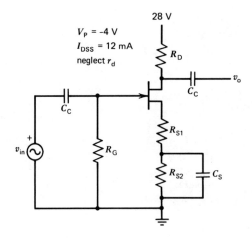

FIGURE 6-69

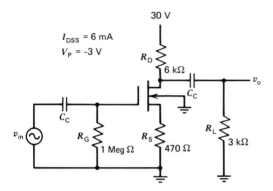

FIGURE 6-70

6-24. A depletion enhancement MOSFET stage is shown in Fig. 6-70. Find:

(a) A_v.
(b) r_{in}.
(c) r_o.

(You must first find the Q-point and g_{mQ}.)

6-25. For the circuit of Fig. 6-71 find:

(a) A_v.
(b) r_{in}.
(c) r_o.

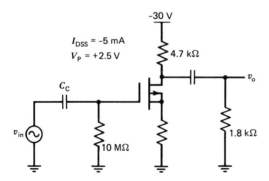

FIGURE 6-71

6-26. Figure 6-72 gives the circuit and ac equivalent circuit of a MOSFET amplifier stage. Find:

(a) A_v.
(b) r_{in}.
(c) r_o.

314 FIELD EFFECT TRANSISTORS (FET)

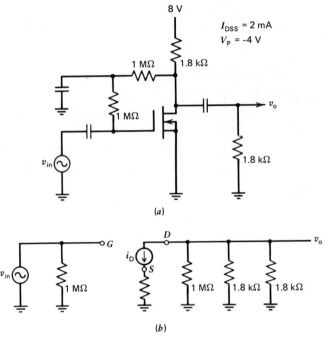

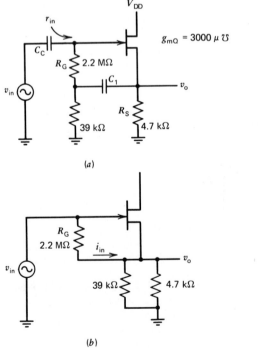

FIGURE 6-72 (a) Circuit. (b) ac equivalent.

FIGURE 6-73 Boot strap circuit. (a) The circuit. (b) Circuit redrawn with C_c and C_1 replaced by shorts (ac eq.).

6-27. The circuit shown in Fig. 6-73 is called a "boot-strap" circuit.
   (a) Find r_{in} in general terms.
   (b) Find r_{in} for the specific values shown.

Hint: Recall that $r_{in} = v_{in}/i_{in}$. Here, i_{in} is the current through R_G which is $i_{in} = \dfrac{v_o - v_{in}}{R_G}$ (the voltage across R_G divided by R_G) and $v_o = A_v \times v_{in}$.

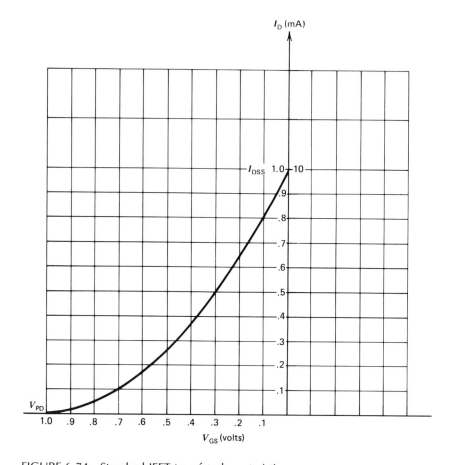

FIGURE 6-74 Standard JFET transfer characteristics.

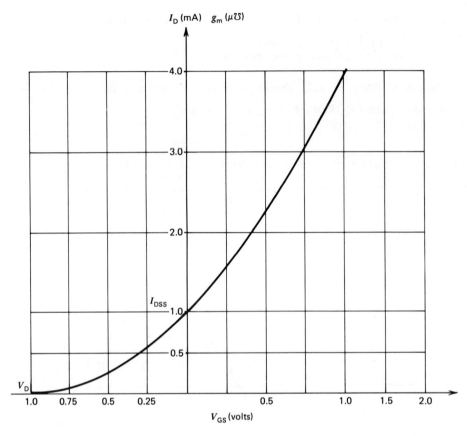

FIGURE 6-75 Standard depletion-enhancement transfer characteristics.

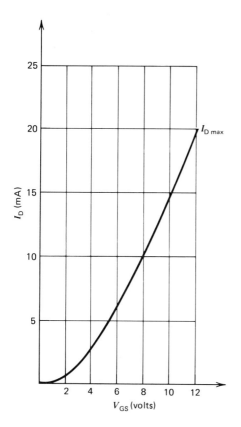

FIGURE 6-76 Typical (nonstandard) MOSFET transfer characteristics (for the 2N4351).

CHAPTER

7

MULTITRANSISTOR AMPLIFIERS

7-1 CHAPTER OBJECTIVES

This chapter is intended to give the student the ability to analyze multitransistor amplifier circuits, and to calculate both dc (biasing) and ac (gain, input and output resistances) parameters.

The student will learn two analytic approaches to the calculation of gain for these circuits: One utilizes the Thevenin's equivalent circuits; the other includes loading effects directly in the computations. Among the specific circuits covered are cascade configurations of both FETs and BJTs, including:

1. common emitter to emitter follower;
2. emitter follower, emitter follower–Darlington pair;
3. source follower–common emitter;
4. common source–emitter follower.

A number of interstage coupling methods are presented:

1. direct or resistive coupling;
2. resistor-capacitor (R–C) coupling;
3. transformer coupling, which includes a brief review of transformer principles;
4. resonant coupling (tuned amplifier).

The first three are presented in detail, giving the student a full understanding of applicable circuit analysis methods. For the resonant coupling only principles are presented.

7-2 INTRODUCTION

The analysis of transistor circuits is done in two separate and distinct parts: the ac analysis, relating to the *signal*-voltage and current, and the dc analysis concerned with transistor biasing. The interconnection of two, or more, transistors may affect both dc and ac circuit behavior. For dc coupling, where only resistors and/or direct lines are used in the interconnection (Fig. 7-1b), *both* dc and ac performance are affected. Both dc and ac analysis, then, must account for the interdependence between the transistor stages. The use of coupling capacitors or transformers (Figs. 7-1c and 7-1 d) isolates the coupled transistors as far as dc is concerned, and the dc analysis may then proceed for each transistor independently.

Throughout the discussion of multitransistor circuits, an attempt will be made to isolate the coupled stages for the purposes of analysis. It will become apparent that the isolation is not always possible.

The computation of voltage gain, for example, can be done for each stage independently and the coupling effects considered separately. The total voltage gain A_T of the two-stage amplifier shown in Fig. 7-1a can be calculated by calculating A_1 and A_2 independently and then considering the voltage dividers $R_s - r_{in1}$, $r_{o1} - r_{in2}$, and $r_{o2} - R_L$. This process, however, cannot always be applied to the computation of r_{in} or r_o of a multitransistor amplifier.

7-3 BIASING

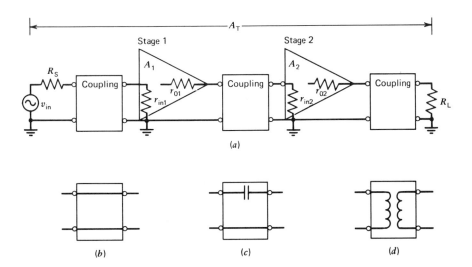

FIGURE 7-1 Cascaded stages. (a) Block diagram. (b) Direct coupling (short or res.). (c) Capacitive coupling. (d) Transformer coupling.

7-3.1 BJT

Only the dc amplifier (direct coupled) is examined here, since the biasing calculations for the ac coupled stages (resistor-capacitor or transformer) follow the descriptions given in chapter 4.

A typical dc two-transistor amplifier is shown in Fig. 7-2a. (This is a cascade connection of CE and EF circuits.) The biasing circuit of Q_1 is a conventional voltage divider biasing circuit. Q_2, however, is biased directly by the collector voltage of Q_1. It is reasonable to assume that $I_{C1} \gg I_{B2}$. Under these conditions, V_{C1} may be found independent of the connection between the transistors.

$$V_{C1} = V_{CC} - I_{C1} \times R_{C1} \tag{7-1}$$

I_{C1} for the conventional voltage divider circuit is given approximately by

$$I_{C1} = \frac{V_{B1} - 0.7}{R_{E1}} \text{ (silicon transistor)} \tag{7-2}$$

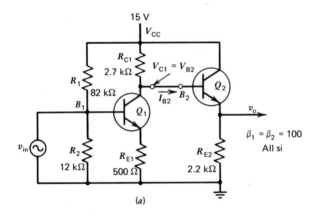

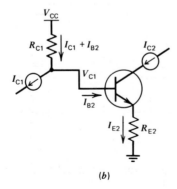

FIGURE 7-2 dc coupled CE-EF stages. (a) Circuit. (b) Q_2 biasing section.

where

$$V_{B1} = \frac{V_{CC} \times R_2}{R_1 + R_2} \qquad (7\text{-}3)$$

Once V_{C1} is found, I_{C2} ($\approx I_{E2}$) is found by applying KVL to the loop containing V_{C1}, V_{BE}, R_{E2}.

$$I_{C2} = \frac{V_{C1} - 0.7}{R_{E2}} \qquad (7\text{-}4)$$

Note that Eq. (7-4) is identical to Eq. (7-2) with V_{B1} replaced by V_{C1}.

Equation (7-1) completely neglects the effects of Q_2 on V_{C1}. A more accurate solution yields

$$V_{C1} = \frac{V_{CC} - I_{C1} \times R_{C1}}{1 + \dfrac{R_{C1}}{\beta_2 R_{E2}}} \qquad (7\text{-}5)$$

In Eq. (7-5) the effects of I_{B2} are considered. The term $R_{C1}/\beta_2 R_{E2}$ is representative of the loading effects of Q_2. Equation (7-5) reverts to Eq. (7-1) for the case in which $R_{C1} \ll \beta_2 R_{E2}$.

EXAMPLE 7-1

Find the operating point of Q_2 (I_{C2}, V_{CE}) in Fig. 7-2.

SOLUTION

To calculate I_{C2} it is first necessary to obtain V_{C1} that, in turn, calls for finding I_{C1}. I_{C1} is found by applying conventional voltage divider biasing analysis.

$$V_{B1} = \frac{V_{CC} \times R_1}{R_1 + R_2} = \frac{15 \times 12\ k\Omega}{82\ k\Omega + 12\ k\Omega}$$
$$= 1.9\ V$$

$$I_{C1} \approx I_{E1} = \frac{V_{B1} - 0.7}{R_{E1}} \qquad (7\text{-}1)$$
$$= \frac{1.9 - 0.7}{500} = 2.4\ mA$$

By KVL applied to a collector circuit of Q_1

$$V_{C1} = V_{CC} - I_{C1} R_{C1} = 15 - 2.4 \times 10^{-3} \times 2.7 \times 10^3 = 8.5\ V$$

Since $V_{C1} = V_{B2} = 8.5\ V$

$$I_{C2} \approx I_{E2} = \frac{V_{B2} - 0.7}{R_{E2}} = \frac{8.5 - 0.7}{2.2\ k\Omega} = 3.55\ mA$$

V_{CE2} is found by KVL applied to Q_2

$$V_{CE2} = 15 - 3.55 \times 10^{-3} \times 2.2 \times 10^3 = \underline{7.2 \text{ V}}$$

(Note that $V_{E2} = I_{E2} \times R_{E2} = 3.55 \times 10^{-3} \times 2.2 \times 10^3 = 7.8$ V.)
A more accurate solution may be obtained by use of Eq. (7-4)

$$V_{C1} = \frac{15 - 2.4 \times 10^{-3} \times 2.7 \times 10^3}{1 + \dfrac{2.7 \text{ k}\Omega}{100 \times 2.2 \text{ k}\Omega}} = \frac{8.5}{1 + .0123}$$

$$= 8.4 \text{ V}$$

$$I_{C2} = \frac{8.4 - 0.7}{2.2 \text{ k}\Omega} = 3.5 \text{ mA}$$

$$V_{CE2} = 15 - 3.5 \times 10^{-3} \times 2.2 \cdot 10^3$$

$$= \underline{7.3 \text{ V}}$$

The error is small, about 1.5%, because the ratio $R_{C1}/\beta_2 R_{E2}$ is small, about 0.015 (0.015 translates into 1.5%).

In the circuit of Fig. 7-2 the stages consisted of CE and EF circuits. It is, of course, also possible to connect a CE stage to a CE stage or an EF stage to a CE stage, etc. The analysis of the various configurations is similar to the one shown above. As an additional example, we consider an EF-CE configuration as shown in Fig. 7-3.

In this configuration $V_{E1} = V_{B2}$ (emitter of Q_1 is connected to base of Q_2).

In order to find the operating point of Q_2, it is first necessary to find $V_{E1} = V_{B2}$. Here, again, it is possible to neglect the effects of Q_2 on V_{E1} if $I_{B2} \ll I_{E1}$. With this assumption, V_{E1} can be found independent of Q_2.

$$V_{E1} = V_{B1} - 0.7 \tag{7-6}$$

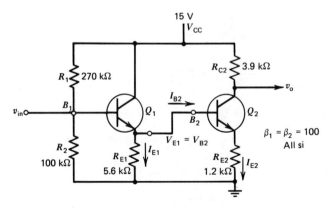

FIGURE 7-3   dc coupled EF-CE stages.

where (using the approximate voltage divider analysis)

$$V_{B1} = \frac{V_{CC} \times R_2}{R_1 + R_2}$$

and

$$I_{C2} = \frac{V_{B2} - 0.7}{R_{E2}} = \frac{V_{E1} - 0.7}{R_{E2}} \quad (7\text{-}7)$$

EXAMPLE 7-2

Find I_{C2}, V_{C2} in the circuit of Fig. 7-3.

SOLUTION

$$V_{B1} = (V_{CC} \times R_2)/(R_1 + R_2) \quad (7\text{-}3)$$

$$= \frac{15 \times 100 \text{ k}\Omega}{270 \text{ k}\Omega + 100 \text{ k}\Omega} = 4.05 \text{ V}$$

$$V_{E1} = 4.05 - 0.7 = 3.35 \text{ V} \quad (7\text{-}6)$$

Since $V_{E1} = V_{B2}$

$$I_{C2} \simeq I_{E2} = \frac{V_{B2} - 0.7}{R_{E2}} = \frac{3.35 - 0.7}{1.2 \text{ k}\Omega}$$

$$= 2.2 \text{ mA}$$

$$V_{C2} = 15 - 2.2 \times 10^{-3} \times 3.9 \times 10^3 = \underline{6.42 \text{ V}} \quad (7\text{-}1)$$

7-3.2 Darlington Pair Biasing

A very popular two-transistor amplifier consists of a cascade connection of two emitter follower stages (Fig. 7-4). The two connected transistors are available commercially in a single three-terminal package (dashed line in Fig. 7-4) known as a *Darlington pair*.

The dc analysis, namely the computation of I_{B1} and I_{E2} (I_B and I_E for the Darlington package), follows the same approach taken in the analysis of the single transistor. KVL for the loop containing V_{CC}, R_B, V_{BE1}, V_{BE2}, and R_E yields

$$V_{CC} - I_B R_B - V_{BE1} - V_{BE2} - I_E R_E = 0 \quad (7\text{-}8)$$

Here, we use $I_B = I_{B1}$ and $I_E = I_{E2}$. Since $I_{E1} = I_{B2}$, the relation between I_B and I_E becomes (for $\beta_1 \times \beta_2 \gg 1$)

$$I_E = \beta_1 \times \beta_2 I_B \; (I_E \simeq I_{C2}) \quad (7\text{-}9)$$

324 MULTITRANSISTOR AMPLIFIERS

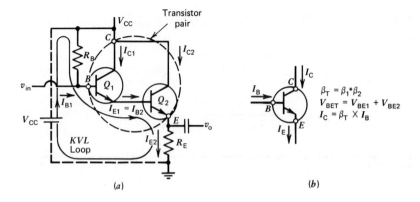

FIGURE 7-4 Darlington pair. (a) Circuit diagram. (b) "Single Transistor" package.

and Eq. (7-8) may be solved for I_B

$$I_B = \frac{V_{CC} - V_{BE1} - V_{BE2}}{R_B + \beta_1\beta_2 R_E} \qquad (7\text{-}10)$$

Assuming that $V_{BE1} = V_{BE2} = V_{BE}$, Eq. (7-10) becomes

$$I_B = \frac{V_{CC} - 2V_{BE}}{R_B + \beta_1\beta_2 R_E} \qquad (7\text{-}10a)$$

Equation (7-10a) indicates that the Darlington pair can be represented, for dc purposes, as a transistor with a $V_{BET} = 2V_{BE}$ and $\beta_T = \beta_1 \times \beta_2$. Equation (7-10a) becomes

$$I_B = \frac{V_{CC} - V_{BET}}{R_B + \beta_T R_E} \qquad (7\text{-}10b)$$

which is identical in form to Eq. (4-15a) developed for the single transistor stage. All biasing configurations, voltage divider, self-bias, etc. may be analyzed using the equivalent of the Darlington shown in Fig. (7-4b).

It is worth noting that the ac analysis of the Darlington pair yields a similar equivalent circuit with the addition of r_{et}. (See Sec. 7-4.1.)

7-3.3 FETs and BJTs

Many multistage amplifiers use both FETs and BJTs. In particular, the FET is often used as the input or first stage to take advantage of its very high input resistance. This combination of a FET followed by a BJT is quite common. With the introduction of the power FET (see Sec. 9-9), "all FET" amplifiers are becoming more and more popular.

BIASING 325

Two examples of bias calculations are given here. The purpose of these examples is to demonstrate the approaches taken rather than serve as a cook-book delineation of the solutions. The equations used throughout these examples are those used in the single stage FET and BJT analysis and/or the direct result of the application of KVL.

EXAMPLE 7-3

Find the operating point of Q_2 in Fig. 7-5—that is, find I_C and V_C.

SOLUTION

By inspection $I_D = I_B$. To find I_D, note that for $I_D > 0$, Q_2 is biased in the active region and $V_B = 0.7$ V, resulting in $V_S = 0.7$ V ($V_B = V_S$). To find I_D we write KVL for the FET input circuit portion (Fig. 7-5b)

$$V_G - V_{GS} - 0.7 = 0$$

since $V_G = 0$ (no current in R_G)

$$V_{GS} = -0.7$$

Using Shockley's equation (or the standard JFET curve with $I_{DSS} = 2$ mA and $V_P = -1.4$ V), we find

$$I_D = 2 \text{ mA} \left(1 - \frac{-0.7}{-1.4}\right)^2 = 0.5 \text{ mA}$$

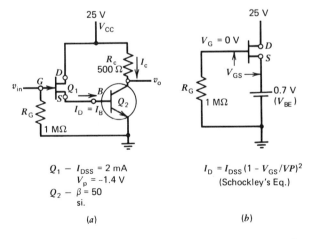

FIGURE 7-5 dc coupled FET-BJT (SF-CE). (a) Circuit diagram. (b) FET biasing portion.

Hence,
$$I_B = 0.5 \text{ mA}$$
$$I_C = \beta \times I_B = 50 \times 0.5 \text{ mA} = \underline{25 \text{ mA}}$$
$$V_C = V_{CC} - I_C R_C = 25 - 25 \times 10^{-3} \times 500 = \underline{12.5 \text{ V}}$$

EXAMPLE 7-4

Find I_C and V_E in the circuit of Fig. 7-6a.

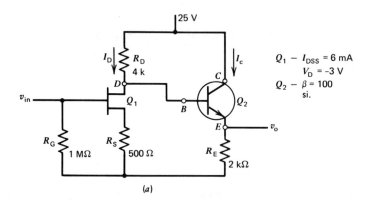

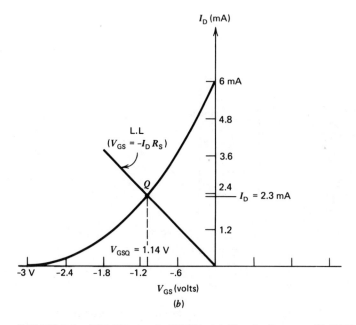

FIGURE 7-6 FET-BJT circuit (CS-EF). (a) Circuit diagram. (b) FET biasing graphic solution.

SOLUTION

Note that $I_C \approx V_E/R_E$. To find V_E we first find V_B. Since $V_B = V_D$, the task is reduced to finding V_D. In order to simplify the solution, we assume that $I_D \gg I_B$. This is a reasonable assumption that will be reexamined once the parameters are evaluated. This assumption permits the evaluation of I_D and V_D for the JFET standing alone, neglecting the loading effects of Q_2. V_D and I_D can be found by the use of Shockley's equation, or, graphically, by the use of the standard transfer characteristics (see Sec. 6-4), as shown in Fig. 7-6b. We get $I_D = 2.3$ mA; hence,

$$V_D = V_{DD} - I_D R_D$$
$$= 25 - 2.3 \times 10^{-3} \times 4 \times 10^3$$
$$= 15.8 \text{ V}$$

and $V_B = 15.8$ V ($V_D = V_B$).

$$V_E = V_B - 0.7 \text{ V} = 15.1 \text{ V}$$
$$I_E = \frac{V_E}{R_E} = \frac{15.1}{2 \text{ k}\Omega} = 7.55 \text{ mA} \approx I_C$$

7-4 SMALL SIGNAL AC ANALYSIS

The ac analysis of the multistage amplifier is essentially an extension of the single stage analysis. Connecting a second stage to the output of the first transistor has the effect of loading the first stage with the input resistance of the second stage. The loaded gain of the first stage may then be calculated by including r_{in2} as the load resistance of that stage.

r_{in} of the complete amplifier is clearly r_{in1} (the input resistance of the first stage), while r_o of the amplifier is r_{o2} (the output resistance of the last stage). Both r_{in} and r_o may be dependent not only on the individual stage (first and last, respectively), but also on other, intermediate stages. As a first example, we analyze the Darlington pair. (The dc analysis has been done in Sec. 7-3.)

7-4.1 The Darlington Pair

Figure 7-7 shows the Darlington EF circuit and its ac equivalent circuit. Before proceeding with the analysis, note that r_{e1} and r_{e2} are related via β_2. Since $I_{C1} = I_{B2}$ and $I_{C2} = \beta_2 I_{B2}$, we get $I_{C2} = \beta_2 \times I_{C1}$. As a result,

$$r_{e2} = \frac{26}{I_{C2}} = \frac{26}{\beta_2 I_{C1}}$$
$$= \frac{1}{\beta_2} \times r_{e1} \text{ (since } \frac{26}{I_{C1}} = r_{e1})$$

328 MULTITRANSISTOR AMPLIFIERS

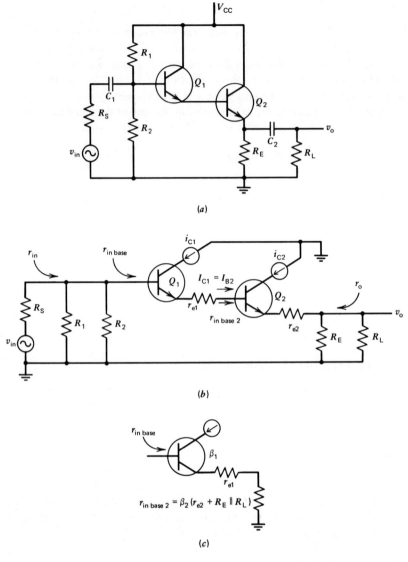

FIGURE 7-7 Darlington pair—ac analysis. (a) Circuit diagram. (b) ac equivalent circuit. (c) Computing $r_{\text{in base}}$.

The relation between r_{e1} and r_{e2} is then given by

$$r_{e1} = \beta_2 \times r_{e2} \tag{7-11}$$

To proceed with the analysis of the Darlington circuit, it is convenient to first obtain an equivalent circuit of the device, which turns out to be much like the

SMALL SIGNAL AC ANALYSIS

equivalent of the single transistor. This equivalent circuit will be obtained by calculating r_{in} at the base and comparing it with that of the single transistor. Calculating $r_{in(base)}$ by referring to Fig. 7-7c yields

$$r_{in(base)} = \beta_2(r_{e1} + r_{in(base\ 2)}) = \beta_1\,[r_{e1} + \beta_2(r_{e2} + R_E\|R_L)]$$

The term $\beta_2 \times (r_{e2} + R_E\|R_L)$ is the input resistance to the base of Q_2, $r_{in(base\ 2)}$. Completing the multiplication, we get

$$r_{in\ (base)} = \beta_1 r_{e1} + \beta_1\beta_2 r_{e2} + \beta_1\beta_2 R_E\|R_L$$

Substituting for r_{e1} from Eq. (7-11) yields

$$r_{in(base)} = \beta_1\beta_2 r_{e2} + \beta_1\beta_2 r_{e2} + \beta_1\beta_2 R_E\|R_L$$
$$r_{in(base)} = \beta_1\beta_2(2r_{e2} + R_E\|R_L) \tag{7-12}$$

If we rename $\beta_1\beta_2 = \beta_T$ and $2r_{e2} = r_{eT}$, Eq. (7-12) becomes

$$r_{in(base)} = \beta_T(r_{eT} + R_E\|R_L) \tag{7-12a}$$

Equation (7-12a) is the typical base input resistance of a single transistor. This leads to a simplified ac equivalent circuit for the Darlington pair (Fig. 7-8). This equivalent circuit may now be used for all computations involving the Darlington pair package. The analysis of the Darlington is then identical to that of the single transistor emitter follower stage.

EXAMPLE 7-5

For the circuit of Fig. 7-9, find
 (a) r_{in}.
 (b) v_o.
 (c) r_o.

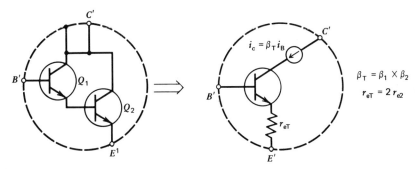

FIGURE 7-8 ac equivalent of Darlington pair.

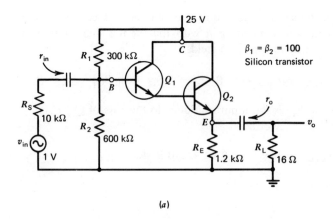

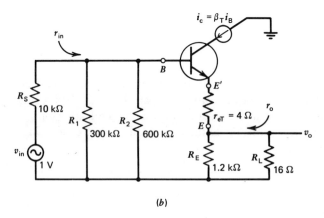

FIGURE 7-9 A Darlington amplifer. (a) Circuit diagram. (b, ac eq. circuit.

SOLUTION

We first find I_{C2} in order to get r_{e2} and r_{eT}. Following the "approximate" analysis of the voltage divider bias circuits (see chapter 4), we get

$$V_B = \frac{25 \times 600 \text{ k}\Omega}{300 \text{ k}\Omega + 600 \text{ k}\Omega} = 16.67 \text{ V (voltage divider)}$$

Hence,

$$V_E = V_B - V_{BET} = 16.67 - 1.4 = 15.27 \text{ V}$$

Recall that $V_{BET} = V_{BE1} + V_{BE2} \approx 1.4$ V for silicon transistor

$$I_C = \frac{V_E}{R_E} = \frac{15.27}{1.2 \text{ k}\Omega} = 12.725 \text{ mA}$$

SMALL SIGNAL AC ANALYSIS

$$r_{e2} = \frac{26}{12.725} = 2.04\,\Omega \approx 2.0\,\Omega$$

$$r_{eT} = 2r_{e2} = 2 \times 2 = 4\,\Omega$$

(a) To find r_{in}, apply the equations used in the emitter follower analysis.

$$r_{in} = R_1 \| R_2 \| [\beta_T \times (r_{eT} + R_E \| R_L)]$$

$$\beta_T = 100 \times 100 = 10{,}000$$

$$R_E \| R_L = 1.2\,k\Omega \| 16\,\Omega \approx 16\,\Omega$$

$$r_{in} = 300\,k\Omega \| 600\,k\Omega \| [10{,}000 \times (4 + 16)]$$

$$= 200\,k\Omega \| 200\,k\Omega = \underline{100\,k\Omega}$$

(b) To find v_o, first find v_b (the ac signal at the base). v_b is a result of voltage division between R_S and r_{in}

$$v_b = \frac{v_{in} \times r_{in}}{R_S + r_{in}} = \frac{1 \times 100}{10 + 100} = 0.91\,V$$

Since $v_b = v_{E'}$ (E' is the emitter of the ideal transistor), we get v_o from another voltage divider between r_{eT} and $R_E \| R_L$

$$v_o = \frac{v_b \times R_E \| R_L}{R_E \| R_L + r_{eT}} = \frac{0.91 \times 16}{16 + 4} = \underline{0.73\,V}$$

Note that the total gain

$$A_v = \frac{v_o}{v_{in}} = \frac{0.73}{1} = \underline{0.73}$$

(c) r_o is found as in the standard emitter follower.

$$r_o = R_E \| \left(r_{eT} + \frac{R_S \| R_1 \| R_2}{\beta_T} \right)$$

$$= 1.2\,k \| \left(4 + \frac{10\,k\Omega \| 300\,k\Omega \| 600\,k\Omega}{10000} \right)$$

$$= 1.2\,k\Omega \| \left(4 + \frac{10\,k\Omega \| 200\,k\Omega}{10{,}000} \right)$$

$$\approx 1.2\,k\Omega \| (4 + 1) \approx 5\,\Omega$$

Note that in the example above $r_o \approx r_{eT} + R_S/\beta_T$, which is quite typical. Furthermore, r_o can be reasonably estimated by r_{eT} alone, $r_o \approx r_{eT}$. The two most important characteristics of the Darlington circuit are very high r_{in} and very low r_o. The voltage gain is approximately unity while the current gain cannot exceed $\beta_T = \beta_1 \times \beta_2$.

332 MULTITRANSISTOR AMPLIFIERS

The analysis of the Darlington is rather atypical of the multistage circuits. We have treated the Darlington pair as if it were a single transistor with a somewhat modified equivalent circuit. This is not the method used in most other multitransistor circuits.

7-4.2 R-C and Direct Coupled Stages

The following examples demonstrate the methods used in the ac analysis of multistage amplifiers. The direct and capacitively coupled stages yield identical ac equivalent circuits, since for the moment, it is assumed that all coupling capacitors constitute a short for ac purposes. To simplify matters, the ac transistor parameters such as r_e, g_{mQ}, etc. will be given (avoiding the need to perform biasing analysis).

EXAMPLE 7-6

For the two-transistor amplifier shown in Fig. 7-10, find
(a) r_{in}.
(b) r_o.
(c) v_o.
(d) A_{VT}.

SOLUTION

r_{in} of the amplifier is r_{in} of the first stage.
(a) $r_{in} = R_{11}\|R_{12}\|\beta_1(r_{e1})$. Note that R_{E1} is shorted ($R_{E1} \to 0$) by the capacitor.

$$r_{in} = 15 \text{ k}\Omega\|4.7 \text{ k}\Omega\|(100 \times 18) = 3.58 \text{ k}\Omega\|1.8 \text{ k}\Omega$$
$$= 1.2 \text{ k}\Omega$$

(b) $r_o = R_{C2} = 1.5 \text{ k}\Omega$
(c) To find v_o we must consider the effect of the coupling networks, as well as

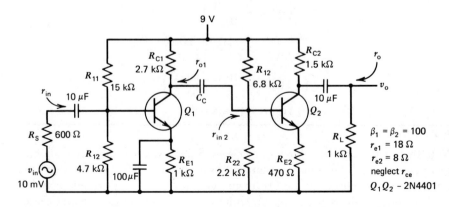

FIGURE 7-10 Capacitive coupled (CE-CE).

SMALL SIGNAL AC ANALYSIS   333

the gains of the transistors. Two methods will be demonstrated. The first is based on the approach outlined in Section 7-2. We treat each stage independently (as if it were not connected to the second stage or the load), and subsequently we evaluate the effect of the coupling networks.

Stage 1

$r_{in1} = r_{in} = 1.2 \text{ k}\Omega$ (from the computation of r_{in} above)

$$r_{o1} = R_{C1} = 2.7 \text{ k}\Omega$$

$$A_{v1} = \frac{-R_C}{r_{e1}} = -\frac{2.7 \text{ k}\Omega}{18}$$

$$= -150 \begin{pmatrix} R_{E1} \text{ is shorted by the} \\ 100 \text{ }\mu\text{f capacitor.} \end{pmatrix}$$

A_{v1} is the gain from the base of Q_1 to the *unloaded* collector of Q_1. (The connection from the collector of Q_1 to the base of Q_2 is opened.) (See Fig. 7-11.)

Stage 2

$$r_{in2} = R_{12} \| R_{22} \| \beta_2 \times (r_{e2} + R_{E2})$$
$$= 6.8 \text{ k}\Omega \| 2.2 \text{ k}\Omega \| [100(8 + 470)]$$
$$= 1.6 \text{ k}\Omega$$
$$r_{o2} = R_o = R_{C2} = 1.5 \text{ k}\Omega$$

$$A_{v2} = -\frac{R_{C2}}{r_{e2} + R_{E2}} = -\frac{1.5}{8 + 470}$$
$$= -3.1$$

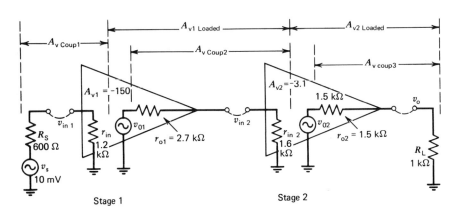

FIGURE 7-11 Computation of A_{VT}.

We now interconnect the amplifier chain (that is, v_s to v_{in1}, v_{o1} to v_{in2}, v_{o2} to v_o). v_{in1} is a result of voltage division between R_S and r_{in}.

$$v_{in1} = \frac{v_s \times r_{in}}{R_S + r_{in}} = \frac{10.0 \text{ mV} \times 1.2 \text{ k}\Omega}{0.6 \text{ k}\Omega + 1.2 \text{ k}\Omega}$$

$$= \underline{6.66 \text{ mV}}$$

$$v_{o1} = A_{v1} \times v_{in1} = -150 \times 6.66$$

$$= \underline{-1000 \text{ mV}}$$

(the negative sign indicates *phase* reversal relative to v_{in1}). To calculate v_{in2}, we again have a voltage divider

$$v_{in2} = \frac{v_{o1} \times r_{in2}}{r_{o1} + r_{in2}} = \frac{-1000 \text{ mV} \times 1.6 \text{ k}\Omega}{2.7 \text{ k}\Omega + 1.6 \text{ k}\Omega}$$

$$= \underline{-372 \text{ mV}}$$

$$v_{o2} = A_{v2} \times v_{in2} = -3.1 \times (-372 \text{ mV})$$

$$= \underline{1153.2 \text{ mV}}$$

$$v_o = \frac{v_{o2} \times R_L}{r_{o2} + R_L} = \frac{1153.2 \text{ mV} \times 1 \text{ k}\Omega}{1.5 \text{ k}\Omega + 1 \text{ k}\Omega}$$

$$= \underline{461 \text{ mV}}$$

(d) The total gain

$$A_{vT} = \frac{v_o}{v_s} = \frac{461.3}{10} = \underline{46.1}$$

This gain is the product of the transistors' gains and the coupling gains (losses)

$$A_{v(COUP1)} = \frac{r_{in}}{R_S + r_{in}} = \frac{1.2 \text{ k}\Omega}{0.6 \text{ k}\Omega + 1.2 \text{ k}\Omega}$$

$$= \underline{0.666}$$

$$A_{v(COUP2)} = \frac{r_{in2}}{r_{o1} + r_{in2}} = \frac{-1.6 \text{ k}\Omega}{2.7 \text{ k}\Omega + 1.6 \text{ k}\Omega}$$

$$= \underline{0.372}$$

$$A_{v(COUP3)} = \frac{R_L}{r_{o2} + R_L} = \frac{1 \text{ k}\Omega}{1.5 \text{ k}\Omega + 1 \text{ k}\Omega}$$

$$= \underline{0.4}$$

The coupling gains (losses) are all due to voltage division.

$$A_{vT} = A_{v1} \times A_{v2} \times A_{v(COUP)} \times A_{v(COUP2)} \times A_{v(COUP3)}$$

$$= -150 \times (-3.1) \times 0.666 \times 0.372 \times 0.4$$

$$= \underline{46.1}$$

SMALL SIGNAL AC ANALYSIS   335

(c,d) In the second method for finding v_o (and A_{vT}), we compute the *loaded* transistor gains.

$$A_{v1(Loaded)} = -\frac{R_{C1}\|r_{in2}}{r_{e1}} = \frac{-2.7\ k\Omega\|1.6\ k\Omega}{18}$$

$$= \underline{-55.8}$$

(r_{in2} was calculated above)

$$A_{v2(Loaded)} = -\frac{R_{C2}\|R_L}{r_{e2} + R_{E2}} = -\frac{1.5\ k\Omega\|1\ k\Omega}{8 + 470} = \underline{-1.25}$$

These two gains combined with the input circuit voltage division (R_S, r_{in}) yield the total gain and v_o.

$$A_{vT} = A_{v(COUP1)} \times A_{v1(Loaded)} \times A_{v2(Loaded)}$$
$$= 0.666 \times (-55.8) \times 1(-.25)$$
$$= \underline{46.4}$$

($A_{v(COUP1)}$) was previously calculated.)

$$v_o = A_{vT} \times V_C = 46.4 \times 10\ mV = \underline{464\ mV}$$

Note that the two methods demonstrated above are different in form only. This is evident from the fact that

$$A_{v1(Loaded)} = A_{v1} \times A_{v(COUP2)}$$

and

$$A_{v2(Loaded)} = A_{v2} \times A_{v(COUP3)}$$

The small difference in v_o between the two methods is due to round-off errors.

EXAMPLE 7-7

For the circuit of Fig. 7-12 find:
(a) r_{in}.
(b) r_o.
(c) $A_v = v_o/v_{in}$.
Neglect r_d and r_{oe}.

SOLUTION

(a) By inspection, $r_{in} = 1\ M\Omega$. To find r_o and A_v, we require g_m (of Q_1) and r_{e2} (of Q_2).

336 MULTITRANSISTOR AMPLIFIERS

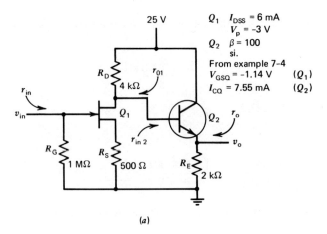

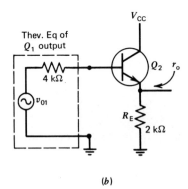

FIGURE 7-12 FET-BJT circuit (redrawing of Figure 7.6a). (a) Circuit diagram. (b) Finding r_o.

Since $V_{GSQ} = -1.14$ V,

$$g_m = g_{mo}\left(1 - \frac{V_{GSQ}}{V_P}\right)$$

$$= \frac{-2 \times 6 \times 10^{-3}}{-3}\left(1 - \frac{-1.14}{-3}\right)$$

$$= 2.48 \text{ m}\mho$$

(This could have been obtained graphically.)

$$r_{e2} = \frac{26}{I_{C2}} = \frac{26}{7.55} = 3.4 \text{ }\Omega$$

(b) Figure 7-12b demonstrates how r_o is found.

$$r_{o1} = 4 \text{ k}\Omega = R_D$$

$$r_o = R_E \| \left(r_{e2} + \frac{r_{o1}}{\beta} \right)$$

$$= 2 \text{ k}\Omega \| \left(3.4 \text{ }\Omega + \frac{4000}{100} \right)$$

$$\simeq \underline{43 \text{ }\Omega}$$

(For details see Sec. 5-6.3)

(c) To obtain A_v, we chose to use the second method demonstrated in Ex. 7-6.

$$A_{v1} = - \frac{g_m \times R_D \| r_{in2}}{1 + g_m \times R_S} \text{ (see chapter 6)}$$

r_{in2} is the load connected to the output of Q_1, effectively in parallel with R_D. We find r_{in2}

$$r_{in2} = \beta(R_E + r_{e2}) = 100(2 \text{ k}\Omega + 3.4)$$

Here, R_{e2} is negligible. $r_{e2} \ll R_E$.

$$r_{in2} = 100 \times 2 \text{ k}\Omega = 200 \text{ k}\Omega$$

$$A_{v1} = \frac{2.48 \times 10^{-3} \times 4 \text{ k}\Omega \| 200 \text{ k}\Omega}{1 + 2.48 \times 10^{-3} \times 500}$$

$$\approx - \frac{2.48 \times 10^{-3} \times 4 \text{ k}\Omega}{1 + 2.48 \times 10^{-3} \times 500}$$

$$(r_{in2} \gg R_D)$$

$$\underline{A_{v1} = -4.4.}$$

$$A_{v2} = \frac{R_E}{R_E + r_{e2}} = \frac{2 \text{ k}\Omega}{2 \text{ k}\Omega + 3.4 \text{ }\Omega}$$

$$\approx 1 \text{ (typical EF gain)}$$

$$A_{vT} = A_{v1} \times A_{v2} = -4.4$$

Should a load R_L be connected across v_o, both r_{in2} and A_{v2} will be affected. The effect on r_{in2} may cause A_{V1} to change. (The condition $r_{in2} \gg R_D$ may not hold anymore.)

7-4.3 Multistage Amplifiers—General Approach

The above two examples point to a general method for analyzing multistage amplifiers. The important parameters, r_{in}, r_o, A_v (and A_i, A_p if necessary), for each grouping of transistors may be calculated as shown. The Thevenin's equivalent

7-4.4 Transformers

In the previous sections, we analyzed the direct and R-C (capacitor) coupled circuits. It was noted that these two yield identical ac performance since, for the time being, the coupling capacitor is considered a short.

The transformer coupling, similar to the R-C circuit, isolates the transistors as far as dc is concerned. The analysis, however, is somewhat more complex.

To facilitate the analysis, let us first review briefly the basic transformer characteristics. We limit ourselves to the ideal (lossless) transformer.

The transformer symbol and the various voltage, current, and impedance relationships are shown in Fig. 7-13. As a memory aid, note that the higher voltage is associated with the higher number of turns in a linear fashion. Similarly, a higher impedance (or resistance, in purely resistive circuits) is associated with the higher number of turns, in proportion to the square of the turns ratio.

EXAMPLE 7-8

In the circuit of Fig. 7-14, find R_p, i_p, v_L, i_L.

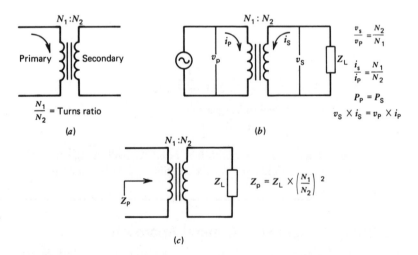

FIGURE 7-13 The transformer. (a) Symbol. (b) Voltage and current relations. (c) Impedance transformation.

SMALL SIGNAL AC ANALYSIS   339

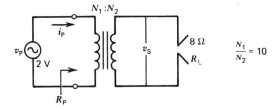

FIGURE 7-14   Transformer circuit.

SOLUTION

Since R_P is associated with the larger number of turns, $R_P > R_L$.

$$R_P = 10^2 \times R_L = 100 \times 8 = \underline{800 \ \Omega}$$

$$i_P = \frac{v_P}{R_P} = \frac{2}{800} = \underline{2.5 \text{ mA}} \text{ (Ohm's law)}$$

$$v_L = \frac{1}{10} v_P = 0.2 \text{ V}$$

($v_L < v_P$ since v_L is associated with the lower number of turns.)

$$i_L = \frac{v_L}{R_L} = \frac{0.2}{8} = \underline{25 \text{ mA}}$$

Note that i_L can be obtained by

$$\frac{i_L}{i_P} = \frac{N_2}{N_1}; \quad i_P = i_L \times \frac{N_1}{N_2}$$

$$= 2.5 \times 10 = 25 \text{ mA}.$$

(i_L is the current in the secondary windings, i_S.)

7-4.5   Transformer Coupling[1]

One of the main purposes of transformer coupling is impedance matching. A simple example will best illustrate this point.

EXAMPLE 7-9

(a) Find the gain of the first stage and v_o of the circuit in Fig. 7-15a.

---

[1] Due to the relatively high cost and bulkiness of transformers, this type of coupling has been largely replaced by appropriate transistor circuits.

340 MULTITRANSISTOR AMPLIFIERS

(b) Repeat for the circuit of Fig. 7-15b.

SOLUTION

(a) The gain

$$A_{v1} = -\frac{R_{C1} \| R_P}{r_{e1} + 470}$$

where R_P is the resistance looking into the primary of the transformer. R_P can be obtained from

$$R_P = \left(\frac{N_1}{N_2}\right)^2 \times r_{in2}$$

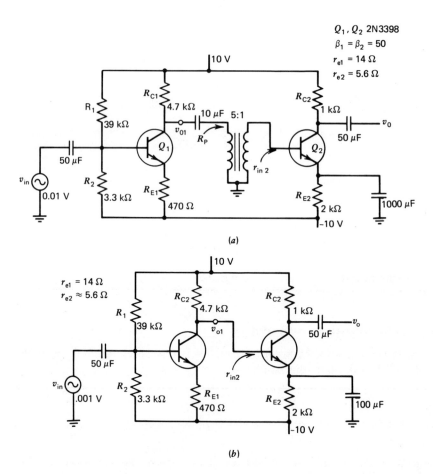

FIGURE 7-15 Transformer coupling. (a) Circuit with transformer coupling. (b) The transformer replaced by a direct connection.

(r_{in2} is the input resistance to the base of Q_2)

$$r_{in2} = \beta_2(r_{e2}) = 50 \times 5.6 = 280 \, \Omega$$

Hence,

$$R_P = \left(\frac{5}{1}\right)^2 \times 280 = 7000 \, \Omega$$

and

$$A_{v1} = -\frac{4700 \| 7000}{470 + 14} = -5.8$$

$$A_{v2} = -\frac{R_{C2}}{r_{e2}} = -\frac{1000}{5.6} = -178.6$$

The total gain must account for the 5:1 "step down" in voltage in the transformer. Hence,

$$A_{vT} = (-5.8) \times (-178.6) \times \frac{1}{5} = 207$$

$$v_o = v_{in} \times A_{vT} = 0.01 \times 207 = 2.07 \, V$$

Had we not used the transformer, but instead, connected the output v_{o1} directly to the base of Q_2 (Fig. 7-15b), A_{v1} then becomes

(b) $A_{v1} = -\dfrac{R_{C1} \| r_{in2}}{r_{e1} + R_{E1}} = -\dfrac{4700 \| 280}{14 + 470}$

$= -0.54$

$A_{vT} = (-0.54) \times (-178.6)$

$= 96$

(the transformer 5:1 step down is not present) $v_o = 0.01 \times 96 = 0.96 \, V$

The impedance transformation of the transformer in Fig. 7-15a has increased the active gain by a factor of 2.16. The input resistance r_{in2} (relatively very low) has been stepped up by a factor $(N_1/N_2)^2$ reducing the loading effect on the output of Q_1.

7-4.6 Other Coupling Methods

Direct, capacitive, and transformer coupling are the three basic coupling methods. Note that the direct coupled circuit is the only one that may be used to amplify very low frequencies—down to dc. (More on that in connection with frequency response and operational amplifiers.)

342 MULTITRANSISTOR AMPLIFIERS

One of the few uses of transformer coupling is in tuned or resonance coupled stages. A typical tuned coupling circuit is shown in Fig. 7-16. Both primary and secondary are made to resonate at approximately the same frequency by means of the tuning capacitors C_1, C_2. As a result of this "tuning," the circuit amplifies only a narrow band of frequencies, around the resonance frequency. Typically, these circuits are used in *Radio Frequency* (RF) amplifiers (the first tuned stage in a broadcast receiver) or *Intermediate Frequency* (IF) amplifiers (typically tuned to 460 kHz in Broadcast AM receivers). The effect of tuning is to allow high voltage gains at (or around) the resonance frequency and low gain, or losses, at all other frequencies.

SUMMARY OF IMPORTANT TERMS

Capacitive coupling The interconnection of two circuits by means of a capacitor.

Darlington (pair) A two-transistor package in which both collectors are interconnected, and the emitter of one is connected to the base of the second. The three terminals available externally are:

1. the collectors;
2. the base of the first;
3. the emitter of the second.

It is associated with a very high r_{in} and a very low r_o.

DC amplifier An amplifier circuit that can be used to amplify signals with as low a frequency as zero—dc. It commonly employs direct coupling.

Direct coupling The interconnection of circuits by means of direct wires or resistors.

IF amplifier An intermediate frequency amplifier, uses a resonant circuit. Associated with the intermediate frequency amplification in radio receivers.

Multistage amplifiers Amplifiers constructed by interconnecting a number (two or more) of individual transistor amplifier stages.

R-C coupling See Capacitive coupling.

Resistive coupling The coupling element is a resistor (see also direct coupling).

RF amplifier A tuned circuit amplifier (resonance circuit) used in selectively amplifying the incoming signal in a radio receiver.

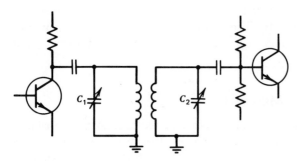

FIGURE 7-16 Tuned coupling.

Transformer coupling The interconnection of two circuits by means of a transformer.
Tuned coupling See IF and RF amplifiers.

PROBLEMS

7-1. A two-stage dc coupled amplifier (Fig. 7-17) has the following parameters:

$r_{in1} = 10$ kΩ, $r_{in2} = 10$ kΩ

$r_{o1} = 4$ kΩ, $r_{o2} = 100$ Ω

Unloaded gains, $A_{v1} = -20$, $A_{v2} = -10$

Source resistance, $R_S = 10$ kΩ

Load resistance, $R_L = 300$ Ω

Find:

(a) r_{in} of the complete amplifiers.
(b) r_o of the complete amplifier.
(c) Loaded total gain (including the effects of R_S).

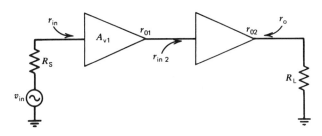

FIGURE 7-17

7-2. In the circuit of Fig. 7-2, $R_1 = 100$ kΩ, $R_2 = 12$ kΩ, $R_{E1} = 330$ Ω, $R_{C1} = 2.2$ kΩ, $R_{E2} = 470$ Ω, $\beta_1 = \beta_2 = 100$, $V_{CC} = 20$ V. Find:

(a) V_{CEQ1}, I_{CQ1}.
(b) V_{CEQ2}, I_{CQ2}.

7-3. In Fig. 7-3, $R_1 = 15$ kΩ, $R_2 = 6.8$ kΩ, $R_{E1} = 4.7$ kΩ, $R_{C2} = 2.0$ kΩ, $R_{E2} = 1$ kΩ, $\beta_1 = 100$, $\beta_2 = 100$, $V_{CC} = 24$ V, and the transistors are silicon. Find:

(a) V_{E1}.
(b) I_{C1}.
(c) V_{E2}.
(d) I_{C2}.

7-4. For the circuit of Fig. 7-18 find:

(a) V_{CQ1}, I_{CQ1}.
(b) V_{CQ2}, I_{CQ2}.

344 MULTITRANSISTOR AMPLIFIERS

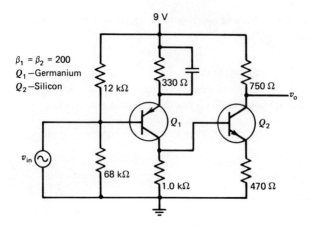

FIGURE 7-18

7-5. For the circuit of Fig. 7-19 find:

(a) I_{C1}.
(b) V_{CQ1}.
(c) I_{C2}.
(d) V_{CQ2}.
(*Hint:* Note that $V_{B2} = V_{C1} - 6.8$ V.)

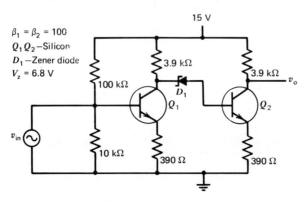

FIGURE 7-19

7-6. For the circuit of Fig. 7-20 find:

(a) I_{B1}.
(b) I_{C1}.
(c) I_{C2}.
(d) V_{C2}.

Assume that v_{in} is an open circuit for dc.
(*Hint:* Use the base emitter loop of Q_1 to find I_{B1}, $I_{C1} = \beta_1 \times I_{B1}$, $I_{C1} = I_{B2}$, $I_{C2} = \beta_2 I_{B2}$.)

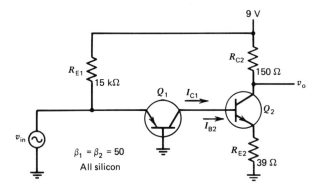

FIGURE 7-20

7-7. In the circuit of Fig. 7-21 find:
 (a) V_{DQ}, I_{DQ} (Q_1).
 (b) V_{CEQ}, I_{CQ} (Q_2).

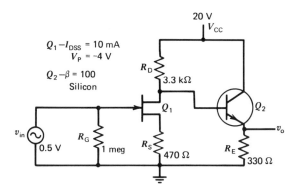

FIGURE 7-21

7-8. In the Darlington circuit of Fig. 7-7a, $R_1 = 220$ kΩ, $R_2 = 220$ kΩ, $R_E = 1$ kΩ, $\beta_1 = \beta_2 = 100$, $V_{CC} = 20$ V. Find:

 (a) V_{E2}.
 (b) I_{C2}.
 (c) I_{C1}.

7-9. In the circuit of Fig. 7-2 with the following values: $R_1 = 47$ kΩ, $R_2 = 6.8$ kΩ, $R_{E1} = 390$ Ω, $R_{C1} = 2.2$ kΩ, $R_{E2} = 470$ Ω, $\beta_1 = \beta_2 = 200$, $V_{CC} = 20$ V, and $R_L = 2.2$ kΩ, R_L is directly coupled to the emitter of Q_2. Find:

 (a) $A_{VT} = \left(\dfrac{v_o}{v_{in}}\right)$ ($r_{e1} = 6$ Ω, $r_{e2} = 1.3$ Ω).

(b) r_{in}.
(c) r_o.

7-10. For the circuit of Fig. 7-3 with $R_1 = 330$ kΩ, $R_2 = 82$ kΩ, $R_{E1} = 4.7$ kΩ, $R_{C2} = 3.3$ kΩ, $R_{E2} = 1$ kΩ, $\beta_1 = \beta_2 = 100$ and $V_{CC} = 24$ V. Find:

(a) A_{VT}.
(b) r_{in}.
(c) r_o.
(Do not neglect βR_{E1} when calculating r_{in}.)

7-11. For the circuit of Fig. 7-18, $r_{e1} = 9$ Ω, $r_{e2} = 5$ Ω, find:

(a) A_{VT}.
(b) r_{in}.
(c) r_o.

7-12. For the circuit of Fig. 7-19 find:

(a) $A_{VT}\left(\dfrac{v_o}{v_{in}}\right)$.

(b) r_{in}.
(c) r_o.
(*Hint:* Note that D_1 is considered a short for ac.)

7-13. For the circuit of Fig. 7-20 $r_{e1} = 47$ Ω, $r_{e2} = 0.94$ Ω. Find:

(a) r_{in}.
(b) r_o.
(c) A_{VT} (v_o/v_{in}). Is A_v positive or negative? Explain.
(*Hint:* The effective value of R_{C1} is $\beta_2 \times R_{E2}$; and, $A_{v1} = R_{C1}/r_{e1}$.)

7-14. For the circuit in Fig. 7-21 find (neglect r_{e2}):

(a) v_o.
(b) r_{in}.
(c) r_o.
(*Hint:* To find r_o, note that the resistance from the base of Q_2 to GND is 3.3 kΩ.)

7-15. In the Darlington circuit of Fig. 7-7a, the values are $R_1 = R_2 = 220$ kΩ, $R_E = 1$ kΩ, $\beta_1 = \beta_2 = 100$, $V_{CC} = 20$ V, $R_L = 100$ Ω and $R_S = 10$ kΩ ($r_{e2} = 3$ Ω). Find:

(a) r_{in}.

(b) $A_{VT}\left(\dfrac{v_o}{v_{in}}\right)$.

(c) $A_i\left(\dfrac{i_o}{i_{in}}\right)$.

(d) r_o.

7-16. In the circuit of Fig. 7-10 substitute an EF circuit for the first stage of the amplifier with $R_1 = 220$ kΩ, $R_2 = 220$ kΩ, $R_E = 5.6$ kΩ.

(a) Sketch the complete amplifier circuit.

(b) Find r_{in}.
(c) Find the total gain.
(d) Find r_o.

7-17. In the circuit of Fig. 7-22 with g_m of Q_1 equal to 2800 $\mu\mho$ find:

(a) A_{VT}.
(b) r_{in}.
(c) r_o.

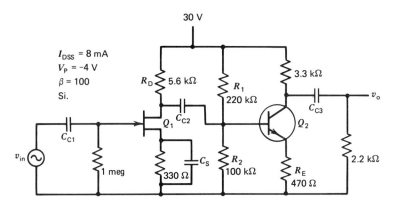

FIGURE 7-22

7-18. In Fig. 7-15 use a 2:1 transformer in place of the 5:1 used. Find:

(a) v_o.
(b) r_{in}.

7-19. For the circuit of Fig. 7-23 find:

(a) The Q point of Q_1.
(b) The Q point of Q_2.
(c) The Q point of Q_3.

7-20. For the circuit in Fig. 7-23 find:

(a) Minimum and maximum gain.
(b) v_o with R_P set to maximum gain.
(c) v_o with R_P set to minimum gain.
(d) P_o (with the given input signal) with R_P set to maximum gain.

348 MULTITRANSISTOR AMPLIFIERS

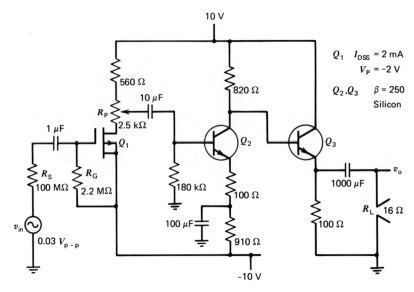

FIGURE 7-23

CHAPTER

8
FREQUENCY RESPONSE

8-1 CHAPTER OBJECTIVES

Completion of this chapter will provide the reader with knowledge related to the frequency dependence of the gain in transistorized circuits. As part of the presentation, a new unit, the decibel (dB), is introduced and its application to power, voltage, and current gain detailed.

The knowledge to compute the two cut-off frequencies, upper cut off (f_{UC}) and lower cut off (f_{LC}), and the amplifier bandwidth (BW) will be acquired. This will entail considerations of coupling capacitors and emitter by pass capacitors as they affect f_{LC}, and stray, interelectrode, and Miller capacitances as they affect f_{UC}. Three distinct equivalent circuits are discussed, the low frequency, midband, and high frequency equivalent circuits. Their use in calculating f_{LC} and f_{UC} will be mastered.

The student will learn how to represent the frequency response graphically, in a frequency versus gain (in dB) plot. Various characteristics of the plot, such as the slope outside the BW, are discussed. The terms "octave" and "decade" are used in the definition of the slope.

Device dependence on frequency is analyzed in terms of f_β and f_T. These terms are defined and their significance related to overall circuit performance.

8-2 INTRODUCTION

In all previous ac analyses, we ignored the effects of operating frequency on the various parameters. It was assumed that the coupling and bypass capacitors were perfect shorts. The β of the transistors were frequency independent, etc. This was

350 FREQUENCY RESPONSE

acceptable as long as we were willing to restrict the analysis to a range of frequencies for which the above assumptions are valid. This frequency range is referred to as "midband": not high enough to involve the frequency limitations of the transistors, and not low enough to invalidate the assumption of $X_C = 0$. If, for example, a coupling capacitor $C = 10\ \mu F$ is used, at a frequency of 10 Hz, $X_C = 1590\ \Omega$, which may markedly affect the performance of the circuit. (It is difficult to consider 1590 Ω as a short unless all other components exhibit *much* higher resistance.)

In the following sections, we investigate the frequency limitations of the various amplifier circuits and quantify their frequency dependence. The frequency spectrum will be divided into three ranges—low, midband and high frequency. Each of these bands is associated with a different equivalent circuit that is used in the analysis. Before proceeding with the analysis, a new gain unit will be introduced.

8-3 BELS AND DECIBELS

It is customary, and convenient, to express power, voltage, and current gains on a logarithmic scale.[1] For this purpose, a new unit, the bel, is introduced. It expresses the power gain $\dfrac{P_o}{P_{in}}$ or any power ratio in logarithmic terms.

$$A_p \bigg|_{ratio} = \frac{P_o}{P_{in}} \qquad (8\text{-}1)$$

$$A_p \bigg|_{bels} = \log \frac{P_o}{P_{in}} = \log A_p$$

The bel may be applied to any power ratio, not necessarily power gain. Thus, if, for example, $P_2 = 2\ W$, $P_1 = 0.2\ W$, the power ratio is $P_2/P_1 = 2/0.2 = 10$. This may be expressed in bels as $\log 2/0.2 = \log 10 = 1$ bel, a power ratio of 1 bel. A smaller unit, the decibel, dB (a tenth of a bel), is defined as

$$\frac{P_2}{P_1}\bigg|_{dB} = 10 \log \frac{P_2}{P_1} \qquad (8\text{-}2)$$

The power ratio P_2/P_1 expressed in decibels (dB) may be found by Eq. (8-2). The 10 to 1 power ratio, which was shown to be the same as 1 bel is also 10 dB.

EXAMPLE 8-1

Given that the power output of a hi-fi amplifier is 10 W and a power input is 20 μW
(a) Find the power gain in terms of a ratio.
(b) Find the power gain in dB.

---

[1] See Appendix F.

SOLUTION

The power gain is given by

(a) $A_p = \dfrac{P_o}{P_{in}} = \dfrac{10}{20 \times 10^{-6}} = 5 \times 10^5$

(b) $A_p|_{dB} = 10 \log A_p = 10 \log (5 \times 10^5) = 10(\log 5 + \log 10^5) = 10(0.699 + 5) = 56.99$ dB

The dB is also used in connection with voltage and current ratios. Since, in general, $P = V^2/R$, we can express the power ratio of P_2/P_1 as $\dfrac{V_2^2/R_2}{V_1^2/R_1}$, where R_1 and R_2 are the resistors associated with P_1 and P_2, respectively.

If it is assumed that both P_1 and P_2 are developed in resistors of the same value, that is, $R_1 = R_2 = R$, then $P_2/P_1 = v_2^2/v_1^2 = (v_2/v_1)^2$. Hence, in dB this becomes

$$10 \log \dfrac{P_2}{P_1} = 10 \log \left(\dfrac{v_2}{v_1}\right)^2 = 2 \times 10 \log \left(\dfrac{v_2}{v_1}\right) \quad (8\text{-}3)$$

$$= 20 \log \dfrac{v_2}{v_1}$$

$$\left.\dfrac{v_2}{v_1}\right|_{dB} = 20 \log \left(\dfrac{v_2}{v_1}\right) \quad (8\text{-}3a)$$

Similarly, since $P = I^2 R$ we get

$$\left.\dfrac{i_2}{i_1}\right|_{dB} = 20 \log \left(\dfrac{i_2}{i_1}\right) \quad (8\text{-}3b)$$

Note that in *all* cases we are actually concerned with the *power ratio* produced by the given voltage or current ratio. The assumption $R_1 = R_2$ implies that when evaluating the voltage or current gain of an amplifier in dB, we must assume that $R_L = r_{in}$. (The power at the input and output must be developed across the same resistance value.) Note that while we may use terms such as voltage gain or current gain in dB, what we actually mean is that power gain in dB expressed in terms of voltage or current ratios.

EXAMPLE 8-2

Assume that $r_{in} = R_L$. Given $A_v = v_o/v_{in} = 10$
 (a) Find the power ratio A_p.
 (b) Find $A_p|_{dB}$.
 (c) Find $A_v|_{dB}$.

SOLUTION

(a) $P_o = \dfrac{v_o^2}{R_L}$, $P_{in} = \dfrac{v_{in}^2}{r_{in}}$

Hence,

$$A_p = \dfrac{P_o}{P_{in}} = \dfrac{v_o^2/R_L}{v_{in}^2/r_{in}}$$

$$A_p = \dfrac{P_o}{P_{in}} = \dfrac{v_o^2}{v_{in}^2} = \left(\dfrac{v_o}{v_{in}}\right)^2$$

$$= 100 \text{ (since } R_2 = r_{in})$$

(b) $A_p|_{dB} = 10 \log \left(\dfrac{P_o}{P_{in}}\right) = 10 \log 100 = 20$ dB

(c) $A_v|_{dB} = 20 \log \left(\dfrac{v_o}{v_{in}}\right) = 20 \log 10 = 20$ dB

Note that in Ex. 8-2, $A_p|_{dB} = 20$ dB that can be gotten from A_v directly using Eq. (8-3a) without having to calculate A_p first. Table 8-1 gives a few voltage and power ratios both as a simple ratio and in dB.

EXAMPLE 8-3

In the circuit of Fig. 8-1, assume $r_{in} = R_L$. The voltage gains of the cascaded amplifiers are given as $A_{v1} = 100$, $A_{v2} = 20$. Find
(a) $A_{v1}|_{dB}$.
(b) $A_{v2}|_{dB}$.
(c) $A_{vT}|_{dB}$, the total gain in dB.

SOLUTION

(a) $A_{v1}|_{dB} = 20 \log 100 = 40$ dB
(b) $A_{v2}|_{dB} = 20 \log 20 = 20 \times 1.3 = 26$ dB

TABLE 8-1 EQUIVALENT POWER AND VOLTAGE RATIOS

V_2/V_1	$V_2/V_1|_{dB}$	P_2/P_1	$P_2/P_1|_{dB}$	dB
0.1	20 log 0.1	0.01	10 log 0.01	-20
0.317	20 log 0.317	0.1	10 log 0.1	-10
0.707	20 log 0.707	0.5	10 log 0.5	$-3\,(-3.010)$
1	20 log 1	1	10 log 4	0
1.414	20 log 1.414	2	10 log 2	3 (3.010)
3.17	20 log 3.17	10	10 log 10	10
10	20 log 10	100	10 log 100	20

LOW FREQUENCY RESPONSE

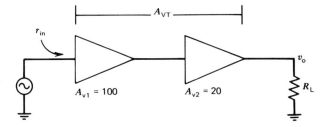

FIGURE 8-1 Cascading amplifiers.

(c) Since $A_{VT} = A_{v1} \times A_{v2} = 100 \times 20 = 2000$

$$A_{VT}|_{dB} = 20 \log 2000 = 66 \text{ dB}.$$

The solution for (c) above can be gotten by

$$A_{VT}|_{dB} = 20 \log (A_{v1} \times A_{v2})$$
$$= 20(\log A_{v1} + \log A_{v2}) \text{ (see Appendix)}$$

$$A_{VT}|_{dB} = 20 \log A_{v1} + 20 \log A_{v2}$$
$$= A_{v1}|_{dB} + A_{v2}|_{dB} = 40 + 26$$
$$= 66 \text{ dB}$$

In general, the total gain in dB is the *sum* of the individual gains in dB.

$$A_{VT}|_{dB} = A_{v1}|_{dB} + A_{v2}|_{dB} + \ldots \qquad (8\text{-}4)$$

Example 8-3 points to some of the advantages of expressing gain in dB. First, it permits a compression of the gain scale, so that a plot of gain versus frequency can be easily accommodated by a standard 8×11 sheet. Instead of dealing with values such as 100 or 2000, the equivalent dB representation is 20 dB or 66 dB, respectively. In addition, the resultant gain when cascaded stages are involved is obtained by the *addition* of the individual gains in dB (as opposed to multiplication for regular gains), simplifying gain calculations and keeping the total gain within the boundaries of the graph paper.

8-4 LOW FREQUENCY RESPONSE

As noted before, we deal with three frequency ranges: low, high, and midband. The midband gain (A_{vmb}) is the maximum gain of the circuit, since the ideal behavior of the components is assumed. (The midband frequency range is precisely that frequency range for which the gain is maximum. In many device-data sheets, 1 kHz is used as a typical midband frequency.) It is reasonable to use A_{vmb} as the reference to which the low and high frequency gains are compared. The question

354 FREQUENCY RESPONSE

that must be dealt with is what are the critical components and/or parameters in terms of high and low frequency behavior and how, quantitatively, do they affect the gain.

8-4.1 Coupling Capacitor

One of the critical components in terms of low frequency behavior is the coupling network, in particular the R-C coupling network. Fig. 8-2 shows the typical capacitive coupling. In the past, C was assumed to be a short, and hence, v_{in} appeared at the base of the transistor. At low frequencies this assumption is invalid and the effects of the reactance X_C must be considered. Clearly, the components C, r_{in} act like a voltage divider, so that the voltage applied to the base is divided down from the source voltage v_{in} by voltage division.

$$v_B = \frac{v_{in} \times r_{in}}{r_{in} - jX_C} = \frac{v_{in} \times r_{in}}{r_{in} - \frac{j}{\omega C}} \tag{8-5}$$

The gain (or attenuation) caused by the coupling network is

$$\frac{v_B}{v_{in}} = \frac{r_{in}}{r_{in} - j/\omega C} \tag{8-6}$$

in terms of magnitude alone

$$\left|\frac{v_B}{v_{in}}\right| = \frac{r_{in}}{\sqrt{R_{in}^2 + \left(\frac{1}{\omega C}\right)^2}}$$

$$= \frac{1}{\sqrt{1 + \left(\frac{1}{r_{in}\omega C}\right)^2}} \tag{8-6a}$$

[Eq. (8-6a) is the magnitude portion of Eq. (8-6)]. At very low frequencies the capacitor is virtually an open circuit ($X_C = \infty$) and the ratio $v_B/v_{in} = 0$, that is $v_B = 0$. No signal voltage is applied to the base. On the other hand, at very high frequencies it may be assumed that $X_C = 0$, and hence, $v_B/v_{in} = 1$, that is, the full voltage v_{in} is applied to the base. Between these two extremes, there is a frequency at which 70.7% of v_{in} is applied to the base, that is, $v_B = 0.707\ v_{in}$ or $v_B/v_{in} = 0.707 = 1/\sqrt{2}$. (For the moment the figure 0.707 seems arbitrary.) The frequency at which this happens is called the *lower cutoff frequency* (f_{LC}). From Eq. (8-6a) we see that when $(1/r_{in}\omega C)^2 = 1$ or when $r_{in}^2 = (1/\omega C)^2$, the denominator becomes $\sqrt{2}$ and $v_B/v_{in} = 1/\sqrt{2}$ (the student should complete the details of this derivation). This condition holds true at a single frequency f_{LC}; hence, f_{LC} can be

LOW FREQUENCY RESPONSE 355

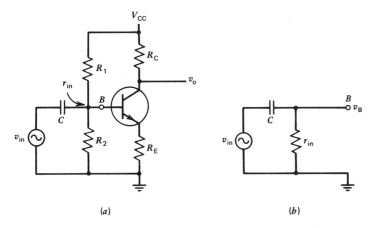

FIGURE 8-2 R-C Coupling. (a) Circuit diagram. (b) Simplified input portion.

obtained from

$$r_{in}^2 = \left(\frac{1}{\omega_{LC}C}\right)^2 = \left(\frac{1}{2\pi f_{LC}C}\right)^2 \tag{8-7}$$

(Recall $\omega_{LC} = 2\pi f_{LC}$.)

$$r_{in} = \frac{1}{\omega_{LC}C}$$

Solving for ω_{LC} and f_{LC}, we get

$$\omega_{LC} = \frac{1}{r_{in}C} \tag{8-8}$$

$$f_{LC} = \frac{1}{2\pi r_{in}C} \tag{8-8a}$$

The circuit of Fig. 8-2b is a typical "high pass" filter circuit encountered in R-C coupled amplifiers. It is the equivalent circuit of the coupling capacitor and input portion of the amplifier circuit. The term "high pass" refers to the fact that the circuit passes, unattenuated, signals of high frequency (while attenuating lower frequency signals).

Figure 8-3 gives the general form for a "high pass" R-C circuit. Here, the source resistance R_S (Thevenin's resistance of the source) is included, since it is practically always present.

The expression for f_{LC} must then be modified to account for R_S. Equation (8-8a) becomes

$$f_{LC} = \frac{1}{2\pi(R_S + R)C} \tag{8-9}$$

356 FREQUENCY RESPONSE

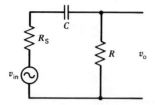

FIGURE 8-3 Typical high-pass R-C Circuit.

In amplifier circuits, R_S is usually r_o of one stage and R is r_{in} of a subsequent stage. [The student should verify Eq. (8-9) by following the procedure used in developing Eq. (8-8a).]

At the frequency f_{LC}, the ratio v_o/v_{in} (the gain of the coupling circuit) has become equal to 0.707. In dB this becomes

$$\left.\frac{v_o}{v_{in}}\right|_{dB} = 20 \log 0.707 = -3 \text{ dB}$$

a drop of 3 dB (from unity gain). Calculating the power delivered to $R(r_{in}$ in Fig. 8-2), we find that at high frequencies it is $P = v_{in}^2/R$, while at f_{LC} it becomes

$$P_{LC} = \frac{(0.707 \times v_{in})^2}{R} = 1/2 \frac{v_{in}^2}{R}$$

At cut off the power to R has become half its maximum (at high frequencies), again a 3 dB drop in power ($-$ 3 dB). The cut-off frequency f_{LC} may be referred to as

a. the low cut-off frequency; or
b. the half power point; or
c. the 3 dB point.

They all mean the same thing.

EXAMPLE 8-4

In Fig. 8-3 $R_S = 500 \, \Omega$, $C = 10 \, \mu F$, $R = 1.5 \, k\Omega$. Find f_{LC}.

SOLUTION

By Eq. (8-9)

$$f_{LC} = \frac{1}{2\pi (R_S + R)C}$$

$$= \frac{1}{2\pi(500 + 1500) \times 10 \times 10^{-6}}$$

$$= 7.96 \text{ Hz}$$

EXAMPLE 8-5

In Fig. 8-2 $R_1 = 22$ kΩ, $R_2 = 2.2$ kΩ, $R_C = 2.7$ kΩ, $R_E = 470$ Ω, $C = 10$ μF, $\beta = 100$. Find
(a) A_{vmb} for the amplifier (neglect r_e).
(b) f_{LC}.
(c) The gain at cut off.

SOLUTION

(a) The midband gain of this CE stage is:

$$|A_{vmb}| = \frac{R_C}{R_E} \text{ (magnitude only)}$$

$$|A_{vmb}| = \frac{2700}{470} = 5.74$$

(b) f_{LC} is a result of C and r_{in}. We find r_{in}.

$$r_{in} = R_1 \| R_2 \| \beta R_E$$
$$= 22 \text{ k}\Omega \| 2.2 \text{ k}\Omega \| 100 \times 470$$
$$= 1.9 \text{ k}\Omega$$

$$f_{LC} = \frac{1}{2\pi \times 1.9 \times 10^3 \times 10 \times 10^{-6}} \qquad (8\text{-}9)$$
$$= 8.4 \text{ Hz}$$

Note that the R-C circuit caused v_B to drop, hence causing v_o to drop, effecting a proportionate reduction in gain.

(c) At f_{LC} the gain has dropped to 0.707 of its maximum; hence,

$$A_{LC} = 0.707 \times 5.74 = 4.06$$

In dB

$$A_{vmb}|_{dB} = 20 \log 5.74 = 15.2 \text{ dB}$$

At cutoff (that is at the cut-off frequency)

$$A_{LC}|_{dB} = A_{vmb}|_{dB} - 3 = 15.2 - 3 = \underline{12.2 \text{ dB}}$$

The last figure can be gotten from

$$A_{LC}|_{dB} = 20 \log A_{LC} = 20 \log 4.06 = \underline{12.2 \text{ dB}}$$

The selection of half power (or 0.707 voltage ratio) is somewhat arbitrary. It, nevertheless, serves as a demarcation between the frequency range for which the

358 FREQUENCY RESPONSE

particular amplifier is suitable and the range (below f_{LC}) for which a substantial loss in gain occurs. The value of f_{LC} gives a quantitative description, for the low frequency end, of the frequency response of the circuit. Note that the use of a 0.707 (or 3 dB) gain drop led to a very simple expression for f_{LC}, which is a good enough reason to select this particular gain drop.

8-4.2 The Emitter By-Pass Capacitor

A very serious low frequency response problem is due to the emitter by-pass capacitor. Figure 8-4a shows a typical CE circuit with an emitter capacitor C_E. This capacitor is usually selected so $X_{CE} \ll R_E$. This is not sufficient. X_{CE} must be selected so that X_{CE} is substantially smaller than r_e at midband, not just R_E. Since, for most circuits $r_e \ll R_E$, it is evident that the choice of $X_{CE} \ll r_e$ implies that $X_{CE} \ll R_E$. The effects of C_E on the frequency response of the circuit are represented by the low cut-off frequency f_{LC}, caused by C_E. To calculate this f_{LC}, note that the voltage gain of the circuit shown in Fig. 8-4 at midband (assuming C_E to be a perfect short) is

$$A_{vmb} = -R_C/r_e \quad \text{(see Ch. 5)}$$

At low frequencies C_E cannot be assumed a short, and hence,

$$A_v = -\frac{R_C}{r_e + R_E \| Z_{CE}} \quad (Z_{CE} = -jX_{CE})$$

Since $X_{CE} \ll R_E$, this becomes

$$A_v = -\frac{R_C}{r_e + (-jX_{CE})} \tag{8-10}$$

$$|A_v| = \frac{R_C}{\sqrt{r_e^2 + X_{CE}^2}} \tag{8-10a}$$

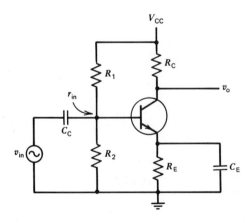

FIGURE 8-4 CE Circuit with emitter by-pass capacitor.

LOW FREQUENCY RESPONSE 359

If we compare Eq. (8-10a) with Eq. (8-6a), it is clear that the low cut-off frequency occurs when $r_e = X_{CE}$ (causing a drop in A_v by a factor of 0.707). Hence,

$$r_e = \frac{1}{\omega C_E} = \frac{1}{2\pi f_{LC} C_E} \quad (8\text{-}11)$$

$$f_{LC} = \frac{1}{2\pi \times r_e \times C_E}$$

It becomes apparent that C_E should be selected so that $X_{CE} \ll r_e$ at midband, rather than $X_{CE} \ll R_E$. (R_E seems to have little effect on the f_{LC} when $r_e \ll R_E$.) To obtain a low f_{LC}, thus expanding the frequency response of the circuits, the product $r_e \times C_E$ must be as large as possible. This is usually done by selecting C_E as large as reasonable.

EXAMPLE 8-6

In Fig. 8-4, $r_e = 5\,\Omega$, $R_E = 200\,\Omega$, $C_E = 100\,\mu F$, $C_C = 10\,\mu F$, $r_{in} = 1.0\,k\Omega$ (r_{in} calculated from $r_{in} = R_1 \| R_2 \| \beta r_e$). Find:

(a) f_{LC} due to the coupling capacitor C_C.
(b) f_{LC} due to the emitter by-pass capacitor C_E.
(c) Compare the two results.

SOLUTION

(a) $f_{LC} = \dfrac{1}{2\pi \times r_{in} \times C_C}$

$= \dfrac{1}{2\pi \times 10^3 \times 10 \times 10^{-6}} = \underline{16\ Hz}$

(b) $f_{LC} = \dfrac{1}{2\pi \times r_e \times C_E}$

$= \dfrac{1}{2\pi \times 5 \times 100 \times 10^{-6}} = \underline{318\ Hz}$

(c) Clearly, the effects of $r_e - C_E$ dominate. The actual cut-off frequency of this stage is approximately 318 Hz (the worse of the two figures), making this stage unsuitable as a quality audio amplifier. (Good hi-fi amplifiers must amplify the full audio frequency range that goes as low as 15 Hz.)

Example 8-6 points to the difficulties introduced by the $r_e - C_E$ network. Even if we made $C_E = 1000\,\mu F$ (100 times larger than C_C), the cut-off frequency would still be $f_{LC} = 31.8\ Hz$.

8-5 HIGH FREQUENCY RESPONSE

The causes of gain loss at high frequencies are substantially different from those for low frequency. The coupling capacitor C_C and emitter by-pass capacitor C_E that were considered a short at midband may certainly be regarded as shorts at higher than midband frequencies. C_C and C_E may now be replaced by shorts. The gain loss at high frequencies is caused by capacitors appearing across (in parallel with) the signal voltage. These capacities are largely due to transistor interelectrode capacities or wiring capacities, etc. Typically, we wind up with an equivalent circuit of the form of Fig. 8-5a. For simplicity, it is temporarily assumed that $r_{in} = \infty$ (r_{in} is open circuited) as shown in Fig. 8-5b. This circuit is referred to as a low pass circuit, because it does not affect low frequencies. At lower frequencies there is a lower voltage drop across R_s. At these low frequencies $v_o = v_{in}$. (The term low here is relative only. It never refers to frequencies below midband.) Clearly, at high (or very high) frequencies, C_{eq} eventually acts like a short ($1/\omega C_{eq} \simeq 0$) and $v_o \simeq 0$. Similar to the high pass circuit, there is a frequency for which $v_o = 0.707 \times v_{in}$ or the gain of the circuit $v_o/v_{in} = 0.707$. In general, for the circuit shown,

$$v_o = \frac{v_{in} \times (-jX_{Ceq})}{R_S - jX_{Ceq}}$$

The magnitude of the gain then is

$$\left|\frac{v_o}{v_{in}}\right| = \frac{X_{Ceq}}{\sqrt{R_S^2 + X_{Ceq}^2}} \tag{8-12}$$

At low frequencies $X_{Ceq} \gg R_S$, thus,

$$\sqrt{R_S^2 + X_{Ceq}^2} \simeq \sqrt{X_{Ceq}^2}$$

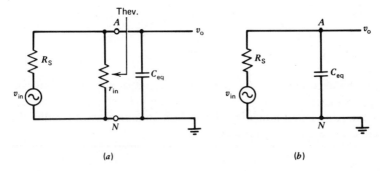

(a) (b)

FIGURE 8-5 High frequency equivalent circuit—low pass. (a) Equivalent circuit. (b) Simplified equivalent circuit ($r_{in} = \infty$—open).

and

$$\left|\frac{v_o}{v_{in}}\right| \simeq \frac{X_{Ceq}}{\sqrt{X_{Ceq}^2}} = 1 \qquad (8\text{-}12a)$$

showing that $v_o \simeq v_{in}$ when $X_{Ceq} \gg R_S$.

Since X_{Ceq} changes with frequency ($X_{Ceq} = 1/2\pi f C_{eq}$), the condition $X_{Ceq} = R_S$ will hold for some frequency. v_o/v_{in} then becomes

$$\left|\frac{v_o}{v_{in}}\right| = \frac{X_{Ceq}}{\sqrt{X_{Ceq}^2 + X_{Ceq}^2}} \qquad (8\text{-}12b)$$

$$= \frac{1}{\sqrt{2}} = 0.707 \ (R_S \text{ replaced by } X_{Ceq})$$

Equation (8-12b) shows that the gain has dropped from a maximum of 1 [Eq. (8-12a)] to 0.707. Consistent with the low frequency cut-off development, the high frequency for which this drop in gain occurs is the upper cut-off frequency, f_{UC}. Since this gain change takes place for $R_S = X_{Ceq}$, f_{UC} can be obtained from

$$R_S = \frac{1}{2\pi f_{UC} \times C_{eq}}$$

and

$$f_{UC} = \frac{1}{2\pi R_S \times C_{eq}} \qquad (8\text{-}13)$$

Returning to Fig. 8-5a and this time not neglecting r_{in}, we can show that

$$f_{UC} = \frac{1}{2\pi (R_S \| r_{in}) \times C_{eq}} \qquad (8\text{-}14)$$

R_S in Eq. (8-13) is replaced by $R_S \| r_{in}$.

To prove Eq. (8-14), we replace the circuit consisting of v_{in}, R_S, r_{in} with a Thevenin's equivalent (point A-N of Fig. 8-5a). Figure 8-6 shows the resultant equivalent circuit. Equation (8-14) is obtained by comparing Fig. 8-6 to Fig. 8-5b. (The circuits have the same form, with R_S replaced by $R_{Th} = R_S \| r_{in}$.)

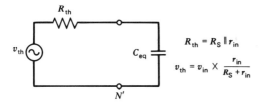

FIGURE 8-6 Applying Thevenin's equivalent to Figure 8-5a.

EXAMPLE 8-7

In Fig. 8-5a $R_S = 600\ \Omega$, $r_{in} = 1.2\ k\Omega$, $C_{eq} = 60\ pF$. Find f_{UC}.

SOLUTION

From Eq. (8-14)

$$f_{UC} = \frac{1}{2\pi(R_S \| r_{in}) \times C_{eq}}$$

$$= \frac{1}{2\pi(600\|1200) \times 60 \times 10^{-12}} = \underline{6.6\ MHz}$$

8-5.1 Stray, Interelectrode, and Miller Capacities

The circuit of Fig. 8-5a is said to be a typical equivalent circuit when considering high frequency response. R_S and r_{in} are the source resistance and input resistance, respectively. The makeup of C_{eq} is somewhat complex. Figure 8-7 shows the elements affecting high frequency performance. The capacitor C_S is a lumped approximation of the wiring or stray capacities. Long wires result in a substantial C_S, degrading the high frequency response. For high frequency amplifiers the physical layout and length of all wires is very important.

In addition to C_S, the transistor interelectrode capacities contribute to high frequency response problems. These capacities are very small, a few picofarads for the small signal transistor. They can be viewed as small parallel plate capacitors in which the internal transistor terminals constitute the two capacitor plates. The collector-base, base-emitter, and collector-emitter terminal pairs are represented by C_{BC}, C_{BE}, and C_{CE}, respectively.

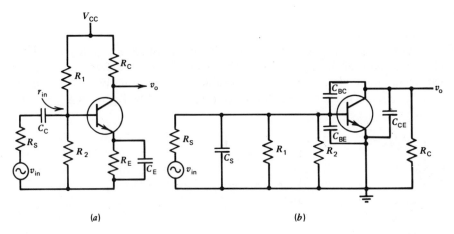

FIGURE 8-7 High frequency equivalent of CE configuration. (a) Circuit diagram. (b) High frequency equivalent.

In our circuits C_S and C_{BE} are simply in parallel. To understand the significance of C_{CB}, we redraw the circuit of Fig. 8-7b in Fig. 8-8a, leaving out all but C_{CB} and the transistor itself, connected to an ideal signal source. We proceed to calculate the current i_c into C_{CB}. This current is part of the input current i_{in} and represents the effect C_{CB} has on the input impedance.

By Ohm's law

$$i_c = \frac{v_{in} - v_o}{X_{CCB}} = \frac{v_{in} - v_o}{\frac{1}{\omega C_{CB}}} \tag{8-15}$$

The voltage *across* the capacitor divided by the reactance of the capacitor. Since $v_o = A_v \times v_{in}$, Eq. (8-15) becomes

$$i_c = \frac{v_{in} - A_v \times v_{in}}{1/\omega C_{CB}} = \frac{v_{in}(1 - A_v)}{1/\omega C_{CB}} \tag{8-16}$$

A_v in Eq. (8-16) is the *base to collector* gain, which is always negative (there is a phase reversal, 180°, between v_o and v_{in}). The magnitude of the gain is denoted by $|A_v|$. The term A_v may then be replaced by $-|A_v|$. (Note that $|A_v|$ is always positive.) Equation (8-16) may be rewritten

$$i_c = \frac{v_{in}(1 - -|A_v|)}{1/\omega C_{CB}} = \frac{v_{in}(1 + |A_v|)}{1/\omega C_{CB}} \tag{8-16a}$$

$$= \frac{v_{in}}{\frac{1}{\omega C_{CB}(1 + |A_v|)}}$$

Equation (8-16a) may be rewritten as

$$i_c = \frac{v_{in}}{1/\omega C_M} \tag{8-17}$$

where $C_M = C_{CB}(1 + |A_v|)$.

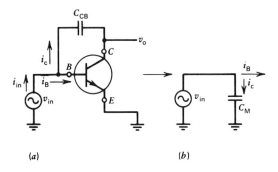

(a)   (b)

FIGURE 8-8 The Miller capacity. (a) C_{CB} considered alone. (b) The equivalent Miller capacity.

364 FREQUENCY RESPONSE

Equation (8-17) indicates that as far as i_c is concerned, it is a result of a voltage v_{in} across a capacity C_M, where the capacity C_M is $(1 + |A_v|)$ times the capacity C_{CB}. In other words, the effect of C_{CB} is multiplied by $(1 + |A_v|)$. Moreover, the collector-base capacitance C_{CB} can be replaced by a capacitor C_M connected from base to ground where $C_M = C_{CB}(1 + |A_v|)$. C_M is referred to as the Miller capacitance. The multiplication effect is called the Miller effect. Note that a capacitor C_M connected from base to ground would have a current i_c through it where

$$i_c = \frac{v_{in}}{-1/\omega C_M}$$

which is the current i_c in C_{CB} as shown by Eq. (8-17), thus justifying the equivalent circuit representing the effects of C_{CB} as shown in Fig. 8-8b.

The total equivalent capacity C_{eq} may now be written as

$$C_{eq} = C_S + C_{BE} + C_M = C_S + C_{BE} + (|A_v| + 1) \times C_{CB} \qquad (8\text{-}18)$$

The capacitor C_{CE} is not represented in Eq. (8-18) since C_{CE} is associated with the transistor output circuitry (collector circuit) and does not affect the input portion of the amplifier circuit.

The Miller effect and Eq. (8-18) as a whole are applicable to FET circuits as well. Clearly, in the case of FETs the interelectrode capacities are referred to as C_{DG}, C_{GS}, C_{DS} (related to the FET terminals drain, gate, source). Thus, Eq. (8-18) becomes

$$C_{eq} = C_S + C_{GS} + C_M = C_S + C_{GS} + (|A_v| + 1) \times C_{DG} \qquad (8\text{-}18a)$$

where A_v is the gate to drain voltage gain.

Note that C_{eq} is the equivalent gate to ground capacity that affects the gate input signals only (in the common source or common drain-source follower configurations).

EXAMPLE 8-8

In the circuit of Fig. 8-7a $R_S = 600\ \Omega$, $r_{in} = 1.2\ k\Omega$ (calculated from $R_1 \| R_2 \| \beta r_e$), $r_e = 15\ \Omega$, $R_C = 2.2\ k\Omega$, $C_{CB} = 2\ pF$, $C_{BE} = 3\ pF$, $C_S = 4\ pF$. Find f_{UC}, the upper cut-off frequency for the base circuit.

SOLUTION

The low pass simplified equivalent circuit is shown in Fig. 8-9. To find C_{eq}, we need the midband base-collector voltage gain

$$A_{vmb} = -\frac{R_C}{r_e} = -\frac{2200}{15} = \underline{-147}$$

Hence, $|A_{vmb}| = 147$.

HIGH FREQUENCY RESPONSE

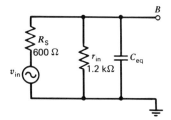

FIGURE 8-9 Input section low pass equivalent.

$$C_M = C_{CB}(|A_{vmb}| + 1)$$
$$= 2 \times (147 + 1) = 296 \text{ pF}$$
$$C_{eq} = C_S + C_{BE} + C_M$$
$$= 4 + 3 + 296 = 303 \text{ pF}$$

(Note that the largest contribution to C_{eq} is due to $C_{CB}(|A_{vmb}| + 1)$. C_S and C_{BE} can easily be neglected *here*.)

$$f_{UC} = \frac{1}{2\pi(R_S \| r_{in}) \times C_{eq}} \quad (8\text{-}14)$$

$$= \frac{1}{2\pi(600\|1200)303 \times 10^{-12}} = \underline{1.3 \text{ MHz}}$$

EXAMPLE 8-9

Figure 8-10a gives a two-stage R-C coupled amplifier. Find:
(a) A_{vmb} of Q_2 (base to collector voltage gain).
(b) f_{LC} for the input section of Q_2 (R_{o1}, C_{C2}, r_{in2}).
(c) f_{LC} for the emitter circuit of Q_2 (C_{E2}, r_{e2}).
(d) f_{LC} for the output portion of Q_2 (r_{o2}, C_{C3}, R_L).
(e) f_{UC} for the input portion of Q_2.

SOLUTION

(a) A_{vmb} of Q_2 can be obtained by

$$A_{vmb} = -\frac{R_C \| R_L}{r_{e2}} = -\frac{3.9 \text{ k}\Omega \| 3.9 \text{ k}\Omega}{25}$$
$$= -78$$
$$\underline{|A_{vmb}| = 78}$$

(b) To obtain f_{LC} for the input section of Q_2, we draw the equivalent circuit

366 FREQUENCY RESPONSE

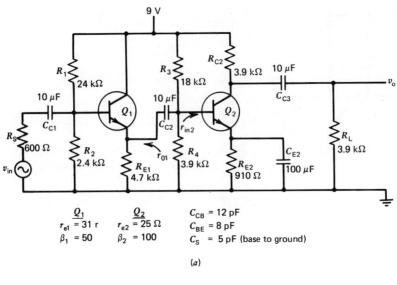

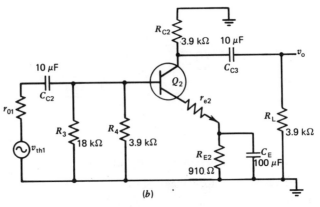

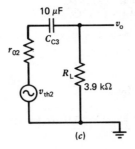

FIGURE 8-10 Two stage R-C coupled amplifier. (a) Circuit diagram. (b) Partial equivalent circuit. (c) Output equivalent of Q_2.

as in Fig. 8-10b. r_{o1} is the output resistance (Thevenin's resistance) of Q_1; hence,

$$r_{o1} = R_{E1} \| \left(r_{e1} + \frac{R_1 \| R_2 \| R_S}{\beta_1} \right)$$

$$= 4.7 \text{ k}\Omega \| \left(31 + \frac{2400 \| 2400 \| 600}{50} \right)$$

$$\approx 39 \, \Omega$$

(output resistance of EF circuit).
r_{o1} is the resistance of the signal source driving Q_2.

$$r_{in2} = R_3 \| R_4 \| \beta_2 r_{e2}$$
$$= 18 \text{ k}\Omega \| 3.9 \text{ k}\Omega \| 100 \times 25 = 1.4 \text{ k}\Omega$$

Hence,

$$f_{LC} = \frac{1}{2\pi(r_{o1} + r_{in2})C_{C2}}$$

$$= \frac{1}{2\pi(39 + 1400) \times 10 \times 10^{-6}} = \underline{11 \text{ Hz}}$$

(c) $f_{LC} = \dfrac{1}{2\pi r_{e2} C_E}$ \hfill (8-12)

$$= \frac{1}{2\pi 25 \times 100 \times 10^{-6}} = 64 \text{ Hz}$$

(d) To calculate f_{LC} for the output portion of Q_2, we draw the equivalent circuit of this portion of the amplifier as shown in Fig. 8-10c. $r_{o2} \approx 3.9 \text{ k}\Omega$. Thus,

$$f_{LC} = \frac{1}{2\pi(r_{o2} + R_L)C_{C3}}$$

$$= \frac{1}{2\pi(3.9 \text{ k}\Omega + 3.9 \text{ k}\Omega) \times 10 \times 10^{-6}} = 2.0 \text{ Hz}$$

The worst low frequency response is due to C_E. The approximate, final, f_{LC} of Q_2 is 64 Hz, the worst of all low cut-off frequencies (the highest f_{LC}).

(e) First compute C_M

$$C_M = C_{CB}(|A_v| + 1) = 12 \times (78 + 1) = 948 \text{ pF}$$

($|A_{vmb}| = 78$ from part (a) above)

$$C_{eq} = C_S + C_{BE} + C_M = 5 + 8 + 948 = 961 \text{ pF} \hfill (8\text{-}18)$$

From Fig. 8-10d

$$f_{UC} = \frac{1}{2\pi(39\|r_{in2})C_{eq}}$$

$$= \frac{1}{2\pi(39\|1400) \times 961 \times 10^{-12}} = 4.2 \text{ MHz}$$

8-6 THE FREQUENCY RESPONSE PLOT

It is often convenient to plot the frequency response curve including both upper and lower frequency cut offs. The plot is usually done in gain in dB versus frequency. If we choose a linear frequency scale (that is, the spacing for 20 Hz is double that for 10 Hz, etc.), it would be impossible to include both f_{UC} and f_{LC} on a single sheet (unless the frequency scale is many feet long). For example, should it be desired to cover the range from 64 Hz to 4.2 MHz (f_{LC} and f_{UC} in Ex. 8-9) in steps of 100 Hz, even if each step is only 0.1 inches long, a frequency scale of more than 300 feet long would be required. In order to compress the frequency scale, we use log (f) as the basic scale. The graduations on the frequency scale are based on the logarithm of the frequency not on the frequency directly. That is, since $\log(10) = 1$ and $\log(100) = 2$, the spacing between the 10 and 100 Hz marks is a unit distance. Similarly, the space between 100 and 1000 Hz is a unit distance. From 10 Hz to 10 MHz (10×10^6 Hz) we require only six unit distances (from 10 to 100, 100 to 1000, etc.). Figure 8-11 gives the frequency response plot of Ex. 8-9. The data used is:

$$A_{vmb} = 78, \quad A_{vmb}|_{dB} = 37.8 \text{ dB}$$
$$f_{LC} = 64 \text{ Hz}, \quad f_{UC} = 4.2 \text{ MHz}$$

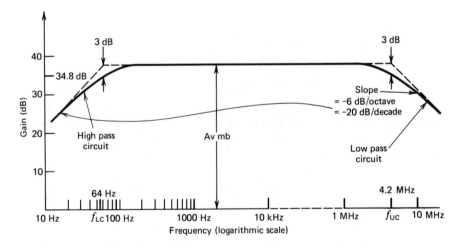

FIGURE 8-11 Frequency response.

Note that at the two cut-off frequencies, the gain is $A_{vmb} - 3$ dB $= 37.8 - 3 = 34.8$ dB. The graph paper is called a semilog paper because one axis is linear (the vertical), while the other is logarithmic (horizontal). It is noteworthy that the graph paper used is a "five-cycle" semilog paper. The five-cycle refers to a total of five unit increments that are displayed on the horizontal axis. To accommodate the 10 Hz to 10 MHz range, it was necessary to do the plot in two parts, 10 Hz to 10 kHz, then starting at 1 to 10 MHz. If we had a six-cycle paper, this maneuver would not be required. To suit our needs, we selected the left-most end as 10 Hz and the right edge as 10 MHz. (Other graph paper such as three- or four-cycle is commercially available.) Note that this type of a scale *cannot* include $f = 0$ (zero frequency) since log(0) is undefined. The frequency assigned to the right-most edge is selected (in powers of ten) to suit the particular plot.

The frequency response plot (Fig. 8-11) serves to give the user of the circuit an appreciation of the useful frequency range of the circuit. This plot, however, is not readily available and it is often desirable to be able to describe the plot, or rather the response itself, without using the plot. There are a number of important terms and concepts that convey the basic ideas presented in the plot. To facilitate the discussion of these concepts, we introduce here two terms, *decade* and *octave*.

A *decade* is a ten-to-one frequency ratio. The change from 10 to 100 Hz is referred to as a one-*decade* increase, while 1 MHz to 1 kHz is a three-decade decrease (1 MHz to 100 kHz, 100 to 10 kHz, 10 to 1 kHz).

An *octave* is a two-to-one frequency ratio; for example, 20 Hz is one *octave* higher than 10 Hz.

8-6.1 Bandwidth

The *bandwidth* (BW) of an amplifier refers to the frequency range between f_{UC} and f_{LC}. This is considered the useful frequency range of the circuit.

$$BW = f_{UC} - f_{LC} \qquad (8\text{-}19)$$

Since in most applications $f_{LC} \ll f_{UC}$, the bandwidth can be approximated by

$$BW \simeq f_{UC} \qquad (8\text{-}19a)$$

(This is true for wide-band amplifiers such as audio, video, and other *non*tuned amplifiers.)

The description and specifications of commercially available amplifier circuits (from the integrated circuit package to the stereo console) will invariably include bandwidth information, since it gives the user very clear and simple applications criteria. The stereo amplifier ought to cover the range of sound frequencies (15 Hz to 15 kHz) or more, while the video amplifier used in TV systems must cover a wider frequency range (0 to 5 MHz). It must be kept in mind that *BW* is but one applications criteria.

8-6.2 Frequency Response Slope

The BW does not by itself fully describe the frequency response. The slopes of the plot outside the BW are an important part of this plot.

We show now that the slope of the frequency response plot below f_{LC} and above f_{UC} is very close to -6 dB/octave or -20 dB/decade. This means that a *decline* of one octave at the low frequency end is associated with a *decrease* of 6 dB in voltage gain. A decline of one decade produces a 20 dB decrease in gain. Similarly, an *increase* above f_{UC} results in similar *decreases* in gain.

Figure 8-2b gives the typical "high pass" circuit that is responsible for the lower cut-off frequency f_{LC}. (For simplicity we refer to r_{in} as R.) The ratio $|v_{B/v_{in}}|$ was given by Eq. (8-6a).

$$\left|\frac{v_B}{v_{in}}\right| = \frac{R}{\sqrt{R^2 + \left(\frac{1}{\omega C}\right)^2}} = \frac{1}{\sqrt{1 + \left(\frac{1}{R\omega C}\right)^2}} \qquad (8\text{-}6a)$$

This equation represents the change in gain (v_B/v_{in}) as a function of frequency (ω). It can be rewritten as:

$$\left|\frac{v_B}{v_{in}}\right| = \frac{1}{\sqrt{1 + \left(\frac{1}{RC}\right)^2 \times \frac{1}{\omega^2}}} = \frac{1}{\sqrt{1 + \left(\frac{\omega_{LC}}{\omega}\right)^2}} \qquad (8\text{-}20)$$

Recall that

$$\omega_{LC} = \frac{1}{RC} \qquad (8\text{-}8)$$

If the circuit of Fig. 8-2b, for which Eq. (8-20) holds, is connected to a transistor with a midband gain of A_{vmb}, then the total gain of the circuit can be expressed as:

$$A_{vL} = \frac{A_{vmb}}{\sqrt{1 + \left(\frac{\omega_{LC}}{\omega}\right)^2}} = \frac{A_{vmb}}{\sqrt{1 + \left(\frac{f_{LC}}{f}\right)^2}} \qquad (8\text{-}20a)$$

A_{vL} is the gain at *any* frequency f, *below* f_{Lc}, while A_{vmb} is a fixed value, the maximum gain associated with the mid-frequency range.

Equation (8-20a) gives the gain as a function of the ratio f_{LC}/f. In other words, the gain below f_{LC} ($f < f_{LC}$) depends on the ratio f_{LC}/f. A very large f_{LC}/f means that we are operating at a frequency well below cut off and, as indicated by Eq. (8-20a), will have a substantial drop in gain.

THE FREQUENCY RESPONSE PLOT

EXAMPLE 8-10

Given $f_{LC} = 200$ Hz, $A_{vmb}|_{dB} = 40$ dB, find the gain at
(a) $f = 20$ Hz.
(b) $f = 10$ Hz.

SOLUTION

We use Eq. (8-20). Since A_{vmb} is given in dB, $A_{vmb}|_{dB} = 40 = 20 \log A_{vmb}$. A_{vmb} = Antilog 40/20 = Antilog (2) ("Antilog" translates to "the number whose logarithm is"; see Appendix G). Hence, $A_{vmb} = 100$ (20 log 100 = 40 dB).

(a) $A_{vL} = \dfrac{A_{vmb}}{\sqrt{1 + \left(\dfrac{f_{LC}}{f}\right)^2}} = \dfrac{100}{\sqrt{1 + \left(\dfrac{200}{20}\right)^2}}$

$= \dfrac{100}{\sqrt{1 + 10^2}} \approx \dfrac{100}{10}$

$\underline{A_{vL} = 10 = 20 \text{ dB}}$

The gain at a frequency one decade below f_{LC} is 20 dB below midband gain (40 dB − 20 dB = 20 dB).

(b) $A_{vL} = \dfrac{100}{\sqrt{1 + \left(\dfrac{200}{10}\right)^2}} = \dfrac{100}{20} = 5$

$\underline{A_{vL} = 5 = 14 \text{ dB}}$

The gain at one octave below 20 Hz is 6 dB lower than the gain at 20 Hz.

Example 8-10 demonstrates the slope of the frequency response plot below f_{LC}. The dashed line in Fig. 8-11 represents this slope. It is possible to obtain an approximate plot by extending the midband gain to the cut off frequency at which point we have a "break" (the cut-off frequency is often referred to as a break point) and continue downward at a −6 dB/octave slope or −20 dB/decade.

The above analysis was done for the low frequency range in which the small capacitors C_M, C_S, C_{BE}, etc. have little effect. A similar analysis for high frequencies, above f_{UC}, where the series capacitor C_C has no effect, yields an equation similar to Eq. (8-20a)

$$A_{vH} = \dfrac{A_{vmb}}{\sqrt{1 + \left(\dfrac{f}{f_{UC}}\right)^2}} \qquad (8\text{-}21)$$

372 FREQUENCY RESPONSE

(Note the "inversion" in the frequency ratio f/f_{UC} compared to f_{LC}/f.) Again, the slope is -6 dB/octave or -20 dB/decade.

If $f/f_{UC} = 10$, a one-decade change, then

$$A_{vH} = \frac{A_v}{\sqrt{1+10^2}} \approx \frac{A_v}{10}$$

a ten-fold drop in gain that equals a 20 dB decrease in gain.

EXAMPLE 8-11

Given $A_{vmb} = 30$ db, $f_{LC} = 100$ Hz, $f_{UC} = 50$ kHz, find:
(a) The gain at 20 Hz.
(b) The gain at 1.0 MHz.
(c) The bandwidth.
(d) Plot the frequency response.

SOLUTION

(a) To get to 20 Hz from $f_{LC} = 100$ Hz, we drop one *decade* to 10 Hz, then go *up* one octave. Hence, the gain decreases by 20 dB and then increases by 6 dB (*down* the slope to 10 Hz, *up* the slope to 20 Hz).

$$A_v \text{ at } 20 \text{ Hz} = 30 - 20 + 6 = \underline{16 \text{ dB}}$$

(b) To get to 1.0 MHz from $f_{UC} = 50$ kHz, we go up one *decade* (*down* the response slope) and up again one *octave* (further down the gain slope).

$$A_v \text{ at } 1\text{M Hz} = 30 - 20 - 6 = \underline{4 \text{ dB}}$$

(c) The bandwidth is obtained from Eq. (8-19) [or (8-19a)]

$$BW = f_{UC} - f_{LC} = 50{,}000 - 100$$
$$= \underline{49.9 \text{ kHz}} \approx 50 \text{ kHz} \approx f_{UC}$$

(d) The plot is shown in Fig. 8-12.

In our introduction to ac analysis, it was noted that the ac performance of the circuit is completely separate from the dc (biasing) operation. Here, the ac analysis itself is separated into three frequency ranges, each yielding a distinctly different equivalent circuit. The ideal situation occurs at *midband*, where all capacitive side-effects are neglected and the gain is maximum. There is the *low frequency range* where the coupling and by-pass capacitors (particularly C_E) are of major importance and the interelectrode capacities are neglected. And, last, there is the *high frequency range* where the interelectrode capacities, particularly C_{CB}, are of the utmost importance, while the coupling and by-pass capacities can easily be considered a short, that is, performing their task perfectly.

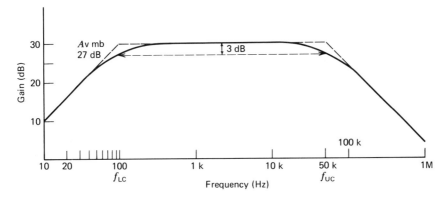

FIGURE 8-12 Frequency response plot for Ex. 8.11.

The three key parameters f_{LC}, f_{UC}, and A_{vmb} give a very good quantitative description of the frequency response of an amplifier. You are certainly *not* going to purchase your hi-fi amplifier if it has f_{LC} = 50 Hz (substantially above the lowest audible sound, such as base drums), f_{UC} = 10 kHz (substantially below the upper limit of audibility of about 16 kHz, of the string section). Its bandwidth does *not* cover the full audible frequency range!

The foregoing frequency considerations were essentially restricted to circuit performance. The intrinsic behaviors of the device itself were not considered.

8-7 TRANSISTOR PARAMETERS, f_β, f_T

It turns out that the β of the transistor *does* vary with frequency; it decreases substantially beyond a certain cut off point. The particular cut-off point depends on the internal structure of the transistor and can be gotten only from the manufacturer's specifications (short of measuring them).

Only two parameters usually given by the manufacturer will be discussed here. f_β is defined as the frequency for which β has dropped by 3 dB from a β midband value (β_{mb} is usually considered β at 1 kHz). This is similar to the upper cut-off frequency. Note that there is no deterioration in β for low frequencies, that is, $\beta_{dc} \simeq \beta_{mb}$. Clearly, it would not be wise to select a transistor with f_β = 100 kHz for use in an amplifier that is expected to amplify signals up to 300 kHz. (The term f_β unfortunately is not universally used. Some manufacturers use the symbol $f_{\alpha e}$ to denote the same quantity.)

Another term often specified is f_T, which is defined as the frequency at which β has become equal to 1 (unity or 0 dB). The term f_T is also called the *gain bandwidth product* of the transistor.

A typical frequency response plot showing both f_T and f_β is shown in Fig. 8-13. Note the similarities of this plot with the frequency response plot of Figs. 8-11 and 8-12, particularly in the upper frequency range.

374 FREQUENCY RESPONSE

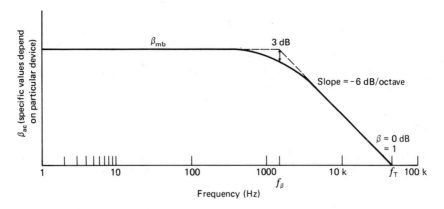

FIGURE 8-13 β versus frequency.

The parameter f_T may be used to obtain a quantitative estimate of *available* gain of the transistor at any frequency below f_T.

$$f_T \simeq GAIN \times BW \qquad (8\text{-}22)$$
$$GAIN = f_T/BW \simeq f_T/f_{UC} \qquad (8\text{-}22a)$$

f_T is the gain-BW product, which is constant for a given transistor. Equation (8-22a) indicates that the available gain of the circuit depends on f_T and the designed BW of the circuit. For example, wth $f_T = 10$ MHz we can obtain a gain of

$$\text{gain} = \frac{10 \text{ MHz}}{100 \text{ kHz}} = 100$$

if the $f_{UC} = 100$ kHz, or a gain of

$$\text{gain} = \frac{10 \text{ MHz}}{1 \text{ MHz}} = 10$$

if $f_{UC} = 1$ MHz. It must be emphasized that the above gives an estimate of the available gain, that is, the best gain that can be hoped for and not necessarily the actual gain of the circuit.

SUMMARY OF IMPORTANT TERMS

A_{vmb} The voltage gain at midband.
Bel A logarithmic representation of a power ratio (see dB).
Cut-off frequency The frequency at which the gain of the circuit drops to 0.707 of its value at midband (see 3 dB point, half power point, f_{LC}, f_{UC}).
Decibel (dB) A tenth of a bel:

$$1 \text{ bel} = 10 \text{ dB}. \left.\frac{P_2}{P_1}\right|_{dB} = 10 \log \frac{P_2}{P_1}$$

f_{LC} **Lower cut-off frequency.** The cut-off frequency on the low end of the frequency response.

Frequency response A description, usually graphic, of the variation of gain with frequency. A plot of gain versus frequency.

f_β The frequency at which the β of the transistor has dropped 3 dB (or become 0.707 of its value at midband).

f_T The frequency of unity gain. In terms of the frequency dependence of β, it's the frequency at which β is unity, $\beta = 1$.

f_{UC} Upper cut-off frequency. (See also f_{LC}, cut-off frequency.)

Half power point See cut-off frequency.

High pass A circuit that exhibits gain deterioration at the low frequency end only. Only f_{LC} exists.

Interelectrode capacities The equivalent of the capacity between the transistor electrodes (base, collector, emitter, C_{BC}, C_{BE}, C_{CE}).

Logarithm (base 10) A method of representing numbers by the power of ten, required to yield that number—for example, $\log 100 = \log 10^2 = 2$.

Low pass A circuit that exhibits gain deterioration at the high frequency end only. Only f_{UC} exists.

Midband The range of frequencies for which all capacitive effects are neglected; they are assumed to have little effect. Coupling capacitors or by-pass capacitors are considered a short. Interelectrode and stray capacitors considered open, nonexistent.

Miller capacity The equivalent of the collector-base interelectrode capacity considering the effects of base to collector gain.

Semilog paper Graph paper in which one axis is marked with a logarithmic scale. For example, the distance between 10 and 100 (10^1 to 10^2) is a unit distance and so is the space between 10^3 to 10^4, etc. Used in frequency response plots.

3 dB point See cut-off frequency.

PROBLEMS

8-1. Express the following power ratios P_2/P_1 in dB:

(a) $\dfrac{P_2}{P_1} = 2$.

(b) $P_1 = 5$ W, $P_2 = 0.5$ W.
(c) $P_1 = 100$ mW, $P_2 = 0.4$ W.

(d) $\dfrac{P_2}{P_1} = 0.02$.

8-2. Express the following voltage ratios $\dfrac{V_2}{V_1}$ or current ratios $\dfrac{I_2}{I_1}$ in dB:

(a) $\dfrac{V_2}{V_1} = 2$.

(b) $V_1 = 0.5$ V, $V_2 = 0.05$ V.

(c) $I_1 = 20 \ \mu A$, $I_2 = 0.05 \ \mu A$.

(d) $\dfrac{I_2}{I_1} = 0.02$.

(e) $\dfrac{V_2}{V_1} = 25$.

8-3. Express the following current, voltage, or power gains in dB:

(a) $A_i = 10$.
(b) $A_i = 0.5$.
(c) $A_v = 100$.
(d) $A_v = -20$.
(e) $A_p = 100$.
(f) $A_p = 10^6$.

8-4. Find the total voltage gain in dB for an amplifier consisting of three stages, each with a gain of $A_v = 20$.

8-5. Three amplifier stages are cascaded to give a total gain of 80 dB. If $A_{v1} = 20$ dB, $A_{v2} = 80$, find A_{v3}.

8-6. Find the low cut-off frequency for the four circuits shown in Fig. 8-14.
Hint: Looking into terminals A-B, obtain the Thevenin's Impedance. Does the 0.1 μ_F have a substantial effect?

8-7. Find f_{LC} for the circuit in Fig. 8-2 with the following values: $R_1 = 80 \ k\Omega$; $R_2 = 20 \ k\Omega$; $R_C = 2 \ k\Omega$; $R_E = 500 \ \Omega$; $\beta = 100$; $C = 1.0 \ \mu F$; $V_{CC} = 10$ V.

8-8. For Problem 8-7, find the gain at midband and at f_{LC}.

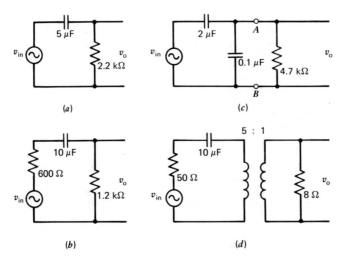

FIGURE 8-14

8-9. The ideal signal source in Problem 8-7 is replaced by a source with a source resistance of 5 kΩ. Find:

(a) f_{LC}.
(b) A_{vmb}.
(c) A_v at the cut-off frequency.

8-10. The circuit in Problem 8-7 is modified to include a *very large* emitter bypass capacitor and $r_e = 16\ \Omega$. Find:

(a) f_{LC}.
(b) A_{vmb}.
(c) A_v at cut off.

8-11. For the circuit of Fig. 8-15 find:

(a) f_{LC} due to C_C.
(b) A_{vmb}.

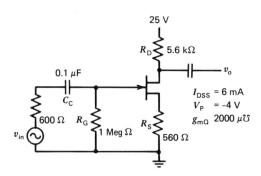

FIGURE 8-15

8-12. In Fig. 8-16 find:

(a) f_{LC} due to the input circuit (C_{C1}).
(b) f_{LC} due to the load coupling capacitor (C_{C2}).
(c) A_{vmb}.
(d) A_v at the *higher* cut-off frequency of the two found in Parts (a) and (b).

8-13. In Fig. 8-17 find:

(a) f_{LC} due to input coupling circuit.
(b) f_{LC} due to interstage coupling circuit.
(c) f_{LC} due to load coupling circuit.

Note: You may neglect r_{e1}, r_{e2}, since these are small relative to R_{E1} and R_{E2}.

8-14. For the circuit in Fig. 8-4: $R_1 = 82\ k\Omega$; $R_2 = 12\ k\Omega$; $R_E = 470\ \Omega$; $R_C = 3.3\ k\Omega$; $C_C = 10\ \mu F$; $C_E = 100\ \mu F$; $V_{CC} = 20\ V$; $\beta = 100$. Find:

(a) f_{LC} due to C_C.
(b) f_{LC} due to C_E.

378 FREQUENCY RESPONSE

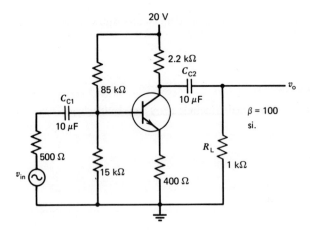

FIGURE 8-16

8-15. Find the upper cut-off frequency for the two circuits of Fig. 8-18. What happens to f_{UC} in Fig. 8-18a if $R_S = 0$?
Hint: Assume C_C is a short. Find f_{UC} and evaluate X_{CC} at f_{UC}. Draw conclusions regarding the significance of C_C here.

8-16. In Fig. 8-7: $R_S = 1$ kΩ; $R_1 = 82$ kΩ; $R_2 = 12$ kΩ; $R_C = 2.7$ kΩ; $R_E = 330$ Ω; $C_C = 100$ μF; $C_E = 1000$ μF; $C_{CB} = 3$ pF; $C_{BE} = 6$ pF; $C_S = 2$ pF; $r_e = 4$ Ω; $\beta = 100$. Find:

(a) f_{UC}.
(b) A_v at the upper cut-off frequency.

8-17. In the circuit of Fig. 8-15: $C_{DG} = 10$ pF (drain to gate capacity); $C_{GS} = 20$ pF (gate to source capacity). Find f_{UC}. (Neglect stray capacities.)

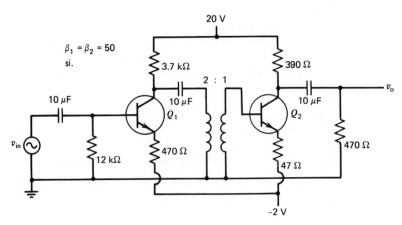

FIGURE 8-17

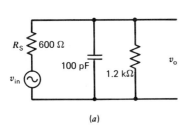
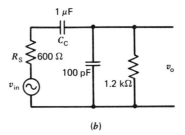

(a)   (b)

FIGURE 8-18

8-18. An amplifier has the following parameters: $A_{vmb} = 1000$; $f_{LC} = 30$ Hz; $f_{UC} = 50$ kHz. The frequency response slope is 6 dB/octave (20 dB/decade). Find in dB and as a ratio:

(a) A_v at 50 Hz.
(b) A_v at 30 kHz.
(c) A_v at 6 Hz.
(d) A_v at 0.5 MHz.
(e) A_v at 1.0 MHz.
(f) Plot the frequency response graph (10 Hz to 1 MHz) on a 5-cycle semilog paper in dB versus frequency.
(g) What is the BW of the amplifier?

8-19. A dc amplifier has an $A_{vmb} = 40$ dB and $f_{UC} = 500$ kHz. Find f_T of the amplifier [f_T is the frequency for which $A_v = 1$ (0 dB)].

8-20. A transistor has $f_\beta = 100$ kHz and β (at 1 kHz) $= 100$. Find:

(a) The value of β at 100 kHz.
(b) f_T.
(c) Estimate β at 400 kHz.

CHAPTER

9

LARGE SIGNAL AMPLIFIERS

9-1 CHAPTER OBJECTIVES

Three major areas are covered in this chapter: (1) large signal circuits–power circuits; (2) heat sinks and thermal considerations; and (3) general transistor specifications. Under items (1) and (2), the student will acquire the knowledge to:

1. Approximate the actual β from β versus I_C plots.
2. Calculate r_e (other than by $r_e = 26/I_E$).
3. Evaluate causes for distortion and their remedies.
4. Calculate efficiency for different classes of operations (class A, class B, class C, and for different circuits, for example, circuits with resistive load, transformer coupled load, complementary symmetry circuits, push pull circuits, etc.).
5. Relate actual power dissipation to the maximum power limitation of the transistor.
6. Analyze the operation of VMOS power amplifiers.
7. Calculate temperature effects on maximum power dissipation (power derating) and the temperature rise resulting from actual transistor power dissipation.
8. Compute allowable power dissipation with and without heat sinks.

Under item (3), general transistor specifications, the student is given a clear definition of most commonly given transistor (BJT, FET) specifications and characteristics. Various maximum voltage, current, and power specifications are discussed, as well as operating characteristics relating to frequency response (BW, f_T, f_β, t_r, t_f, C_{BC}, C_{BE}) and electrical noise figure (NF). The student will also gain an understanding of how these specifications affect component selection (which transistor is suitable for what application) and how to use the data in calculating various amplifier parameters.

9-2 INTRODUCTION

The discussions so far were restricted to "small" signals. This means that the signal excursion around the bias point I_{CQ}, V_{CEQ} is small. This, however, does not hold for all circuits. The ac output current i_{o1} in Fig. 9-1 deviates very little from I_{CQ}, while i_{o2} deviates by ± 0.4 A from $I_{CQ} = 0.6$ A ($i_{o2(p-p)} = 0.8$ A). i_{o1} may be considered a "small" signal, while i_{o2} is a "large one." Since the transistor behavior depends on I_{CQ}, that is, $r_e = 26/I_{CQ}$ and β varies with I_{CQ}, it is not possible to simply calculate r_e by $r_e = 26/I_{EQ}$. (This expression may serve only as a crude estimate.) In the case of "large" signals, the instantaneous collector current varies over a wide range, resulting in large changes in r_e, as well as β.

9-3 r_e AND β FOR LARGE SIGNALS

From Fig. 9-1, for i_{o2}, we note that at the peak current i_{o2} reaches 1 A; hence, $r_e = 26/1000 = 0.026$ Ω. While at its minimum i_{o2} reaches 0.2 A; hence, $r_e = 26/200 = 0.13$ Ω, a 5 to 1 variation in r_e. A simple way of handling this problem is to use the *average slope* of the V_{BE} versus I_C (or I_E) plot (Fig. 9-2).

Recall that the equation for r_e, $r_e = 26/I_E$ was obtained from the slope of this plot evaluated at a point, particularly at I_{CO} (I_{EO}). The average slope (mathematically

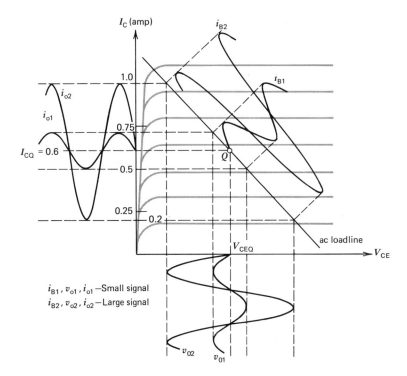

FIGURE 9-1 Small signals–large signals.

382 LARGE SIGNAL AMPLIFIERS

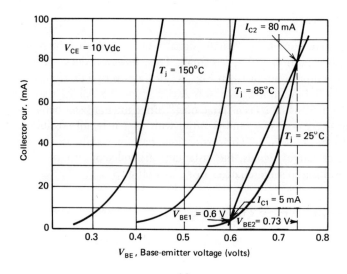

(a)

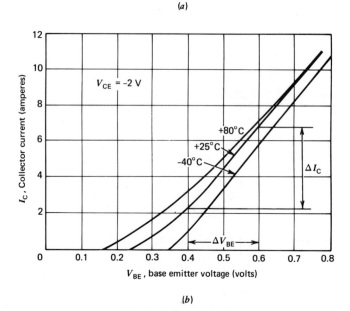

(b)

FIGURE 9-2 I_C versus V_{BE}. (a) For power transistor MJ 2251-2N3738. (v) For power transistors 2N2075-2N2078. (Courtesy Motorola, Inc.)

speaking, we actually use here the inverse of the slope) can be found by

$$\text{Slope} = \frac{\Delta V_{BE}}{\Delta I_C} = \frac{V_{BE2} - V_{BE1}}{I_{C2} - I_{C1}} = r'_e \tag{9-1}$$

(The notation r'_e is used in order to distinguish this r_e from the small signal r_e.)

In Fig. 9-2a r'_e is evaluated for a signal excursion from $I_{C1} \simeq 5$ mA to $I_{C2} = 80$ mA corresponding to $V_{BE1} = 0.6$ V and $V_{BE2} = 0.73$ V. Hence,

$$r'_e = \frac{V_{BE2} - V_{BE1}}{I_{C2} - I_{C1}} = \frac{0.73 - 0.6}{80 \times 10^{-3} - 5 \times 10^{-3}}$$

$$= \frac{0.13}{75} \times 10^3 = 1.73 \, \Omega$$

Note that neither $26/I_{C1}$ or $26/I_{C2}$ gives a good approximation for r'_e.

In many instances, the I_C versus V_{BE} is reasonably linear (Fig. 9-2b). In this case it is possible to compute r'_e by taking any interval ΔV_{BE} divided by the corresponding ΔI_C.

$$r'_e = \frac{\Delta V_{BE}}{\Delta I_C} \tag{9-2}$$

In Fig. 9-2b we have

$$\Delta I_C = 6.8 - 2.2 = 4.4 \text{ A}$$
$$\Delta V_{BE} = 0.6 - 0.4 = 0.2 \text{ V}$$

and

$$r'_e = \frac{0.2}{4.4} = 0.045 \, \Omega$$

A second transistor parameter that depends somewhat on I_C is β. To simplify matters, it is common practice to use the smallest β corresponding to the range of I_C. Figure 9-3a gives two graphs of β_{dc} (or h_{FE}) as a function of I_C for the 2N3715. If the 2N3715 is used with a swing of 0.2 A to 1 A (as in Fig. 9-1), the β corresponding to $I_C = 1$ A, $\beta = 80$, will be used in the circuit analysis, since it is the lowest β in the operating range. (For $I_C = 0.2$ A, $\beta = 100$.) On the other hand, for the 2N3447 (Fig. 9-3b) operating in the same range a $\beta \simeq 75$ corresponding to $I_c = 0.2$ A will be used. (The plot of β at 25°C was used in these examples.)

In the "small signal" transistor analysis we have, almost always, neglected r_{ce} (r_{oe} from the hybrid parameter approach). For power transistors, this parameter may be as low as a few hundred ohms (usually larger current transistors have lower values of r_{ce}) and materially affects the performance of the transistor.

The analysis of large signal circuits follows the same procedure used for small signal computations; hence, all the equations developed in the last few chapters hold. The value of r_e to be used is that given by Eqs. (9-1) or (9-2) and the value for β is as indicated above. It is also necessary to be somewhat more careful in evaluating the effects of r_{ce} to see whether r_{ce} may be neglected or must be included in the computations.

9-4 LINEARITY AND DISTORTION

As long as the input signal to a transistor circuit is small, the essentially nonlinear behavior of the semiconductor devices may be assumed linear. For large signals,

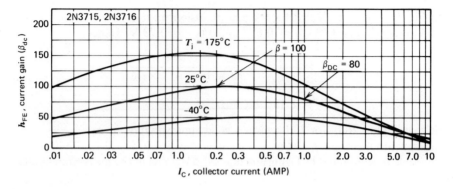

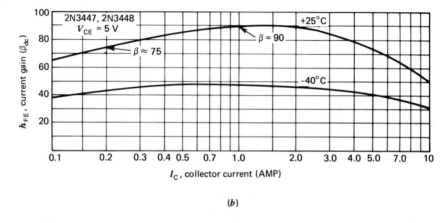

FIGURE 9-3 β_{dc} versus collector current. (a) For transistors 2N3715-16. (b) For transistors 2N3447-48. (Courtesy Motorola Inc.)

nonlinearities become more pronounced and introduce distortion. Figure 9-4 demonstrates this linearity problem for a JFET. (See also Section 6-6.) The upper portion of i_D for the large signal (Fig. 9-4a) is substantially larger than the lower portion; hence, i_D cannot be considered a sinusoid anymore. This is due to the curvature in the I_D versus V_{GS} curve. In contrast to the nonlinearity of the JFET, the VMOS FET has a linear I_D versus V_{GS} plot, and hence, minimal distortion even for large signals (Fig. 9-4b).

To combat this source of distortion in power amplifiers–large signal amplifiers, the manufacturers have produced linear power transistors (mainly bipolar devices and very recently the VMOS), as well as developed special circuits that overcome, to a great extent, the linearity problem.

POWER CONSIDERATIONS—EFFICIENCY AND MAXIMUM POWER RATING

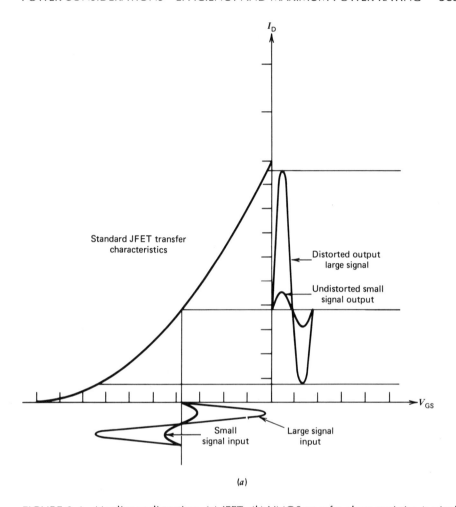

FIGURE 9-4 Nonlinear distortion. (a) JFET. (b) VMOS transfer characteristics (typical for 2N6656, VN30).

9-5 POWER CONSIDERATIONS—EFFICIENCY AND MAXIMUM POWER RATING

The power delivered to the speakers in your hi-fi system is on the order of watts. For example, assume that the system delivers 9 W to an 8 Ω speaker. This translates to a voltage of 24 V p-p and a current of 3 A p-p, produced by a transistorized power stage. This example demonstrates the need for large voltage and current swings. It becomes evident that for large-signal amplifiers, we are concerned with the *power* delivered to the load, and hence must be concerned with the *efficiency* of the amplifier. *Efficiency* (η) is defined as the ratio of ac output power to total

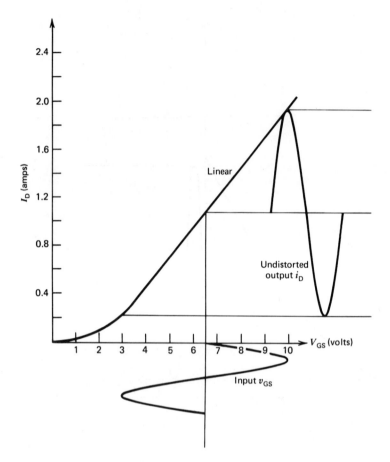

FIGURE 9-4 (Continued)

power dissipated by the circuit. For most power circuits this translates to the ratio of signal output power to dc power.

$$\eta \text{ (efficiency)} = \frac{P_{ac}}{P_{dc}} \qquad (9\text{-}3)$$

$$\% \, \eta = \frac{P_{ac}}{P_{dc}} \times 100\% \qquad (9\text{-}3a)$$

It is self-evident that low efficiency means waste. If, for example, an amplifier requires 36 W dc power to deliver 9 W signal power to the load, an efficiency of 25%, the power wasted is four times the power driving the load. In most high power circuits, this waste is intolerable. (High efficiency circuits are discussed in Sec. 9-8.)

Besides this concern with the efficiency of the circuit, it is necessary to make sure that the semiconductor device is operating within its permissible power range.

POWER CONSIDERATIONS—EFFICIENCY AND MAXIMUM POWER RATING

Every transistor has a specified maximum power limitation. If operated beyond its maximum power, the transistor may be permanently damaged. The significance of this maximum power limitation can best be demonstrated by obtaining a plot of maximum power on an I_C and V_{CE} (or I_D and V_{DS} for the FET) coordinate system. This plot can be superimposed on the collector (or drain) characteristics of the device, since these characteristics use the desired I_C and V_{CE} set of coordinates. The plot is obtained by first realizing that the power dissipated by the transistor P_T is given by

$$P_T = I_C V_{CE} \tag{9-4}$$

Hence, the equation representing the maximum allowable power dissipation P_D is:

$$P_D = I_C V_{CE}$$

For example, the Motorola power transistor MJ2252 has a specified total power dissipation P_D (maximum permissible dissipation) of $P_D = 10$ W. The plot of $10 = I_C V_{CE}$ (or $V_{CE} = 10/I_C$) is shown in Fig. 9-5 superimposed on a family of collector characteristics. (For example, for $V_{CE} = 200$ V, we have $I_C = 50$ mA and $P_D = 200 \times 50 \times 10^{-3} = 10$ W, etc.).

The portion below and to the left of the P_D graph represents the permissible operating range for which $P_T < P_D$. The actual power dissipated in the transistor P_T is less than the maximum allowable power P_D. The transistor may not operate in the region above and to the right of the P_D plot. In this region $P_T > P_D$. It should be noted that a similar presentation can be made for other devices such as JFET, VMOS, etc. (Fig. 9-6).

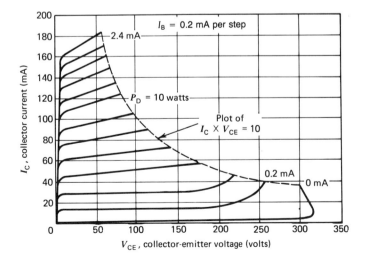

FIGURE 9-5 Maximum power plot for the MJ2252 power transistor (Motorola, Inc.).

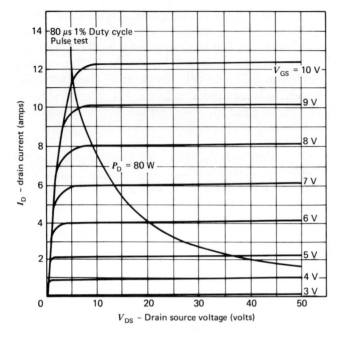

FIGURE 9-6 Maximum power plot for the VN 64GA-VMOS power transistor (Siliconix, Inc.).

EXAMPLE 9-1

A 2N3715 power transistor is used in a power circuit. i_c is given as $i_{c(p-p)} = 5$ A. $I_{CQ} = 3$ A. $V_{CEQ} = 25$ V.
 (a) Find whether the transistor operates within its maximum power limitations.
 (b) Find r'_e (average r_e).
 (c) Determine the β to be used in the calculations.

SOLUTION

 (a) From the specifications we find that $P_D = 150$ W (maximum power). The actual dc power dissipated by the transistor is $P_T = I_{CQ} \times V_{CEQ} = 3 \times 25 = 75$ W. Since $P_T < P_D$, the transistor is well within its power specifications.
 (b) We use the V_{BE} versus I_C plot for this transistor (Fig. 9-7). (Note that the vertical scale I_C is logarithmic, giving rise to the unusual shape of the curve.) The maximum and minimum values of I_C for the given peak to peak $i_c = 5$ A are

$$I_{c2} = 3 + \frac{5}{2} = 5.5 \text{ A}$$

$$I_{c1} = 3 - \frac{5}{2} = 0.5 \text{ A}$$

$$(i_{c(p-p)} = I_{c2} - I_{c1} = 5.5 - 0.5 = 5 \text{ A.})$$

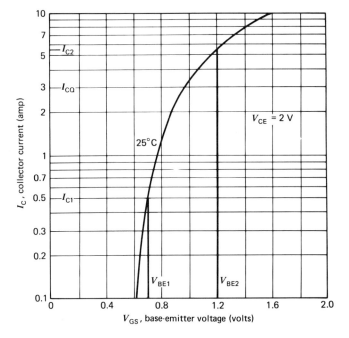

FIGURE 9-7 I_c versus V_{BE} for the 2N3713 through 2N3716 (Motorola, Inc.).

The corresponding V_{BE} values are:

$$V_{BE2} = 1.2 \text{ V}$$
$$V_{BE1} = 0.7 \text{ V}$$

$$r'_e = \frac{V_{BE2} - V_{BE1}}{I_{c2} - I_{c1}} = \frac{1.2 - 0.7}{5.5 - 0.5} \quad (9\text{-}1)$$

$$= \frac{0.5}{5} = 0.1 \text{ }\Omega$$

(c) Figure 9-3a gives the plot of β versus I_C. At $I_C = 5.5$ A (peak current), $\beta = 30$ (the plot for 25°C). For $I_C = 0.5$ A, $\beta = 80$. We use $\underline{\beta = 30}$.

9-6 CLASSES OF OPERATION

In the previous chapters, the quiescent operating point, the Q-point, was consistently placed in the "middle" of the collector (or drain) characteristics. It was also tacitly assumed that no clipping takes place, that is, the active device never exceeds saturation or cut off (see Fig. 5-30). This type of operation is called a *Class A* operation. In Class A circuits the transistor conducts for 360°, the full input cycle, and never saturates or cuts off. Figure 9-1 gives a graphic signal presentation of a Class A amplifier. For most efficient operation we set $V_{CEQ} = \frac{1}{2} V_{CC}$ so that the output voltage v_o (at the collector) can attain a maximum value, namely it can

390 LARGE SIGNAL AMPLIFIERS

swing from $V_C = V_{CEQ}$ to $V_C = 0$ and from $V_C = V_{CEQ}$ to $V_C = V_{CC}$ (Fig. 9-8). The maximum peak to peak output voltage is then $v_{o(max)} = V_{CC}$, that is, the largest peak to peak output voltage is V_{CC}. If, say, $V_{CC} = 20$ V, then $v_{o(max)} = 20$ V p-p. (In order to actually produce such an output, a sufficiently large input signal is necessary.) The output current swings from $I_C = 0$ to $I_C = I_{C(sat)}$, yielding $i_{c(max)} = 2I_{CQ}$, that is, the largest peak to peak output current possible is double I_{CQ}. If $I_{CQ} = 50$ mA, we have $i_{c(max)} = 100$ mA p-p.

A *Class B* operation occurs when the bias point is at cut off. That means that conduction takes place for only 180° (half the input cycle). Figure 9-9 shows the ac load line with the resulting waveforms. Here, $I_{CQ} = 0$, $V_{CEQ} = V_{CC}$, $v_{o(max)} = V_{CC(peak)}$, and $i_{o(max)} = I_{C(sat)}$. Here, the output current and voltages consist of *half* the input wave, that is, 180°.

To obtain a *Class C* operation the transistor is biased well beyond cut off. Figure 9-10 shows the resulting waveforms. Since the quiescent operating point is well beyond cut off, it cannot be shown in the $I_C - V_{CE}$ plot. The transistor conducts for a duration well below 180°. Usually the angle of conduction is about 90°.

Audio applications require operation over a full 360° so that the full sinusoid is reproduced. This may lead to the conclusion that only Class A may be used in audio amplifiers. It is, however, possible to obtain full 360° operation also with *two* transistors operating in Class B. Class C is certainly not usable for audio applications. It is commonly used in tuned amplifiers in which a resonance circuit produces the full sinusoid. The short Class C pulses that drive the resonance circuit

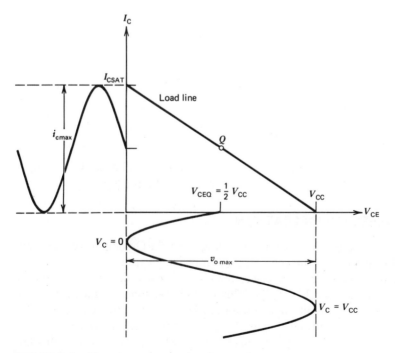

FIGURE 9-8 Class A maximum signal excursions.

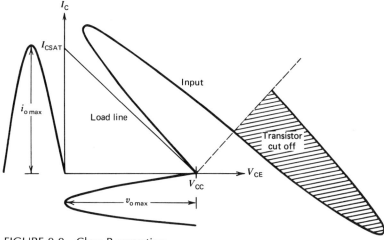

FIGURE 9-9 Class B operation.

serve to supplement the losses in the resonant circuit and maintain a full sinusoidal waveform.

9-7 CLASS A POWER AMPLIFIER

9-7.1 Resistive Load

Figure 9-11 shows a typical Class A amplifier with a resistive load. Note that this circuit is nothing more than the typical CE circuit.

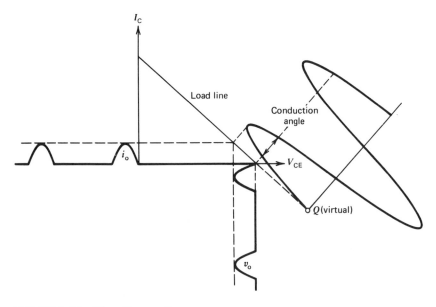

FIGURE 9-10 Class C operation.

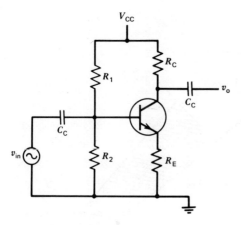

FIGURE 9-11   CE class A.

The parameters r_{in}, r_o, A_v, A_i, and A_p are computed in the same way they have been in the previous chapters. The dynamic resistance r_e to be used in the computations must be obtained by Eq. (9-1) or (9-2).

An additional parameter, efficiency (η), is of importance in power amplifiers. Recall that

$$\eta = \frac{P_{ac}}{P_{dc}} = \frac{P_o}{P_{dc}} \qquad (9\text{-}3)$$

($P_{ac} = P_o$ the signal output power.)

To calculate P_o we refer to Fig. 9-1 and note that

$$P_o = i_{o(rms)} \times v_{o(rms)} = \frac{v_{o(rms)}^2}{R_L}$$

since

$$i_{o(rms)} = \frac{i_{o(p\text{-}p)}}{2\sqrt{2}}$$

and

$$v_{o(rms)} = \frac{v_{o(p\text{-}p)}}{2\sqrt{2}}$$

we find that

$$P_o = \frac{v_{o(p\text{-}p)} \times i_{o(p\text{-}p)}}{2\sqrt{2} \times 2\sqrt{2}} \qquad (9\text{-}5)$$

$$= \frac{v_{o(p\text{-}p)} \times i_{o(p\text{-}p)}}{8} = \frac{v_{o(p\text{-}p)}^2}{8R_L}$$

(using the relation $i_{o(p\text{-}p)} = v_{o(p\text{-}p)}/R_L$).

P_{dc} is the power delivered by the supply voltage. V_{CC} supplies the quiescent current; consequently,

$$P_{dc} = V_{CC} \times I_{CQ} \qquad (9\text{-}6)$$

CLASS A POWER AMPLIFIER   393

The actual efficiency can be calculated only when v_o and i_o are available (clearly, we also require V_{CC} and I_{CQ}). These depend on the input voltage and the gain. In order to obtain some idea of the possible values of η, it is useful to evaluate the *maximum possible* efficiency. This occurs when v_o and i_o are maximum and when I_{CQ} and V_{CEQ} are perfectly "placed," that is, $I_{CQ} = \frac{1}{2}I_{C(sat)}$, $V_{CEQ} = \frac{1}{2}V_{CC}$ (in the center of the load line). From the previous discussion and Fig. 9-8 we get $v_{o(max)} = V_{CC}$ (peak to peak output voltage); $i_{o(max)} = 2I_{CQ}$ (peak to peak output current). By substitution into Eq. (9-5), we get

$$P_{o(max)} = \frac{V_{CC} \times 2I_{CQ}}{8} = \frac{V_{CC} \times I_{CQ}}{4}$$

$$\eta_{max} = \frac{P_o}{P_{dc}} = \frac{V_{CC} \times I_{CQ}}{4} \bigg/ V_{CC} \times I_{CQ} = \frac{1}{4} \quad (9\text{-}3)$$

$$\%\eta_{max} = \frac{1}{4} \times 100\% = 25\%$$

We see that the *best* that can be hoped for, in *this case*, is an efficiency of 25%, rather very poor.

EXAMPLE 9-2

For the circuit of Fig. 9-11 $R_C = R_L = 390 \ \Omega$, (R_C is the load resistance R_L). $R_1 = 1.5 \ k\Omega$, $R_2 = 100 \ \Omega$, $R_E = 39 \ \Omega$, $V_{CC} = 50 \ V$, $v_{in} = 1 \ V$ rms, $\beta_{min} = 25$; r_{ce} may be neglected.
Find:
(a) $|v_o|$.
(b) η.
(c) The necessary input voltage in order to obtain maximum efficiency.

SOLUTION

(a) From CE gain equations

$$A_v = -\frac{390}{r'_e + 39}$$

To *estimate* r'_e we use $26/I_{CQ}$. I_{CQ} can be found by the exact voltage divider biasing analysis (the use of Thevenin's equivalent).

$$V_{Th} = \frac{50 \times 0.1}{1.5 + 0.1} = 3.125 \ V$$

$$R_{Th} = 1.5 \ k\Omega \| 0.1 \ k\Omega = 94 \ \Omega$$

$$I_{CQ} = \frac{V_{Th} - 0.7}{R_{Th}/\beta + R_E} = \frac{3.125 - 0.7}{\frac{94}{25} + 39} = 57 \ mA$$

An estimate of r'_e is $26/57 = 0.46 \ \Omega$. It is reasonable here to neglect r'_e, since $r'_e \ll R_E$,

$$|A_v| = \frac{390}{39} = 10$$

$$|v_o| = |A_v| \times v_{in} = 10 \times 1 = 10 \text{ V rms}$$

(b) $P_o = \dfrac{v_{o(rms)}^2}{R_L} = \dfrac{10^2}{390} = 0.26 \text{ W}$

$P_{dc} = V_{CC} \times I_{CQ}$
$= 50 \times 57 \times 10^{-3} = 2.85 \text{ W}$

$\eta = \dfrac{0.26}{2.85} = 0.09, \quad \% \eta = 9\%$

This compares with a possible maximum of 25%.

(c) To get maximum efficiency, the output voltage must be 50 V p-p ($v_{o(max)} = V_{CC(p-p)}$). This translates into

$$v_{o(rms)} \simeq \frac{50}{2\sqrt{2}} = 17.7 \text{ V rms}$$

With a gain of 10, v_{in} must be

$$v_{in} = \frac{17.7}{10} = 1.77 \text{ V rms}$$

Note that the actual input voltage is less than that required for maximum efficiency; as a result, efficiency is well below its maximum value. (The actual maximum v_o is somewhat less than 50 V due to the voltage drop across R_E.)

9-7.2 Transformer Coupled

Figure 9-12a shows a typical CE transformer-coupled stage. To calculate the various parameters, we convert the circuit as shown in Fig. 9-12b. R_L has been transformed to the collector side, becoming R_L', by

$$R_L' = R_L \times \left(\frac{N_1}{N_2}\right)^2 \qquad (9\text{-}7)$$

The voltage gain of the equivalent circuit (Fig. 9-12b) is

$$|A_v'| = \frac{v_o'}{v_{in}} = \frac{R_L'}{r_e} \qquad (9\text{-}8)$$

(The gain magnitude is used since the 180° phase shift, usually involved, may be present or eliminated by interchanging the two primary or secondary wires.) To find the gain of the original circuit, note that

$$v_o = v_o' \times \frac{N_2}{N_1} \qquad (9\text{-}8a)$$

CLASS A POWER AMPLIFIER

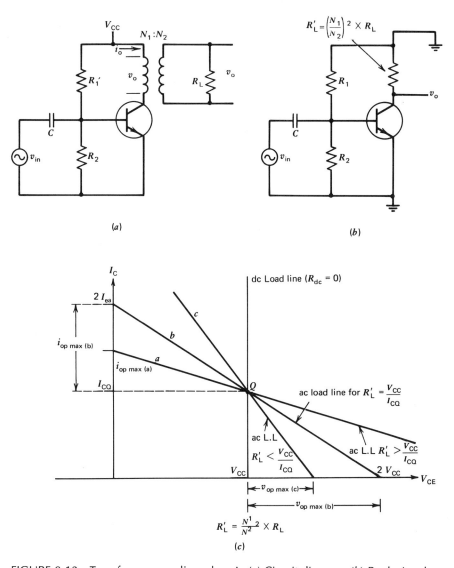

FIGURE 9-12 Transformer coupling–class A. (a) Circuit diagram. (b) Replacing the transformer with R'_L (ac equivalent). (c) Load lines $R'_L = (N_1/N_2)^2 \times R_L$.

Combining Eqs. (9-7), (9-8) and (9-8a) we get the expression for the gain

$$A_v = \frac{v_o}{v_{in}} = \frac{N_1}{N_2} \times \frac{R_L}{r_e} \tag{9-8b}$$

Equation (9-8b) indicates that the gain increases for $N_1 > N_2$ or $R'_L > R_L$. The transformer is usually used to transform a low R_L into a higher R'_L so that the overall gain is increased.

396 LARGE SIGNAL AMPLIFIERS

The value of R'_L [Eq. (9-7)] determines the *ac* loadline. By selecting $R'_L = V_{CC}/I_{CQ}$, we obtain an ac loadline that extends from $V_{CE} = 2V_{CC}$, $I_C = 0$ to $V_{CE} = 0$, $I_C = 2I_{CQ}$, allowing for the maximum possible signal amplitude. Figure 9-12c gives three loadlines: $R'_L > V_{CC}/I_{CQ}$ (graph a), $R'_L = V_{CC}/I_{CQ}$ (graph b), and $R'_L < V_{CC}/I_{CQ}$ (graph c). Graph b permits a maximum i'_{op} and v'_{op}, while in graph a i'_{op} maximum is smaller, and in c, v'_{op} maximum is smaller. Theoretically, R'_L can always be selected to permit a maximum signal output by carefully selecting $N_1:N_2$.

As demonstrated by graph b in Fig. 9-12c, the maximum output voltage is $v'_{o(p-p)} = 2V_{CC}$. As a result, $\eta_{max} = 50\%$. This compares with a maximum signal of $v_{o(p-p)} = V_{CC}$ and $\eta_{max} = 25\%$ for a similar transformerless circuit.

EXAMPLE 9-3

In Fig. 9-12a $V_{CC} = 30$ V, $I_{CQ} = 0.5$ A, $r'_e = 0.3$ Ω, $N_1/N_2 = 2$, $R_L = 8$ Ω, $v_{in} = 0.1$ V rms. The transistor used is 2N3766 with a maximum power dissipation $P_D = 20$ W. Find

(a) $|A_v|$.
(b) $|v_o|$.
(c) η.
(d) Check total power dissipation.

SOLUTION

To find v_o and A_v we find R'_L

$$R'_L = \left(\frac{N_1}{N_2}\right)^2 \times R_L = (2)^2 \times 8 = 32 \ \Omega$$

(a) $|A'_v| = \dfrac{R'_L}{r'_e} = \dfrac{32}{0.3} = 107$

$$|A_v| = A'_v \times \frac{N_2}{N_1} = 107 \times \frac{1}{2} = \underline{53.5} \qquad (9\text{-}8a)$$

(b) $|v_o| = |A_v| \times v_{in} = 53.5 \times 0.1 = 5.35$ V
Alternately

$$|v'_o| = |A'_v| \times v_{in} = 107 \times 0.1 = 10.7 \text{ V}$$

$$|v_o| = |v'_o| \times \frac{N_2}{N_1} = 10.7 \times \frac{1}{2} = 5.35 \text{ V} \qquad (9\text{-}8)$$

(c) First, find P_{dc} and P_o

$$P_{dc} = V_{CC} \times I_{CQ} = 30 \times 0.5 = 15 \text{ W}$$

$$P_o = \frac{v_o^2}{R_L} = \frac{5.35^2}{8} = 3.6 \text{ W}$$

$$\eta = \frac{3.6}{15} = 0.24, \quad \% \, \eta = 24\%$$

Note that $\eta_{max} = 50\%$.

(d) The allowable dissipation is 20 W. At the quiescent point $V_{CEQ} = V_{CC} = 30$ V, $I_{CQ} = 0.5$ A (see dc load line).

$$P_T = V_{CEQ} \times I_{CQ} = 30 \times 0.5 = 15 \text{ W}$$

$P_T < P_D$. The transistor is operating within the allowable power range.

9-8 CLASS B AUDIO AMPLIFIERS

In the discussion of Class A amplifiers, it was found that the maximum efficiency for this class of amplifiers is 50% for the transformer coupled type. Transformers, however, are notoriously bulky, expensive, and have a poor frequency response. To avoid transformers one is forced to use low efficiency, 25% at best, Class A resistive circuits. It is this low efficiency that stimulated the development of high efficiency Class B circuits that can be used in audio applications.

9-8.1 The Concept

Since in Class B operation only half the sinusoid is reproduced, and since audio applications require the full 360° of the waveform, it becomes necessary to combine the properly phased outputs of two amplifier circuits in order to reproduce the full sinusoidal waveform. This concept is shown in block diagram form in Fig. 9-13. Each of the circuits, 1 and 2, amplify only half the waveform. The two outputs are combined ("added") to produce v_o, the full sinusoid. To avoid distortion both

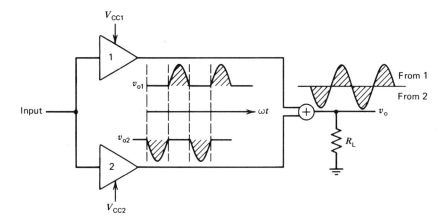

FIGURE 9-13 Class B concept.

398 LARGE SIGNAL AMPLIFIERS

circuits must have the exact same gain; otherwise, one of the two halves may have a larger amplitude, which translates into distortion.

Figure 9-13 leads one to think in terms of a NPN-PNP transistor pair. Circuit 1 amplifying the positive half may consist of an NPN transistor stage, while circuit 2 consists of a PNP stage. A typical circuit utilizing this complementary symmetry is shown in Fig. 9-14. As will be seen later, this circuit produces a somewhat distorted output; it is used here to simplify the explanation. Each transistor is biased at cutoff, $I_B = 0$. Q_1, an NPN transistor, operates as an emitter follower for the *positive* half cycle, while Q_2 acts as an EF for the *negative* half. The total output current flows in R_L, producing an output voltage that is the sum of the two transistor outputs.

9-8.2 Crossover Distortion

The distortion mentioned in the previous section is called "crossover" distortion and is caused by improper biasing. In the circuit of Fig. 9-14, both transistors are biased at $V_{BE} = 0$. Zero voltage applied to the base emitter junction results in $I_B = 0$, namely cut off. Since conduction requires that $V_{BE} \geq 0.7$ V (for silicon), conduction starts only after the input signal exceeds 0.7 V (-0.7 V for the PNP transistor). For the input signal below 0.7 V in magnitude (Fig. 9-15), none of the transistors conduct and thus there is no output for this interval, resulting in what is called, "crossover distortion."

To cure crossover distortion, it is necessary to provide external bias voltages such that $|V_{BE}| \simeq 0.7$ V for both transistors. Two suitable circuits are shown in Fig. 9-16. In order to provide 0.7 V for each base-emitter junction, we have only to provide $2 \times 0.7 = 1.4$ V from B_1 to B_2 (see polarities of the V_{BE} voltages in Fig. 9-16). In Fig. 9-16a this 1.4 V is provided by the voltage divider R_1, R_2, R_3, where R_2 is made adjustable so that the biasing voltage can closely match the actual forward diode drops of the BE junctions. In Fig. 9-16b the drop across D_1, D_2 provides the necessary bias voltage. D_1 and D_2 should be selected to match the base-emitter voltage drops of Q_1 and Q_2 (Germanium diodes for germanium transistors, etc.).

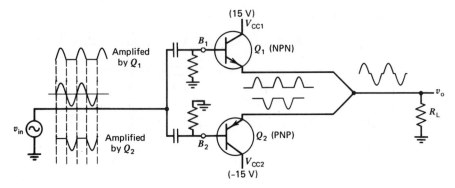

FIGURE 9-14 Complementary symmetry EF.

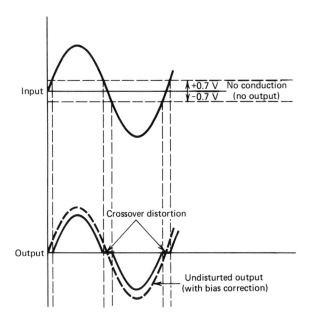

FIGURE 9-15 Crossover distortion.

Note that in Fig. 9-16b, the input is applied to the transistor bases B_1 and B_2 through the forward biased diodes D_1 and D_2. This is possible because the forward biased diodes represent a short for ac signals.

There are many other circuits, besides those shown in Figs. 9-16a and 9-16b that are designed to operate as Class B audio amplifiers. Some of these circuits will be discussed later. Since improved efficiency was the main aim of these circuits, it is worthwhile calculating the maximum efficiency of a Class B audio circuit. Fortunately, the calculations can be generalized and may be applied to all Class B audio circuits.

9-8.3 Class B Efficiency

The ac output power can be easily calculated once v_o is found. From Eq. (9-5)

$$P_o = \frac{v_{o(p-p)}^2}{8R_L} = \frac{(2v_{op})^2}{8R_L} = \frac{v_{op}^2}{2R_L} \tag{9-5a}$$

Note that $v_{o(p-p)} = 2v_{op}$. To find v_o we apply standard analysis techniques to either transistor (both have identical gains).

The total dc power dissipated by each transistor is

$$P_{dc} = V_{CC} \times I_{cAV} \tag{9-9}$$

This expression differs from the one given in Eq. (9-6) because here $I_{CQ} = 0$

400 LARGE SIGNAL AMPLIFIERS

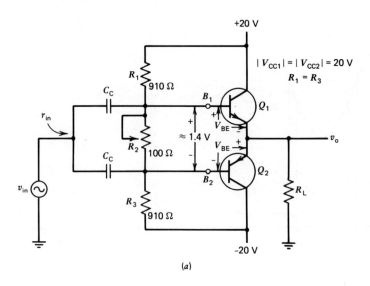

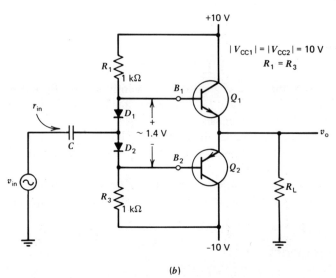

FIGURE 9-16 Complementary symmetry class B circuits. (a) Voltage divider bias. (b) Diode bias.

and the dc power is a function of the *average* collector current I_{cAV}. Since each transistor conducts for only 180°, half a cycle, the output produced by a single transistor is a "half-wave rectified" waveform (Fig. 9-17). In terms of peak current (not peak to peak) we have, for each transistor,

$$i_{op} = \frac{v_{op}}{R_L} \tag{9-10}$$

CLASS B AUDIO AMPLIFIERS

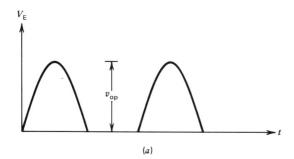

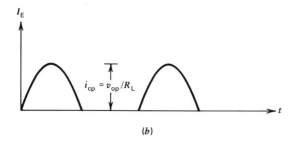

FIGURE 9-17 Calculating I_{cAV} (single transistor). (a) v_{op}. (b) (I_E).

The average current in the half wave is (Eq. 2-11)

$$I_{cAV} = \frac{i_{op}}{\pi} \qquad (2\text{-}11)$$

From Eqs. (9-10) and (2-11), we get

$$I_{cAV} = \frac{v_{op}}{\pi \times R_L}$$

The dc power for *each* transistor is

$$P_{dc1} = \frac{v_{op}}{\pi \times R_L} \times |V_{CC1}| \qquad (9\text{-}11)$$

V_{CC1} is the supply voltage for Q_1. The total dc power, for both transistors (since I_{cAV} is the same for both transistors)

$$P_{dc} = \frac{v_{op}}{\pi \times R_L} (|V_{CC1}| + |V_{CC2}|) \qquad (9\text{-}11a)$$

In most cases, $|V_{CC1}| = |V_{CC2}| = V_{CC}$ and $|V_{CC1}| + |V_{CC2}| = 2V_{CC}$; hence,

$$P_{dc} = \frac{v_{op}}{\pi R_L} \times 2V_{CC} \qquad (9\text{-}11b)$$

402 LARGE SIGNAL AMPLIFIERS

Please note that V_{CC} is the magnitude of the supply voltage of *either* transistor for the case in which the two supplies have the same magnitude.

The efficiency then becomes

$$\eta = \frac{P_o}{P_{dc}} = \frac{v_{op}^2}{2R_L} \bigg/ \frac{v_{op} \times 2V_{CC}}{\pi \times R_L} = \frac{v_{op}}{V_{CC}} \times \frac{\pi}{4} \qquad (9\text{-}12)$$

From Eq. (9-12) we see that in order to reach the maximum efficiency η_{max}, we must provide a large enough input signal to make $v_{op} = V_{CC}$ (v_{op} may not be larger than V_{CC} lest clipping will occur). For the case in which $v_{op} = V_{CC}$, yielding maximum efficiency,

$$\eta_{max} = \frac{v_{op}}{V_{CC}} \times \frac{\pi}{4} = \frac{\pi}{4} = 0.785 \qquad (9\text{-}12a)$$

$$\% \ \eta_{max} = 78.5\% \qquad (9\text{-}12b)$$

9-8.4 Class B Circuits

Two typical Class B complementary symmetry circuits are shown in Fig. 9-16. Both are of the EF type. Figure 9-18 shows an EF circuit using a single supply voltage. To get an appropriate bias point, we get $R_1 = R_2$. The two transistors are *barely* on. Since these are matched transistors, it is reasonable to expect that $V_{CE1} = V_{CE2} = \frac{1}{2} V_{CC}$. Each transistor has half the supply voltage across it under

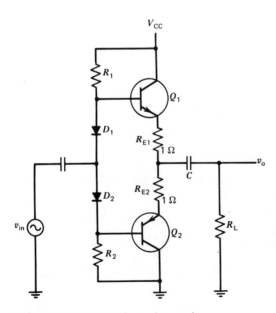

FIGURE 9-18 EF with single supply.

quiescent operating conditions. This implies that the maximum output *peak* voltage, produced by each transistor separately, is $V_{CC}/2$ (maximum $v_{o(p-p)} = V_{CC}$). For this circuit

$$\eta = \frac{V_{op}}{V_{CC}/2} \times \frac{\pi}{4} \qquad (9\text{-}12c)$$

(We replaced the supply voltage in Eq. (9-12) by half the supply voltage from the circuit in Fig. 9-18.) The small emitter resistors (here we use 1 Ω) serve to equalize the transistor gains and provide some protection against a dc short.

Figure 9-19 shows a CE Class B configuration. Each emitter is connected to ac ground (the -20 V and $+20$ V sources are both ac ground). The resistive networks R_1, R_2 are designed to produce a $V_{BE} = 0.7$ V across both base-emitter junctions. Quiescently, $I_C \approx 0$. (In many instances Class AB is used, yielding a small collector current.) Q_1 amplifies the negative half cycle (it is a PNP) and Q_2 the positive half (NPN). Gain, input resistance, etc. are calculated the same way as for the regular CE circuit. It is important to note that only *one* transistor is operative at a time (the other providing no output at all) so that the overall gain is that of a single transistor. To obtain an undistorted output, the transistors must be matched, that is $\beta_1 = \beta_2$, $r'_{e1} = r'_{e2}$. Thus,

$$A_{v1} = A_{v2} = A_v = -\frac{R_L}{r'_e} \qquad (9\text{-}13)$$

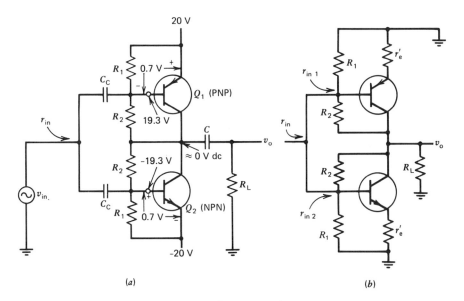

FIGURE 9-19 CE Class B circuit. (a) The circuit. (b) ac equivalent.

LARGE SIGNAL AMPLIFIERS

where

$$r'_e = r'_{e1} = r'_{e2}$$

It is important to remember that r'_e is an average r_e and not based on $r_e = 26/I_{EQ}$. This latter expression would yield $r_e = \infty$ (r_e equals infinity) since $I_{EQ} = 0$.

To calculate r_{in}, the input resistance to each individual base must first be examined. There are three parallel branches: R_1, the effect of R_2, and $\beta r'_e$. Since R_2 is not connected to ground but from base to collector (output), its effect on r_{in} is similar to that of the Miller capacitance, namely $R_2/|A_v| + 1$. The three parallel terms for each base are then (Fig. 8-19b)

$$r_{in1} = r_{in2} = R_1 \left\| \frac{R_2}{|A_v| + 1} \right\| \beta r'_e$$

At first glance, since two such bases are involved, r_{in} appears to be the parallel combination of r_{in1} and r_{in2}. This is erroneous, since the term $\beta r'_e$ is due to the *active* transistor only. Since only *one* transistor is active at a time, $\beta r'_e$ can appear only once. (The base of the off transistor is open.) The input resistance then is

$$r_{in} = R_1 \left\| \frac{R_2}{|A_v| + 1} \right\| R_1 \left\| \frac{R_2}{|A_v| + 1} \right\| \beta r'_e$$

$$= \frac{R_1}{2} \left\| \frac{R_2/2}{|A_v| + 1} \right\| \beta r'_e \qquad (9\text{-}14)$$

In many practical power amplifier circuits, the term $\beta r'_e$ is smaller than the other terms in Eq. (9-14) so that

$$r_{in} \simeq \beta r'_e \qquad (9\text{-}14a)$$

Figure 9-20 shows the typical transformer coupled "push pull" amplifier. It is a CE dual transistor amplifier; again, each transistor amplifies only half the waveform.

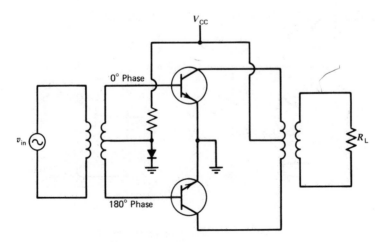

FIGURE 9-20   Push-pull circuit.

CLASS B AUDIO AMPLIFIERS   405

Since both are NPN transistors, amplifying the positive half only, it is necessary to provide a 180° phase shift between the input signals to the two bases. The two output currents, from the two transistors, are "summed" in the transformer reproducing the full input sinusoid across R_L.

The transformer coupled power amplifier has seen very little use recently, in particular in audio applications, because of:

1. Substantial weight of the transformer.
2. High cost.
3. Poor frequency response.

EXAMPLE 9-4

In Fig. 9-16b, $R_1 = R_3 = 1\ k\Omega$, $R_L = 8\ \Omega$, $r_{e1} = r_{e2} = 0.5\ \Omega$, $\beta_1 = \beta_2 = 25$. $v_{in} = 10$ V p-p. Find:
(a) v_o.
(b) r_{in}.
(c) η.

SOLUTION

(a) A_v for the EF circuit is

$$A_v = \frac{R_L}{r_e + R_L} = \frac{8}{0.5 + 8} = 0.94$$

$$v_o = v_{in} \times 0.94 = 10 \times 0.94 = 9.4\ V_{p\text{-}p}$$

(b) $r_{in} = R_1 \| R_3 \| (\beta \times (r_e + R_L))$ (see CE analysis, Ch. 5)
$= 1\ k\Omega \| 1\ k\Omega \| (25(8.5)) = 149\ \Omega$

D_1 and D_2 are ac shorts, placing R_1 in parallel with R_3.

(c) $\eta = \dfrac{V_{op}}{V_{CC}} \times \dfrac{\pi}{4}$ (9-12)

$$\eta = \frac{9.4/2}{10} \times \frac{\pi}{4} = 0.369$$

$\%\eta = 36.9\%$

Example 9-4 could have been applied to the circuit of Fig. 9-16a with identical results. In Fig. 9-16a the two bases are interconnected for ac purposes by the two coupling capacitors C_C, effectively shorting R_2. (The setting of R_2 becomes irrelevant to the ac analysis.)

EXAMPLE 9-5

In the circuit of Fig. 9-19 $\beta_1 = \beta_2 = 25$, $r_{e1} = r_{e2} = 0.5\ \Omega$, $R_1 = 220\ \Omega$, $R_2 = 5.6\ k\Omega$, $R_L = 16\ \Omega$. Find:
(a) The voltage gain v_o/v_{in}.
(b) r_{in}.

SOLUTION

(a) Since the circuit is of the CE type (neglecting the feedback effects of the two R_2 resistors), $|A_v| = R_L/r_e$. This is the gain of either transistor. Since only one transistor is active at a time, this is also the total circuit gain.

$$|A_v| = \frac{16}{0.5} = \underline{32}$$

(b) To find r_{in} we note that v_{in} is connected by *two* R_1 resistors to ac ground (the $+20$ and -20 V) and to only *one* active base, with a *base* input resistance of βr_e (Fig. 9-21).
In addition, there are two R_2 resistors, effectively in parallel, connected from base to collector. These are equivalent to a resistance of $R_2/|A_v|+1 \| R_2/|A_v|+1$ from base to ground.

$$r_{in} = R_1 \| R_1 \left\| \frac{R_2}{|A_v|+1} \right\| \left\| \frac{R_2}{|A_v|+1} \right\| \beta r_e \quad (9\text{-}14)$$

$$= 220 \| 220 \left\| \frac{5600}{33} \right\| \left\| \frac{5600}{33} \right\| 25 \times 0.5 = \underline{10\ \Omega}$$

The term βr_e is usually the smallest of all parallel terms and may be used as a preliminary *estimate* of r_{in}.

$$r_{in} \approx \beta r_e = 12.5\ \Omega \quad (9\text{-}14a)$$

Examples 9-4 and 9-5 point to the basic differences between the EF (Fig. 9-16) and CE (Fig. 9-19) circuits. The EF has a high r_{in} (149 Ω compared to 10 Ω) and a low A_v (0.94 compared to 32). This, of course, was expected.

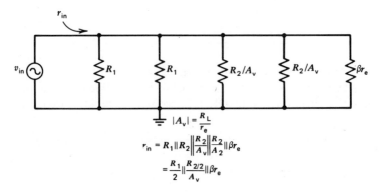

FIGURE 9-21 r_{in} for the circuit of Fig. 9-19.

9-9 VMOS POWER AMPLIFIERS

Some of the basic problems associated with the BJT power amplifier have been practically eliminated by the VMOS. The relatively low input resistance (r_{in} = 149 Ω of the EF in Ex. 9-4 is still a very low r_{in}) of the BJT is replaced by the very high (practically an open circuit) r_{in} of the FET. The nonlinearities of the BJT give way to the relatively linear characteristics of the VMOS, minimizing distortion. In addition, the likelihood of "thermal-run-away" does not exist for the VMOS (see Sec. 6-12). The frequency response of the VMOS is as wide or wider than that of the BJT. All these advantages lead to the conviction that VMOS will rapidly replace the BJT in most, if not all, power applications.

A single Class A VMOS power circuit is shown in Fig. 9-22a using the 2N6657 VMOS. The transistor is biased at

$$V_{GSQ} = \frac{V_{DD}R_2}{R_1 + R_2} = \frac{30 \times 1}{4.5 + 1} = 5.5 \text{ V}$$

$I_{DQ} \simeq 0.8$ A (Fig. 9-22b)

To calculate the voltage gain, we first find g_m at the operating point:

$$g_m = \frac{\Delta I_D}{\Delta V_{GS}} = \frac{0.92 - 0.68}{6 - 5} = 240 \text{ m}\mho$$

Since the I_D versus V_{GS} curve (Fig. 9-22b) is very nearly linear at the Q-point, g_m may be calculated by using increments ΔI_D and ΔV_{GS} of any size. (There is no need to use derivatives where $\Delta I_D \rightarrow 0$ and $\Delta V_{GS} \rightarrow 0$.) The voltage gain becomes (see Ch. 6)

$$A_v = -g_m R'_L = -240 \times 10^{-3} \times 16 = -3.84 \qquad (6\text{-}12a)$$

(Here R'_L is the ac load resistance at the drain side. Because the transformer has a 1:1 turns ratio, $R_L = R'_L$.)

The zener diode connected between gate and source serves to protect against static charge that may damage the transistor. This zener is usually constructed as an integral part of the transistor. In many circuits an additional external zener is sometimes used to protect against excessive signal swing that may exceed the maximum permissible V_{GS} voltage.

A Class B VMOS circuit is shown in Fig. 9-23a. The portion of the circuit used to drive the two VMOS transistors Q_1 and Q_2, is not shown to avoid the complexities of the drive circuit. The biasing of the VMOS is set by properly setting the dc levels in the drive circuit.

The two output transistors Q_1, Q_2 are both N-channel (at this time only N-channel VMOS are available). In order to obtain a push pull type operation a "quasi complementary" circuit functioning much like the complementary circuit is used. The waveforms in Fig. 9-23b describe the operation of the circuit. Since Q_1 is a source follower, it affects no inversion, no 180° phase shift, while Q_2 is a

408 LARGE SIGNAL AMPLIFIERS

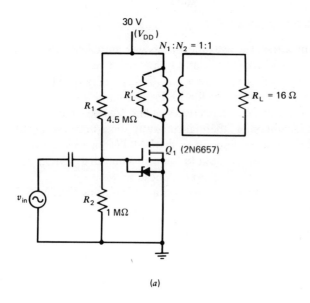

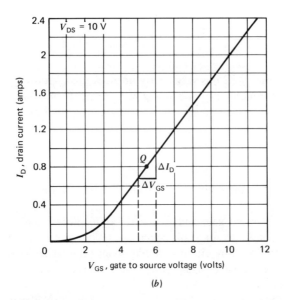

FIGURE 9-22 VMOS class A circuit. (a) Circuit diagram. (b) Transfer characteristics for 2N6657 (Siliconix, Inc.).

common source configuration producing a 180° phase shift. The output v_o is composed of the unshifted source follower signal combined with the shifted common source. The inputs to Q_1 and Q_2 are 180° apart so that each transistor operates on a different half cycle (Fig. 9-23) as it is in the BJT push-pull circuit. At the output the contributions from each transistor are combined, in the proper phase, to produce the full sinusoid. The 9 V zeners have been included for overvoltage protection.

R_2 is used to set the gain of Q_2 compared to Q_1 to produce a signal at v_o of equal positive and negative swing. (The effective gains must be the same.) Since the gain of the SF (Q_1) is less than one, the gain of the CS (Q_2) is also set to less than one.

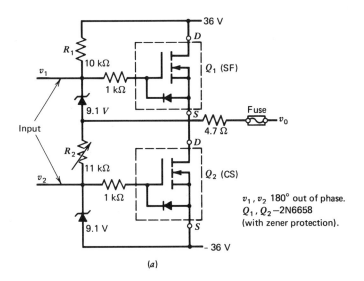

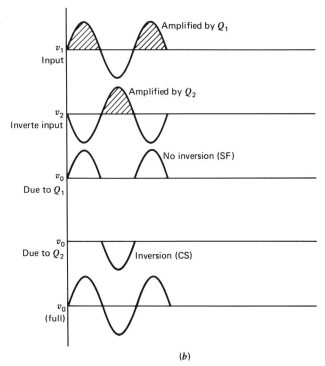

FIGURE 9-23 Class B VMOS. (a) Circuit diagram. (Complete circuit can be found in "VMOS Power FETs—Applications" Siliconix handbook, 1980.) (b) Waveforms.

9-10 CLASS C AMPLIFIERS

The discussion of Class C amplifiers belongs more appropriately in a communications text since the Class C circuit is the mainstay of a goodly number of communication systems. Our discussion here is, therefore, brief and concerned largely with principles rather than analytical details.

To produce a sinusoidal output signal, the Class C stage incorporates a resonance circuit. The resonance circuit converts the short Class C current pulses into a full sinusoid.

Most Class C circuits are biased beyond cut off by the signal itself, rather than by a separate reverse bias supply. Figure 9-24a shows a voltage divider circuit with a negative supply voltage used to produce a V_{BQ} (base operating voltage) that substantially reverse biases the B-E junction. Only the peaks of the input signal produce collector current. The various waveforms of the circuit are shown in Fig. 9-24b. The ac collector current i_c consists of very short current pulses. The conduction angle is substantially less than 180° that is typical for the Class C circuit. The bias voltage V_{BQ} in this circuit must be selected to match the input signal amplitude in order to obtain a proper operation. Too small an input signal may result in no collector current at all.

The circuit of Fig. 9-25 remedies this dependence on the input amplitude. Here, the cut-off bias is produced by the signal itself due to the clamping action of the B-E junction. With no signal applied, the circuit is biased *at* cut off and we *may* expect a 180° conduction angle. The application of an input signal, however, produces an additional reverse bias and results in a Class C operation.

During the positive half of the input signal there is a base current, the B-E junction is turned on (approximating a short), and the coupling capacitor is charged to approximately the peak input voltage with the polarity shown in Fig. 9-25b. That charge is held, during the rest of the input cycle producing a reverse bias V_{BQ} approximately equal to the peak input voltage. A small amount of base current is needed to replenish the capacitor charge. This small base current pulse produces a collector current pulse. The waveforms are much the same as in Fig. 9-24b. The advantage of this circuit is that the bias voltage automatically adjusts to the input amplitude. As a result, this circuit operates in Class C for *any* input amplitude. (It must be larger than the forward V_{BE} in order to produce the capacitive charge and the current pulses.)

The load consists of an L-C resonance circuit tuned to the frequency of the input signal, f_1. Recall that the resonance frequency is given by

$$f_o = \frac{1}{2\pi\sqrt{LC}} \tag{9-15}$$

In order to obtain "clean" undistorted sinusoids, it is necessary to use a high Q resonance circuit, $Q > 10$. The Q of the resonance circuit at the resonance frequency is given by

$$Q = \frac{\omega_o L}{R_{dc}} = \frac{2\pi f_o L}{R_{dc}} \tag{9-16}$$

CLASS C AMPLIFIERS 411

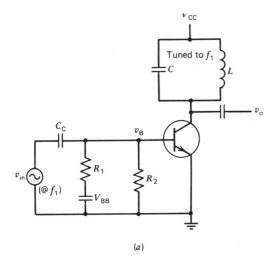

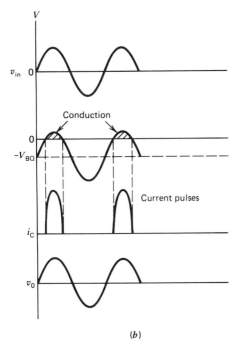

FIGURE 9-24 Class C with reverse bias supply. (a) Circuit diagram. (b) Waveforms.

where R_{dc} is the internal resistance of the inductor (Fig. 9-26). (This may represent not only the dc resistance of the wire, but also a variety of other power loss effects.)

For purposes of analysis, we replace the resonance circuit by an equivalent resistance R_o, where R_o represents the impedance (purely resistive) of the circuit at resonance

$$R_o = Q \times X_L = Q \times \omega_o L \qquad (9\text{-}17)$$

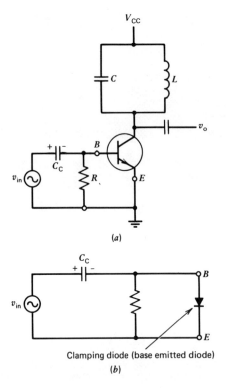

FIGURE 9-25 Class C with self bias. (a) Circuit diagram. (b) Clamping effect.

(ω_o is the resonant frequency.) The above relationships assume a lossless capacitor and a high Q coil.

As noted previously, the maximum power efficiency of a Class C circuit approaches 100% since dc power is supplied only when the transistor conducts current, a very small conduction angle results in a high efficiency. A small conduction angle, however, requires that the resonance circuit have a high Q. This is because

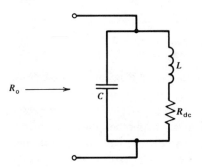

FIGURE 9-26 Resonance circuit.

the current pulse must compensate for any losses in the resonance circuit. A low Q implies larger losses and requires a larger (or longer duration) current pulse.

9-11 POWER RATINGS AND DEVICE TEMPERATURE

One of the main failure causes of power semiconductors is excessive heat at the junction. Above certain temperatures, the junction structure may disintegrate (burn up). The maximum permissible operating temperature is largely a function of the material used. Silicon devices can operate at temperatures up to about 175°C to 200°C, while germanium devices have a temperature limit of about 100°C to 110°C. Most specification sheets give specific temperature limits for the devices described, and for the most part these are given in terms of the maximum *junction* temperature, T_j. (We refer to this temperature as $T_{j(max)}$, while T_j represents the junction temperature for any given condition. Specification sheets use T_j to denote maximum allowable temperature.) For safe operation, the temperature of the junction must at all times be kept below its maximum permissible value. Since the junction temperature is directly dependent on the power dissipated by the transistor, it is necessary to limit this power so that the maximum temperature is not exceeded. The manufacturer usually gives the relation between power and temperature in terms of *"thermal resistance,"* which is defined as a temperature increase in °C per watt of dissipation (°C/W). The thermal resistance (θ_{jc}) between the junction and the case, the transistor housing, gives the temperature difference between the junction and the case per watt dissipated by the junction. Similarly, θ_{cA}, the thermal resistance between the case and the ambient (the surrounding temperature), gives the temperature difference, per watt of dissipation, between the case and its surrounding. Since we must guard against unacceptable junction temperatures, compared to the surroundings, we, therefore, must consider the thermal resistance θ_{jA} junction to ambient that is given by

$$\theta_{jA} = \theta_{jc} + \theta_{cA} \qquad (9\text{-}18)$$

For example, the 2N4900 has a thermal resistance $\theta_{jc} = 7°C/W$, which means that if, for example, the transistor dissipates 5 W, the junction temperature will be $7 \times 5 = 35°C$ *above* the *case* temperature. If we assume, for the sake of the example, that $\theta_{cA} = 5°C/W$, then for the same 5 W of power dissipated by the transistor we get a junction temperature that is 60°C *above* the *ambient* 5(W) × (7+5) = 60°C. At an ambient temperature of 25°C (a typical reference temperature), we find that for 5 W of power the temperature of the junction becomes 60°C + 25°C = 85°C. In general, junction temperature T_j is given by

$$T_j = P_T(\theta_{jc} + \theta_{cA}) + T_A = P_T(\theta_{jA}) + T_A \qquad (9\text{-}19)$$

where P_T is the power dissipated by the transistor and T_A is the ambient temperature. The maximum permissible dissipation P_D is then related to the maximum permissible junction temperature by

$$T_{j(max)} = P_D(\theta_{jc} + \theta_{cA}) + T_A = P_D(\theta_{jA}) + T_A \qquad (9\text{-}20)$$

414 LARGE SIGNAL AMPLIFIERS

If we reference our equation to T_c, the case temperature, then,

$$T_{j(max)} = P_D(\theta_{jc}) + T_c \tag{9-20a}$$

Equation (9-20) indicates that a higher $T_{j(max)}$ would permit a higher maximum power. It also shows that if the ambient temperature is high, the P_D given for an ambient of 25° must be derated, reduced, so that the total temperature does not exceed $T_{j(max)}$. The specification sheet usually gives this derating factor in W/°C. A derating factor of 0.143 W/°C (for the 2N4900) referenced to 25°C means that the P_D that is 25 W at 25°C must be decreased by 0.143 W for every degree above 25°C. Thus, at 200°C, 175°C above 25°C, the maximum power becomes

$$P_{D(200°C)} = 25 - 0.143 \times 175 = 0 \text{ W}$$

Indeed, the specifications give 200°C as the maximum temperature. Any power dissipated would now cause an increase above 200°C, possibly damaging the transistor. In general, the maximum permissible power at some case temperatures, T_c, is given by

$$P_D \text{ (at } T_c) = P_D \text{ (at } T_c = 25°C) - (D.F.) \times (T_c - 25) \tag{9-21}$$

where D.F. is the derating factor in W/°C above 25°C. Equation (9-21) gives the maximum power at any temperature, T_c (P_D at T_c), as the maximum power at 25°C (usually given by the manufacturer) less the derating necessary. For every °C above 25°C the maximum permissible power is reduced by an amount, in watts, given by the derating factor, D.F.

Equation (9-21) is usually shown graphically in the specification sheet (Fig. 9-27).

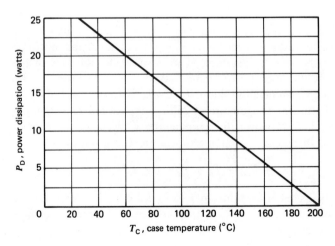

FIGURE 9-27 Power-temperature derating curve for 2N4900 (Motorola, Inc.).

EXAMPLE 9-6

The following specifications are given for the 2N4234.
P_D (at $T_A = 25°C$) is 1 W
Derate above 25°C = 5.7 mW/°C
Maximum operating temperature $T_{j(max)} = 200°C$
Thermal resistance $\theta_{jC} = 29°C/W$
Find:
(a) θ_{cA}.
(b) P_D at $T_c = 25°C$ (P_D was given for $T_A = 25°C$, not $T_c = 25°C$).
(c) P_D for $T_A = 100°C$.
(d) If the power actually being dissipated by the transistor is 0.7 W and $T_A = 35°C$, what is T_j? Is it above the $T_{j(max)}$?

SOLUTION

(a) We first find θ_{jA}, noting that for $T_A = 25°C$ the maximum power is 1 W, implying that for this power $T_j = 200°C$ the maximum permissible, we get

$$T_{j(max)} = P_D \times \theta_{jA} + 25 \quad (9\text{-}19)$$
$$200 = 1 \times \theta_{jA} + 25$$
$$\theta_{jA} = \frac{200 - 25}{1} = 175 \ °C/W$$

From Eq. (9-18), we get

$$\theta_{cA} = \theta_{jA} - \theta_{jc} = 175 - 29 = \underline{146 \ °C/W}$$

(b) P_D is always associated with $T_{j(max)}$. Hence, if $T_c = 25°C$ and $\theta_{jc} = 29°C/W$,

$$T_{j(max)} = P_D(\theta_{jc}) + T_c \quad (9\text{-}20a)$$
$$P_D = \frac{T_{j(max)} - T_c}{\theta_{jc}} = \frac{200 - 25}{29} = \underline{6 \ W}$$

(c) Based on a derating factor of 5.7 mW/°C, we get

$$P_D \text{ (at } T_A) = P_D \text{ (at } T_A = 25°C) - D.F.(T_A - 25)$$
$$P_D \text{ (@ } T_A = 100°C) = 1 - 5.7 \times 10^{-3}(100 - 25)$$
$$= 1 - 0.43 = \underline{0.57 \ W}$$

The last expression is identical to Eq. (9-21) with T_A replacing T_c. This is done because we are given $T_A = 100°C$ and D.F. is given referenced to $T_A = 25°C$.

(d) $T_j = P_T(\theta_{jA}) + T_A \quad (9\text{-}19)$
$T_j = 0.7(175) + 35 = \underline{157.5°C}$

416 LARGE SIGNAL AMPLIFIERS

This junction temperature is below the maximum; hence, 0.7 W dissipation is well within the permissible power.

In order to ascertain whether power specifications are being exceeded, it is necessary to calculate the largest power the device is expected to dissipate. The following equations give these "largest possible"[1] power dissipations for different audio power amplifiers discussed in the previous sections.

For Class A transformer coupled:

$$P_{T(max)} = I_{CQ} \times V_{CC} \qquad (9\text{-}22)$$

Note that here $V_{CQ} \simeq V_{CC}$. For Class B two transistor circuits, each transistor dissipates

$$P_{T(max)} = \frac{(V_{CC1} + V_{CC2})^2}{4\pi^2 R_L} \qquad (9\text{-}23)$$

Equation (9-23) can be modified. If $|V_{CC1}| = |V_{CC2}| = V_{CC}$, then

$$P_{T(max)} = \frac{(2V_{CC})^2}{4\pi^2 R_L} = \frac{V_{CC}^2}{\pi^2 R_L} \qquad (9\text{-}23a)$$

For a single power supply circuit (Fig. 9-18), we set one of the supplies to zero, say $V_{CC2} = 0$, and get

$$P_{T(max)} = \frac{(V_{CC1})^2}{4\pi^2 R_L} \qquad (9\text{-}23b)$$

The equations given above are derived using derivatives and maxima point computations. (See Appendix H.) $P_{T(max)}$ given above must always be less than the maximum power specified for the device.

EXAMPLE 9-7

The 2N4234 is used in a Class B two transistor stage, driving a 50 Ω load with $V_{CC1} = V_{CC2} = 20$ V (two supplies). The ambient temperature (T_A) is 80°C. Check whether it is operating below its maximum allowable power dissipation, P_D.

SOLUTION

Using the specification in Ex. 9-6, we get at $T_A = 80$°C

$$P_D = 1 - 5.7 \times 10^{-3} \times (80 - 25) = 0.687 \text{ W}$$

---

[1] We use the term "largest possible" to denote the largest possible power the transistor *may* be called upon to dissipate under certain conditions.

The largest possible power the transistor may dissipate is

$$P_{T(max)} = \frac{V_{CC}^2}{\pi^2 R_L} = \frac{20^2}{\pi^2 \times 50} = 0.81 \text{ W} \qquad (9\text{-}23a)$$

We are exceeding permissible dissipation because $P_{T(max)} > P_D$. To remedy the problem we may:

1. Reduce V_{CC}; or
2. increase R_L; or
3. select another transistor with a higher power rating; or
4. use a heat sink.

9-12 HEAT SINKS

The maximum operating temperature of the junction cannot be increased since it is essentially a characteristic of the material used. In order to permit higher power dissipation, it is necessary to remove as much heat from the junction as possible, thus keeping T_j as low as possible. This can be accomplished by mounting the transistor on a "heat sink." Figure 9-28 shows a number of different heat sinks. Practically all are black or grey and made of aluminum. The dark color helps radiate heat and the highly heat conducting aluminum helps to conduct heat away from the transistor case.

The function of the heat sink is essentially to reduce the term θ_{cA} in Eq. (9-20)

FIGURE 9-28 A variety of power transistor heat sinks (Motorola, Inc.).

and thus permit a higher P_D. An examination of Ex. 9-6, part (a), shows how important θ_{cA} is. Had we been able to reduce θ_{cA} from 140°C/W to, say, 50°C/W by the use of a heat sink, we would get $\theta_{jA} = \theta_{jc} + \theta_{cA} = 29 + 50 = 79°C/W$, and the corresponding P_D (at 25°C) then becomes

$$P_D = \frac{T_{j(max)} - 25}{\theta_{jA}} = \frac{200 - 25}{79} = \underline{2.2 \text{ W}}$$

This P_D is double the 1 W specified by the manufacturer (who assumed no heat sink). When a heat sink is used, its thermal resistance is substituted for the θ_{cA} of the transistor without the heat sink. Clearly, to be useful, the thermal resistance of the heat sink must be substantially less than the θ_{cA} of the bare transistor.

The heat sinks shown in Fig. 9-28 are described in the specification sheet by their weight, size, etc. and most important, by their thermal resistance. This thermal resistance can then replace θ_{cA} of the bare transistor and a new P_D may be calculated (P_D with heat sink), for the specific *ambient* temperature involved, using Eq. (9-20). It is noteworthy that the power transistor itself has a large case designed to promote heat transfer away from the junction and away from the large, well-exposed housing.

Most P_D values in the specifications are given for $T_c = 25°C$, not for $T_A = 25°C$. P_D for $T_A = 25°C$ is less than that for $T_c = 25°C$, the difference depending on θ_{cA}. Given P_D (@$T_c = 25°C$), we can use Eq. (9-20) to get P_D (at $T_A = 25°C$) if we know θ_{jc} and θ_{cA}.

EXAMPLE 9-8

The 2N1120 (germanium) has the following specifications: $T_{j(max)} = 100°C$, P_D (@$T_c = 25°C$) $= 90$ W, $\theta_{jc} = 0.8°C/W$. The transistor is mounted on a heat sink H1 with $\theta_{cA} = 1.1°C/W$. Find P_D (at $T_A = 25°C$).

SOLUTION

$$P_D \text{ (at } T_A = 25°C) = \frac{T_{j(max)} - 25}{\theta_{jc} + \theta_{cA}}$$

$$= \frac{100 - 25}{0.8 + 1.1} = 39 \text{ W}$$

We substituted the thermal resistance 1.1°C/W of the heat sink for θ_{cA}. Without the heat sink, θ_{cA} may be much higher and consequently, P_D much lower.

Note that in this example the maximum power of 39 W at $T_A = 25°C$ is less than half the 90 W specified for the *case* temperature of 25°C. Once P_D for $T_A = 25°C$ is found, we may then proceed with the power calculations as described in Section 9-11.

9-13 TRANSISTOR SPECIFICATIONS

Throughout this text, reference has often been made to various transistor parameters and characteristics. It is the purpose of this section to summarize and further explain the significance of these parameters. Since most of these characteristics are given in the manufacturer's data sheet, the presentation will, essentially, consist of an elaboration on and clarification of the data sheet.

The general comments made in Section 3-7 in connection with diode specifications apply here, as well. In particular, we note that the specifications are given together with a detailed description of conditions under which the parameters were measured. Appropriate adjustments must be made if the device is to be operated in conditions other than those specified. For example, β_{dc} may be given for a specific I_C. For collector currents other than those specified, β_{dc} may be different.

In the following discussion we define the parameters usually included in the transistor data sheets (dealing separately with BJT and FETs). These parameters are not necessarily found on a single data sheet, since the specifications given by the manufacturer are geared to the applications category. An audio frequency transistor data sheet is not likely to contain switching data such as "turn on" or "turn off" times, while the switching transistor data sheet may not contain the small signal current gain h_{fe} (β_{ac}).

9-13.1 BJT—Maximum Ratings

The notation used here is essentially that used by Motorola Inc. in the "maximum ratings" table (see Fig. 9-29).

V_{CEO} (BV_{CEO})   The maximum allowable collector emitter voltage with the base open. The BV_{CEO} given in the section on electrical characteristics (Fig. 9-29) is essentially the same as V_{CEO}. It is referred to as the collector emitter breakdown voltage. Clearly, if a breakdown may occur at this voltage, we ought not operate the transistor beyond that level. (Note that a minimum value is given for BV_{CEO}, since this is the *lowest* voltage at which breakdown *may* occur.)

V_{CB} (BV_{CBO})   The maximum collector base voltage. (BV_{CBO} is the collector base breakdown voltage with $I_E = 0$, emitter open.)

V_{EB} Maximum base emitter voltage (reverse).

I_C Maximum collector current.

P_D The maximum average power that may be dissipated by the transistor.

T_j, T_{stg} The maximum junction temperatures (lowest and highest) operating and storage (T_{stg}), respectively.

θ_{jA} Thermal resistance, junction to ambient (see Sec. 9-12).

θ_{jc} Thermal resistance, junction to case.

420 LARGE SIGNAL AMPLIFIERS

2N5088 (SILICON)
2N5089

CASE 29 (1) (TO-92)

NPN silicon annular transistors designed for low-level, low-noise amplifier applications.

MAXIMUM RATINGS

Rating	Symbol	2N5088	2N5089	Unit
Collector-Emitter Voltage	V_{CEO}	30	25	Vdc
Collector-Base Voltage	V_{CB}	35	30	Vdc
Emitter-Base Voltage	V_{EB}	4.5		Vdc
Collector Current	I_C	50		mAdc
Total Device Dissipation @ T_A = 25°C Derate above 25°C	P_D	310 2.81		mW mW/°C
Operating and Storage Junction Temperature Range	T_J, T_{stg}	-55 to +135		°C

THERMAL CHARACTERISTICS

Characteristic	Symbol	Max	Unit
Thermal Resistance, Junction to Ambient	θ_{JA}	0.357	°C/mW

FIGURE 9-29 Transistor data sheet (Motorola, Inc.).

9-13.2 Electrical Characteristics

(See Fig. 9-29—electrical characteristics.)

BV_{CEO} See V_{CEO}.
BV_{CBO} See V_{CB}.
I_{CBO} The collector base leakage current while the emitter is open (with a specified reverse collector base voltage).
I_{EBO} The emitter base leakage with an open collector (with a specified reverse emitter base voltage).

2N5088, 2N5089 (continued)

ELECTRICAL CHARACTERISTICS ($T_A = 25°C$ unless otherwise noted)

Characteristic		Symbol	Min	Typ	Max	Unit
OFF CHARACTERISTICS						
Collector-Emitter Breakdown Voltage ($I_C = 1.0$ mAdc, $I_B = 0$)	2N5088	BV_{CEO}	30	-	-	Vdc
	2N5089		25	-	-	
Collector-Base Breakdown Voltage ($I_C = 100$ μAdc, $I_E = 0$)	2N5088	BV_{CBO}	35	-	-	Vdc
	2N5089		30	-	-	
Collector Cutoff Current ($V_{CB} = 20$ Vdc, $I_E = 0$)	2N5088	I_{CBO}	-	-	50	nAdc
($V_{CB} = 15$ Vdc, $I_E = 0$)	2N5089		-	-	50	
Emitter Cutoff Current ($V_{EB(off)} = 3.0$ Vdc, $I_C = 0$)		I_{EBO}	-	-	50	nAdc
($V_{EB(off)} = 4.5$ Vdc, $I_C = 0$)			-	-	100	
ON CHARACTERISTICS						
DC Current Gain ($I_C = 100$ μAdc, $V_{CE} = 5.0$ Vdc)	2N5088	h_{FE}	300	-	900	
	2N5089		400	-	1200	
($I_C = 1.0$ mAdc, $V_{CE} = 5.0$ Vdc)	2N5088		350	-	-	
	2N5089		450	-	-	
($I_C = 10$ mAdc, $V_{CE} = 5.0$ Vdc)	2N5088		300	-	-	
	2N5089		400	-	-	
Collector-Emitter Saturation Voltage ($I_C = 10$ mAdc, $I_B = 1.0$ mAdc)		$V_{CE(sat)}$	-	-	0.5	Vdc
Base-Emitter On Voltage ($I_C = 10$ mAdc, $V_{CE} = 5.0$ Vdc)		$V_{BE(on)}$	-	-	0.8	Vdc
DYNAMIC CHARACTERISTICS						
Current-Gain – Bandwidth Product ($I_C = 500$ μAdc, $V_{CE} = 5.0$ Vdc, $f = 20$ MHz)		f_T	50	175	-	MHz
Collector-Base Capacitance ($V_{CB} = 5.0$ Vdc, $I_E = 0$, $f = 100$ kHz, emitter guarded)		C_{cb}	-	1.8	4.0	pF
Emitter-Base Capacitance ($V_{BE} = 0.5$ Vdc, $I_C = 0$, $f = 100$ kHz, collector guarded)		C_{eb}	-	4.0	10	pF
Small-Signal Current Gain ($I_C = 1.0$ mAdc, $V_{CE} = 5.0$ Vdc, $f = 1.0$ kHz)	2N5088	h_{fe}	350	-	1400	-
	2N5089		450	-	1800	
Noise Figure ($I_C = 100$ μAdc, $V_{CE} = 5.0$ Vdc, $R_S = 10$ k ohms, $f = 10$ Hz to 15.7 kHz)	2N5088	NF	-	-	3.0	dB
	2N5089		-	-	2.0	

FIGURE 9-29 (Continued)

Maximum values are given for the above since they represent worst case conditions.

h_{FE}–β_{dc} The dc current gain—the minimum and maximum values are usually given.

$V_{CE(sat)}$ The collector emitter saturation voltage.

$V_{BE(ON)}$ The base emitter voltage drop for the ON transistor.

All of the above are dc characteristics.

$C_{cb}(C_{ob})$ The collector-base capacities (at specified conditions). This parameter is used in the computation of Miller capacitance.

C_{eb} Emitter-base capacities (at specified conditions).

Both C_{cb} and C_{eb} depend somewhat on the exact voltage applied to the junction.

422 LARGE SIGNAL AMPLIFIERS

C_{oe} The collector-emitter capacity.

The above three are interelectrode capacitances contributing to the degradation of the high frequency response. C_{cb} is particularly detrimental.

h_{fe} (β_{ac}) The small signal (ac) CE current gain i_c/i_b (at specified conditions) usually given at midband frequency.

h_{fb} (α) The small signal CB current gain i_c/i_e usually given at midband frequency.

f_T β_{ac} × bandwidth gain bandwidth product. In Fig. 9-29 the typical f_T is given as 175 MHz. We note that this value of f_T was found for specific I_C and V_{CE} conditions and at 20 MHz. This means that the β_{ac} measured at 20 MHz was 175/20 = 8.75, yielding a gain bandwidth product of 20 × 8.75 = 175 MHz. f_T can also be defined as the frequency at which β_{ac} = 1 (see Sec. 8-6).

$f_{\alpha b}$ The frequency at which α has dropped to 0.707 of its value at midband.

$f_{\alpha e}$, f_{β} The cut off frequency related to β. The frequency at which β dropped to 0.707 of its value at midband.

NF Noise figure is a quantitative description of the electrical noise power contributed by the device itself (due to various random currents). It is defined as the ratio in dB of noise power at the output to that at the input over a specified bandwidth. It is given for specific operating conditions (it is very dependent on I_C and I_B). Source resistance is usually specified since it strongly affects noise power at the input.

9-13.3 FET Specifications

Most of the data given in Fig. 9-30 is self-explanatory. We add here brief comments relating to some of the parameters.

I_{GSS} Since this is a MOS transistor, V_{GS} may be positive *or* negative and still only leakage current will flow.

$V_{DS(ON)}$ Essentially drain source saturation voltage.

$r_{ds(ON)}$ This should not be confused with r_d, the ac drain source, equivalent resistance as discussed in Chapter 6. $r_{ds(ON)}$ is the series drain source resistance for an ON transistor, measured either under dc or ac conditions. It is usually relatively small.

Y_{fs} (g_m, g_{fs}) For the MOSFET this parameter is specified at *some* I_D, while for junction FETs it is usually specified at $V_{GS} = 0$, $I_D = I_{DSS}$ and is equivalent to g_{mo}. [The use of Y instead of g is intended to account for reactive components in this parameter. g is real, while Y is complex (G + jB).]

C_{vss} (C_{dg}) The feedback capacitance between drain and gate (used in Miller capacitance calculations).

C_{iss} (C_{gs}) Gate to source capacitance. Usually considered to be connected between gate and ground.

TRANSISTOR SPECIFICATIONS

——— Field-Effect Transistors ———

2N4351 (SILICON)

CASE 20(2)
(TO-72)

$V_{DS} = 25\,V$
$I_D = 30\,mA$
$P_D = 800\,mW$

Silicon N-channel MOS field effect transistors, designed for enhancement-mode operation in low power switching applications. The 2N4351 is complementary with type 2N4352.

MAXIMUM RATINGS ($T_A = 25°C$ unless otherwise noted)

Rating	Symbol	Value	Unit
Drain-Source Voltage	V_{DS}	25	Vdc
Drain-Gate Voltage	V_{DG}	-25	Vdc
Gate-Source Voltage	V_{GS}	±15	Vdc
Drain Current	I_D	30	mAdc
Power Dissipation at $T_A = 25°C$ Derate above 25°C	P_D	300 1.7	mW mW/°C
Power Dissipation at $T_C = 25°C$ Derate about 25°C	P_D	800 4.56	mW mW/°C
Operating Junction Temperature	T_J	175	°C
Storage Temperature	T_{stg}	-65 to +200	°C

HANDLING CONSIDERATIONS:

MOS field-effect transistors, due to their extremely high input resistance, are subject to potential damage by the accumulation of excess static charge. To avoid possible damage to the devices while handling, testing, or in actual operation, the following procedure should be followed:

1. The leads of the devices should remain wrapped in the shipping lead washer or foil except when being tested or in actual operation to avoid the build-up of static charge.
2. Avoid unnecessary handling; when handled, the devices should be picked up by the case instead of the leads.
3. The devices should not be inserted or removed from circuits with the power on as transient voltages may cause permanent damage to the devices.

FIGURE 9-30 FET Data Sheet (Motorola, Inc.).

424 LARGE SIGNAL AMPLIFIERS

ELECTRICAL CHARACTERISTICS ($T_A = 25°C$ unless otherwise noted)
Substrate connected to source.

Characteristic	Figure No.	Symbol	Min	Max	Unit
OFF CHARACTERISTICS					
Drain-Source Breakdown Voltage ($I_D = 10~\mu A$, $V_{GS} = 0$)	—	$V_{(BR)DSS}$	25	—	Vdc
Gate Leakage Current ($V_{GS} = \pm 15$ Vdc, $V_{DS} = 0$)	—	I_{GSS}	—	± 10	pAdc
Zero-Gate-Voltage Drain Current ($V_{DS} = 10$ V, $V_{GS} = 0$)	—	I_{DSS}	—	10	nAdc
ON CHARACTERISTICS					
Gate-Source Threshold Voltage ($V_{DS} = 10$ V, $I_D = 10~\mu A$)	—	$V_{GS(TH)}$	1.0	5	Vdc
"ON" Drain Current ($V_{GS} = 10$ V, $V_{DS} = 10$ V)	3	$I_{D(on)}$	3	—	mAdc
Drain-Source "ON" Voltage ($I_D = 2$ mA, $V_{GS} = 10$ V)	—	$V_{DS(on)}$	—	1.0	Vdc
SMALL SIGNAL CHARACTERISTICS					
Drain-Source Resistance ($V_{GS} = 10$ V, $I_D = 0$, $f = 1$ kHz)	4	$r_{ds(on)}$	—	300	ohms
Forward Transfer Admittance ($V_{DS} = 10$ V, $I_D = 2$ mA, $f = 1$ kHz)	1	$|y_{fs}|$	1000	—	μmho
Reverse Transfer Capacitance ($V_{DS} = 0$, $V_{GS} = 0$, $f = 140$ kHz)	2	C_{rss}	—	1.3	pF
Input Capacitance ($V_{DS} = 10$ V, $V_{GS} = 0$, $f = 140$ kHz)	2	C_{iss}	—	5.0	pF
Drain-Substrate Capacitance ($V_{D(SUB)} = 10$ V, $f = 140$ kHz)	—	$C_{d(sub)}$	—	5.0	pF
SWITCHING CHARACTERISTICS					
Turn-On Delay	6, 10	t_{d1}	—	45	ns
Rise Time	7, 10	t_r	—	65	ns
Turn-Off Delay	8, 10	t_{d2}	—	60	ns
Fall Time	9, 10	t_f	—	100	ns

($I_D = 2.0$ mAdc, $V_{DS} = 10$ Vdc, $V_{GS} = 10$ Vdc) (See Figure 10; Times Circuit Determined)

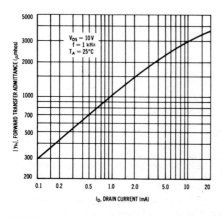

FIGURE 1 — FORWARD TRANSFER ADMITTANCE

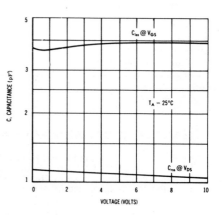

FIGURE 2 — CAPACITANCE

FIGURE 9-30   (Continued)

Figure 9-30 includes a number of graphs that give more detailed data for some of the parameters. For example, the forward transfer admittance graph gives the continuous relation between y_{fs} (g_m) and I_D.

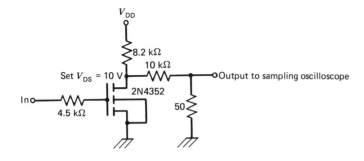

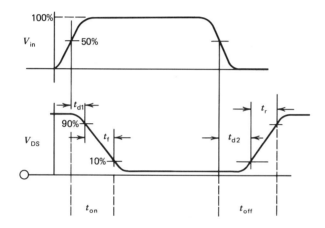

FIGURE 9-31 Switching waveforms and circuit.

9-13.4 BJT and FET Switching Characteristics (Fig. 9-31)

The following group of parameters describes how fast the device can respond to a switching signal. The data sheet usually gives worst case values, for example, maximum delay or maximum t_r, as well as typical values. It also provides schematics of the circuits used in obtaining the data. This gives a more careful and precise definition of the parameters and is helpful in using the data for uncommon applications.

t_{d1} (or $t_{d(on)}$) The turn-on delay represents the time it takes the device to start responding to an input voltage that will turn the transistor on. Defined as the interval between 50% input voltage level to a 10% change in output voltage V_{CE} (or V_{DS}), usually from $V_{CE} = V_{CC}$ (or V_{DD}) to $V_{CE} = 90\%$ of V_{CC} *on the way* to saturation. (For FETs use corresponding symbols.)

t_r Rise time is the time it takes the transistor output voltage to change from 10 to 90% of its final amplitude in response to an appropriate input signal.

t_f — Fall time is the 90 to 10% interval in the output waveform (high to low portion of output waveform).

t_{d2} ($t_{d(off)}$) — The delay in turn off. In BJT this is usually the storage time t_s, which results from the excess carriers in the base region during saturation. t_{d2} is defined much like t_{d1} for a transistor being turned *off*. The time interval from a 50% input signal level to a 10% change in output voltage on the way to cut off.

t_{on} — Turn-on time. The time it takes the device to reach 90% saturation current (or drop to 10% voltage in output waveform) from the 50% level at the input waveform $t_{on} \approx t_{d1} + t_f$.

t_{off} — Turn-off time. Similar to t_{on}, but applied to turn-off conditions. $t_{off} \approx t_{d2} + t_r$.

SUMMARY OF IMPORTANT TERMS

Class A Transister stage with an operating point at the center of the collector characteristics. A conduction angle of 360° and no saturation or cut off.

Class B Operating point at cut off. Conduction angle of 180° with no saturation.

Class C Operating point below cut off. Conduction angle less than 180°, approximately 90°.

Complementary symmetry A power amplifier circuit in which one NPN and one PNP (or N channel and P channel) transistor, each operating at about Class B, combine to produce a 360° output (the full sinewave).

Conduction angle The part of a full cycle (360°), in degrees, during which the transistor is operating in the active region. $I_C \neq 0$.

Crossover distortion Waveform distortion occurring in Class B push-pull or complementary symmetry circuits due to a lack of pre-bias. Occurs at the point where one transistor turns on and the other turns off (around $V_{BE} = 0$).

Efficiency (power) The ratio of signal power output to total power input. It is very nearly P_o/P_{dc}, the ac power out over the power from the dc source (or sources). (Ac input power is neglected.)

Heat Sink A structure, usually metallic colored black or grey, used to help remove heat from the transistor, thus increasing its maximum power dissipation.

Impedance matching The use of transformers or active devices to match load resistance to source (Thevenin's) resistance (usually in order to obtain maximum power transfer from source to load).

Large signal (versus small signal) The operation of an active device with signals that produce large voltage–current excursions at the output, usually associated with power amplifiers.

Maximum efficiency The *best* efficiency that can be expected from a circuit (under the most favorable conditions).

Maximum power rating (P_D) The maximum power the device may dissipate at a given temperature (usually 25°C) as specified by the manufacturer (without the use of a heat sink).

Nonlinear distortion Distortion in the waveform caused by an amplifier due to the fact that the transfer characteristics (collector current versus base voltage or drain current versus gate-source voltage) are not linear (typically, β changes with I_c or g_m changes with I_D).

Power derating The reduction of the maximum power rating to account for higher temperatures (above that for which P_D was given).

Push pull A Class B two-transistor circuit (usually using transformers) that produces a full 360° at the output. (See complementary symmetry.)

Quasi-complementary When two similar types of transistors (for example, both N channels) are used to accomplish what the complementary symmetry circuit does.

r_e average (r'_e) $\Delta V_{BE}/\Delta I_C$. Here, both ΔI_C and ΔV_{BE} are substantial intervals.

Self bias (Class C) When Class C biasing is generated by the signal voltage itself.

Thermal resistance θ The increase in temperature accompanying a unit increase in power dissipated by the device usually given in °C/W.

VMOS A MOS device constructed to produce a vertical flow of I_D, *down* the structure rather than across the surface, end to end.

PROBLEMS

9-1. The MJ2251 transistor is operating with $I_{CQ} = 40$ mA; $I_{C(p\text{-}p)} = 60$ mA ($\pm$ 30 mA) at 85°C. Find the average r_e. (Use the graphs on Fig. 9-2a.)

9-2. A transistor is operating in the range $I_C = 0.5$ A to $I_C = 1.5$ A ($I_C = 1.0 \pm 0.5$ A). What value of β would you use in the analysis if the transistor is a

(a) 2N3715;
(b) 2N3447.
(Use Fig. 9-3 at 25°C.)

9-3. An MJ2252 transistor is operating Class A at Q-points as listed below. For each calculate the power dissipated by the transistor and state whether it is within allowable range.

(a) $I_{CQ} = 100$ mA; $V_{CEQ} = 75$ V.
(b) $I_{CQ} = 40$ mA; $V_{CEQ} = 260$ V.
(c) $I_{CQ} = 150$ mA; $V_{CEQ} = 50$ V.
(d) $I_{CQ} = 30$ mA; $V_{CEQ} = 150$ V.
(Use the MJ2252 data sheet or Fig. 9-5.)

9-4. A Class A amplifier (transformerless) operates with a single supply voltage $V_{CC} = 50$ V at a Q-point $I_{CQ} = 0.5$ A; $V_{CEQ} = 30$ V. Find:

(a) The maximum undistorted output voltage (v_{ce}) swing.
(b) The maximum undistorted output current (i_c) swing.
Hint: Obtain the load line from the Q-point and $V_{CE} = 50$ V, $I_C = 0$.

9-5. In the circuit of Fig. 9-32 $I_{CQ} = 0.5$ A, $V_{CEQ} = V_{CC} = 75$ V, $R_L = 8$ Ω.

(a) What is the maximum output voltage swing across the primary (using a suitable transformer turns ratio)?
(b) Find $N_1:N_2$ to attain that maximum.
(c) What is the maximum swing of v_o?
Hint: The maximum ΔI_C (i_c) is 0.5 A (from the Q point to $I_C = 0$), must produce the maximum ΔV_{CE} or peak output voltage across R'_L.

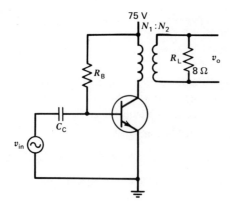

FIGURE 9-32

9-6. Calculate P_{dc} and P_o and η for the circuit of Fig. 9-32 with $I_{CQ} = 1$ A; $V_{CEQ} = 75$ V; $N_1:N_2 = 2$; $R_L = 8\ \Omega$.

  (a) for $v_{in} = 0.1$ V p-p
  (b) for $v_{in} = 0.2$ V p-p
       use $r_e = 0.2\ \Omega$.
  (c) With the given values can you ever reach maximum efficiency by increasing v_{in}? Explain.

9-7. Find P_{dc}, P_o, and η for the circuit in Fig. 9-33 with

  (a) $v_{in} = 1$ V rms.
  (b) $v_{in} = 7$ V rms (neglect the effect of r_e).
  Assume that r_e and R_{dc} of the transformer may be neglected.

9-8. Define the three classes of operation: Class A, Class B, Class C.

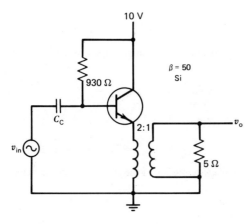

FIGURE 9-33

9-9. Calculate the efficiency of a complementary symmetry EF circuit using two supplies of $+20$ V and -20 V and driving an 8 Ω load for:

(a) $v_{in} = 5$ V rms.
(b) $v_{in} = 10$ V rms.
(c) Find v_{in} that yield maximum efficiency (neglect r_e).

9-10. Explain the terms:

(a) Nonlinear distortion.
(b) Crossover distortion.
(c) Tuned amplifier.

9-11. Find r_{in} for the circuit in Fig. 9-19 with $R_1 = 100$ Ω; $R_2 = 2.7$ kΩ; $R_L = 8$ Ω; $r_e = 0.5$ Ω; $\beta_1 = \beta_2 = 100$.

9-12. Sketch and explain the operation of the following Class B amplifiers:

(a) Push-pull with transformer (CE).
(b) Complementary symmetry EF.
(c) Quasi-complementary symmetry VMOS.

9-13. The 2N2526 is a germanium power transistor with the following maximum ratings: $T_j = 110°C$; P_D (at $T_c = 25°C$) $= 85$ W. The typical $\theta_{jc} = 1°C/W$. The transistor is to operate at $T_A = 40°C$.

(a) What is the maximum allowable power if $\theta_{cA} = 29°C/W$?
(b) The transistor may be dissipating as much as 20 W at $T_A = 40°C$. Specify the thermal resistance of a suitable heat sink.

9-14. The 2N3618 has the following *maximum* ratings: $T_j = 110°C$; P_D (at $T_c = 25°$) $= 77$ W; P_D (at $T_A = 25°C$) $= 2.6$ W. The derating curves are given in Fig. 9-34.

(a) Find θ_{jc}.
(b) Find θ_{cA}.
(c) Find P_D at $T_A = 50°C$ (use a T_A plot and a T_A scale).
(d) Find P_D at $T_c = 50°C$ (use a T_c plot and a T_c scale).
(e) Specify the thermal resistance of a heat sink to permit the use of the transistor at $T_A = 50°C$ with a power dissipation of 20 W.

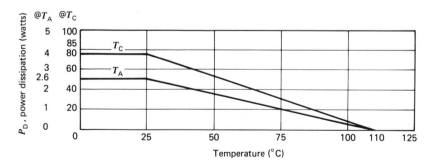

FIGURE 9-34 Power-temperature derating curves for 2N3615-18 (Motorola, Inc.).

CHAPTER

10

DIFFERENTIAL AMPLIFIERS

10-1 CHAPTER OBJECTIVES

This chapter will introduce the student to the basic ideas of the differential amplifier (an amplifier designed to amplify the *difference* of two signals) and to some of its more important characteristics. The Common Mode Rejection Ratio (CMRR) that describes (and assigns numerical value to) the ability of the amplifier to distinguish between *difference* voltages and other signals will be discussed and its effect on amplifier performance numerically analyzed.

As part of the analysis of the differential amplifier (DA) circuit, the student will be presented with various circuits that "behave like" current sources (current source circuits). These constitute an important part of the basic DA circuit.

Other parameters of the DA, such as voltage gain and input resistance, will also be discussed and numerically evaluated.

10-2 INTRODUCTION

In Chapters 2 through 9 the emphasis was placed on discrete circuits, that is, circuits that are constructed from discrete components such as resistors, capacitors, transistors, etc. With the advent of Integrated Circuits (I.C.) technology, a large number of circuits have become available in integrated circuit form. The user for the most part is not interested in the details of the circuit (even though these details usually reveal important features of the device) but rather in its performance. It is however useful, particularly in connection with differential and operational amplifiers, to have an understanding of the circuits used before interpreting the functional specifications.

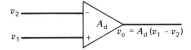

FIGURE 10-1 Differential amplifier (DA)—block diagram.

The differential-operational amplifier has been one of the first linear devices to become available in an I.C. package. This is due to its extreme popularity in the everyday life of the electronics practitioner. The differential amplifier (DA) plays an essential role in most, if not all, electronic control systems.

Typically, the DA produces an output that is related to the *difference* between two inputs. In Fig. 10-1 the term A_d represents the differential gain, that is, the amplification factor applied to $(v_1 - v_2)$ to produce v_o. Note that in this ideal differential amplifier, if $v_1 = v_2$, the output becomes 0. (The (+) and (−) signs at the input terminals indicate whether inversion (180° phase shift) takes place from the input to the output, (−) input, or no inversion occurs, (+) input. Most control systems require such a comparison between some reference signal and a signal representing the controlled element. Figure 10-2 is a block diagram of such a control system incorporating the DA. When $v_1 = v_2$, the table has reached its desired position (represented by v_2), $v_o = 0$ and the motor stops. The motor is energized only as long as $v_1 \neq v_2$, that is, as long as the table position is *not* the desired position. Figure 10-2 is representative of a large number of control systems.

10-3 DIFFERENTIAL AMPLIFIER (DA) FUNCTIONAL DESCRIPTION

As noted, the differential amplifier is designed to produce an output, v_o, proportional to the *difference* between two input voltages, v_1 and v_2. Should $v_1 = v_2$, v_o must be zero. The real DA, as opposed to the idealized version, however, does not perform this task perfectly. It does produce an output, though very small, even when $v_1 = v_2$. This output (for $v_1 = v_2$) is dependent on the magnitude of the

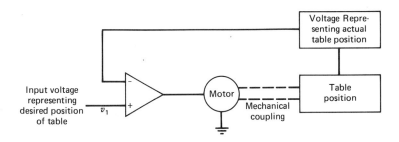

FIGURE 10-2 Typical application of the DA.

input and is the result of what is referred to as "common mode gain." The output of the real DA may then be represented by:

$$v_o = A_d(v_1 - v_2) + A_{cm}(v_1 + v_2)/2 \qquad (10\text{-}1)$$

where A_d is the differential gain, amplifying the *difference* input voltage, and A_{cm} is the common mode gain amplifying the *average* input. Note that for $v_1 = v_2 = v$ the second term in Eq. (10-1) is $A_{cm} \times v$.

Figure 10-3 demonstrates the common mode and differential input signals as they appear in the normal DA connection. Numerical values are given for clarity only. As shown in Fig. 10-3a, the differential voltage v_d is $v_d = v_1 - v_2 = 6 - 5 = 1.0$ V. In Fig. 10-3b this v_d is split into two equivalent sources v' and v'' connected in series. The common mode input voltage $v_{cm} = (v_1 + v_2)/2 = (5 + 6)/2 = 5.5$ V is shown to be part of both v_1 and v_2; hence, both inputs are referenced to v_{cm}. They fluctuate above and below v_{cm}. Note that the circuit of Figure 10-3b is equivalent to Figure 10-3a. Since in Fig. 10-3b $v_1 = v_{cm} + v' = 5.5 + 0.5 = 6$ V

$$v_2 = v_{cm} - v'' = 5.5 - 0.5 = 5 \text{ V}$$

which agrees with Fig. 10-3a.

A simple way of measuring A_{cm} is to apply only a common mode signal as shown in Fig. 10-4a. Here, $A_{cm} = v_o/v_{cm}$. The differential input is zero since the two input terminals are interconnected. In theory, A_d can be measured by applying differential input only (Fig. 10-4b), and $A_d = v_o/v_d$. This method has a basic drawback. The signal v_d must be extremely small, since A_d is very large, in the order of 10^4 to 10^5. v_d larger than about 200 μV may very well saturate the amplifier, making the measurement meaningless. Typically, $A_d \gg A_{cm}$. The ratio A_d/A_{cm} determines the quality of the DA in terms of its ability to distinguish between

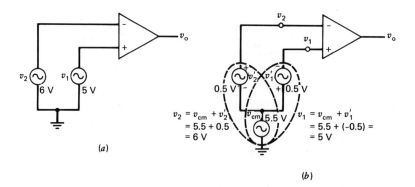

FIGURE 10-3 Differential and common mode signals.

DIFFERENTIAL AMPLIFIER (DA) FUNCTIONAL DESCRIPTION

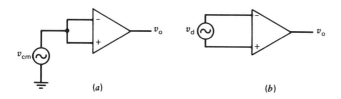

FIGURE 10-4 Measuring A_{CM} and A_d. (a) Measuring A_{CM}. (b) Measuring A_d.

differential and common mode signals. It is called the *common mode rejection ratio*, CMRR.

$$CMRR = A_d/A_{cm} \tag{10-2}$$

For example, a typical DA may have CMRR = 10^5. That means that the differential signal would be amplified 10^5 times more than the common mode input signal. CMRR is also given in dB where

$$CMRR|_{dB} = 20 \log CMRR = 20\log\frac{A_d}{A_{cm}} \tag{10-3}$$

(In some texts CMRR is defined as A_{cm}/A_d. This results in CMRR's with negative powers of ten or negative dB values. In either case $A_d \gg A_{cm}$ and the significance of CMRR is the same.)

EXAMPLE 10-1

A DA has CMRR = 80 dB, $A_d = 10^4$; the inputs are $v_1 = 10.001$ V. $v_2 = 10$ V.
Find v_o.

SOLUTION

To use Eq. (10-1) we must first find A_{cm}. From Eq. (10-3)

$CMRR|_{dB} = 20\log\dfrac{A_d}{A_{cm}}$

$80 = 20 \log (10^4/A_{cm})$

$4 = \log 10^4 - \log A_{cm} = 4 - \log A_{cm}$

$\log A_{cm} = 0$

$A_{cm} = \text{Antilog } 0 = \underline{1}$

From Eq. (10-1)

$$v_o = (v_1 - v_2) \times A_d + \frac{(v_1 + v_2)}{2} A_{cm}$$
$$= (10.001 - 10) \times 10^4 + 10.0005 \times 1$$
$$\approx 10 + 10 = 20 \text{ V}$$

434 DIFFERENTIAL AMPLIFIERS

The results of Ex. 10-1 show that the output voltage v_o contains two parts: 10 V as a result of the differential gain and another 10 V resulting from common mode gain. This is hardly a *differential* amplifier since only half of the output is a result of differential gain, even though $A_d \gg A_{cm}$. Thus, a CMRR of 80 dB is, in some cases, quite unsuitable. (The values selected in Ex. 10-1 may be somewhat exaggerated to make the point. They nevertheless are encountered in many sophisticated control systems.) With a $CMRR = 120$ dB in Ex. 10-1, we would find that the common mode output voltage would be 0.1 V compared to a 10 V differential output. Here, the differential output (10 V) is 100 times larger than the common mode output (0.1 V).

EXAMPLE 10-2

Given $CMRR = 120$ dB, $v_1 = 1.01$ V, $v_2 = 1.0$ V. Find the ratio of the common mode output to the differential mode output.

SOLUTION

From Eq. (10-1)

$$v_o = \underbrace{(v_1 - v_2)A_d}_{\text{differential}} + \underbrace{\frac{(v_1 + v_2)}{2}A_{cm}}_{\text{common mode}}$$

We are interested in the ratio

$$(v_1 - v_2) \times A_d \Big/ \frac{(v_1 + v_2)}{2} A_{cm} = (v_1 - v_2)/\{v_1 + v_2/2\} \times \frac{A_d}{A_{cm}}$$

$$= (v_1 - v_2)/\{v_1 + v_2/2\} \times CMRR$$

Since 120 dB means 10^6, the desired ratio is

$$(1.01 - 1.0)/\{(1.01 + 1)/2\} \times 10^6 \approx \frac{(0.01 \times 10^6)}{1}$$

$$= \underline{10^4}$$

Clearly, the amplifier in Ex. 10-2 is a quality differential amplifier, in terms of common mode rejection.

To obtain high CMRR it is of the utmost importance that the transistors, in particular the transistors in the input differential stage, be matched in their parameters. The pair of transistors must produce identical gains for as large a voltage input as possible. They must have matched V_{BE} versus I_C characteristics, the same value for β over the operating range, etc.

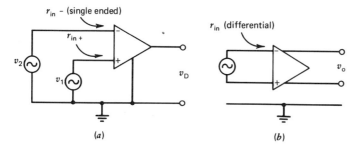

FIGURE 10-5 DA configurations. (a) Single-ended in–single-ended out. (b) Differential in–differential out.

10-4 DA CONFIGURATIONS

In general, the DA can be designed for operation in a number of configurations. Two of them are shown in Fig. (10-5). "Single ended input" is a configuration in which the input signals are referenced to ground (Fig. 10-5a). In the "differential" mode, the signal is applied directly to the two differential input terminals of the amplifier. There is no ground reference for the input signal (Fig. 10-5b). This definition applies also to "single-ended" and "differential" outputs. It stands to reason that additional configurations that mix single-ended input or output with differential output or input also exist. The DA can also be used as a simple amplifier in a single-input, single-output configuration. An inverting amplifier, when input is applied to the $(-)$ terminal, while the $(+)$ terminal is grounded, or noninverting with input connections reversed, as shown in Fig. 10-6. (More details of these configurations are given in Sec. 10-10.)

10-5 THE DIFFERENTIAL AMPLIFIER (DA) CIRCUIT BIASING

A typical, if somewhat simplified, differential amplifier circuit is shown in Figure 10-7. As far as the bias (dc) currents are concerned, it is reasonable to expect that $I_{CQ1} = I_{CQ2} = I_{CQ}$. This is based on the assumption that the two transistors are *matched* for β, V_{BE}, and other parameters. Since both transistors are produced simultaneously

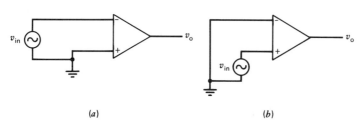

FIGURE 10-6 Single input configurations. (a) Inverting. (b) Noninverting.

436 DIFFERENTIAL AMPLIFIERS

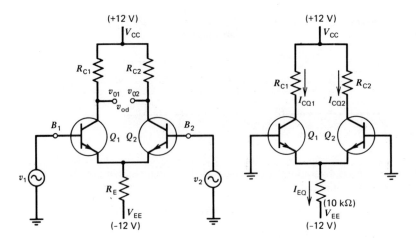

FIGURE 10-7 Basic DA circuit. (a) Circuit diagram (simplified). (b) Biasing (dc) equivalent ($e_1 = e_2 = 0$).

by the same integrated circuit process, they are indeed reasonably well-matched. (See the discussion of CMRR.) The collector bias current I_{CQ} can be calculated by applying KVL to one of the base loops and realizing that $I_{EQ} \approx 2I_{CQ}$

$$-V_{BE} - I_{EQ} \times R_E - V_{EE} = 0$$

$$-V_{EE} - V_{BE} - 2I_{CQ} \times R_E = 0$$

$$I_{CQ} = \frac{-V_{EE} - V_{BE}}{2R_E} \qquad (10\text{-}4)$$

For the values given in Fig. 10-5 and if we assume that Q_1 and Q_2 are silicon transistors, $I_{CQ} = [-(-12) - 0.7]/[2 \times 10 \times 10^3] = [12 - 0.7]/[2 \times 10 \times 10^3] = 0.565$ mA. (Note the double negative associated with the negative V_{EE}.)

It is desirable in these differential amplifier circuits to keep I_{EQ} independent of the transistors themselves and at the same time make the effective value of R_E as large as possible. The reasons for this will become evident from the ac analysis. This task is accomplished by use of a current source in the emitter circuit. Figure 10-8 shows an ideal current source when its internal resistance is infinite. This makes the effective value of R_E infinite. Real current sources have a finite internal resistance, usually very high and thus the effective R_E is very high.

10-6 CURRENT SOURCES

There are many circuits that exhibit the characteristics of a current source. In Chapter 4 we noted that the collector current in a transistor behaves like and indeed is represented by a current source. I_C does not, for the most part, depend on the

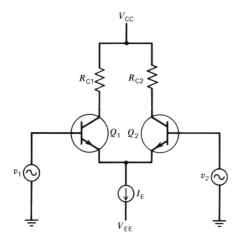

FIGURE 10-8 Current source drive.

collector circuit. This leads to the simple current source circuit shown in Figure 10-9. Here,

$$I_C \approx \frac{|V_{EE}| - 0.7}{\frac{R_B}{\beta} + R_E} \text{ (silicon transistor)}$$

and is not dependent, to any large extent, on the collector circuit, whether it is a resistor or a differential amplifier (as long as the transistor is operating in the active region).

An improved circuit that also has the advantage of using only one resistor is shown in Fig. 10-10. Again, it is reasonable to expect that the two transistors will

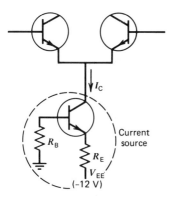

FIGURE 10-9 Simple current source.

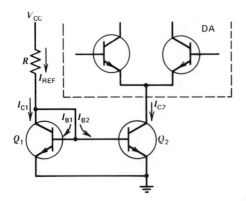

FIGURE 10-10 Improved current source ("Mirror" type).

be "matched." In particular, their transfer characteristic V_{BE} versus I_C must be the same. (This simply means that for the same V_{BE} the two transistors produce the same I_C.) Since V_{BE} for both transistors is the same (they are connected in parallel), it follows that

$$I_{C1} \simeq I_{C2} \tag{10-5}$$

Applying KVL to the Q_1 collector circuit and realizing that $V_{CE1} = V_{BE}$ (the collector of Q_1 is connected to base), we get

$$V_{CC} - I_{REF}R - V_{BE} = 0$$

$$I_{REF} = \frac{V_{CC} - V_{BE}}{R} \tag{10-6}$$

Summing currents at the collector of Q_1 yields

$$I_{REF} - I_{C1} - I_{B1} - I_{B2} = 0 \tag{10-7}$$

For transistors with high β, I_{B1} and I_{B2} are small relative to I_{C1} and may be neglected, so that

$$I_{REF} - I_{C1} \simeq 0$$
$$I_{REF} \simeq I_{C1} \tag{10-8}$$

and from Eqs. (10-8), (10-6), and (10-5)

$$I_{C1} \simeq I_{REF} = \frac{V_{CC} - V_{BE}}{R} \simeq I_{C2} \tag{10-9}$$

As in the simple current source, I_{C2} is independent of the collector circuit of Q_2. Equation (10-9), which gives I_{C2}, contains no component or supply associated with the Q_2 collector circuit. Equation (10-9) must be modified in the case in which an emitter supply voltage V_{EE} is used

$$I_{REF} = \frac{V_{CC} - V_{BE} - V_{EE}}{R} = I_{C2} \qquad (10\text{-}9a)$$

Note that since usually V_{EE} is *negative* (when using NPN transistors), Eq. (10-9a) can be rewritten as

$$I_{REF} = \frac{|V_{CC}| + |V_{EE}| - V_{BE}}{R} = I_{C2} \qquad (10\text{-}9b)$$

EXAMPLE 10-3

It is desired to design a current source with a current of 1.0 mA using the circuit of Fig. 10-10. Assume that β_1 and β_2 are very high and Q_1 and Q_2 are silicon transistors ($V_{BE} = 0.7$ V), $V_{CC} = 30$ V. Find R.

SOLUTION

Solving Eq. (10-9) for R, we get

$$R = \frac{V_{CC} - V_{BE}}{I_{C2}} = \frac{30 - 0.7}{1 \times 10^{-3}}$$
$$= 20.3 \text{ k}\Omega$$

Example 10-3 points to one of the difficulties with this type of current source. To obtain small currents, for example, 1.0 mA or less, R becomes relatively large and not suitable for I.C. processing. (Large resistors require large die area and hence are wasteful in terms of the complete I.C. package.) It is also noteworthy that most first stages of differential amplifiers operate with currents well below 1.0 mA requiring resistances in the hundreds of kΩ.

A circuit that solves this problem is the Widlar circuit that is a modification (and rather simple at that) of the circuit of Figure 10-10. A resistor R_E, in the order of a few kΩ, is inserted in the emitter of Q_2 (Fig. 10-11). Table 10-1 gives some typical values of R_E for various values of I_{C1} and I_{C2}.

TABLE 10-1 WIDLAR CURRENT SOURCE R_E FOR DIFFERENT VALUES OF I_{C1}, I_{C2}

I_{C1}	R_E for $I_{C2} = 10$ μA	R_E for $I_{C2} = 50$ μA	R_E for $I_{C2} = 200$ μA
0.5 mA	10.2 kΩ	1.2 kΩ	0.12 kΩ
1.0 mA	12 kΩ	1.6 kΩ	0.21 kΩ
2.0 mA	13.8 kΩ	1.9 kΩ	0.3 kΩ

440 DIFFERENTIAL AMPLIFIERS

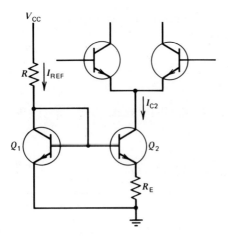

FIGURE 10-11   "Widlar" current source.

The values of R_E given in Table 10-1 are based on theoretical calculations and are only an approximation of circuit behavior. To find I_{C2} we calculate $I_{C1}(I_{REF})$ from Eq. (10-9) and then use Table 10-1 to find I_{C2}.

The circuits presented here are but a fraction of the current source circuits used, in particular those in the I.C. industry. It is important to remember that besides the fact that the current is *independent* of the circuit the current source is connected to, it also has a very high internal resistance. In terms of the DA circuit, this means that the emitter resistance in the differential stage (R_E in Figure 10-7) is replaced by the very high internal resistance of the current source.

10-7   THE DA CIRCUIT—AC ANALYSIS

The DA circuit consists of two CE stages, with separate input signals, connected to a *common* emitter resistance. Since two signal sources are involved, the use of superposition suggests itself. In addition, it is reasonable to neglect R_E, because, as noted before, R_E is very large and hence assumed to be an open circuit ($R_E \simeq \infty$). These two premises lead to the equivalent circuits shown in Figure 10-12 that represent the first step in the application of superposition, that is, v_{in2} set to zero. Note that it is assumed that the transistors are matched; hence, $r_{e1} = r_{e2} = r_e$ and we use $R_{C1} = R_{C2} = R_C$. Figure 10-12b shows the equivalent circuit resulting from setting $R_E = \infty$ (open circuit) and applying step one of superposition (eliminate source v_{in2}). We can now calculate v'_{o1} and v'_{o2}, for the first step of superposition, as shown from Figs. 10-12c and 10-12d. The resistance r_{e2} appears as part of the calculation of v'_o as well as v''_o. r_{e2} in Fig. 10-12c is returned to ground through the ideal base emitter junction of Q_2.

$$v'_{o1} = -v_{in1} \times R_C/(r_e + r_e) = -v_{in1} \times R_C/2r_e \qquad (10\text{-}10)$$

Equation (10-10) represents the output of a CE stage.

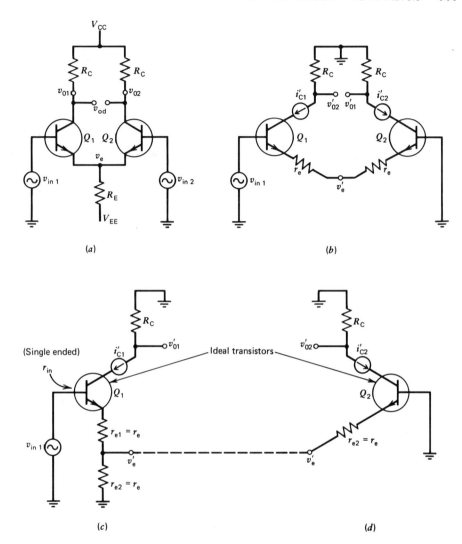

FIGURE 10-12 DA circuit analysis using superposition. (a) Circuit diagram. (b) Eq. circuit superposition step 1: eliminate v_{in2} ($v_{in2} \to 0$) and $R_E \to \infty$. (c) Computing v'_{o1}. (d) Finding v'_{o2}.

$$v'_e = v_{in1} \times r_e/(r_e + r_e) = v_{in1}/2 \tag{10-11}$$

To calculate v'_{o2}, note that this voltage is the output of a CB (input coupled to the emitter) stage with an input v'_e. We first find v'_e, which is essentially the output of an EF circuit as shown in Fig. 10-12c.

Hence,

$$v'_{o2} = v'_e \times \frac{R_C}{r_e} \quad \text{(CB output)} \tag{10-12}$$

Substituting for v'_e from Eq. (10-11), we get

$$v'_{o2} = \frac{v_{in1}}{2} \times \frac{R_C}{r_e} = v_{in1} \times \frac{R_C}{2r_e} \qquad (10\text{-}13)$$

Applying the second step of superposition (eliminating the source v_{in1}) yields

$$v''_{o1} = v_{in2} \times \frac{R_C}{2r_e} \qquad (10\text{-}14)$$

$$v''_{o2} = -v_{in2} \times \frac{R_C}{2r_e} \qquad (10\text{-}15)$$

The last equations can be obtained directly from Eq. (10-13) and (10-10), respectively, by realizing the symmetry of the two steps of superposition. Equation (10-14) is obtained from Eq. (10-13) by replacing v_{in1} with v_{in2} and v'_{o2} with v''_{o1} and Eq. (10-15) from Eq. (10-10) by similar substitutions. The last step of superposition, the summation of voltages, yields:

$$v_{o1} = v'_{o1} + v''_{o1} = -v_{in1} \times \frac{R_C}{2r_e} + v_{in2} \times \frac{R_C}{2r_e} \qquad (10\text{-}16)$$

$$= -\frac{R_C}{2r_e}(v_{in1} - v_{in2})$$

$$v_{o2} = v'_{o2} = v_{in1} \times \frac{R_C}{2r_e} - v_{in2} \times \frac{R_C}{2r_e} \qquad (10\text{-}17)$$

$$= \frac{R_C}{2r_e}(v_{in1} - v_{in2})$$

The differential input to the DA is $v_{in1} - v_{in2} = v_{ind}$, (the difference between two voltages). We calculate the gains referred to this input for the various configurations. For a *single-ended* output, we calculate output v_o taken from one collector to ground.

$$|A_v| = \left|\frac{v_o}{v_{ind}}\right| = \frac{R_C}{2r_e} \qquad (10\text{-}18)$$

(v_o may refer to v_{o1} or v_{o2}.)
For differential output, in which the output is taken between the two collectors, Eqs. (10-16) and (10-17), yield

$$v_{od} = v_{o1} - v_{o2}$$

$$= -\frac{R_C}{2r_e}(v_{in1} - v_{in2}) - \frac{R_C}{2r_e}(v_{in1} - v_{in2})$$

$$= -\frac{R_C}{r_e}(v_{in1} - v_{in2})$$

and since $v_{in1} - v_{in2} = v_{ind}$

$$|A_v| = \left|\frac{v_{od}}{v_{ind}}\right| = \frac{R_C}{r_e} \quad (10\text{-}19)$$

(the gain for differential output to differential input). It is important to note that both Eqs. (10-18) and (10-19) represent the *differential* gains since they relate to the *differential* input signal.

EXAMPLE 10-4

In Fig. 10-13 find

$$\left|\frac{v_{o1}}{v_{ind}}\right| \text{ and } \left|\frac{v_{od}}{v_{ind}}\right| \quad (v_{ind} = v_{in1} - v_{in2})$$

SOLUTION

We first find the value of the current source. From Eq. (10-9b)

$$I_{C2} = \frac{10 + 10 - 0.7}{39 \times 10^3} = 0.495 \approx 0.5 \text{ mA}$$

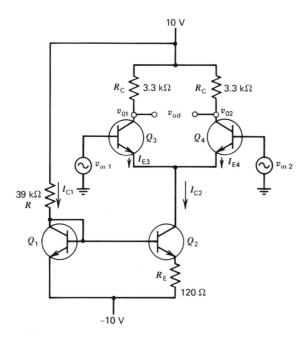

FIGURE 10-13 DA stage with current source.

444 DIFFERENTIAL AMPLIFIERS

From Table-1 for $I_{C1} = 0.5$ mA and $R_E = 0.12$ kΩ, we find $I_{C2} = 200$ μA. This current divides evenly between the two transistors Q_3 and Q_4, giving $I_{E3} = I_{E4} = 0.1$ mA (100 μA) Hence,

$$r_e = \frac{26}{0.1} = 260 \; \Omega$$

From Eq. (10-18)

$$\left|\frac{v_{o1}}{v_{ind}}\right| = \frac{R_C}{2r_e} = -\frac{3300}{520} = \underline{6.35}$$

and from Eq. (10-19)

$$\left|\frac{v_{od}}{v_{ind}}\right| = \frac{R_C}{r_e} = \frac{3300}{260} = \underline{12.7}$$

A circuit sometimes used in IC DAs consists of *Darlington pairs* in a differential amplifier configuration. Typically, it serves to increase the input resistance of the DA circuit. A basic differential Darlington stage is shown in Fig. 10-14. Each pair Q_1, Q_2 and Q_3, Q_4 replaces a single transistor of Fig. 10-8. Gain calculations are the same as in the regular case with the understanding that $r_e = r_{eT}$ where r_{eT} is the dynamic resistance of the Darlington *pair*. (For a Darlington stage $r_{eT} = 2r_{e2}$ double the r_e for the second transistor in the Darlington pair.) Thus, the dynamic resistance in Fig. 10-14 can be found by finding $I_{E2} = I_{E4}$ (equals half the current source value) and $r_e = r_{eT} = 2 \times 26/I_{E2}$

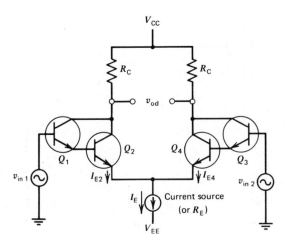

FIGURE 10-14 A Darlington DA stage.

10-8 INPUT AND OUTPUT RESISTANCE

Corresponding to the two types of input signals "single-ended" and "differential," there are two types of input resistance to be considered: "single-ended" (or common mode) and "differential" input resistance. The "single-ended" is the input resistance "seen" by the signal source applied to either one of the input terminals. Matching of the two input sections makes $r_{in-} = r_{in+}$ (see Figs. 10-5a and 10-12c). The "differential" input resistance is that "seen" by a signal source applied to the two inputs differentially (Fig. 10-5b). The value of r_{in}, differential or single-ended, depends on the β and r_e of the input transistors. In particular, the differential r_{in}, r_{ind} is approximately given by

$$r_{ind} \approx \beta \times (2r_e) \qquad (10\text{-}20)$$

This expression can be obtained directly from the differential equivalent circuit Fig. 10-15. Note that R_E (or the current source) is absent since this resistance is very large.

An idea of the magnitude of r_{ind} can be obtained by using typical parameter values for the two transistors involved. Typically, the dc collector currents are about 10 μA (a 20 μA current source divided between the two transistors), while β is about 200. The $I_C = 10$ μA yields

$$r_e = \frac{26 \text{ mV}}{10 \times 10^{-3} \text{ mA}} = 2.6 \text{ k}\Omega$$

so that

$$r_{ind} \simeq 200 \times 2 \times 2.6 \text{ k}\Omega = 1.04 \text{ M}\Omega$$

A substantially larger r_{in} is attained by the use of the Darlington differential stage, since β_T of the Darlington is so much larger than β of a single transistor. (r_{eT} of the Darlington is also somewhat larger than r_e of a single transistor.) The input resistance specified by the manufacturers of the integrated DA is indeed as high as

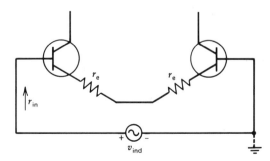

FIGURE 10-15 ac equivalent circuit of differential input stage.

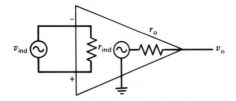

FIGURE 10-16 DA equivalent.

10 MΩ for some of these circuits. To obtain even higher values of r_{in}, FET circuits are used. These yield r_{in} as high as 10^{12} Ω. It is worth noting that the single-ended input resistance is often larger than r_{ind}. Manufacturers of I.C. DA circuits usually specify only r_{ind}, which is simply referred to as r_{in} (see Ch. 11 for the manufacturer's specifications).

As far as output resistance is concerned, it is usually given for a single-ended output and is typically 50 to 100 Ω. Some DA devices are produced with integral buffer amplifier stages (these are relatively high current, low output resistance circuits), bringing r_o down as low as 0.3 Ω typically. For example, the r_o is typically 75 Ω for the μA741 device and 0.3 Ω for the NE5533. Figure 10-16 shows an equivalent circuit of a DA similar to the circuits used in discussing r_{in} and r_o in Ch. 5 (see Fig. 5-1). It further clarifies the meaning of r_{ind} and r_o.

10-9 THE TYPICAL I.C. DA

Most I.C. DA circuits are very complex and difficult to analyze in detail. The circuit shown in Fig. 10-17 for the μA709 I.C. amplifier is relatively simple. Instead of a detailed analysis, we divide the circuit into its basic blocks, each of which has been previously discussed in this text.

The transistors $Q_1 - Q_2$ constitute the basic differential stage with Q_{10} as its current source. The two outputs of the last differential stage (collectors of Q_1 and Q_2) are coupled to a Darlington differential stage $Q_4Q_6 - Q_3Q_5$.

The signal from this Darlington circuit is coupled to the complementary symmetry output stage $Q_{13} - Q_{12}$ via Q_8 and Q_9 with Q_{11}, a CE circuit driving the output stage.

One of the reasons for the large number of coupled stages is the importance of a high voltage gain from input to output. This will become more evident from the discussion of operational amplifiers in the next chapter.

SUMMARY OF IMPORTANT TERMS

Common mode gain (A_{cm}) Amplification applied only to the common mode signal.
Common mode rejection ratio (CMRR) The ratio A_d/A_{cm} in which A_d is the differential voltage gain and A_{cm} the common mode voltage gain.

SUMMARY OF IMPORTANT TERMS 447

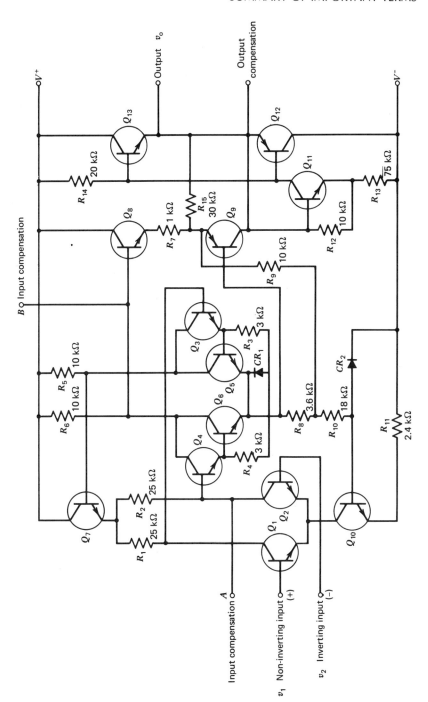

FIGURE 10-17 The µA709 schematic.

448 DIFFERENTIAL AMPLIFIERS

$$CMRR = \frac{A_d}{A_{cm}}$$

$$CMRR_{dB} = 20 \log \frac{A_d}{A_{cm}}$$

Common mode signal A signal (or its equivalent) applied simultaneously to both input terminals of a differential amplifier.

Common mode r_{in} See single-ended r_{in}.

Current source circuit A circuit (usually using active devices) that produces a current in one of its legs (for example, collector current) that is independent (to a large degree) of the circuit connected to that leg. The current behaves like that of a current source.

Differential amplifier (DA) A circuit, with two inputs, that is primarily designed to amplify only the *difference* between two signals applied to its inputs.

Differential gain Amplification applied to the differential signal only.

Differential signal A signal applied *between* the two input terminals of a DA. A signal not referenced to ground (floating).

Differential r_{in} (r_{ind}) The input resistance seen by a signal connected to the two input terminals (no ground reference necessary).

Inverting amplifier An operational-differential amplifier configuration that produces an output that is 180° out of phase with the input signal.

Inverting input (−) The terminal to which the input is applied in an inverting amplifier.

Noninverting amplifier A differential amplifier configuration that produces no phase shift (0°) between input and output.

Noninverting input (+) The input terminal used in the noninverting amplifier configuration.

Operational amplifier The general class of dc amplifiers that may be configured to produce mathematical operations such as addition, subtraction, multiplication, integration, and differentiation.

Single-ended signal A signal with one of its legs connected to ground. Ground referenced signal.

Single-ended r_{in} The input resistance seen by a signal applied from either input terminal to GND.

PROBLEMS

10-1. Find the output of a DA with the following characteristics:

$$A_d = 20{,}000.$$
$$A_{cm} = 0.5.$$
$$v_1 = 12.1 \text{ mV}.$$
$$v_2 = 12.0 \text{ mV}.$$

10-2. Find the *CMRR* and the $CMRR|_{dB}$ for the DA of Problem 10-1 above.

10-3. In Fig. 10-4a, a measurement of v_o/v_{cm} yielded 0.2, while v_o/v_d in Fig. 10-4b yielded 10,000.

    (a) Find $CMRR|_{dB}$.
    (b) The amplifier is used with the signals as in Problem 10-1. Find v_o.
    (c) Which DA, the one in Problem 10-1 or the one in Problem 10-3, is the better one? Explain.

10-4. A DA has a CMRR of 80 dB. It is used with inputs such that $v_d = v_{cm}/20$. (The differential input is a twentieth of the common mode input.) Find the relative magnitudes of the output due to v_d and v_{cm} (output due to v_d)/(output due to v_{cm}).

10-5. Find I_C in Fig. 10-9 with $R_B = 1\ M\Omega$, $R_E = 12\ k\Omega$, $\beta = 200$, $V_{EE} = -12\ V$, silicon transistors.

10-6. Find I_{C2} in the circuit of Fig. 10-10 with $R = 10\ k\Omega$, $V_{CC} = 20\ V$. Both transistors are silicon.

10-7. Find I_{C2} (approx.) in Fig. 10-11, with $R = 5.6\ k\Omega$, $V_{CC} = 12\ V$, $R_E = 1.8\ k\Omega$. All transistors are silicon. (Use Table 10-1.)

10-8. In Fig. 10-12a, $R_{C1} = R_{C2} = 27\ k\Omega$, $R_E = 22\ k\Omega$, $V_{EE} = -10\ V$. Find:

    (a) $v_{o1}/(v_{in1} - v_{in2})$
    (b) $v_{od}/(v_{in1} - v_{in2})$

10-9. Find $v_{od}/(v_{in1} - v_{in2})$ in Fig. 10-14, with $R_C = 18\ k\Omega$, $I_E = 1.0\ mA$.

10-10. Find $v_{od}/(v_{in1} - v_{in2})$ in Fig. 10-13 with the following resistance: $R = 5.6\ k\Omega$, $R_E = 1.8\ k\Omega$, $R_C = 100\ k\Omega$. (I_{C2} has been calculated in Problem 7.)

10-11. The current source circuit of Problem 10-7 is used as the current source in a two-transistor differential stage, with $\beta_1 = \beta_2 = 200$. Find r_{ind}.

10-12. Find r_{ind} if in Problem 10-11 a Darlington differential stage is used. All βs are 200.

CHAPTER

11

OPERATIONAL AMPLIFIERS

11-1 CHAPTER OBJECTIVES

The student is introduced to basic negative feedback principles in this chapter, particularly as they apply to operational amplifiers. Following this introduction, the student will learn to analyze the following operational amplifier configurations:

1. adders;
2. subtracters;
3. integrators;
4. differentiators.

A number of applications are presented, giving the student an understanding of, and an analytical insight into these circuits. The applications presented are:

1. audio mixers;
2. bridge amplifiers;
3. active R.C. filters.

The last portion of the chapter gives the student the ability to interpret manufacturer's specifications and relate them to specific applications. A discussion of static characteristics (such as bias current, offset current, and offset voltage) and dynamic characteristics (such as slew rate and bandwidth) gives the student a very practical knowledge of OA performance.

11-2 INTRODUCTION

The term "operational" refers to the fact that the amplifier can be used to perform mathematical operations, for example, addition (adding two voltages), subtraction, multiplication, and division (by a constant), integration, and differentiation (integrals and derivatives). The operational amplifier (OA) can be configured to perform the various operations by appropriate external circuitry (circuitry that is not part of the amplifier itself). The same basic amplifier may be used as an adder, subtracter, or integrator. This flexibility makes the OA circuit very versatile and extremely useful. It is used in most control systems and is the essential component of the analog computer (a computer that performs mathematical operations using analog input signals).

As described later in this chapter, the OA uses the basic DA circuit in a negative feedback arrangement. It is, consequently, useful to describe some of the important features of negative feedback in general before proceeding to discuss the OA circuits.

11-3 NEGATIVE FEEDBACK

A feedback circuit is one in which part or all the output signal is fed back (or added) to the input signal. In Fig. 11-1 v_1 is the signal to the overall amplifier (with the feedback in place) and $v_f = \beta v_o$ is the part of v_o fed back to the input. β represents the fraction of v_o that is fed back to the input. v_f is added to v_1 to yield the input to the amplifier circuit (smaller triangle in Fig. 11-1), so that

$$v_{in} = v_1 + v_f = v_1 + \beta v_o \tag{11-1}$$

Note that two gain terms are involved in Fig. 11-1. The open loop gain A is the gain of the amplifier itself without any additional circuitry, and by definition,

$$A = \frac{v_o}{v_{in}} \tag{11-2}$$

Similarly, the closed loop gain A_{CL} is the gain from v_1 to v_o of the complete circuit, including the effect of the feedback,

$$A_{CL} = \frac{v_o}{v_1} \tag{11-3}$$

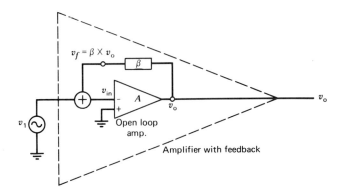

FIGURE 11-1 Feedback amplifier.

452 OPERATIONAL AMPLIFIERS

The latter gain will be evaluated shortly and is one of the most important features of the OA circuit.

In Fig. 11-1 the open loop gain is negative, since the input is applied to the $(-)$ terminal of the amplifier. Recall, that by definition, this means that the output is 180° (or negative) with respect to the input. The feedback that results from this configuration is *negative* feedback. The signal fed back from the output *opposes* (has the reverse polarity of) the input signal and tends to reduce the effective input to the open loop amplifier.

11-3.1 Voltage Gain A_{CL}

The closed loop gain can be calculated from the block diagram in Fig. 11-1. v_o is given by Eq. (11-2)

$$v_o = A \times v_{in} \qquad (11\text{-}2a)$$

If we use v_{in} from Eq. (11-1), Eq. (11-2a) becomes

$$v_o = Av_{in} = A(v_1 + \beta v_o) \qquad (11\text{-}2b)$$

Solving for v_o, we get

$$v_o = \frac{Av_1}{1 - A\beta} \qquad (11\text{-}3)$$

and

$$\frac{v_o}{v_1} = A_{CL} = \frac{A}{1 - \beta A} \qquad (11\text{-}3a)$$

Since A is negative, that is, $A = -|A|$ ($|A|$ is the magnitude of A and is always positive), Eq. (11-3a) becomes

$$A_{CL} = \frac{v_o}{v_1} = -\frac{|A|}{1 + \beta|A|} \qquad (11\text{-}3b)$$

Both A and A_{CL} in the above calculation refer to midband gains. (A and A_{CL} should be referred to as A_{mb} and A_{CLmb}, although this was not done here in order to avoid cumbersome notation). For the case in which $|\beta A| \gg 1$, the closed loop gain A_{CL} becomes

$$A_{CL} = \frac{v_o}{v_1} \approx -\frac{|A|}{|\beta A|} = -\frac{1}{\beta} \qquad (11\text{-}4)$$

The assumption $|\beta A| \gg 1$ is quite reasonable, since $|A|$ is in the order of 20,000 and up while β is rarely less than 1/100. These figures result in $|\beta A| = 20{,}000 \times 1/100 = 200 \gg 1$.

NEGATIVE FEEDBACK

The voltage v_{in}, applied to the $(-)$ terminal in the closed loop configuration, is very small. This can be seen by substituting $v_o = -|A| \times v_{in}$ in Eq. (11-1) and solving for v_{in} as follows:

$$v_{in} = \frac{v_1}{1 + \beta|A|} \approx \frac{v_1}{\beta|A|} \qquad (11\text{-}5)$$

Since it was assumed that $\beta|A|$ is large, and that v_1 is usually limited to a few volts, v_{in} is in the order of no more than a few mV (if that much). As a result, the $(-)$ terminal is referred to as *virtual ground*. It is "virtually" zero volts. Extending this result to the DA, we find that in closed loop applications, with a feedback network in place, $v_d \approx 0$. The differential voltage between the $(+)$ and $(-)$ inputs is very small.

A very important result of Eq. (11-4) is that the closed loop gain, A_{CL}, depends solely on the feedback network, and not the gain A. This network can usually be set accurately to obtain precise gains with very high stability.

EXAMPLE 11-1

Find A_{CL} for an OA circuit for

(a) $|A| = 20{,}000$, $\beta = \dfrac{1}{100}$ $\left(\dfrac{1}{100} \text{ of } v_o \text{ is fed back}\right)$.

(b) $|A| = 20{,}000$, $\beta = 1$ (all of v_o is fed back).

SOLUTION

(a) $A_{CL} = -\dfrac{|A|}{1 + \beta|A|} = -\dfrac{20{,}000}{1 + \dfrac{1}{100} \times 20{,}000} = \underline{-99.5}$

If we use

$A_{CL} = -\dfrac{1}{\beta} = \dfrac{1}{1/100} = \underline{-100}$

(b) $A_{CL} = -\dfrac{20{,}000}{1 + 1 \times 20{,}000} = \underline{-1}$

and

$A_{CL} = -\dfrac{1}{\beta} = \underline{-1}$

Example 11-1 points to the fact that when all the output is fed back ($\beta = 1$), the closed loop gain is 1. (This will be the case for the unity gain OA circuit discussed later.) As this example indicates, the closed loop gain may be varied by changing β only (assuming A is very large). There is a substantial loss in gain from part (a) to (b) in Ex. 11-1. Why then use the low gain? Fortunately, what is lost in gain is recaptured in frequency response and distortion.

11-3.2 Frequency Response

The lower the gain of an OA circuit, the wider the frequency bandwidth. Since the OA circuit is a dc amplifier, it amplifies signals down to dc (zero frequency), and the bandwidth of the amplifier is given by the upper cut-off frequency f_{UC}. (In many data sheets it is referred to simply as f_2.) The upper cut-off frequency for the amplifier *with feedback* can be obtained from the cut-off frequency of the amplifier alone from

$$f_{2CL} = (1 + \beta|A|) \times f_2 \qquad (11\text{-}4)$$

where f_{2CL} is the closed loop cut-off frequency and f_2 is the open loop (no feedback) cut-off frequency. From Eq. (11-3b) (considering magnitudes only), we get

$$1 + \beta A = \frac{A}{A_{CL}}$$

so that Eq. (11-4) becomes

$$f_{2CL} = \frac{A}{A_{CL}} \times f_2 \qquad (11\text{-}4a)$$

Equation (11-4a) shows that the BW is increased by the same proportion that the gain is decreased, namely A/A_{CL}. The μA741 IC OA circuit, for example, has an $f_2 \approx 6$ Hz with $A \approx 1.5 \times 10^5$. At a closed loop gain of $A_{CL} = 100$ the cut-off frequency becomes, from Eq. (11-4a),

$$f_{2CL} = \frac{1.5 \times 10^5}{100} \times 6 = 9.0 \text{ kHz}$$

a marked increase in the BW. f_{2CL} for $A_{CL} = 1$, the unity gain BW, is often given by the data sheet. It is given as the unity gain BW or the gain-BW product (see the discussion of f_T in Ch. 8). It can, of course, be calculated from Eq. (11-4a) as follows: For $A_{CL} = 1$,

$$GBW = \frac{1.5 \times 10^5}{1} \times 6 = 0.9 \text{ MHz.}$$

GBW is the gain BW product. The manufacturer of the μA741 device gives this data in a graph shown in Fig. 11-2. Point *a* on the plot gives $f_2 = 9$ kHz for a gain of 100, and at point *b* $f_2 = 0.9$ MHz for unity gain consistent with the above calculations. The last frequency, 0.9 MHz, is of course the gain-bandwidth-product (GBW). With GBW given, f_{2CL} for a particular closed loop gain (which represents the expected bandwidth for the particular gain) can be found from

$$GBW = A_{CL} \times f_{2CL} \qquad (11\text{-}5)$$

$$f_{2CL} = \frac{GBW}{A_{CL}} \qquad (11\text{-}5a)$$

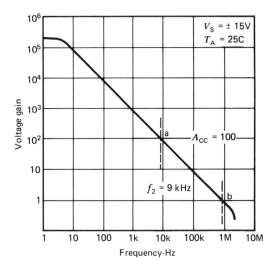

FIGURE 11-2 Frequency response of the μ741 (Courtesy Signetics Corp.)

11-3.3 Distortion

Much like the bandwidth, distortion is improved, or reduced, as the closed loop gain is decreased. To avoid some mathematical details, we define distortion as deviation from a pure sinewave. For low A_{CL} this distortion is minimal, and as a result, the output signal is much closer to a sinewave waveform (assuming the input was a perfectly pure sinewave). In other words, the amplifier itself introduces less distortion if its gain is lower. As before, the distortion improvement is a fraction of the ratio between the open loop and closed loop gains. For a given closed loop gain, A_{CL}, the distortion D_{CL} is given by

$$D_{CL} = \frac{D}{1 + A\beta} = D \times \frac{A_{CL}}{A} \qquad (11\text{-}6)$$

Where D_{CL} is the distortion of the overall amplifier with gain A_{CL}, D is the distortion of the internal amplifier (gain A). For the μA741 with a given $A = 1.5 \times 10^5$, the improvement, or reduction, in distortion at a closed loop gain of 150 would be

$$D_{CL} = D \times \frac{150}{1.5 \times 10^5} = D \frac{1}{1000} \qquad (11\text{-}6)$$

for a 1000:1 improvement.

11-4 VOLTAGE GAIN OF AN I.C. OPERATIONAL AMPLIFIER (OA)

In the years before the I.C. revolution the OA consisted of a very high-gain dc amplifier in a feedback configuration. With the advent of the I.C. technology, the basic differential amplifier circuit, such as the μA709 or μA741, is used in a feedback arrangement, yielding the typical operational amplifier performance. The noninverting input to the DA (the (+) input terminal) is connected to ground (0 V), and the inverting input (the (−) terminal) is used in the feedback circuit. Figure 11-3 gives a typical OA configuration using an I.C. DA circuit, for example, the μA709 or μA741. Some of the detailed circuit connections for offset adjustment or frequency compensation have not been included in order to maintain simplicity. The feedback network consists of the two resistors R_f and R_1. In terms of Eq. (11-4), it can be shown that $\beta = -R_1/R_f$.

This is based on the following assumptions:

1. $r_{in(-)}$ is very high.
2. $A \times \beta \gg 1$.
3. The source resistance of v_1 and the Thevenin's resistance of v_o are small.
4. The use of superposition for the two sources v_1 and v_o at the (−) terminal, to find $v_{in(-)}$.

From this feedback factor and Eq. (11-4), we derive the following closed loop gain for the circuit of Fig. 11-3:

$$A_{CL} = \frac{v_o}{v_1} = -\frac{1}{\beta} = -\frac{R_f}{R_1} \tag{11-7}$$

This result may be obtained in a somewhat different manner as follows: In section 11-3 it was shown that v_i of the OA is approximately zero. This means that in Fig. 11-3, $v_{in(-)} \simeq 0$ as well as $v_d \simeq 0$ (the differential input). Thus, $i_{in(-)} \simeq 0$; otherwise, the current $i_{in(-)}$ will produce a voltage drop across the input resistance $r_{in(-)}$ resulting in $v_{in(-)}$ having a substantial value. Using KCL at the (−) terminal (Fig. 11-3), we get

$$i_1 + i_f = i_{in(-)} \simeq 0 \tag{11-8}$$

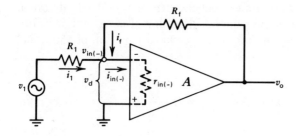

FIGURE 11-3 The DA as an operational amp (OA).

VOLTAGE GAIN OF AN I.C. OPERATIONAL AMPLIFIER (OA)

Since the voltage at the $(-)$ terminal is approximately zero,

$$i_1 = \frac{v_1}{R_1} \tag{11-9}$$

and

$$i_f = \frac{v_o}{R_f} \tag{11-10}$$

(v_1 is dropped across R_1. v_o is dropped across R_f.) Substituting in Eq. (11-8), we arrive at

$$\frac{v_1}{R_1} + \frac{v_o}{R_f} = 0$$

Solving for v_o/v_1 yields

$$A_{CL} = \frac{v_o}{v_1} = -\frac{R_f}{R_1} \tag{11-11}$$

Extending Eq. (11-11) to the general sinusoidal case, we get:

$$A_{CL} = \frac{v_o}{v_1} = -\frac{Z_f}{Z_1} \tag{11-11a}$$

where R_f and R_1 have been replaced by the more general terms Z_f and Z_1.

Equation (11-11a) shows that the closed loop gain, the gain of the OA with the external resistors, is dependent on the two impedances Z_1 and Z_f (or R_1 and R_f), making it very easy to obtain any desired gain (as long as $A_{CL} \ll A$). Note that Eq. (11-11a) also indicates that an inversion takes place: the output is 180° out of phase with the input. In terms of dc signals, a *positive* input voltage produces a *negative* output voltage. The circuit (Fig. 11-3) is connected as an *inverting* amplifier.

EXAMPLE 11-2

Find v_o in the circuit of Fig. 11-4 where $Z_f = 20$ kΩ,

$$Z_1 = 5 \text{ k}\Omega, \qquad v_1 = -2 \text{ V}.$$

SOLUTION

Using Eq. (11-11)

$$A_{CL} = -\frac{Z_f}{Z_1} = -\frac{20 \text{ k}\Omega}{5 \text{ k}\Omega} = -4$$

$$v_o = A_{CL} \times v_1 = (-4) \times (-2) = 8 \text{ V}.$$

Note the polarities of the input and output.

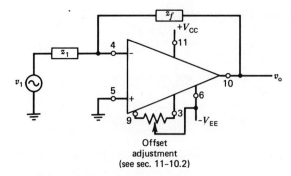

FIGURE 11-4 OA circuit (μA741CA).

EXAMPLE 11-3

In the circuit of Fig. 11-4, Z_f consists of a 10 mH inductor, $Z_1 = 1.0$ kΩ, $v_1 = 1$ V rms.
(a) Find v_o, if v_1 has a frequency of 10 kHz ($v_1 = 1.4 \sin(2\pi\ 10{,}000t)$.
(b) Find v_o for a 100 kHz frequency.

SOLUTION

(a) To use Eq. (11-11a), first find $Z_f = j\omega L = j2\pi fL = j2\pi \times 10{,}000 \times 10 \times 10^{-3} = j628$ Ω

$$A_{CL} = -\frac{Z_f}{Z_1} = -\frac{j \times 628}{1 \times 10^3}$$

$$= -j \times 0.628$$

$$A_{CL} = 0.628\ \angle -90° \quad (|A_{CL}| = 0.628)$$

The circuit produces a $-90°$ phase shift.

$$|v_o| = |A_{CL}| \times v_1 = 0.628 \times 1 = 0.628\ \text{V}$$

In sinusoidal terms,

$$v_o = 0.628\ \angle -90° \times 1 = 0.628\angle -90°\ \text{V}$$
$$v_o = 0.88 \sin(\omega t - 90) \quad (0.88 = 1.41 \times 0.628)$$

(b) For $f = 100$ kHz $\omega L = 6280$ Ω

$$A_{CL} = -\frac{Z_f}{R_1} = -j6280/10^3 = 6.28\angle -90°$$
$$v_o = A_{CL} \times v_1 = 6.28\angle -90° \times 1 = 6.28\angle -90°\ \text{V}$$
$$v_o = 8.8 \sin(\omega t - 90) \quad (\omega = 2\pi \times 10^5)$$

11-5 z_{in} OR r_{in} AND z_o (r_o) OF THE INVERTING OA

As in the treatment of other amplifiers, we are again concerned with some of the basic parameters, such as z_{in} (r_{in}) and z_o (r_o). z_{in} is the impedance "seen" by the signal source. For the circuit of Fig. 11-4 the signal source is v_1. Since the $(-)$ node is virtual ground, the source v_1 "sees" an impedance Z_1 to ground (to virtual ground); hence,

$$z_{in} = Z_1 \tag{11-12}$$

Equation (11-12) is valid for the particular configuration shown. It is important to understand that z_{in} is not the same as, or even related to, $r_{in(-)}$ or $r_{in(+)}$, which are the resistances "looking into" the respective terminals. These depend on the OA circuit itself (see Sec. 10-8), while z_{in} depends on the feedback network.

The output impedance, z_o (usually purely resistive, hence the use of r_o), is largely a function of the internal circuitry of the amplifier. It is specified by the manufacturer of the I.C., and it is in the order of 100 Ω, or less.

11-6   NONINVERTING DA

In the previous OA circuits the input was connected to the inverting input $(-)$, yielding an inverting OA. Since the OA circuit is actually a differential amplifier, which has both $(-)$ and $(+)$ input terminals, it is possible to obtain a noninverting amplifier circuit by connecting the input signal to the $(+)$ terminal. A typical noninverting circuit is shown in Fig. 11-5.

To compute the gain of this configuration, note that $v_d \simeq 0$ (see Sec. 11-3); this means that

$$v_{(-)} = v_{(+)}$$

In this circuit

$$v_{(+)} = v_1 = v_{(-)} \tag{11-13}$$

and

$$v_{(-)} = v_o \times \frac{R_1}{R_f + R_1} \quad \text{(voltage division)} \tag{11-14}$$

From Eqs. (11-13) and (11-14), we get

$$v_1 = v_o \times \frac{R_1}{R_1 + R_f} \tag{11-15}$$

Solving for v_o/v_1 yields

$$A_{CL} = \frac{v_o}{v_1} = \frac{R_1 + R_f}{R_1} = 1 + \frac{R_f}{R_1} \qquad (11\text{-}15a)$$

The output v_o is *in phase* with v_1, the source signal. The input resistance seen by v_1, r_{in} is the input resistance of the (+) terminal, which is very high (specified by the manufacturer):

$$r_{in} = r_{in(+)}$$

Note the difference in r_{in} for the inverting (Fig. 11-3) and noninverting (Fig. 11-5) circuits.

EXAMPLE 11-4

For the circuit of Fig. 11-5, $R_f = 10\text{ k}\Omega$ and $R_1 = 2\text{ k}\Omega$. The I.C. chip is the μA741. Find:
(a) the closed loop gain A_{CL}.
(b) the input resistance.

SOLUTION

(a) From Eq. (11-15)

$$A_{CL} = 1 + \frac{R_f}{R_1} = 1 + \frac{10\text{ k}\Omega}{2\text{ k}\Omega} = \underline{6}$$

(b) The data sheet of the μA741 gives $r_{in} = 2.0\text{ M}\Omega$, typically. (This is the input resistance of the device itself with no external components.) Thus

$$r_{in} = \underline{2.0\text{ M}\Omega}, \text{ typically.}$$

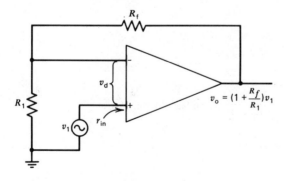

FIGURE 11-5 Noninverting amp.

ADDERS, SUBTRACTORS, AND MULTIPLIERS 461

A unity gain noninverting amplifier is often used to increase the signal current available to the load. Since the output resistance of the amplifier is low it can drive low-resistance loads. Figure 11-6 shows a unity gain amplifier. Again, since $v_d \simeq 0$, $v_{(+)} = v_{(-)}$, $v_{(+)} = v_o$, and $v_{(-)} = v_1$, we get

$$v_1 = v_o$$

$$A_{CL} = \frac{v_o}{v_1} = 1 \tag{11-16}$$

11-7 ADDERS, SUBTRACTORS, AND MULTIPLIERS

The following circuits are extensions of the circuits discussed in the previous sections. Figure 11-7 shows a typical inverting adder (or mixer in sound studio terminology). Using superposition, we find that

$$v_o = -v_1 \times \frac{R_f}{R_1} - v_2 \times \frac{R_f}{R_2} - v_3 \times \frac{R_f}{R_3} - \ldots \tag{11-17}$$

The first step of superposition is shown in Fig. 11-7b. Here, we get $v_o = -v_1 \times R_f/R_1$ [Eq. (11-11)]. Note that the other input resistors (R_2, R_3, etc.) have no effect on the circuit since the $(-)$ terminal is a virtual ground, effectively *isolating all inputs from one another*. Each "sees" its own input resistor and nothing else. Repeating the above superposition step for every input and summing yields Eq. (11-17). Equation (11-17) can be written as

$$v_o = -\left(v_1 \frac{R_f}{R_1} + v_2 \frac{R_f}{R_2} + v_3 \frac{R_f}{R_3} + \ldots\right) \tag{11-17a}$$

showing that v_o is the inverted sum of the weighted inputs. Each input is amplified by a selected factor (R_f/R_1 for v_1, etc.) and added. To obtain a simple nonweighted inverting adder, simply set

$$R_1 = R_2 = R_3 = \ldots = R$$

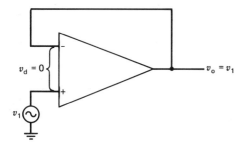

FIGURE 11-6 Unity gain–noninverting.

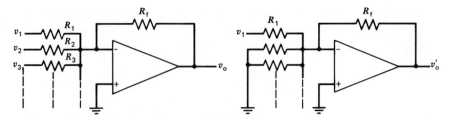

FIGURE 11-7   Adder (mixer). (a) Schematic diagram. (b) Step 1 of superposition.

Equation (11-17a) becomes

$$v_o = -\frac{R_f}{R}(v_1 + v_2 + v_3 + \ldots) \qquad (11\text{-}17b)$$

Thus, the inputs are summed, and the sum is amplified by the factor $-R_f/R$. Setting $R_f = R$ ($R_f = R_1 = R_2 = R_3 = \ldots$), we get

$$v_o = -(v_1 + v_2 + v_3 + \ldots) \qquad (11\text{-}17c)$$

EXAMPLE 11-5

The circuit in Fig. 11-7a has the following values: $R_1 = 2$ kΩ, $R_2 = 4$ kΩ, $R_3 = 5$ kΩ, $R_f = 10$ kΩ, $v_1 = -3$ V dc, $v_2 = 2$ V dc, $v_3 = 3$ V dc.
(a) Find v_o.
(b) Find the input resistance for each of the signal sources.

SOLUTION

(a) From Eq. (11-17a), we get

$$v_o = -\left(-3 \times \frac{10\text{ k}}{2\text{ k}} + 2\frac{10\text{ k}}{4\text{ k}} + 3 \times \frac{10\text{ k}}{5\text{ k}}\right)$$
$$= -(-15 + 5 + 6) = \underline{+4\text{ V dc}}.$$

(b) r_{in} for each source is the respective input resistor. For v_1, $r_{in1} = 2$ kΩ; for v_2, $r_{in2} = 4$ kΩ, $r_{in3} = 5$ kΩ.

EXAMPLE 11-6

In Ex. 11-5 the inputs are three sinusoidal signals with different frequencies, ω_1 for v_1, ω_2 for v_2, and ω_3 for v_3. That is, $v_1 = 0.5 \sin \omega_1 t$, $v_2 = 2 \sin \omega_2 t$, $v_3 = 0.5 \sin \omega_3 t$. Find v_o.

SOLUTION

From Eq. (11-17a), we get

$$v_o = -\left(v_1 \frac{10\ k}{2\ k} + v_2 \frac{10\ k}{4\ k} + v_3 \frac{10\ k}{5\ k}\right)$$
$$= -(v_1 \times 5 + v_2 \times 2.5 + v_3 \times 2)$$
$$= -(2.5 \sin \omega_1 t + 5 \sin \omega_2 t + 1 \sin \omega_3 t)$$

Since the frequencies are different, no further simplification is possible. All three frequencies with their respective amplitudes are present in the output voltage.

A subtracter circuit may be obtained directly by the use of both (+) and (−) terminals or indirectly by the use of a cascade arrangement of two amplifiers. Figure 11-8a shows a weighted subtracter using a single DA. Applying superposition we obtain v_o as follows: step 1, setting v_2 to GND (0 V), $v'_o = -v_1 \times R_f/R_1$; step 2, setting v_1 to GND, yielding the noninverting configuration, and $v''_o = v_2(1 + R_f/R_1)$. Summing the partial voltages yields

$$v'_o + v''_o = -v_1 \frac{R_f}{R_1} + v_2\left(1 + \frac{R_f}{R_1}\right)$$

$$v_o = v_2 \frac{(R_1 + R_f)}{R_1} - v_1 \frac{R_f}{R_1} \qquad (11\text{-}18)$$

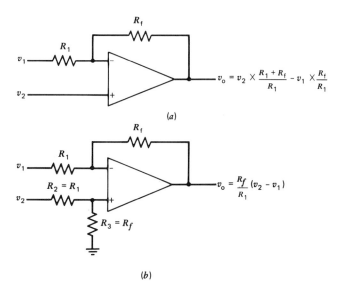

FIGURE 11-8 Single amp. subtractors. (a) Subtractor with unbalanced weight. (b) Balanced subtractors.

464 OPERATIONAL AMPLIFIERS

The subtraction is applied to the scaled (amplified) signals.

To overcome the difference in gain for the two inputs we use voltage division on v_2, as shown in Fig. 11-8b. Applying superposition and realizing that the input to the (+) terminal is $v_2 \times R_3/(R_2 + R_3)$, we get:

$$v_o = -v_1 \frac{R_f}{R_1} + v_2 \left(\frac{R_3}{R_2 + R_3}\right)\left(\frac{R_1 + R_f}{R_1}\right) \quad (11\text{-}19)$$

By setting $R_2 = R_1$, $R_3 = R_f$ we arrive at

$$v_o = -v_1 \frac{R_f}{R_1} + v_2 \left(\frac{R_f}{R_1 + R_f}\right)\left(\frac{R_1 + R_f}{R_1}\right)$$

$$= -v_1 \frac{R_f}{R_1} + v_2 \frac{R_f}{2R_1}$$

$$v_o = \frac{R_f}{R_1}(v_2 - v_1) \quad (11\text{-}19a)$$

Note that the voltage applied to the (+) input does *not* become inverted. It has *no* negative sign associated with it. The gain factor R_f/R_1 is applied to the result of the subtraction, namely $v_2 - v_1$. For the case in which $R_f = R_1$, we have:

$$v_o = v_2 - v_1 \quad (11\text{-}19b)$$

a simple subtracter.

The circuit of Fig. 11-9 executes a subtraction by using two amplifiers. For the sake of simplicity all resistors were chosen to have the same value—R.

To find v_o we first find v_{o1}:

$$v_{o1} = -v_2 \times \frac{R}{R} = -v_2 \quad (11\text{-}20)$$

The adder stage yields:

$$v_o = -\left(v_{o1} \times \frac{R}{R} + v_1 \frac{R}{R}\right) = -(v_{o1} + v_1) \quad (11\text{-}21)$$

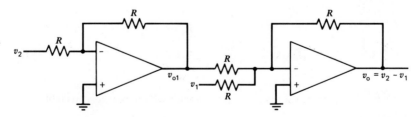

FIGURE 11-9 Subtractor–two stage.

ADDERS, SUBTRACTORS, AND MULTIPLIERS   465

Substituting Eq. (11-20) into (11-21), we get

$$v_o = -(-v_2 + v_1) = v_2 - v_1 \qquad (11\text{-}22)$$

The following examples will require the use of the various gain relationships discussed so far:

EXAMPLE 11-7

Find v_o for the circuit of Fig. 11-10.

SOLUTION

Using Eq. (11-19a) (since here $R_2 = R_1$, $R_f = R_3$), we get:

$$v_{o1} = \frac{R_f}{R_1}(3 - 3.2) = \frac{10\ k}{2\ k}(-0.2)$$
$$= -1.0\ V$$

If we use the expression for the adder,

$$v_o = -\frac{5\ k}{5\ k} \times v_{o1} - \frac{5\ k}{5\ k} \times v_3$$
$$= -(-1.0) - 1.0 = 0\ V$$

EXAMPLE 11-8

Find v_o in the circuit of Fig. 11-11.

SOLUTION

We note that A_1 is an inverting stage, A_2 a noninverting, and A_3 an adder; so we get for v_{o1}

$$v_{o1} = -0.1 \times \frac{12\ k}{2\ k} = -0.1 \times 6 = -0.6$$

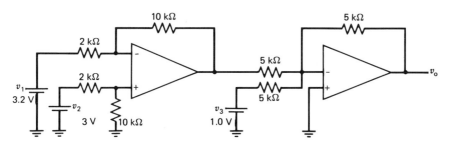

FIGURE 11-10 Circuit for Ex. 11-7.

466 OPERATIONAL AMPLIFIERS

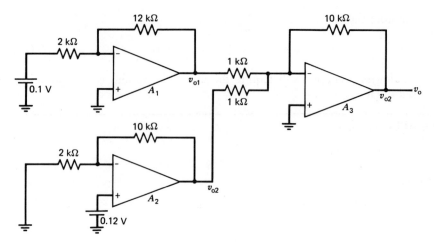

FIGURE 11-11 Circuit for Ex. 11-8.

From Eq. (11-15) we derive

$$v_{o2} = 0.12\left(1 + \frac{10\,k}{2\,k}\right) = 0.72$$

The application of Eq. (11-17) to A_3 results in

$$v_o = -v_{o1} \times \frac{10\,k}{1\,k} - v_{o2}\frac{10\,k}{1\,k}$$
$$= -(-0.6) \times 10 - 0.72 \times 10 = \underline{-1.2\ V}$$

Any of the OA circuits can be considered a multiplier in the sense that they all provide an output that is a product of an input times some gain factor. The multiplier is the gain factor of the particular circuit.

11-8 DIFFERENTIATOR AND INTEGRATOR

Equation (11-11a) gives the gain of an inverting OA circuit for steady state inputs, that is, for dc or sinusoidal signals. It does not account for sharp discontinuities in the input signal, such as step voltages (unless only resistive elements are used, in which Eq. (11-11) holds).

To obtain the general expression for the gain of the OA with capacitors and/or inductors, it is necessary to use differential and integral calculus. A solution obtained in this manner (see Appendix I) is valid for all inputs, dc, sinewave, step voltage, etc. The expression for v_o that will be obtained here, however, is for one type of circuit only and is limited to step voltage inputs.

Before proceeding, it is useful to review basic relations of voltage current and charge in the capacitor. For a capacitor being charged by a dc current I_C (Fig. 11-12)

$$Q = Cv_C \qquad (11\text{-}23)$$

Dividing by t (time)

$$\frac{Q}{t} = I_C = C\frac{v_C}{t} \qquad (11\text{-}24)$$

$$v_C = \frac{I_C t}{C} \qquad (11\text{-}24a)$$

The last expression is valid only for dc current, that is for the case in which I_C is constant. Note that v_C [Eq. (11-24a)] keeps changing linearly with time. As t increases so does v_C. Equation (11-24) may now be used to obtain the gain (v_o/v_1) of the integrator circuit shown in Fig. 11-13a. Note that $v_o = v_C$, the output voltage appears across C because the $(-)$ terminal acts like a ground. Similarly, the input voltage v_1 appears across R_1. Thus, from Eq. (11-24)

$$I_C = C \times \frac{v_C}{t} = C \times \frac{v_o}{t}$$

Since $I_C = I_f$, the feedback current

$$I_f = C \times \frac{v_o}{t} \qquad (11\text{-}25)$$

Similarly,

$$I_1 = \frac{v_1}{R_1} \qquad (11\text{-}26)$$

As noted before, in order to maintain the zero potential (virtual ground) at the $(-)$ terminal

$$I_f + I_1 = 0 \qquad (11\text{-}8)$$

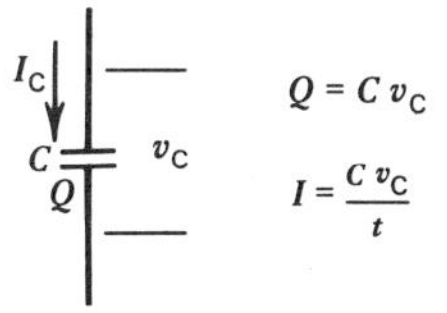

FIGURE 11-12 Voltage current relations in a capacitor.

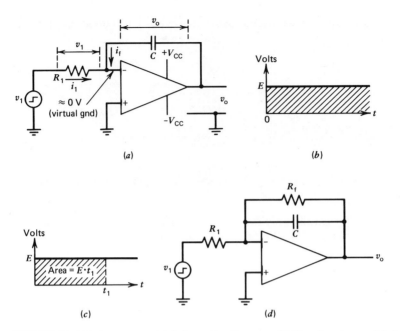

FIGURE 11-13 Integrator. (a) Schematic diagram. (b) Step input—v_1. (c) Area of input step. (d) The practical integrator.

Thus, from Eqs. (11-25) and (11-26)

$$I_1 + I_f = \frac{v_1}{R_1} + C \times \frac{v_o}{t} = 0$$

Solving for v_o yields

$$v_o = -\frac{v_1}{R_1 C} \times t \qquad (11\text{-}27)$$

The last equation is valid *only* when v_1 is a step voltage, independent of t (Fig. 11-13b). This is necessary so that I_C (as well as I_1) be dc values, since the current voltage relations in the capacitor given in Eq. (11-24) were limited to dc current. The step voltage is indeed constant (for all t larger than 0) so taht I_C is also a constant. For Fig. 11-13a, since $v_1 = E$, *the amplitude of the step is E*, Eq. (11-27) becomes

$$v_o = -\frac{1}{R_1 C} \times E \times t \qquad (11\text{-}27a)$$

As expected from the basic current voltage relations in the capacitor (Eq. 11-24), v_o that is the same as v_C increases linearly with time. As t increases (a longer period of time, a larger t) so does v_o. Clearly, this increase cannot continue indefinitely.

Indeed, it stops when either cut off or saturation is reached. In practice, the magnitude of v_o cannot exceed either supply voltage $+v_{CC}$ or $-v_{CC}$. The negative sign in Eq. (11-27) represents the inversion introduced by the amplifier. For a positive input step there will be a negative going output and vice-versa.

The circuit of Fig. 11-13 is called an *integrator* because the output voltage represents the integral of the input. Stated differently, the output at any particular instant of time represents the area under the input waveform up to that instant. Note that $E \times t_1$ in Fig. 11-13c is the area of the input step up to point t_1. The coefficient $-1/RC$ in Eq. (11-27) is essentially only a gain factor.

The circuit of Fig. 11-13a, while theoretically correct, does not guard against the effects of the dc offset current (see sec. 11-10). This offset current tends to charge the capacitor, driving the amplifier into cut off. To eliminate this problem it is necessary to include a resistor connected across the capacitor (Fig. 11-13d). The resistor R_f serves to discharge the capacitor, avoiding any voltage buildup due to offset current. The value of R_f should be as large as possible, yet small enough to provide a relatively fast and full discharge. A good rule is to make $R_f = 20R_1$.

EXAMPLE 11-9

For the circuit in Fig. 11-13a, $R_1 = 10$ kΩ, $C = 0.1$ μF, v_1 is a 5 V step as shown in Fig. 11-14. Plot v_o.

SOLUTION

If we use Eq. (11-27a)

$$v_o = -\frac{1}{10^4 \times 0.1 \times 10^{-6}} \times 5 \times t$$
$$= -5000t$$

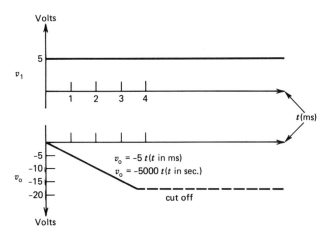

FIGURE 11-14 Input (v_1) and output v_o for Ex. 11-9.

470 OPERATIONAL AMPLIFIERS

This last expression represents a voltage changing linearly with time. For each millisecond ($\Delta t = 1$ ms), v_o changes by -5.0 V. A plot of v_o is shown in Fig. 11-14. The slope of v_o (up to cut off) is -5000 V/s (-5 V/ms). For this example, a feedback resistor R_f should be added where R_f is about 200 kΩ or larger to eliminate the effect of offset currents.

The output voltage of the *differentiator* (Fig. 11-15) can be obtained following similar methods used in the development of Eq. (11-27) for the integrator. Here, however, the output is dependent upon the rate of change of the input, rather than its amplitude. For fast changing inputs the output will be larger. For a linear (ramp) input signal, in which the rate of change is constant, the output is

$$v_o = -R_f C \times \frac{\Delta v_1}{\Delta t} \tag{11-28}$$

where $\Delta v_1/\Delta t$ represents the rate of change of the input voltage (assumed constant, otherwise differential notation must be used. See Appendix I).

EXAMPLE 11-10

In Fig. 11-15 $R_f = 1$ MΩ, $C = 0.05$ μF, v_1 is shown in Fig. 11-16a. Plot v_o.

SOLUTION

We first find the rate of change of v_1. We note that v_1 in Fig. 11-15a changes linearly from 0 to 10 V in 100 ms; hence,

$$\frac{\Delta v_1}{\Delta t} = \frac{10 \text{ V}}{100 \text{ ms}} = 100 \text{ V/s}$$

$\Delta v_1/\Delta t$ has the units of volts/seconds. v_o must be found in two parts. For the period 0–100 ms, v_1 changes at 100 V/s, from 100 ms on, v_1 is constant. Hence, its rate of change is zero and for this portion of time $\Delta v_1/\Delta t = 0$. From Eq. (11-28), for 0 to 100 ms

$$v_o = \underbrace{-10^6}_{R_f} \times \underbrace{0.005 \times 10^{-6}}_{C} \times \underbrace{100}_{\frac{\Delta v_1}{\Delta t}} = -5.0 \text{V}$$

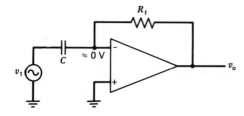

FIGURE 11-15 The differentiator circuit.

DIFFERENTIATOR AND INTEGRATOR

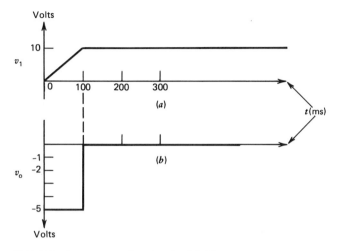

FIGURE 11-16 v_1 and v_o for Ex. 11-10.

For the period 100 ms on

$$v_o = -10^6 \times 0.005 \times 10^{-6} \times \underbrace{0}_{\frac{\Delta v_1}{\Delta t}} = 0$$

A plot of v_o is shown in Fig. 11-16b.

The circuits of Figs. 11-13 and 11-15 produce 90° phase shifts when used with sinusoidal signals. From Eq. (11-11a) we get for the circuit in Fig. 11-13:

$$v_o = -v_1 \times \frac{z_f}{z_1} = -v_1 \times \frac{-j\omega C}{R_1}$$

$$= v_1 \times \frac{j}{\omega C R_1}$$

$$= v_1 \frac{1}{\omega C R_1} \angle 90° \qquad (11\text{-}29)$$

The gain factor is $1/\omega C R_1$ and the phase shift from input to output is 90°. Similarly, Fig. 11-15 results in:

$$v_o = v_1(-j\omega C R_f) = v_1 \times \omega C R_f \angle -90° \qquad (11\text{-}30)$$

The gain in these circuits depends on the frequency while the phase shift is 90° *regardless* of frequency (within the frequency response limits of the OA).

472 OPERATIONAL AMPLIFIERS

EXAMPLE 11-11

In the circuit of Fig. 11-15 $R_f = 10$ kΩ, $C = 0.1$ μF, and $v_1 = 0.3\angle 0°$ V at 1 kHz. Find v_o in phasor form and plot the sinusoidal wave forms for v_1 and v_o.

SOLUTION

From Eq. (11-30) for the differentiation circuit

$$v_o = v_1 \omega C R_f \angle -90°$$
$$= 0.3\angle 0° \times 2\pi \times 10^3 \times 0.1 \times 10^{-6} \times 10^4 \angle -90°$$
$$= 6.28 \times 0.3 \angle -90°$$
$$= \underline{1.88\angle -90° \text{ V}}$$

The plots are shown in Fig. 11-17.

Both the differentiator and the integrator circuits may be designed using inductors instead of capacitors. Figure 11-18 gives the typical circuits. For the integrator (Fig. 11-18a) with a step input of amplitude E

$$v_o = -\frac{R_f}{L} \times Et \tag{11-31}$$

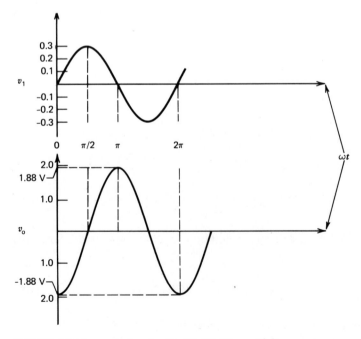

FIGURE 11-17 v_1 and v_o for Ex. 11-10 (Sinusoidal).

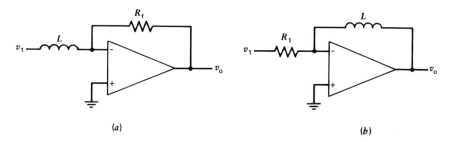

FIGURE 11-18 Integrator and differentiator using inductors. (a) Integrator. (b) Differentiator.

for a sinusoidal input v_1

$$v_o = -\frac{R_f}{j\omega L} \times v_1 \qquad (11\text{-}31a)$$

For the differentiator circuit with v_1 having a constant rate of change (linear ramp)

$$v_o = -\frac{L}{R_1} \times \frac{\Delta v_1}{\Delta t} \qquad (11\text{-}32)$$

($\Delta v_1/\Delta t$ is the rate of change of v_1) and with v_1 sinusoidal

$$v_o = -\frac{j\omega L}{R_1} \times v_1 \qquad (11\text{-}32a)$$

11-9 OA-DA APPLICATIONS

11-9.1 The Audio Mixer

In studio work, it is often necessary to mix sounds from different sources: background music from a recorder with speaking voices from a number of microphones, and so on. The levels of these sound sources must be controlled so that the effects of sound fade-in and fade-out may be accomplished. The OA lends itself perfectly to such an application. As shown in Fig. 11-19, the preamplified signals from the various sources are added in a typical OA adder. To facilitate level control, each input resistor (R_1, R_2, . . .) is made up of two resistors, a fixed resistor R_{1a} and a variable resistor R_{1p}. The gain, hence the level, of each signal is controlled by adjusting the respective input resistance. The overall level of the mixed signal is set by the feedback resistance R_f. The switches S_1, S_2, etc. were included to permit turning off completely any of the sources.

The preamplication is usually used to eliminate the effects of the large variations in source resistance of the different signal sources. (The output resistances of all preamps driving the mixer are about the same.)

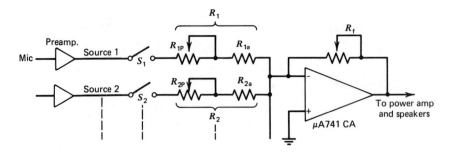

FIGURE 11-19   Audio mixer.

11-9.2   High Input Impedance Differential Amplifier

In measuring voltages or currents from sources with high internal resistance it is essential that the measuring instrument present a very high input resistance; otherwise, loading effects may introduce substantial errors. It is also often necessary to make a "floating" measurement not referenced to ground. Figure 11-20 shows a suitable high input impedance differential amplifier. v_{in} is connected differentially (both terminals of v_{in} are floating, not referenced to ground) to the (+) terminals of the two DAs. As noted previously, the input resistance $r_{in(+)}$ is very high, so that v_{in} sees a very high input resistance.

To evaluate v_o, with the resistances as shown, we note that $v_A = v_1$ and $v_B = v_2$. This is a result of the fact the v_d in both DAs is zero. The current through R_2, i_o is:

$$i_o = (v_A - v_B)/R_2 = (v_1 - v_2)/R_2 \qquad (11\text{-}33)$$

For $R_2 = aR_1$, that is R_2 is given relative to R_1 (for example, if $R_2 = 2$ kΩ and $R_1 = 1$ kΩ, then $a = 2/1 = 2$), we get $i_o = (v_1 - v_2)/aR_1$. v_{od} is the voltage drop across the $R_1 - R_2 - R_1$ combination, and i_o is the current through these

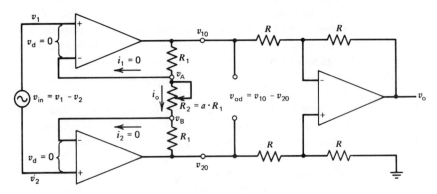

FIGURE 11-20   High input-resistance DA.

resistors. By Ohm's law

$$v_{od} = i_o \times (R_1 + R_2 + R_1) = i_o \times (2R_1 + aR_1). \quad (11\text{-}34)$$

Substituting for i_o from Eq. (11-33a), we get

$$v_{od} = \frac{(v_1 - v_2)}{aR_1} \times (2R_1 + aR_1) = (v_1 - v_2) \times \frac{(2 + a)}{a} \quad (11\text{-}34a)$$

To obtain v_o we refer to the circuit of Fig. 11-8b (with all resistors having the same value, R). From Eq. (11-19b)

$$v_o = (v_{20} - v_{10}) = -v_{od} \quad (11\text{-}35)$$

v_{od} is given by Eq. (11-34a), so Eq. (11-35) becomes

$$v_o = -v_{od} = -(v_1 - v_2) \times \frac{(2 + a)}{a} = -v_{in} \times \frac{(2 + a)}{a} \quad (11\text{-}35a)$$

and the voltage gain is

$$A_v = \frac{v_o}{v_{in}} = -\frac{(2 + a)}{a} = -(2/a + 1) \quad (11\text{-}36)$$

Note that selecting the factor a by adjusting R_2 determines the gain of the complete circuit. For $a = 4$ (that is, $R_2 = 4R_1$) we have, from Eq. (11-36),

$$\frac{v_o}{v_{in}} = -\left(\frac{2}{a} + 1\right) = -1.5$$

For $a = \frac{1}{5}$

$$A_v = -\left(\frac{2}{a} + 1\right) = -11$$

It is apparent that the gain is large for $R_2 < R_1$ ($a < 1$) and approaches unity as "a" approaches infinity (R_2 open circuit, $R_2 \gg R_1$). With R_2 open, the circuit reverts to two separate unity gain noninverting amplifiers.

11-9.3 Bridge Amplifier

Various parameters such as weight, temperature, etc. can be measured indirectly by measuring resistance changes (in a transducer) caused by changes in these parameters. A resistive temperature transducer (a device that represents changes in temperature by resistive changes) may be used to measure temperature. At a reference temperature T_R (usually $T_R = 25°C$) the transducer resistance is R. As the temperature changes by ΔT so does R change by ΔR. Figure 11-21 shows a circuit

476 OPERATIONAL AMPLIFIERS

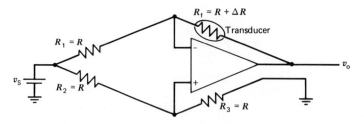

FIGURE 11-21 Bridge amplifier.

that may be used to convert the resistance change ΔR into an output voltage v_o. Thus, v_o represents ΔT indirectly. R_1, R_2, R_3 were selected to equal the transducer resistance R at the reference temperature. Note that this circuit is identical to the circuit of Fig. 11-8b with $V_1 = V_2 = V_s$. Equation (11-19a) developed for this circuit can be used here. At the reference temperature, when $R_1 = R_2 = R_3 = R_f = R$,

$$v_o = (v_2 - v_1)\frac{R_f}{R}$$
$$= (v_s - v_s)\frac{R_f}{R} = 0$$

The output is zero when no temperature deviation, from the reference, occurred. For a transducer resistance of $R + \Delta R$ (at a temperature other than the reference) we get, from Eq. (11-19)

$$v_o = -v_s \times \frac{R_f}{R_1} + v_s \times \left(\frac{R_3}{R_2 + R_3}\right)\left(\frac{R_1 + R_f}{R_1}\right)$$
$$= -v_s\left(\frac{R + \Delta R}{R}\right) + v_s\left(\frac{R}{2R}\right)\left(\frac{R + R + \Delta R}{R}\right)$$
$$= v_s\left(\frac{2R + \Delta R}{2R}\right) - \left(\frac{R + \Delta R}{R}\right)$$
$$= -v_s \times \frac{1}{2} \times \frac{\Delta R}{R} \qquad (11\text{-}37)$$

v_o depends on the excitation voltage v_s and on the *relative* change of R, $\Delta R/R$. It does not depend on the magnitude of R itself. By selecting a large excitation voltage v_s, v_o can be made responsive to very small changes in R, and hence, it will measure small temperature variations.

EXAMPLE 11-13

In Fig. 11-21, $R = 10 \text{ k}\Omega$. The temperature coefficient of the transducer is $0.5\%/°C$; that is, the resistance changes by 0.5% for each degree C change. Use $v_s = 40$ V. Find v_o for a 10°C increase in temperature.

SOLUTION

To find $\Delta R/R$ for a 10°C change, note that $\Delta R/R = 0.5/100$ (0.5%) for a 1° change. Thus, $\Delta R/R$ for a $\Delta T = 10°C$ is

$$\left.\frac{\Delta R}{R}\right|_{(\Delta T = 10°C)} = \frac{0.5}{100} \times 10 = 5 \times 10^{-2}\ \Omega/\Omega$$

$$v_o = -v_s \times \frac{\frac{1}{2}\Delta R}{R} = -40 \times \frac{1}{2} \times 5 \times 10^{-2}$$
$$= -100 \times 10^{-2} = \underline{-1.0\ V}$$

A relatively small change in R produces a substantial v_o. Note that here $\Delta R/R$ is positive, because ΔT is positive, producing a negative v_o. A negative ΔT (decrease in temperature) will produce a positive v_o.

11-9.4 Active R-C Filters

The OA is very useful in filter circuits. It eliminates the use of inductors and still provides frequency responses otherwise attainable only with L-C circuits. The active filters, as the OA filter circuits are called, use only resistive and capacitive components.

The details of the design are complex and will not be presented. Instead, we discuss a relatively simply low pass filter ("passing" frequencies below cut off and attenuating higher frequencies). We encountered the simple R-C filter in our discussion of the frequency response of amplifiers. In this simple R-C circuit the "fall-off," that is, the rate at which attenuation increases above cut off, was 20 dB/decade.

In filter applications it is often necessary to have the "fall-off" substantially higher, so that the demarcation between the "pass band" (frequency range passed by the circuit) and "stop band" (heavily attenuated frequency range) is sharper. Figure 11-22 demonstrates two "fall-off" rates, the simple R-C filter with 20 dB/decade and an active filter response (one of many circuits available) with 40 dB/decade. The circuit that yields the 40 dB/decade is given in Fig. 11-23. The components are given in such a way that the cut-off frequency f_{CU} is given by:

$$f_{CU} = \frac{1}{2\pi RC}, \qquad \omega_{CU} = \frac{1}{RC} \qquad (11\text{-}38)$$

and the gain of the circuit as a function of frequency is:

$$|A_v| = \left|\frac{v_o}{v_1}\right| = \frac{1}{\sqrt{1 + (RC)^4 \omega^4}} = \frac{1}{\sqrt{1 + \left(\dfrac{\omega}{\omega_{CU}}\right)^4}} \qquad (11\text{-}39)$$

478 OPERATIONAL AMPLIFIERS

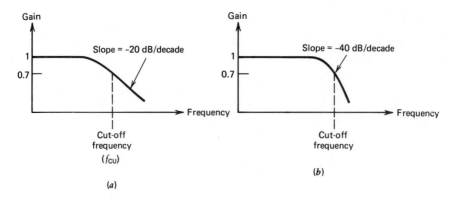

FIGURE 11-22 Low pass (LP) filter responses (normalized). (a) Simple R-C. (b) Sharper fall off—active filter.

Thus, for $C = 1000$ pF we have $C_1 = C \times 3/2 = 1500$ pF and we must set $C_2 = C \times 2/3 = 1000 \times 0.667 = 667$ pF. If we set $R = 10$ kΩ the cut off becomes:

$$f_{CU} = \frac{1}{2\pi \times 10 \times 10^3 \times 1000 \times 10^{-12}}$$
$$= 15.9 \text{ kHz}$$

Note that all the resistors marked R must have the value (10 kΩ in this instance).

A large variety of filter responses can be produced with active filter circuits. These include the low pass, high pass, band pass, and band reject. The detailed response, fall off, bandwidth, etc. depend on the exact circuit used—and again a large variety of responses can be accommodated.

11-10 PRACTICAL OA-DA—THE DATA SHEET

In all previous discussions, an ideal model of OA and DA was assumed. Open loop gain was assumed practically infinite, setting no limits on closed loop gain.

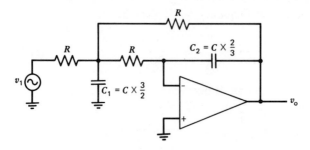

FIGURE 11-23 Active low pass filter.

Input resistance was assumed extremely large and dc bias current effects were neglected. Frequency response was completely ignored.

Since nothing is ideal, an understanding of the OA and DA limitations is both useful and necessary. These depend to a large extent on the detailed construction of the circuit and are usually given in the data sheet. As a typical data sheet, we use the one for the µA741 shown in Fig. 11-24.

11-10.1 A_{CL} Versus A

In section 11-3, the closed loop gain A_{CL} of a feedback amplifier was given as:

$$A_{CL} = \frac{v_o}{v_i} = -\frac{|A|}{1 + \beta |A|} \tag{11-3b}$$

It was noted later that the feedback factor for the OA is $\beta = R_1/R_f$ (Section 11-4). This leads to a more precise expression for the gain of the OA:

$$A_{CL} = -\frac{|A|}{1 + \frac{R_1}{R_f} \times |A|} \tag{11-40}$$

Equation (11-40) reduces to the common expression for A_{CL} for an OA [Eq. (11-7)] when the denominator is approximated by $R_1/R_f|A|$ (neglecting the 1). This approximation introduces an error in A_{CL} that directly depends on the relative magnitude of A with respect to A_{CL}, or A with respect to R_f/R_1. The error introduced by Eq. (11-7) can be approximated by:

$$\% \text{ error} \approx \frac{R_f/R_1}{A} \times 100\% \tag{11-41}$$

Note that R_f/R_1 is the approximate expression [Eq. (11-7)] for A_{CL}. The error is then approximately:

$$\% \text{ error} = \frac{A_{CL}}{A} \times 100\% \tag{11-41a}$$

EXAMPLE 11-13

Given an OA with $|A| = 10000$ and $R_f/R_1 = 1000$ (Fig. 11-3), find:
(a) The exact $|A_{CL}|$.
(b) Approximate $|A_{CL}|$.
(c) The exact error.
(d) The error using Eq. (11-41a).

signetics

GENERAL PURPOSE OPERATIONAL AMPLIFIER μA741

LINEAR INTEGRATED CIRCUITS

DESCRIPTION
The μA741 is a high performance operational amplifier with high open loop gain, internal compensation, high common mode range and exceptional temperature stability. The μA741 is short-circuit protected and allows for nulling of offset voltage.

FEATURES
- INTERNAL FREQUENCY COMPENSATION
- SHORT CIRCUIT PROTECTION
- OFFSET VOLTAGE NULL CAPABILITY
- EXCELLENT TEMPERATURE STABILITY
- HIGH INPUT VOLTAGE RANGE
- NO LATCH-UP

ABSOLUTE MAXIMUM RATINGS

	μA741C	μA741
Supply Voltage	±18V	±22V
Internal Power Dissipation (Note 1)	500mW	500mW
Differential Input Voltage	±30V	±30V
Input Voltage (Note 2)	±15V	±15V
Voltage between Offset Null and V⁻	±0.5V	±0.5V
Operating Temperature Range	0°C to +70°C	−55°C to +125°C
Storage Temperature Range	−65°C to +150°C	−65°C to +150°C
Lead Temperature (Solder, 60 sec)	300°C	300°C
Output Short Circuit Duration (Note 3)	Indefinite	Indefinite

Notes
1. Rating applies for case temperatures to 125°C; derate linearly at 6.5mW/°C for ambient temperatures above +75°C.
2. For supply voltages less than ±15V, the absolute maximum input voltage is equal to the supply voltage.
3. Short circuit may be to ground or either supply. Rating applies to +125°C case temperature or +75°C ambient temperature.

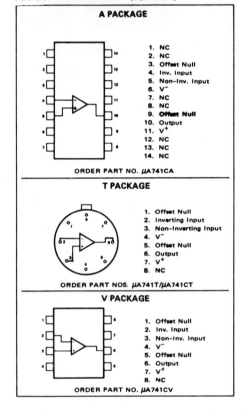

EQUIVALENT CIRCUIT

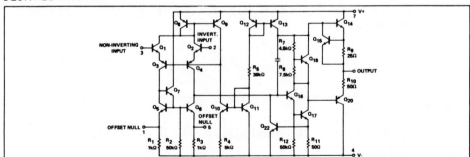

FIGURE 11-24 The μA741 data sheets (courtesy Signetics, Inc.).

SIGNETICS GENERAL PURPOSE OPERATIONAL AMPLIFIER ■ μA741

ELECTRICAL CHARACTERISTICS ($V_S = \pm 15V$, $T_A = 25°C$ unless otherwise specified)

PARAMETER	MIN.	TYP.	MAX.	UNITS	TEST CONDITIONS
μA741C					
Input Offset Voltage		2.0	6.0	mV	$R_S \leq 10k\Omega$
Input Offset Current		20	200	nA	
Input Bias Current		80	500	nA	
Input Resistance	0.3	2.0		MΩ	
Input Capacitance		1.4		pF	
Offset Voltage Adjustment Range		±15		mV	
Input Voltage Range	±12	±13		V	
Common Mode Rejection Ratio	70	90		dB	$R_S \leq 10k\Omega$
Supply Voltage Rejection Ratio		10	150	μV/V	$R_S \leq 10k\Omega$
Large-Signal Voltage Gain	20,000	200,000			$R_L \geq 2k\Omega$, $V_{out} = \pm 10V$
Output Voltage Swing	±12	±14		V	$R_L \geq 10k\Omega$
	±10	±13		V	$R_L \geq 2k\Omega$
Output Resistance		75		Ω	
Output Short-Circuit Current		25		mA	
Supply Current		1.4	2.8	mA	
Power Consumption		50	85	mW	
Transient Response (unity gain)					$V_{in} = 20mV$, $R_L = 2k\Omega$, $C_L \leq 100pF$
Risetime		0.3		μs	
Overshoot		5.0		%	
Slew Rate		0.5		V/μs	$R_L \geq 2k\Omega$
The following specifications apply for $0°C \leq T_A \leq +70°C$					
Input Offset Voltage			7.5	mV	
Input Offset Current			300	nA	
Input Bias Current			800	nA	
Large-Signal Voltage Gain	15,000				$R_L \geq 2k\Omega$, $V_{out} = \pm 10V$
Output Voltage Swing	±10	±13		V	$R_L \geq 2k\Omega$
μA741					
Input Offset Voltage		1.0	5.0	mV	$R_S \leq 10k\Omega$
Input Offset Current		10	200	nA	
Input Bias Current		80	500	nA	
Input Resistance	0.3	2.0		MΩ	
Input Capacitance		1.4		pF	
Offset Voltage Adjustment Range		±15		mV	
Large-Signal Voltage Gain	50,000	200,000			$R_L \geq 2k\Omega$, $V_{out} = \pm 10V$
Output Resistance		75		Ω	
Output Short Circuit Current		25		mA	
Supply Current		1.4	2.8	mA	
Power Consumption		50	85	mW	
Transient Response (unity gain)					$V_{in} = 20mV$, $R_L = 2k\Omega$, $C_L \leq 100pF$
Risetime		0.3		μs	
Overshoot		5.0		%	
Slew Rate		0.5		V/μs	$R_L \geq 2k\Omega$
The following specifications apply for $-55°C \leq T_A \leq +125°C$					
Input Offset Voltage		1.0	6.0	mV	$R_S \leq 10k\Omega$
Input Offset Current		7.0	200	nA	$T_A = +125°C$
		20	500	nA	$T_A = -55°C$
Input Bias Current		0.03	0.5	μA	$T_A = +125°C$
		0.3	1.5	μA	$T_A = -55°C$
Input Voltage Range	±12	±13		V	
Common Mode Rejection Ratio	70	90		dB	$R_S \leq 10k\Omega$
Supply Voltage Rejection Ratio		10	150	μV/V	$R_S \leq 10k\Omega$
Large-Signal Voltage Gain	25,000				$R_L \geq 2k\Omega$, $V_{out} = \pm 10V$
Output Voltage Swing	±12	±14		V	$R_L \geq 10k\Omega$
	±10	±13		V	$R_L \geq 2k\Omega$
Supply Current		1.5	2.5	mA	$T_A = +125°C$
		2.0	3.3	mA	$T_A = -55°C$
Power Consumption		45	75	mW	$T_A = +125°C$
		45	100	mW	$T_A = -55°C$

FIGURE 11-24 (Continued)

SOLUTION

(a) Using Eq. (11-40) the exact A_{CL} is

$$A_{CL} = \frac{10000}{1 + \frac{1}{1000} \times 10000} = \frac{10000}{1 + 10}$$

$$= 909$$

(b) Using Eq. (11-7) $|A_{CL}| = R_f/R_1 = 1000$

(c) % error $= \frac{1000 - 909}{1000} \times 100\% = 9.1\%$

(d) % error $= \frac{1000}{10000} \times 100\% = 10\%$ \hfill (11-41a)

The lesson to be learned from Ex. 11-13 is that the closed loop gain A_{CL} must always be *substantially* lower than the open loop gain A. For the μA741 (not μA741C) with a minimum open loop gain of 50,000 (as given in the data sheet) it is not advisable to select R_f/R_1 larger than about 200 (allowing A to be at least 250 times larger than A_{CL}). For higher closed loop gains, two OAs can be connected in cascade (Fig. 11-25).

11-10.2 Input Bias Current and Offset Voltage and Current

Since the typical differential amplifier (for example, μA741) is often used in dc applications, it is important to evaluate some of the more significant dc parameters.

The *input bias current* is the biasing current at the input terminals that keeps the circuit at a proper operating point. Since we are dealing with two currents, $I_{B(-)}$ and $I_{B(+)}$ (Fig. 11-26), which are usually not exactly equal, the manufacturer

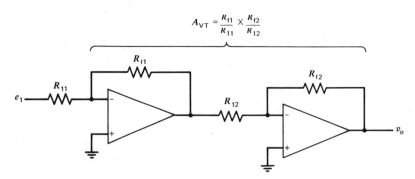

FIGURE 11-25 Cascading for gain.

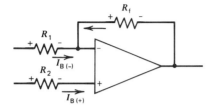

FIGURE 11-26 Input bias circuit.

specifies the bias current I_B as the average of the two:

$$I_B = \frac{I_{B(-)} + I_{B(+)}}{2} \tag{11-42}$$

The bias current may produce spurious (unwanted) dc voltage drops at the input terminals, which in turn produce a spurious dc output voltage. This can best be demonstrated by an example:

EXAMPLE 11-14

The unity gain (voltage follower) circuit shown in Fig. 11-27 uses a µA741 that has a maximum bias current of 500 nA (10^{-9} A). Find v_o for $v_1 = 0$.

SOLUTION

The circuit is shown in Fig. 11-27a and the effects of the bias current are shown in Fig. 11-27b. Since $I_B = 500$ nA we find that $V_{off} = I_B \times R_1 = 500 \times 10^{-9} \times 10^6 = 500$ mV. The offset voltage V_{off} acts *like* an input signal (Fig. 11-27c), producing an output $V_o = V_{off} = 500$ mV (unity gain). Note there is no input, yet there is an output.

The effect of I_B can be substantially reduced by balancing the resistance at the two input terminals. In Ex. 11-14, this would mean replacing the feedback short (R_f

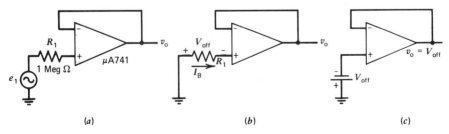

FIGURE 11-27 Bias currents in voltage follower circuit. (a) Circuit diagram. (b) Effect of bias current with $e_1 = 0$. (c) Equivalent input offset voltage.

484 OPERATIONAL AMPLIFIERS

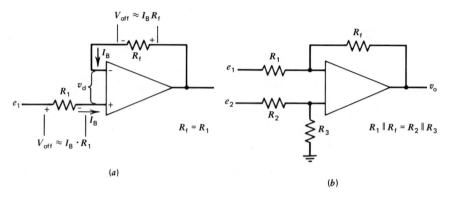

FIGURE 11-28 Compensating for bias current. (a) For the voltage follower. (b) For a complex circuit.

= 0) with $R_f = R_1 = 1$ MΩ. As shown in Fig. 11-28a, the introduction of $R_f = 1$ MΩ will cause similar offsets at both input terminals, and hence the differential offset $V_{OFF(+)} - V_{OFF(-)}$ is near zero. (Only if $I_{B(-)} = I_{B(+)}$ will the offset actually be nulled out.) A very low offset is then present at the output. Figure 11-28b shows a more complex DA circuit. Here, bias current compensation requires that

$$R_f \| R_1 = R_2 \| R_3 \tag{11-43}$$

This relation is a result of the fact that the bias current $I_{B(+)}$ flows in the parallel combination of R_2 and R_3, while $I_{B(-)}$ flows in $R_f \| R_1$. If $I_{B(+)} \simeq I_{B(-)}$, the resulting voltage drops are equal, keeping the differential offset voltage at zero.

The above compensation method is useful to the extent that $I_{B(+)} = I_{B(-)} = I_B$. However, as defined, I_B is an average value, implying that $I_{B(+)} \neq I_{B(-)}$ ($I_{B(+)}$ is not equal to $I_{B(-)}$). The difference between $I_{B(+)}$ and $I_{B(-)}$ is defined as the *input current offset*

$$I_{OFF} = I_{B(+)} - I_{B(-)} \tag{11-44}$$

This offset current contributes to a spurious output voltage, an offset voltage, that *cannot* be compensated for as shown by Eq. (11-43).

EXAMPLE 11-15

Find the offset voltage at the output in the circuit of Fig. 11-29a.

SOLUTION

The input offset current for the μA741 is given as 200 nA maximum (to meet the "worst case" design criteria we use the maximum, worst case, rather than the typical value). Since in the circuit shown, the criteria of Eq. (11-43) were met, there is no offset caused by I_B. Only I_{OFF} produces an offset voltage. The equivalent circuit,

PRACTICAL OA-DA—THE DATA SHEET

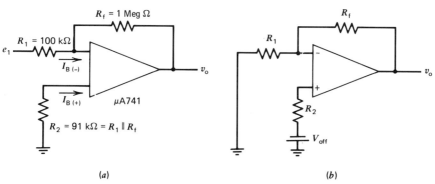

FIGURE 11-29 Offset current. (a) Circuit. (b) V_{OFF} due to I_{OFF}.

Fig. 11-29b, represents this offset, where $V_{OFF} = I_{OFF} \times R_f \| R_1 = I_{OFF} \times R_2 = 200 \times 10^{-9} \times 91 \times 10^3 = 18.2$ mV.

Using the expression for noninverting amplifier gain,

$$v_o = V_{OFF} \times \left(1 + \frac{R_f}{R_1}\right)$$
$$= -18.2 \times 10^{-3} \times 11 = -200.2 \text{ mV}$$

In Ex. 11-15, the offset voltage of the input due to I_{OFF} is 18.2 mV, while as referred to the output it is 200.2 mV. The assumption that V_{OFF} is applied to the (+) terminal, as shown in Fig. 11-29b, results in the "worst case" output offset. For a balanced circuit ($R_1 \| R_f = R_2$) this leads to an offset voltage at the *output* given by

$$V_{OFF} \text{ (at output)} = I_{OFF} \times \frac{R_1 \times R_f}{R_1 + R_f}\left(1 + \frac{R_f}{R_1}\right) \qquad (11\text{-}45)$$
$$= I_{OFF} \times R_f$$

The offset voltage, produced by I_{OFF}, can be a serious problem, particularly in applications involving small dc signals. The remedy is, first, keep $R_f \| R_1$ as small as possible. Second, select an amplifier with as low an offset current as economically feasible. (The low offset circuit is usually substantially more expensive.) Third, use the external offset null circuitry that is discussed below.

In addition to current offset there is a direct *voltage offset* that results from internal mismatches in the IC circuit. This offset voltage is present at the output even when all input resistors are shorted (Fig. 11-30a). Since the circuit shown has unity gain from the (+) terminal, we can represent this output offset as shown in Fig. 11-30b. The μA741, for example, has a maximum offset voltage referred to the input of 6.0 mV (at 25°C).

486 OPERATIONAL AMPLIFIERS

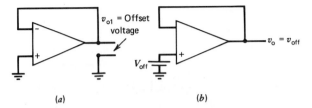

FIGURE 11-30 Offset voltage. (a) Circuit. (b) Equivalent input offset.

We have used the same symbol V_{OFF} to denote the offset voltages from all the different sources, those caused by I_B, I_{OFF} and the internal circuit mismatches. This is reasonable because ultimately the total V_{OFF} is the algebraic sum of all offset sources. This offset can be nulled, for a particular circuit configuration, by the offset null circuit shown in Fig. 11-31a (for the μA741). The configuration chosen is a simple inverting amplifier. The pin-out and a sketch of the IC package are presented for reference only. As shown, all inputs are grounded while adjusting the offset (10 kΩ) potentiometer, R_p, until $v_o = 0$. Because of variations in offset parameters from chip to chip, it is necessary to readjust the null every time the chip is replaced and certainly when the circuit is changed. Most specification sheets give an expected range of offset voltage that can be adjusted (for the μA741, it is typically ± 15 mV). It is important to emphasize that the dc offsets discussed above have no effect in strictly ac applications, in which capacitive coupling blocks any dc voltages and thus eliminates the effects of dc offset.

11-10.3 Frequency Response

The low end of the frequency response of the OA and DA is dc. The bandwidth is then dc to some upper cut-off frequency. The data sheet for the μA741, for example, does not specify the upper cut-off frequency; instead, it gives the *rise time* and *slew rate*, which are frequency related.

Rise time is defined in terms of a pulse response to a step input voltage. We assume a perfect input step (*no* time required to reach its full value). The *rise time* t_r is the time it takes the output signal to rise from 10% to 90% of final value (Fig. 11-32). The *rise time* t_r is related to the upper cut-off frequency or the bandwidth BW by

$$BW = \frac{0.35}{t_r} \tag{11-46}$$

where the BW is in Hz, and t_r in seconds. The specified, typical *rise time* for the μA741 (see Fig. 11-24) is 0.3 μs; its BW is:

$$BW = \frac{0.35}{0.3 \times 10^{-6}} = 1.167 \text{ MHz}$$

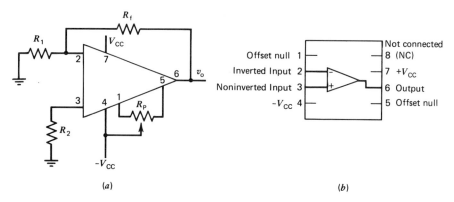

FIGURE 11-31 Offset nulling circuit for μA741cV (pin numbers are for the V package). (a) The circuit with all inputs grounded. (b) The I.C. package.

As noted in the data sheet, the $t_r = 0.3$ μs is given for unity gain; hence, the BW of 1.167 MHz is also for unity gain. It is then not the 3 dB bandwidth but rather more like f_T (unity gain frequency) or the gain-bandwidth product. It is worth noting that other specifications sheets such as the Burr Brown data sheet for their 3542 OA (Fig. 11-33) give the unity-gain frequency or gain-bandwidth product directly ($f_T = 1.0$ MHz). Clearly, the circuit is useless beyond that frequency and it cannot be expected to produce output waveforms with t_r less than that specified. Since f_T is the gain-bandwidth product, the amplifier when used in a unity gain configuration may yield a total $BW = f_T$. In a configuration yielding a closed loop gain A_{CL} (more than unity), the BW becomes

$$BW = \frac{f_T}{A_{CL}} \qquad (11\text{-}47)$$

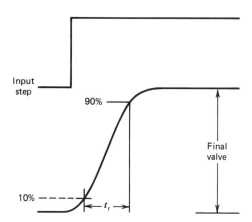

FIGURE 11-32 Rise time.

488 OPERATIONAL AMPLIFIERS

Specifications typical at 25°C and ±15 Vdc Power Supply unless otherwise noted.

MODEL	3542J	3542S
OPEN LOOP GAIN, dc		
no load	100 dB	
1 kΩ, load, min.	88 dB	
RATED OUTPUT		
Voltage, min.	±10V	
Current, min.	±10 mA	
Output Impedance	75 Ω	
FREQUENCY RESPONSE		
Unity Gain, Open Loop	1 MHz	
Full Power Response	8 kHz	
Slew Rate	0.5 V/μsec	
INPUT OFFSET VOLTAGE		
Initial Offset, 25°C, max.	±20 mV	
vs. Temp (0° to 70°C)	±10 typ, ±50 max. μV/°C	
vs. Supply Voltage	±50 μV/V	
vs. Time	±100 μV/mo	
INPUT BIAS CURRENT		
Initial bias, 25°C	-10 typ, -25 max. pA	
(doubles every +10°C)		
vs. Supply Voltage	1 pA/V	
INPUT DIFFERENCE CURRENT		
Initial difference, 25°C	±2 pA	
INPUT IMPEDANCE		
Differential	10^{11} Ω	
Common Mode	10^{11} Ω	
INPUT NOISE		
Voltage, .01 Hz - 10 Hz, p-p	2 μV	
10 Hz - 1 kHz, rms	3 μV	
Current, .01 Hz - 10 Hz, p-p	0.3 pA	
10 Hz - 1 kHz, rms	0.6 pA	
INPUT VOLTAGE RANGE		
Common Mode Voltage	±($|V_S|$ -5 V)	
Common Mode Rejection	80 dB	
Max. Safe Input Voltage	±V_S	
POWER SUPPLY		
Rated Voltage	±15 VDC	
Voltage Range, derated	±5 to ±20 VDC	
Current, quiescent	±4 mA	
TEMPERATURE RANGE		
Specification	0° to +70°C	-55° to +125°C
Operating	-25° to +85°C	-55° to +125°C
Storage	-65° to +150°C	
PRICE [1]		
1-24	6.45	11.50
25-99	5.00	9.75
100	4.25	8.50

(1) Check with factory for current prices.

FIGURE 11-33 Data sheet for the OA3542 (courtesy Burr Brown, Inc.).

So that the Burr Brown 3542 OA with $f_T = 1.0$ MHz will have a BW of 100 kHz when used with an $A_{CL} = 10$

$$BW = \frac{1.0 \text{ MHz}}{10} = 100 \text{ kHz}$$

The *slew rate* is defined as the rate of change of the output voltage at full load (Fig. 11-34). It is related to both the frequency response *and* the particular signal amplitude. The maximum rate of change of a sinewave of amplitude E_m and

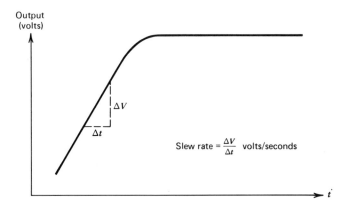

FIGURE 11-34 Slew rate.

frequency f is:

$$\text{Maximum slew rate of sine wave} = 6.28 \, E_m \times f \qquad (11\text{-}48)$$

[Eq. (11-48) is arrived at by use of differential calculus.] Equation (11-48) indicates that there is no clear relation between maximum slew rate and the frequency unless we assume a specific signal amplitude. For a typical slew rate of 0.5 V/μs $= 0.5 \times 10^6$ V/s (that of the μA741), if the output amplitude is 0.1 V the upper cut off frequency becomes from Eq. (11-48)

$$f_{CU} = \frac{\text{slew rate}}{6.28 \times E_m} = \frac{0.5 \times 10^6}{6.28 \times 0.1} = 796 \text{ kHz}$$

At higher amplitudes the frequency response is proportionately narrower (a lower f_{CU}).

The slew rate is usually important in large signal applications and in systems where "ramp" signals are used (digital voltmeters, etc.).

11-10.4 Frequency Compensation

In many of the OA IC circuits, the manufacturers provide for the inclusion of external components to improve the frequency response. (In a goodly number, this compensation is internal, included within the IC package.) For example, the Signetics Inc. OA type NE5533 requires a frequency compensating capacitor C_C to be connected between pins 5 and 8. The specific value of C_C depends on component values and circuit configuration. To assist in this determination, the manufacturer gives various graphs relating the BW for particular circuits to the values of C_C. A typical plot of the slew rate versus C_C is shown in Fig. 11-35a. Note that the BW can be determined from the slew rate and the signal amplitude (for sinewaves).

Another OA requiring frequency compensation is the NE531. A compensating

490 OPERATIONAL AMPLIFIERS

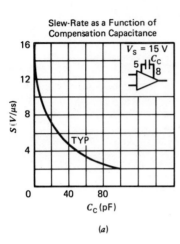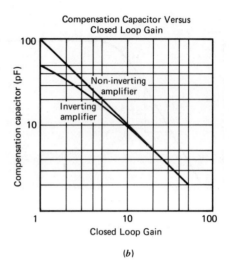

FIGURE 11-35 Frequency compensation (courtesy Signetics, Inc.). (a) Slew rate versus C_C for the NE5533. (b) Gain versus C_C for the NE531.

capacitor C_C is to be connected between pin 8 and the output. The value of this capacitor can be determined from the graph in Fig. 11-35b given by the manufacturer. Here, C_C is given as a function of gain, both in an inverting and noninverting configuration.

In many cases C_C must be selected experimentally. The manufacturer's data can then serve as a starting point in this process of "trial and error."

11-10.5 Miscellaneous Parameters

Most parameters listed in Figs. 11-24 and 11-33, other than the ones discussed above, are self-explanatory. Input resistance can be given "single ended"—terminal to ground or "differential" between terminals.

Output voltage swing is usually related to supply voltage. For the μA741 it is specified for $V_{supply} = \pm 15$ V. The *supply voltage rejection ratio* gives the expected changes in offset voltage as a function of changes in supply voltage. The value of 10 μV/V means that for a 1.0 V change in supply voltage we may expect, typically, a change of 10 μV in the offset voltage (maximum of 150 μV).

11-10.6 Parameter Variations with Temperature and Frequency

Practically all the OA-DA parameters are affected by temperature and frequency. It must be self-evident that, in the foregoing discussion, both temperature and frequency were assumed constant at some convenient value—temperature at 25°C and frequency at some midband frequency. The *CMRR* and the open loop gain, for example, change with frequency. Input resistance, and offset and bias currents, for example, vary with temperature. The details of these relations are given graphically in Fig. 11-36. (A number of other graphs are given for the sake of completeness. The advanced student may wish to interpret these other graphs by himself.)

SUMMARY OF IMPORTANT TERMS

Active filter A filter using active circuits such as operational amplifiers.

Active R-C filter An active filter with resistors and capacitors (no inductors).

Adder An operational amplifier configuration that performs the function of adding input voltages.

Bridge amplifier A circuit that uses an OA in a configuration similar to that of a Wheatstone bridge. It is designed to respond to changes in resistance.

Closed loop gain The gain of an OA circuit that accounts for the effects of feedback.

Differentiator A circuit, using an OA, that performs differentiation. The output voltage is the derivative of the input signal.

Fall off The rate of increase in attenuation with changes in frequency outside of the BW of the circuit (away from cut off), usually given in dB/octave or dB/decade.

Gain-bandwidth product (GBW, f_T) The bandwidth at unity gain.

Input bias current (I_B) The average dc bias current in the two input terminals ($+$) and ($-$) of a differential amplifier.

$$I_B = \frac{I_{B(+)} + I_{B(-)}}{2}$$

Integrator An OA configuration that performs integration. The output voltage is the integral of the input signal.

Inverting amplifier An amplifier in which the output is 180° out of phase with the input.

Mixer A circuit used to combine signals from different sources.

Negative feedback A circuit using an OA (in the context of this chapter) in which a part of the output is fed back (added) to the input is said to have feedback. The word "negative" indicates that there is a 180° phase difference between the input and the output. The fed back signal reduces the effective input signal.

Offset current (I_{OFF}) (referred to the input) The difference between the bias current in the two input terminals of an OA.

$$I_{OFF} = I_{B(+)} - I_{B(-)}$$

Offset voltage The dc output voltage of an OA with zero input to both input terminals. The ($+$) and ($-$) terminals are grounded. Usually, a result of internal mismatches. This offset voltage is generally given for unity gain.

Operational amplifier (OA) A high gain dc amplifier capable of being configured to perform various mathematical operations (add, subtract, and so on).

Pass band (in band pass filter) The range of frequencies with minimum attenuation (or maximum gain) up to the two cut-off frequencies (the 3 dB points).

492 OPERATIONAL AMPLIFIERS

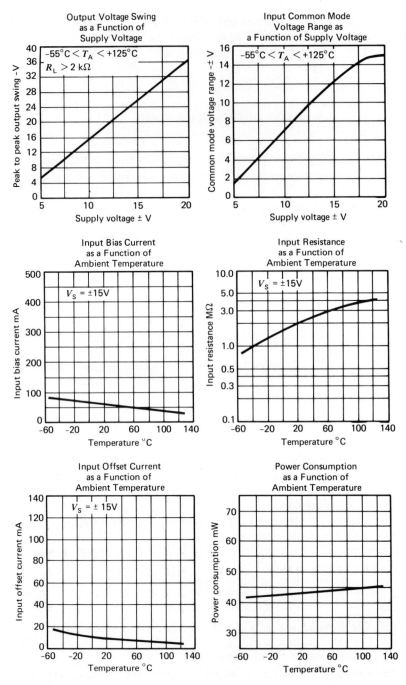

FIGURE 11-36 Typical characteristics of µA741 (courtesy Signetics. Inc.).

SUMMARY OF IMPORTANT TERMS 493

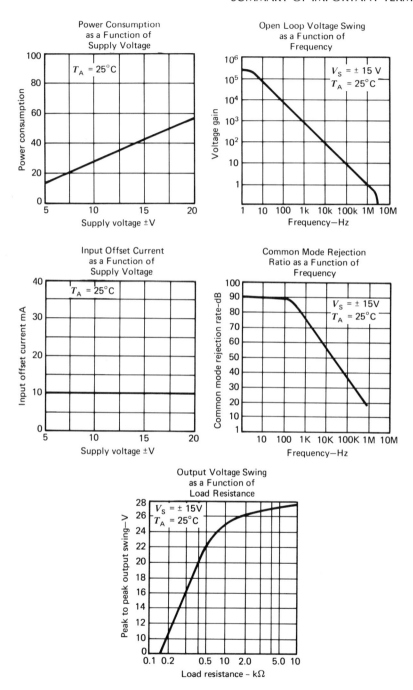

FIGURE 11-36 (Continued)

494 OPERATIONAL AMPLIFIERS

Rise time (pulse characteristic) The time it takes for a pulse or square wave to change from 10 to 90% of its final (maximum) value.

Slew rate (of an amplifier) The linear rate of change of voltage that the amplifier can accommodate, in V/s.

Stop band The range of frequencies for which the filter introduces the highest attenuation, relative to that of the pass band.

Subtractor An amplifier, usually an OA or a DA, that performs the subtraction of two input voltages.

Virtual ground The $(-)$ terminal in the inverting amplifier configuration. The voltage at this point is virtually zero.

f_T See gain-bandwidth product.

PROBLEMS

11-1. Find A_{CL} for Fig. 11-1, with $A = 5000$, $\beta = 1/100$.

   (a) Use the approximate expression for A_{CL}.
   (b) Use the exact expression.
   (c) Find the error in part (a).

11-2. The NE538 OA circuit has an open loop gain of 200,000 and a cut-off frequency f_2 of 20 Hz.

   (a) Find the BW when the NE538 is used in a closed loop circuit with a gain of 100.
   (b) Find the gain-bandwidth product (unity gain BW).

11-3. The gain-BW product of the μA741 is 1 MHz.

   (a) Find the BW for $A_{CL} = 20$.
   (b) Find the BW for $A_{CL} = 100$.
   (c) What is the open loop BW if the open loop gain is 20,000?

11-4. The Burr Brown Inc. OA Model 3542J has an open loop gain of 88 dB and a gain-BW product of 1 MHz. Find:

   (a) The open loop BW.
   (b) The gain available for a BW of 20 kHz.

11-5. Find v_o for the circuit shown in Fig. 11-3, with $v_1 = 0.1$ V dc, $R_1 = 2.2$ kΩ, $R_f = 12$ kΩ.

11-6. (a) Find A_{CL} and v_o (in complex form) for the circuit of Fig. 11-4.

   $v_1 = 0.5 \sin 6280t$, $Z_1 = 100$ Ω, $Z_f = 0.1$ H.

   (b) What is the phase shift involved?

11-7. Find v_o for the circuit shown in Fig. 11-37. $v_1 = 0.1$ V rms at 10 kHz.

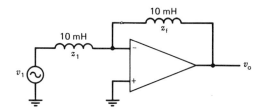

FIGURE 11-37

11-8. In the circuit of Fig. 11-37 make the following changes:

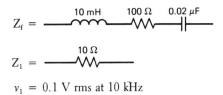

$v_1 = 0.1$ V rms at 10 kHz

(a) Find v_o.
(b) Find the phase shift from input to output.

11-9. In Problem 10-8 find the resonant frequency of Z_f and A_{CL} when operating at that frequency.

11-10. Find A_{CL} and v_o for Fig. 11-15 with $R_1 = 2.2$ kΩ, $R_f = 22$ kΩ and $v_1 = 0.2$ V.

11-11. Find v_o for the circuit shown in Fig. 11-38.

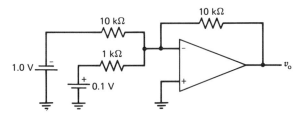

FIGURE 11-38

11-12. Find v_o in Fig. 11-39.

11-13. Find v_o in Fig. 11-40.

11-14. Plot v_o for the first 11 seconds, for the circuit of Fig. 11-41, with v_1 as given.

11-15. For Fig. 11-13d, plot v_o for the first 50 ms, with a step input of 4 V, $R_1 = 100$ kΩ, $C = 0.1$ μF. (Note that cut off occurs at $v_o = \pm 15$ V since $V_{CC}^+ = 15$ V and $V_{CC}^- = -15$ V.) Neglect R_f.

496 OPERATIONAL AMPLIFIERS

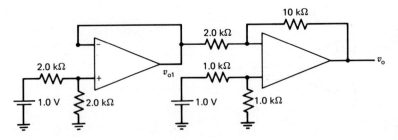

FIGURE 11-39

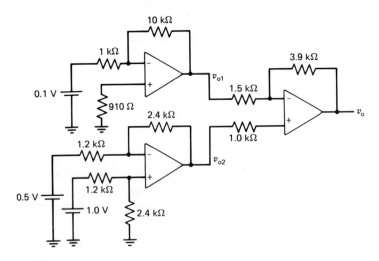

FIGURE 11-40

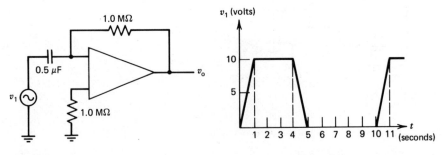

FIGURE 11-41

11-16. In the circuit of Fig. 11-18a $L = 0.1$ H, $R_f = 10$ kΩ. Find A_{CL} when operating at the following frequencies:

    (a) 100 Hz.
    (b) 1 kHz.

(c) 10 kHz.
(d) 100 kHz.

11-17. In Problem 11-16 v_1 is a 5-volt step. Plot v_o for at least 50 μs. (Note that $V_{CC} = \pm 15$ V.)

11-18. In Fig. 11-20, $v_{in} = 0.1$ V, $R_1 = 10$ kΩ, $R_2 = 1$ kΩ, $R = 2.2$ kΩ. Find v_o.

11-19. The gain of an inverting OA is designed so that $R_f/R_1 = 1000$. The amplifier has an open loop gain, $A = 20,000$.

(a) Find the *exact* gain.
(b) Find the error in gain when using $|A_{CL}| = R_f/R_1$.

11-20. Find the offset output voltage in Fig. 11-42 due to the bias current of 10 μA. Suggest a way to reduce this offset.

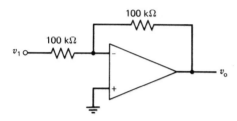

FIGURE 11-42

11-21. The offset bias current is 200 nA. Find the worst case offset voltage at the output in the circuit of Fig. 11-43.

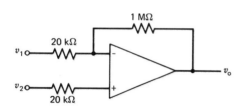

FIGURE 11-43

11-22. The Burr Brown OA #3508J has the following specifications:

Gain-BW product	100 MHz (at $A_{CL} = 100$)
Full power BW	600 kHz (at $v_o = 24$ V p-p, $i_o = 36$ mA p-p)
Slew rate	35 V/μs
Rise time	17 ns

For which of the following is this amplifier unsuitable? Explain.

(a) A gain of 100 for a 2 MHz signal.

498 OPERATIONAL AMPLIFIERS

(b) Produce $v_o = 24$ V p-p, $i_o = 36$ mA p-p for a 1 MHz signal.
(c) Yield an output of 0.1 V p-p and 1 mA p-p for a signal frequency of 2 MHz.
(d) A control system in which the input signal to the amplifier changes at a rate of 10^6 V/sec.
(e) Amplify a pulse with $t_r = 22$ ns.

11-23. Find the maximum slew rate associated with the waveform given by $v = 10 \sin 10{,}000t$. (Note, amplitude = 10 V, $\omega = 10{,}000$.)

11-24. The maximum slew rate of an amplifier is 0.5 V/ms. The full load maximum output swing is 10 V p-p (sinusoid). What is the BW for full output?

11-25. Show that the current in the ammeter in Fig. 11-44 is given by $v_1/2$ kΩ and is independent of the resistance of the meter, R_m.

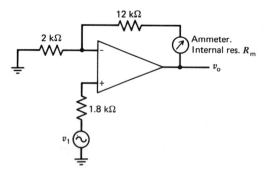

FIGURE 11-44

11-26. A resistive strain gauge (its resistance changes with pressure or weight) is used in an OA bridge configuration in a weighing system. The resistance of the gauge with no weight is 50 Ω. The resistance changes by 0.01 Ω for each pound of weight.

(a) Sketch the bridge circuit with all values shown and all pin numbers. Use the µA741.
(b) For your bridge, with an excitation voltage of 10 V, find v_o for a 10 lb. weight.
(c) If cut off occurs at $v_o = 5$ V, what is the maximum measurable weight with your system?

CHAPTER

12
THYRISTORS

12-1 CHAPTER OBJECTIVES

The purpose of this chapter is to give the student a working knowledge of the various devices classified as thyristors. Structure, equivalent circuits, and performance characteristics are discussed. The student will gain an understanding of the significance of the device specifications, as given by the manufacturer, and how to interpret the data sheets. Some specific applications are discussed, with particular emphasis on the relation between applications and device parameters.

12-2 INTRODUCTION

The term thyristor is used to denote a class of multilayer semiconductor devices that exhibit *positive feedback* (or regenerative feedback) and hence are useful largely in switching applications. (Positive or regenerative feedback means that a part or all of the output signal is returned to the input so as to strengthen the effect of the input signal.)

Typically, a short pulse is enough to turn the device on. The inherent positive feedback provides the holding current to keep the device on with the trigger pulse gone. The power dissipated in the device is very small relative to the power the device is controlling, resulting in a very efficient method of controlling power delivered to a load.

The class of thyristors includes a large number of devices that vary in their characteristics and applications. We present here a detailed discussion of the silicon controlled rectifier (SCR), the triac, and only a brief functional description of a number of other devices in the class.

500 THYRISTORS

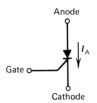

FIGURE 12-1 The SCR symbol.

12-3 THE SILICON CONTROLLED RECTIFIER (SCR)

The SCR is a three-terminal device, as shown in Fig. 12-1. It is functionally a diode (anode and cathode) that can be turned on by a signal applied to the gate terminal. It is very much like a switch between anode and cathode, which is closed by the voltage applied to the gate. To open the anode-cathode path, it is necessary to reduce the anode current to below a minimum holding current. (It cannot readily be turned off by gate signals.) Note that just a small current at the gate is capable of controlling very high anode (load) currents. This is probably the most important feature of the SCR.

Structurally, the SCR consists of four layers, PNPN (Fig. 12-2a). As shown in Figs. 12-2b and 12-2c these layers are a composite of two interconnected transistors, a PNP (Q_1) and an NPN (Q_2). The base of Q_1 is connected to the collector of Q_2, and similarly, the base of Q_2 and the collector of Q_1 are interconnected. Note that the gate in Fig. 12-2 is the base of Q_2, while the base of Q_1 is unused. As we see later, this single gate structure permits one only to turn the device on. Another device, the silicon controlled switch (SCS), makes use of both bases; it has two gate terminals and is capable of controlled turnon as well as turnoff.

12-3.1 SCR Operation and Characteristics

The operations of the SCR can best be explained in terms of the two-transistor equivalent structure (Fig. 12-2). This equivalent circuit is redrawn in Fig. 12-3 showing the various currents and voltages.

We first assume that $V_{GC} = 0$ and hence $I_G = 0$, no gate current is applied. (The student is cautioned not to confuse the gate of the SCR with that of the FET. The SCR is *current* operated, while the gate of the FET requires *no current*.) It is further assumed that the anode-cathode voltage, V_{AC}, is less than some specified breakover voltage V_{FOB} (see next section), with the polarity shown in Fig. 12-3. Under these conditions both transistors are off. As shown $I_{B2} = 0$, making $I_{C2} = 0$, and hence, $I_{B1} = 0$ and $I_{C1} = 0$. By inspection we see that the total device current I_A is the same as $I_{C1} + I_{C2}$.

$$I_A = I_{C1} + I_{C2} \tag{12-1}$$

THE SILICON CONTROLLED RECTIFIER (SCR)

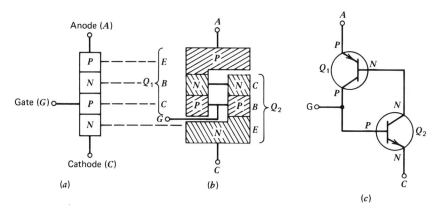

FIGURE 12-2 SCR structure. (a) Basic structure. (b) Structure showing two transistors. (c) Transistor equivalent.

Since both $I_{C1} = 0$ and $I_{C2} = 0$, clearly $I_A = 0$. Under these conditions only leakage currents flow in the device. It is worth noting that since most SCRs are high current devices, 1 A and higher, the leakage current ranges from about 100 μA and up. Typically for a 2 A SCR a 100 μA maximum leakage current is specified.

To turn the device on it is necessary to provide a base current to Q_2, I_{B2}, to produce some collector current I_{C2}. With $I_{C1} = 0$, I_{B2} is the gate current, $I_{B2} = I_G$. (By KCL, $I_{B2} = I_{C1} + I_G$.) Q_2 can be turned "on" with an external trigger signal applied to the gate, V_{GC}, producing the current I_{B2} (Fig. 12-3). Turning on Q_2 results in a substantial I_{C2} that in turn means that I_{B1} (since $I_{C2} = I_{B1}$) is substantial enough to turn Q_1 on. Once Q_1 is conducting, the current $I_{C1} = I_{B2}$ is sufficient to keep Q_2 "on" even if V_{GC} is removed, that is, $I_G = 0$. The feedback

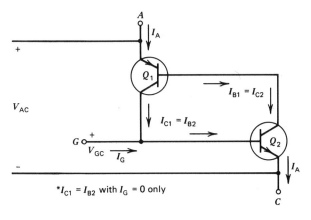

FIGURE 12-3 Two-transistor equivalent.

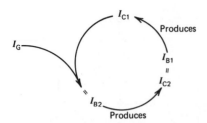

FIGURE 12-4 SCR turnon representation.

in the device keeps the device *latched*[1] in the "on" state after the trigger signal is removed. Consequently, only a short duration positive trigger pulse is necessary for turnon (the exact minimum pulse duration depends on the response time of the transistors and is usually specified by the manufacturer). The turnon operation is shown symbolically in Fig. 12-4. If we start with I_G and move counterclockwise, the initial I_G equals I_{B2} (with the absence of any I_{C1}), which produces an I_{C2}. $I_{C2} = I_{B1}$ in turn produces I_{C1} that replaces I_G as the drive (turnon) signal. Clearly, if a negative signal is applied to the gate, initially, it will serve to cut off Q_2 or rather keep it off, and *not* turn the device on.

If the device latches in the on state, how then is it turned off? The only turnoff mechanism is to reduce V_{AC} to zero (or below zero; that is, reverse bias the device). To be more precise, to turn the device off, it is necessary to reduce the current through the device I_A to below a specified "holding" current I_{HO}. This can be done by reducing V_{AC} to about zero, or by simply opening the circuit. Once turned off, V_{AC} can be reapplied. The device will stay off as long as V_{GC} is below the trigger level ($I_G \approx 0$) and V_{AC} is kept below the breakover voltage V_{FOB}.

The device can be turned on by exceeding V_{FOB} even if $I_G = 0$. This turnon mode, which is *not* the common operating mode, results from the increase in leakage in I_{C1} ($I_{C1} = I_{B2}$) sufficient to produce conduction in Q_2 and allow the feedback to turn the device on (Q_2 and Q_1).

In the reverse bias condition, the device follows essentially the characteristics of a reverse-biased diode, including the zener (or avalanche) breakdown. Figure 12-5a is a V-I plot of a typical SCR. For clarity, it is drawn for $I_G = 0$ and does not show gate-operated turnon. Figures 12-5b shows the test circuit used. The parameter I_{FO} in the forward-biased quadrant is the forward leakage current with the gate open, $I_G = 0$. Note that even though it is a forward current, it is a leakage current, since the device is "off" in this range. As V_{AC} is increased to $V_{AC} = V_{FOB}$, the device turns on, and I_A is increased dramatically causing V_{AC} to drop. (Most of the supply voltage is now dropped as $I_A R_L$.) Further increases in V_{AC} follow the typical forward characteristics of a diode, with I_F and V_F the forward current and voltage. A drop in V_{AC} below that necessary to maintain the holding current I_{HO} will turn the device off.

---

[1] The term "latch" means that the device turns on and stays on regardless of the input signal.

THE SILICON CONTROLLED RECTIFIER (SCR)

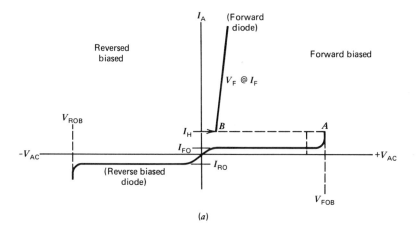

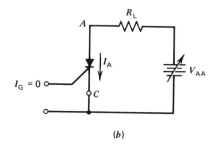

FIGURE 12-5 SCR characteristics. (a) The V-I Plot, ($I_G = 0$). (b) Test circuit.

In the reverse-biased state, that is anode negative with respect to cathode (reversing the polarity of V_{AA} in Fig. 12-5b), I_{RO} denotes the reverse leakage current and V_{ROB} the reverse voltage breakdown. This portion of the characteristics is identical in form to that of a diode. V_{ROB} is essentially a zener or avalanche breakdown, not necessarily destructive.

The characteristics shown in Fig. 12-5a do not show the effect of gate current, that is, the effect of input trigger voltage. Here, the SCR is turned on (the transition from point A to point B) only by the increase in V_{AC} while $I_G = 0$. This is not the normal operating mode. Figure 12-6 shows the effect of gate current. With a given I_{G1} (nonzero I_G) the "turnon" occurs at a $V_{AC} = V_{AC1}$ lower than the breakover voltage V_{FOB}. Increasing I_G, $I_{G2} > I_{G1}$, further reduces the anode-cathode voltage required for turnon. With $I_G = I_{G4}$ the device behaves very much like a diode with a forward bias voltage of about 1.3 to 1.8 V. It should be noted that the holding current varies somewhat with I_G (after all I_G is part of the total current I_A). This relation has been neglected in Fig. 12-5 since it is usually insignificant.

The interpretation of Fig. 12-6 takes on a somewhat different significance for ac and dc applications. In dc applications, in which V_{AA}, the supply voltage, is dc, the trigger current must be made large enough to turn the device on with the given

504 THYRISTORS

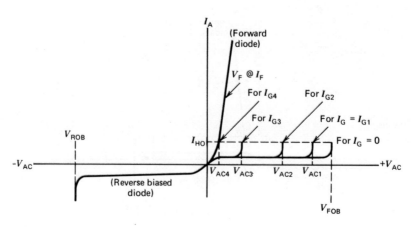

With $I_G =$	0	I_{G1}	I_{G2}	I_{G3}	I_{G4}
Turnon occurs at $V_{AC} =$	V_{FOB}	V_{AC1}	V_{AC2}	V_{AC3}	$V_{AC4}(\approx 1.3$ to 1.8 V)

V_{FOB}—Forward breakover, $I_G = 0$, the device turns on.
V_{ROB}—Reverse breakover—zener breakdown in nature.
I_{Ho} —Holding current device does not stay on if current drops below I_{Ho}.
I_{Fo} —Forward blocking current forward leakage.
I_{Ro} —Reverse leakage.

FIGURE 12-6 Effects of I_G on turnon.

V_{AA} (as long as the device is off $V_{AA} \simeq V_{AC}$). For a $V_{AC} = V_{AC1}$ the trigger current must be larger than I_{G1} (Fig. 12-6). A trigger current less than that may not turn the device on. Similarly for $V_{AC} = V_{AC2}$, I_G must be larger than I_{G2} to trigger the device.

For ac applications, where V_{AC} varies with time, as a result of V_{AA} being an ac voltage, a given I_G such as I_{G1} will trigger the device as V_{AC} reaches V_{AC1}. Figure 12-7 shows this dependence between I_G and the trigger point for a sinusoidal supply voltage. The larger I_G (up to I_{G4}), the larger the conduction angle, that part of the cycle during which the device is on. The device turns off at $V_{AC} \simeq 0$, since I_A drops below the holding current I_{HO}. Note that in ac applications, turnoff is a direct result of the variation in the supply voltage and no special turnoff provisions are necessary. Clearly, the larger the conduction angle, the more power is delivered to the load. (The load is in series with the SCR; see Sec. 12-7.) As Fig. 12-7 shows,

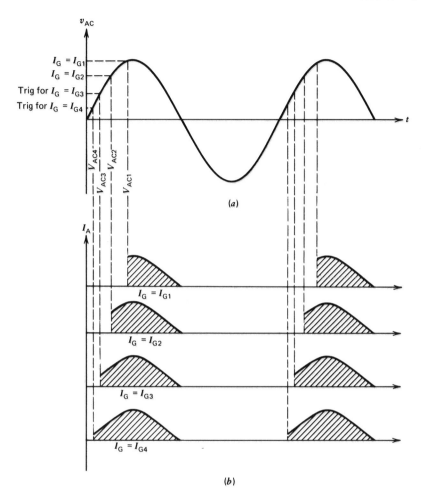

FIGURE 12-7 Trigger points for ac applications (see Fig. 12-6 for notations). (a) Trigger points. (b) Conduction angles (shaded portion).

the device may be on for, at most, half the cycle. This results in reduced power delivered to the load. Special circuits using two SCRs as well as a bidirectional SCR, also called a Triac, have been developed to allow control during both halves of the ac cycle.

12-4 THE TRIAC

The triac structure, as the symbol in Fig. 12-8a implies, consists of two parallel complementary SCRs with a common gate: A composite, NPNP and PNPN structure. The NPNP part conducts and controls the negative half of the cycle while the PNPN does the same for the positive half. (Recall the PNP-NPN complementary

506 THYRISTORS

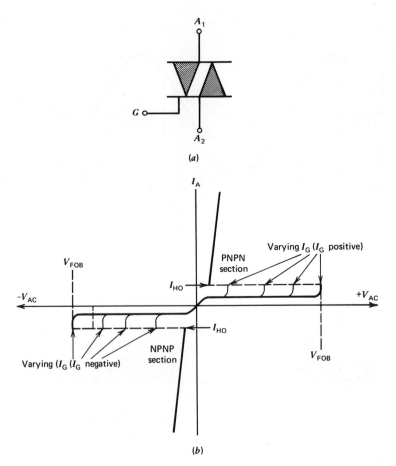

FIGURE 12-8 Triac symbol and characteristics. (a) Triac symbol. (b) V-I plot.

symmetry circuits of Ch. 8.) The device characteristics are shown in Fig. 12-8b. Since the device operates in both directions, we refer to the terminals as anode 1 and anode 2, A_1, A_2, rather than anode and cathode. The operation in either polarity is identical to that of the SCR and needs no further elaboration.

12-5 SCR AND TRIAC SPECIFICATIONS

Figure 12-9 gives the maximum ratings and the electrical characteristics of the 2N2573 to 2N2579SCRs. The notation is that used by Motorola, Inc.

The maximum ratings give parameter values that may not be exceeded in operation. Most of the parameters listed are self-explanatory; nevertheless, a brief explanation with an application-oriented interpretation will be given.

V_{ROM} Referring to Fig. 12-5a, this is the largest reverse voltage, V_{AC}, that is guaranteed not to cause reverse breakover (zener or avalanche). Clearly,

2N2573 thru 2N2579

$I_f = 25$ A RMS
$V_{ROM(rep)} = 25\text{-}500$ V

Industrial-type, silicon controlled rectifiers in a "diamond" package for applications requiring a high surge-current rating or low thermal resistance.

CASE 61 (TO-41) **CASE 54** (TO-3)

For units with pins (TO-3) specify devices MCR649AP-1(2N2573) thru MCR649AP-7(2N2579).

MAXIMUM RATINGS ($T_J = 125°C$ unless otherwise noted)

Rating		Symbol	Value	Unit
Peak Reverse Blocking Voltage*	2N2573	$V_{ROM(rep)}$*	25	Volts
	2N2574		50	
	2N2575		100	
	2N2576		200	
	2N2577		300	
	2N2578		400	
	2N2579		500	
Forward Current RMS (all conduction angles)		I_f	25	Amp
Circuit Fusing Considerations ($T_J = -65°$ to $+125°C$, $t \leq 8.3$ ms)		I^2t	275	A^2s
Peak Surge Current (One Cycle, 60 Hz, $T_J = -65$ to $+125°C$)		$I_{FM(surge)}$	260	Amp
Peak Gate Power - Forward		P_{GFM}	5	Watts
Average Gate Power - Forward		$P_{GF(AV)}$	0.5	Watt
Peak Gate Current - Forward		I_{GFM}	2	Amp
Peak Gate Voltage - Forward		V_{GFM}	10	Volts
Reverse		V_{GRM}	5	
Operating Junction Temperature Range		T_J	-65 to +125	°C
Storage Temperature Range		T_{stg}	-65 to +150	°C

*V_{ROM} for all types can be applied on a continuous dc basis without incurring damage.

V_{ROM} ratings apply for zero or negative gate voltage.

FIGURE 12-9 SCR data sheet (courtesy Motorola Inc.)

V_{ROM} is less than the actual breakover voltage V_{ROB} to allow for a safety margin and for variation among devices (of the same type). Exceeding specified V_{ROM} may cause breakover, heavy conduction and if power limitations are exceeded, permanent damage. The "O" in the subscript refers to open gate conditions for this parameter.

2N2573 thru 2N2579 (continued)

ELECTRICAL CHARACTERISTICS ($T_C = 25°C$ unless otherwise noted)

Characteristic	Symbol	Min	Typ	Max	Units
Peak Forward Blocking Voltage* ($T_J = 125°C$)	V_{FOM}*				Volts
2N2573		25	—	—	
2N2574		50	—	—	
2N2575		100	—	—	
2N2576		200	—	—	
2N2577		300	—	—	
2N2578		400	—	—	
2N2579		500	—	—	
Peak Forward Blocking Current (Rated V_{FOM} with gate open, $T_J = 125°C$)	I_{FOM}	—	0.6	5.0	mA
Peak Reverse Blocking Current (Rated V_{ROM}, $T_J = 125°C$)	I_{ROM}	—	0.6	5.0	mA
Gate Trigger Current (Continuous dc) (Anode Voltage = 7 Vdc, $R_L = 100\ \Omega$)	I_{GT}	—	20	40	mA
Gate Trigger Voltage (Continuous dc) (Anode Voltage = 7 Vdc, $R_L = 100\ \Omega$)	V_{GT}	—	1.0	3.5	Volts
(Anode Voltage = Rated V_{FOM}, $R_L = 100\ \Omega$, $T_J = 125°C$)	V_{GNT}	0.3	—	3.5	
Forward On Voltage ($I_F = 20$ Adc)	V_F	—	1.1	1.4	Volts
Holding Current (Anode Voltage = 7 Vdc, Gate Open)	I_{HO}	—	20	—	mA
Turn-On Time ($t_d + t_r$) ($I_G = 50$ mA, $I_F = 10$A)	t_{on}	—	1.0	—	µs
Turn-Off Time ($I_F = 10$ A, $I_R = 10$ A, $dv/dt = 20$ V/µs, $T_J = 125°C$) (V_{FXM} = rated voltage) (V_{RXM} = rated voltage)	t_{off}	—	30	—	µs
Forward Voltage Application Rate (Gate Open, $T_J = 125°C$)	dv/dt	—	30	—	V/µs
Thermal Resistance (Junction to Case)	θ_{JC}	—	1.0	1.5	°C/W

*V_{FOM} for all types can be applied on a continuous dc basis without incurring damage.

V_{FOM} ratings apply for zero or negative gate voltage.

FIGURE 12-9 *(Continued)*

I_f The continuous forward current that may not be exceeded. I_A must be less than this specified value for dc or repetitive sinewave operation. $I_A < I_F$.

I_t^2 This is a measure of the energy that may be handled by the device, under nonrepetitive conditions. (Pulse period < 8.3 ms, which is the 60 Hz, half-cycle duration.) For example with a single 2 ms pulse applied

SCR AND TRIAC SPECIFICATIONS   509

to the anode, turning the device on, and with the given I^2t of 275 A²s (Ampere square times seconds), we find that since

$$275 = I^2t = I^2 \times 2 \times 10^{-3}$$

$$I = \sqrt{\frac{275}{2 \times 10^{-3}}} = 370 \text{ A}$$

The device can carry a maximum of 370 A for this short duration. The I^2t limit must be considered when operating in the pulse mode.

$I_{\text{FM(surge)}}$ The peak anode current (I_A) during each conducting half cycle of a 60 Hz sinewave. This is a peak value, not rms or average. With sinusoidal voltage the actual peak current through the device, I_{AP}, can be calculated by

$$I_{AP} = \frac{E_m}{R_L} \tag{12-2}$$

Where E_m is the peak supply voltage and R_L is the driven load resistance. Hence, we must satisfy

$$\frac{E_m}{R_L} < I_{\text{FM(surge)}} \tag{12-3}$$

P_{GFM} The maximum instantaneous power that may be dissipated in the gate, while turning the device on. The peak instantaneous positive gate voltage times the resultant gate current may not exceed this power rating.

$P_{\text{GF(ave)}}$ The average power that may be dissipated by the gate, dc or average ac power.

I_{GFM} Maximum instantaneous forward gate to cathode current.

V_{GFM} Maximum allowable gate to cathode voltage in the forward mode (that is, during turnon).

V_{GRM} Maximum allowable gate to cathode voltage in the reverse mode. This voltage is applied in a polarity opposite to that required for turnon.

In contrast with maximum ratings, the electrical characteristics do not set limits of operation, but rather they give performance characteristics. For example, the holding current I_{HO} in Fig. 12-9 is given as 20 mA typical. This means that for the most part (*not* guaranteed for all devices) the device will turn off if the anode current becomes less than 20 mA.

In many instances the manufacturer will give values that are guaranteed, that is, worst case values. The values for V_{FOM} in the data sheet are given under the minimum column. (V_{FOM} is the anode-cathode voltage below which the device stays blocked, that is, off, with $I_G = 0$, no gate trigger.) That means that under no condition will the device unblock below the given value. If V_{FOM} is specified at 25 V (for the 2N2573), then it is guaranteed *not* to turn on with anode-cathode voltages below 25 V (with $I_G = 0$).

In the following, we give the basic definitions and explanations of the various functional parameters. Some have been discussed previously.

V_{FOM} The anode-cathode voltage below which the device will *not* turn on (with $I_G = 0$). Clearly, $V_{FOM} < V_{FOB}$, where V_{FOB} is the actual breakover voltage.

I_{FOM} The forward "leakage" current under the conditions of the device being "off" with a forward V_{AC} voltage. As shown in Fig. 12-9, the typical value is 0.6 mA while the worst case, maximum value is 5.0 mA. At worst, this leakage will never exceed 5.0 mA. The parameter is given for rated V_{FOM}; that is, $V_{AC} = V_{FOM}$.

I_{ROM} Leakage current with a reverse V_{AC} voltage (similar to I_{FOM} with a reversed polarity V_{AC}).

I_{GT} This gives the gate current required to turn the device on. (Under the given conditions $V_{AC} = 7$ V dc, $R_L = 100\ \Omega$.) Again, *typically* a gate current of 20 mA will turn the device on, but *never* will it be necessary to have a gate current larger than 40 mA (again for the conditions shown).

V_{GT}, V_{GMT} Gate voltage required to produce turnon. V_{GT} and V_{GMT} differ only in the operating conditions. $V_G = 3.5$ V (gate voltage of 3.5 V) will always guarantee turnon. When designing the circuit, we use this worst case value rather than the typical values.

V_F Anode-cathode voltage when the device is on will *never* exceed 1.4 V (under the given conditions of $I_F = 20$ A dc).

I_{HO} The holding current. Reducing I_A below this value will turn the device off.

t_{ON} This is the turnon time. The time it takes from the application of the trigger pulse to the time the load current reaches 90% of its final "on" value. In words, it gives the delay from trigger pulse to turnon.

t_{OFF} Similar to t_{ON}, except that it refers to turnoff. Here, I_A must be brought to below I_{HO} and then within 30 μs. Typically, the device will be off (this implies that I_A must be kept below I_{HO} for 30 μs typically, to allow full turnoff).

dv/dt   The SCR (or Triac) may be turned on by anode voltages that change too rapidly. This parameter gives the value of the rate of change of V_{AC} that the device can typically operate with. Recall that the sinusoidal voltage has a rate of change of $E_m \times 2\pi f$ at the zero crossing point. Thus, in sinusoidal applications the value $E_m \times 2\pi f$ may not exceed the given *dv/dt*.

θ_{jc} See Chapter 8.

It should be noted that the symbols shown above are those used by the Motorola Inc. Other manufacturers may use different symbols to describe essentially the same parameters.

12-6 SCR AND TRIAC CIRCUIT

The SCR and Triac are used mostly in AC power control. This is due to the fact that these devices can be turned off only by reducing the anode current to near zero. In ac applications, the supply voltage itself crosses zero, reducing anode current to well below I_{HO} and turning the device off.

12-6.1 Direct (dc) Control

To preserve simplicity and clarity, a simple, rarely used, dc circuit will be analyzed first. The circuit is shown in Fig. 12-10. The trigger signal V_{trig} is a single pulse used to turn the SCR "on," connecting the 20 V battery to the load R_L. The SCR can be turned off by opening switch S_1, returning I_A to zero, well below the I_{HO}. (I_{HO}, from the data sheet of Fig. 12-9 is 20 mA typically.)

To turn the SCR on, i_G (ac, lowercase symbols are used since the current is really an ac quantity, it is present for a short duration only. Its value is essentially the amplitude of i_G. must equal or exceed the trigger current, the gate current that will guarantee turnon. In Fig. 12-9 this is given as I_{GT} = 20 mA typically, and I_G = 40 mA maximum. The maximum value here guarantees that *all* 2N2573 SCRs will be turned on if this maximum gate current is present (not just the "typical" 2N2573). In the circuit of Fig. 12-10, V_{trig} must be large enough to produce the 40 mA I_G. Again, this is given in Fig. 12-9 as V_{GT} = 3.5 V.

12-6.2 ac Control

Most SCR and Triac circuits involve ac (or pulsed dc) applications. Figure 12-11a is a simplified SCR circuit used to control ac current to the load R_L.

For the purposes of explanation, it is assumed that the SCR is turned on by a V_{GC} = 3.0 V (gate to cathode voltage). Figure 12-11b shows the waveforms for three different R_p settings. The turnon occurs always at V_{GC} = 3.0 V. Note that $V_{GC} = V_{GN}$ as long as the SCR is "off," since there is no voltage drop across R_L. (Both sides of R_L are at the same potential.) The conduction angles shown in Fig. 12-11b increase with increases in V_{GN}; this means that the total power supplied to the load that depends on the conduction angle can be varied by varying the potentiometer setting. It is important to realize that in this circuit, power is delivered to the load for a maximum of approximately 180°. There is no load current during

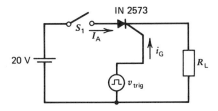

FIGURE 12-10 Simple dc SCR control circuit.

512 THYRISTORS

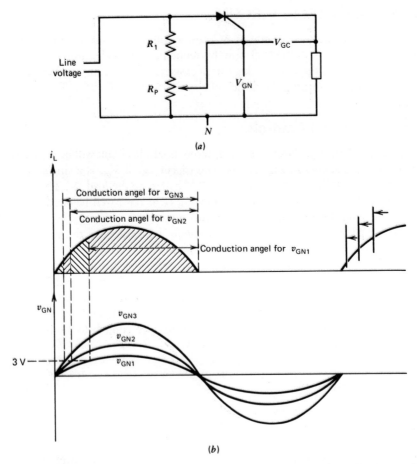

FIGURE 12-11 ac control. (a) The circuit. (b) Waveforms and conduction angles.

the negative portion of the line voltage. A full 360° control range (or full wave control) can be attained by using two SCRs in an inverted parallel connection (Fig. 12-12a), or by using a Triac circuit that is functionally equivalent to the two-SCR connection (Fig. 12-12b). Figure 12-12c shows the waveforms for a symmetrical setting of R_{P1} and R_{P2}, $V_{G1} = V_{G2}$ in Fig. 12-12a.

The Triac circuit (Fig. 12-12c) operates in a similar fashion. The diodes D_1 and D_2 provide the gate current of the appropriate polarity. For the positive half cycle, D_1 conducts, triggering the Triac to allow positive current through the load. D_2 conducts during the negative half cycle, turning the Triac "on" in the opposite direction. The setting of R_P determines the precise gate voltage that will produce sufficient trigger current, thus establishing the trigger points (triggers 1 and 2 in Fig. 12-12c).

SCR AND TRIAC CIRCUIT

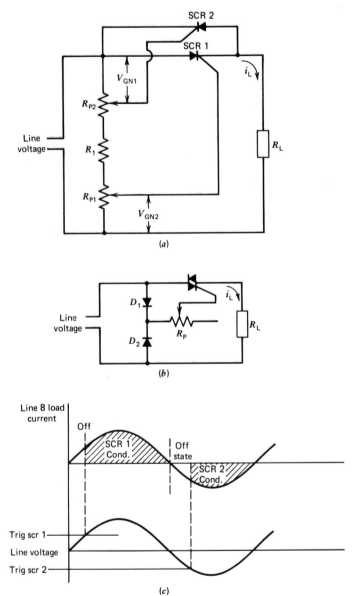

FIGURE 12-12 360° full wave control. (a) Using two SCRs. (b) Using a triac. (c) Waveforms for one setting R_{P_1} and R_{P_2} (in Fig. 12-12a).

12-6.3 Phase Control

The turnon point and, consequently, the conduction angle of the SCR and Triac may be controlled by varying the phase between the gate voltage V_{GC} and the anode to cathode voltage V_{AC}. In practical circuits both the amplitude and phase of V_{GC} are varied.

A simple circuit that uses this mode of operation is shown in Fig. 12-13a. The amplitude of v_C varies as R is varied. This results in shifts in the "turnon" point much like those shown in Fig. 12-10. In addition, as R is varied, the phase between v_C and v_{Line} is varied. The effect of this phase shift alone is shown in

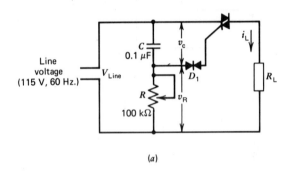

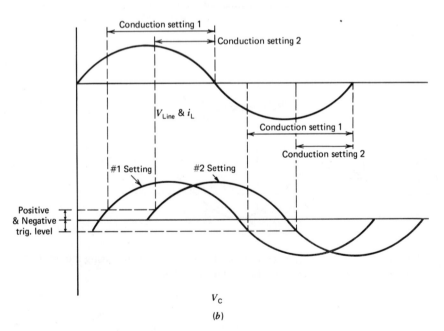

FIGURE 12-13 Phase shift control. (a) The circuit. (b) Waveforms for two settings of R (v_C shifted in phase from v_{Line} by different amounts).

Fig. 12-13b. As the phase shift between v_C and v_{Line} increases, the conduction angle decreases, supplying less power to the load. The diode D_1 is a 3-layer device operating in a back-to-back zener mode. To turn D_1 on, a substantial voltage is required, the specific value depending, of course, on the particular device selected. This voltage plus the Triac trigger voltage (about 3 V) is the total voltage v_C that will turn the Triac on.

All the circuits discussed in the preceding two sections are designed to control the power delivered to a load. The SCR or Triac is usually in series with the load controlling the *duration* of the current through the load, thereby controlling the average load power. One is tempted to replace these sophisticated circuits with a simple variable resistor in series with the load. The problem with this simple approach is that the power dissipated in the control resistor is very high. Moreover, the power from the line does not change substantially, even when the control is set to minimize the power to the load. The power used in the load may be small and the power taken from the line, and paid for, quite substantial since a good portion of it is lost in the series variable resistor. In contrast, the power lost in the SCR or Triac is minimal. The result is a very efficient power control method.

12-7 MISCELLANEOUS THYRISTORS

A brief description of the structure, characteristics, and symbols of a number of other thyristors will be presented here.

Four-Layer Diode (Shockley Diode)

The symbol of the four-layer diode is shown in Fig. 12-14. It is essentially a low current SCR without the gate terminal. The device is turned "on" when V_{AC} exceeds the forward breakover voltage, V_{FOB}. The V-I plot is that shown in Fig. 12-5. It is practically the plot for the SCR with $I_G = 0$. Once "on," the four-layer diode can be turned off by returning the anode current below I_{HO}, the holding current. A typical data sheet for the M4L-3052 is shown in Fig. 12-15. The various parameters are explained in the discussion of the SCR data sheet (Sec. 12-6).

FIGURE 12-14 Four-layer diode symbol.

M4L3052 thru M4L3054
NOW 1N5158 thru 1N5160

$V_{(BR)F} = 8\text{-}12V$
$V_{RM(rep)} = 10\text{-}12V$
$I_F = 150\ mA$
$P_D = 150\ mW$

CASE 51
(DO-7)

PNPN 4-layer diodes, two-terminal, fast switching devices specifically designed for low voltage applications such as logic circuits, pulse generators, memory and relay devices, relay replacement, alarm circuit, multivibrators, ring counters, and signal switching circuits. These devices feature low breakover (switching) voltage, fast switching speeds, low junction capacitance, low breakover currents, and sub-miniature package.

MAXIMUM RATINGS ($T_A = 25°C$ unless otherwise noted)

Rating	Symbol	Value	Unit
Peak Reverse Blocking Voltage M4L3052 M4L3053 M4L3054	$V_{RM(rep)}$	10 11 12	Volts
Continuous Forward Current	I_F	150	mA
Steady State Power Dissipation @ $T_A = 50°C$ Derate above $50°C$	P_D	150 1.5	mW mW/°C
Peak Pulse Current (50 μs maximum pulse width)	I_{pulse}	10	Amp
Operating Junction Temperature Range	T_J	-65 to +150	°C
Storage Temperature Range	T_{stg}	-65 to +175	°C

ELECTRICAL CHARACTERISTICS ($T_A = 25°C$ unless otherwise noted)

Characteristic	Symbol	Min	Typ	Max	Unit
Forward Breakover (Switching) Voltage M4L3052 M4L3053 M4L3054	$V_{(BR)F}$	8 9 10	— — —	10 11 12	Volts
Forward Breakover (Switching) Current	$I_{(BR)F}$	—	5	50	μA
Forward Blocking Current (Measured at 75% of $V_{(BR)F}$)	I_{FM}	—	1	5	μA
Reverse Blocking Current (Measured at rated $V_{RM(rep)}$)	I_{RM}	—	2	10	μA
Holding Current	I_{HO}	1	4	20	mA
Forward On Voltage ($I_F = 150$ mAdc)	V_F	—	1.0	1.5	Volts
Junction Capacitance (AC Voltage = 10 mV, $V_F = 0$, $f = 100$ kHz)	C_J	—	42	—	pF
Turn-On Time*	t_{on}	—	50*	—	ns
Turn-Off Time*	t_{off}	—	100*	—	ns

*Time depends on a wide variety of circuit conditions. Consult manufacturer for further information.

FIGURE 12-15 Four-layer diode data sheet (courtesy Motorola Co., Inc.).

Bilateral Four-Layer Diode

The four-layer diode, as its symbol indicates, can be turned "on" in one direction only (it is unidirectional). The connection of two four-layer diodes in an inverse parallel connection produces a device that can be turned "on" in either direction. The bilateral four-layer device is indeed an integrated pair of four-layer diodes. The symbol shown in Fig. 12-16 is a composition of two inverted four-layer diode symbols.

The SUS and SBS

The SUS, the silicon unilateral switch, and the SBS, the silicon bilateral switch, are corresponding four-layer diodes (unilateral and bilateral) with gate terminals added, permitting an operation much like a low-current SCR or Triac, respectively. The symbols are shown in Fig. 12-17.

Other Thyristors (See Symbols in Fig. 12-18.)

Other thyristors include the *light activated SCR (Lascr)* that is turned on by the combined effects of light and gate voltage and the *silicon controlled switch (SCS)* that is essentially an SCR with two gate terminals (both bases of the equivalent circuit, Fig. 12-3, are accessible). The device can be turned on by either gate. Another thyristor is the *Diac*, a Triac with no gate terminal. The V-I plot is that of a Triac with $I_G = 0$ (see Fig. 12-8). In addition, there is the *gate turnoff thyristor (GTO)* that is an SCR constructed to allow both turnon and turnoff via the gate. The turnoff gate current is negative and substantially larger than the required turnon current.

SUMMARY OF IMPORTANT TERMS

Conduction angle The portion of the input waveform, in degrees, for which conduction takes place in a particular circuit.

Halfwave control An SCR circuit in which conduction can take place for a maximum of 180°.

FIGURE 12-16 Bilateral four-layer diode.

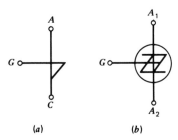

FIGURE 12-17 SUS and SBS.

518 THYRISTORS

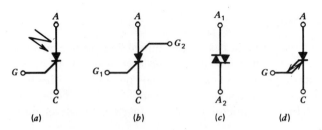

FIGURE 12-18 Miscellaneous thyristor symbols.
(a) Lascr (b) SCS.
(c) Diac (d) GTO

Holding current, I_{HO} The current I_A below which the SCR or Triac turns off.

Fullwave control An SCR or Triac circuit in which the maximum conduction angle is 360°. The exact conduction angle depends on circuit values.

Latching A circuit in which the action initiated by an input signal is maintained by the device itself. (Exists in SCR, Triac due to regenerative feedback.)

Regenerative feedback (positive) As applied to SCR and Triac, it refers to the fact that the input current initially present due to an input turnon signal is replaced by, or supplemented by, an internal device current, keeping the device in the ON state.

Trigger point The condition (V_{GG}) that turns the SCR or Triac on.

The following terms are listed and defined in Sec. 12-5:

dv/dt, I_F, I_{FM}, I_{FOM}, I_{GFM}, I_{GT}, I_{HO}, I_{ROM}, P_{GF}, t_{OFF}, t_{ON}, V_F, V_{FOM}, V_{GFM}, V_{GRM}, V_{GT}, V_{ROM}.

PROBLEMS

12-1. Describe briefly the latching operation of the SCR.

12-2. An SCR circuit is shown in Fig. 12-19. What should R_1 be set at in order to guarantee that the SCR is turned on? (Refer to data sheet Fig. 12-9 for specs.)
Hint: I_G must be 40 mA while V_{GT} = 3.5 V. Assume I_L = 0 before turnon.

12-3. Suggest two distinctly different ways the SCR in Fig. 12-19 can be turned off (circuit modifications using additional components).

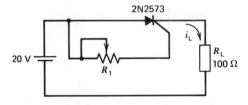

FIGURE 12-19

12-4. In Fig. 12-20, D_2 is a three-layer diode that requires 20 V across it to conduct (firing voltage), $R_1 = 1$ kΩ. $I_C = 30$ mA for turnon. Draw the waveform for i_L and approximate the conduction angle. (Neglect V_{GT}.)
Hint: The trigger level is approximately $20 + 1$ k$\Omega \times 30$ mA.

12-5. Draw the waveform for i_L in Fig. 12-20 with D_2 shorted.

12-6. In Fig. 12-20 the i_L waveform was found to be as shown in Fig. 12-21. What is the likely cause? (What has failed and how?)

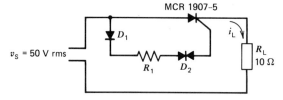

FIGURE 12-20

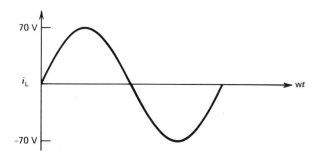

FIGURE 12-21

12-7. In Fig. 12-22 assume that the SCR trigger voltage (gate to cathode) is about zero (it is typically 0.7 V for the 2N4442). The firing voltage of D_1 is 40 V. Plot the waveform of i_L and v_L. (Notice the polarity of D_2.)

12-8. If I_{HO} for the 2N4442 in Fig. 12-22 is $I_{HO} = 40$ mA, what is the voltage across R_L at the moment the SCR turns "off"?

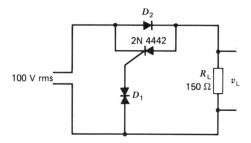

FIGURE 12-22

520 THYRISTORS

12-9. If the waveform of v_L for Fig. 12-22 is as shown in Fig. 12-23, give two possible failures that would give rise to this waveform.

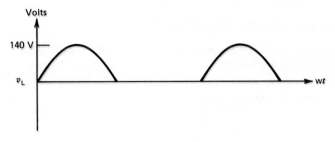

FIGURE 12-23

12-10. Explain the function of D_1 and D_2 in the circuit of Fig. 12-24.
Hint: Replace the Triac with two inverted parallel SCRs and consider the operation of each set of diodes and SCRs separately.

12-11. The circuit in Fig. 12-24 is not functioning properly. It is activating the line circuit breaker (too much current drawn). What is the probable cause (the load is *not* shorted)?

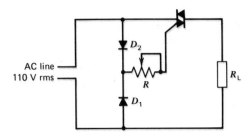

FIGURE 12-24

12-12. Specify the minimum V_{ROM} and I_f for the SCRs and Triac in the following circuits:
  (a) Figure 12-19.
  (b) Figure 12-20.
  (c) Figure 12-22.
  (d) Figure 12-24. ($R_L = 10\ \Omega$)

CHAPTER

13
MISCELLANEOUS SEMICONDUCTOR DEVICES

13-1 CHAPTER OBJECTIVES

It is the purpose of this chapter to acquaint the student with a number of semiconductor devices that have not been described in the text so far. The student will become familiar with standard symbols (if they exist) of the devices and with their function. The student is not expected to be able to analyze circuits using these devices, but rather gain an appreciation of the device operation and some of its more obvious applications. The devices that will be described include light sensitive devices, voltage sensitive capacitors, and a three-terminal device, the unijunction transistor, that like the SCR is a triggerable device; it is either "on" or "off." For the latter device the student will gain a somewhat more detailed analytical understanding.

13-2 INTRODUCTION

The list of semiconductor devices is very long and growing every year. It is impossible to discuss details of all or even most of these devices in a single text. In order that they be most useful, the brief discussions presented here will concentrate on basic functional descriptions and on some of the more significant characteristics of the devices.

13-3 THE UNIJUNCTION TRANSISTOR (UJT)

Some of the more common applications of the Unijunction Transistor (UJT) are in timing circuits, SCR trigger circuits, and signal generators (oscillators), specifically the "relaxation" oscillator that produces a sawtooth waveform.

Lest the name be misunderstood, the UJT is very *unlike* the BJT. The term *Unijunction* implies a single *p-n* junction compared to the Bipolar Junction Transistor (BJT) that contains two *p-n* junctions. A simplified UJT structure is shown in Fig. 13-1a. It consists of a bar of N material (B_2 to B_1) with a *p-n* junction between the emitter terminal and the bar of N material. *A note of caution.* The terms base and emitter are used here even though they bear no relation or resemblance to the same terms used in conjunction with the BJT. The symbol for the UJT is shown in Fig. 13-1b.

13-3.1 UJT Characteristics

The operation of the UJT can best be understood by referring to its equivalent circuit (Fig. 13-2). External voltage sources are included in Fig. 13-2 to facilitate the discussion. The two resistances r_{B2} and r_{B1}[1] represent the N bar. While r_{B2} does not vary as the operating conditions change, r_{B1} drops substantially when the UJT is turned "on." Turnon is accomplished by forward biasing the diode D. To get D to be forward biased, it is necessary to make V_E more positive than V_A (Fig. 13-2b), enough to provide for the forward drop of the diode. Thus, the turnon condition is

$$V_E > V_D + V_A \qquad (13\text{-}1)$$

$$V_E > V_P \qquad (13\text{-}1a)$$

where $V_P = V_D + V_A$. For turnon V_E must be larger than V_P. Since this condition is applied to the device in the "off" state, ($I_E = 0$), V_A can be found by

$$V_A = \frac{V_{B2B1} \times r_{B1}}{r_{B1} + r_{B2}} \qquad (13\text{-}2)$$

Note that only current through the "bar" is considered here since $I_E = 0$. This is simply a voltage division of the voltage V_{B2B1}, the voltage across $r_{B1} + r_{B2}$.

The ratio $r_{B1}/(r_{B1} + r_{B2})$, called the *intrinsic stand-off ratio*, η, is given by the manufacturer.

$$\eta = \frac{r_{B1}}{r_{B1} + r_{B2}} \quad \text{(intrinsic stand-off ratio)} \qquad (13\text{-}3)$$

---

[1] Many texts and data sheets use capital letters, R_{B1}, R_{B2}. Lowercase symbols are used here to be consistent with the rest of the text, in which equivalent representation used lowercase letters, such as r_e of the transistor.

THE UNIJUNCTION TRANSISTOR (UJT)   523

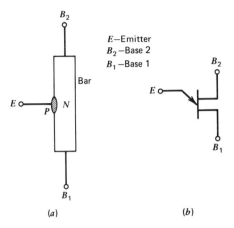

FIGURE 13-1   The UJT. (a) Simplified structure. (b) Symbol.

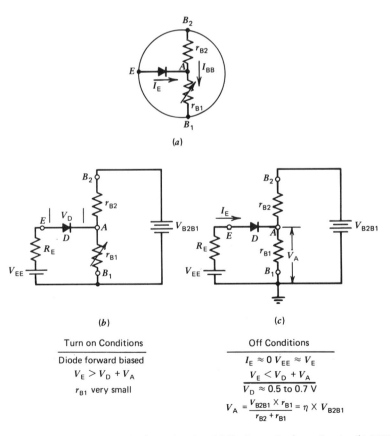

Turn on Conditions

Diode forward biased
$V_E > V_D + V_A$
r_{B1} very small

Off Conditions

$I_E \approx 0 \quad V_{EE} \approx V_E$
$V_E < V_D + V_A$
$V_D \approx 0.5 \text{ to } 0.7 \text{ V}$

$V_A = \dfrac{V_{B2B1} \times r_{B1}}{r_{B2} + r_{B1}} = \eta \times V_{B2B1}$

FIGURE 13-2   UJT equivalent circuits. (a) Basic equivalent circuit. (b) UJT on. (c) UJT off.

The voltage V_A then becomes

$$V_A = \eta \times V_{B2B1} \qquad (13\text{-}2a)$$

Manufacturers often define η in terms of V_P and the forward diode drop V_D. Motorola Inc., for example, in its data sheet defines η as

$$\eta = \frac{V_A}{V_{B2B1}} = \frac{V_P - V_D}{V_{B2B1}} \qquad (13\text{-}4)$$

The expression is directly obtainable from Eq. (13-2a) since $V_A = V_P - V_D$. To turn the UJT "on," either V_E must be increased or V_A decreased so that Eq. (13-1) is satisfied. V_A can be decreased by decreasing V_{B2B1}. It must be emphasized that V_E is the emitter to base 1 voltage and V_{B2B1} is the voltage across the N bar, base 2 to base 1. When the UJT is "on," I_E becomes substantial while when it is "off," $I_E \approx 0$. The plot of V_E versus I_E is shown in Fig. 13-3. As long as V_E is less than V_P (recall $V_P = V_D + V_A$), the device is "off" and $I_E \approx 0$. The only emitter current under these conditions is a leakage current, usually less than 1 μA. (In Fig. 13-3, the abscissa is drawn with two different scales to permit a reasonably expanded view of the cut-off region. In the cut-off region, the scale of I_E is in μA, while for the "on" region, the scale is in mA.)

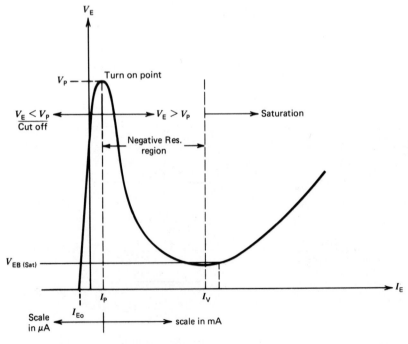

FIGURE 13-3 V_E versus I_E for UJT.

I_{EO} in Fig. 13-3 is the emitter leakage current for $V_E = 0$. As V_E is increased and reaches V_P (somewhat larger than V_P) the current I_E becomes substantial. The fact that I_E is not zero anymore causes the resistance r_{B1} (this is an equivalent resistance rather than an actual resistor) to decrease drastically. As a result, V_A drops suddenly, allowing a sharp increase in I_E. The increase in I_E causes V_E to drop ($V_E = V_{EE} - I_E R_E$). Note that I_E increases while V_E decreases. This behavior can be described as that of a *negative resistance*.

Recall that resistance = $\Delta V / \Delta I$. V and I must both change in the same direction, V and I both positive or both negative. In the case of the turnon of the UJT, the change in V_E is negative (decreasing), while the change in I_E is positive (increasing), making the ratio $\Delta V/\Delta I$ *negative*. The range from the turnon point $I_E = I_P$ to $I_E = I_V$ (Fig. 13-3) is thus called the *negative resistance* region. It is worth noting that the device stays on even though V_E decreases in this range. This is due to the fact that V_A also decreases, keeping $V_E > V_D + V_A$ as required for turnon. It should be noted that turnon is maintained as long as I_E is larger than I_P (Fig. 13-3). I_P, usually given by the manufacturer, is in the order of a few μA. It stands to reason that in order to turn the UJT off, it is necessary to reduce I_E below I_P (by lowering V_E or increasing V_A).

The turnon and turnoff conditions for the UJT can be summarized as follows:

TURNON

$$V_E > V_P$$
$$V_P = \eta \times V_{B2B1} + V_D$$

with I_E maintained above I_P.

η is usually between 0.55 and 0.85 and $V_D \approx 0.5$ V. V_{B2B1} is the base 2 to base 1 voltage, usually dependent on the supply voltage.

TURNOFF

$$I_E < I_P$$

I_E reduced below I_P.

EXAMPLE 13-1

Determine whether the UJT in Fig. 13-4 is "on" or "off," and find I_E and V_E. The specifications of the 2N4851 are given in Fig. 13-5. Use minimum values of η, R_{BB}, etc. Use $V_D \approx 0.5$ V.

SOLUTION

To determine whether the UJT is on or off, it is necessary to find V_P and compare it with V_E. Note that R_{B2} affects V_P since it changes the voltage division between r_{B1} and r_{B2}. The calculations here are based on the assumption that the UJT is

526 MISCELLANEOUS SEMICONDUCTOR DEVICES

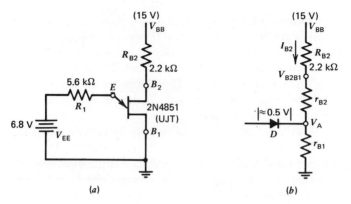

FIGURE 13-4 Circuits for Ex. 13-1. (a) The circuit. (b) Finding V_p.

"off." V_{B2B1} can be found by voltage division.

$$V_{B2B1} = \frac{V_{BB} \times (r_{B1} + r_{B2})}{R_{B2} + (r_{B1} + r_{B2})}$$

$r_{B1} + r_{B2} = r_{BB}$ and is given in the data sheet (as r_{BB}), $r_{BB} = 4.7$ kΩ (minimum value)

$$V_{B2B1} = \frac{15 \times 4.7 \text{ k}\Omega}{2.2 \text{ k}\Omega + 4.7 \text{ k}\Omega} = 10.2 \text{ V}$$

(Please note the distinction between the external resistor R_{B2} and r_{B2}.) V_P is obtained from $V_P = V_A + V_D = \eta \times V_{B2B1} + V_D = 0.56 \times 10.2 + 0.5 = 6.2$ V.

Since $V_E > V_P$ (for the assumed "off" UJT where $I_E = 0$ and $V_E = V_{EE}$), the UJT is "on." The calculation of I_E and V_E must be based on the fact that the UJT is indeed on.

When "on," r_{B1} is very small (well below 100 Ω) and for convenience will be assumed equal to zero. To find I_E, we apply KVL to the E–B$_1$ circuit.

$$V_{EE} - I_E \times R_E - V_D = 0$$

$$I_E = \frac{V_{EE} - V_D}{R_E} = \frac{6.8 - 0.5}{5.6} = \underline{1.1 \text{ mA}}$$

$V_E \approx V_D = \underline{0.5 \text{ V}}$ (approximately the forward diode voltage drop)

13-4 UJT APPLICATIONS

One of the more common applications of the UJT is the relaxation oscillator (Fig. 13-6a). This circuit is an oscillator (a signal generator) that produces a sawtooth waveform (Fig. 13-6b). The circuit operates as follows. While the UJT is "off," I_E

UJT APPLICATIONS 527

2N4870 (SILICON)
2N4871

$V_{BB} = 35$ V
$I_e = 50$ mA RMS

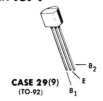

CASE 29(9)
(TO-92)

PN unijunction transistors designed for use in pulse and timing circuits, sensing circuits and thyristor trigger circuits.

MAXIMUM RATINGS ($T_A = 25°C$ unless otherwise noted)

Rating	Symbol	Value	Unit
RMS Power Dissipation*	P_D*	300	mW
RMS Emitter Current	I_e	50	mA
Peak-Pulse Emitter Current**	i_e**	1.5	Amp
Emitter Reverse Voltage	V_{B2E}	30	Volts
Interbase Voltage†	V_{B2B1}†	35	Volts
Operating Junction Temperature Range	T_J	-65 to +125	°C
Storage Temperature Range	T_{stg}	-65 to +150	°C

*Derate 3.0 mW/°C increase in ambient temperature.
**Duty cycle ≤ 1%, PRR = 10 PPS (see Figure 5).
†Based upon power dissipation at $T_A = 25°C$.

ELECTRICAL CHARACTERISTICS ($T_A = 25°C$ unless otherwise noted)

Characteristic	Figure No.	Symbol	Min	Typ	Max	Unit
Intrinsic Standoff Ratio* ($V_{B2B1} = 10$ V) 2N4851 2N4852, 2N4853	4, 8	η*	0.56 0.70	—	0.75 0.85	—
Interbase Resistance ($V_{B2B1} = 3.0$ V, $I_E = 0$)	11, 12	R_{BB}	4.7	—	9.1	k ohms
Interbase Resistance Temperature Coefficient ($V_{B2B1} = 3.0$ V, $I_E = 0$, $T_A = -65$ to $+125°C$)	12	αR_{BB}	0.2	—	0.8	%/°C
Emitter Saturation Voltage** ($V_{B2B1} = 10$ V, $I_E = 50$ mA)		$V_{EB1(sat)}$**	—	2.5	—	Volts
Modulated Interbase Current ($V_{B2B1} = 10$ V, $I_E = 50$ mA)		$I_{B2(mod)}$	—	15	—	mA
Emitter Reverse Current ($V_{B2E} = 30$ V, $I_{B1} = 0$) 2N4851, 2N4852 2N4853	7	I_{EB2O}	—	—	0.1 0.05	μA
Peak-Point Emitter Current ($V_{B2B1} = 25$ V) 2N4851, 2N4852 2N4853	9, 10	I_P	—	—	2.0 0.4	μA
Valley-Point Current** ($V_{B2B1} = 20$ V, $R_{B2} = 100$ ohms) 2N4851 2N4852 2N4853	13, 14	I_V**	2.0 4.0 6.0	—	—	mA
Base-One Peak Pulse Voltage 2N4851 2N4852 2N4853	3, 17	V_{OB1}	3.0 5.0 6.0	—	—	Volts
Maximum Frequency of Oscillation	5	$f_{(max)}$	1.0	1.25	—	MHz

* η, Intrinsic standoff ratio, is defined in terms of the peak-point voltage, V_P, by means of the equation: $V_P = η V_{B2B1} + V_F$, where V_F is about 0.49 volt at 25°C @ $I_F = 10$ μA and decreases with temperature at about 2.5 mV/°C. The test circuit is shown in Figure 4. Components R_1, C_1, and the UJT form a relaxation oscillator; the remaining circuitry serves as a peak-voltage detector. The forward drop of Diode D_1 compensates for V_F. To use, the "cal" button is pushed, and R_S is adjusted to make the current meter, M_1, read full scale. When the "cal" button is released, the value of η is read directly from the meter, if full scale on the meter reads 1.0.

** Use pulse techniques: PW = 300 μs, duty cycle ≤ 2.0% to avoid internal heating, which may result in erroneous readings.

FIGURE 13-5 Data sheet for 2N4851-53 UJT (courtesy Motorola, Inc.).

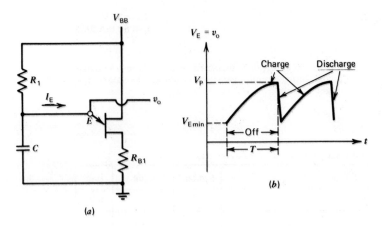

FIGURE 13-6 Relaxation oscillator. (a) Circuit. (b) V_E waveform.

≈ 0, the capacitor C is being charged toward V_{BB} through the resistor R_1. As soon as V_E reaches the turnon value, V_P, conduction takes place, and the capacitor is rapidly discharged by the large I_E current. The discharge path is through the drastically lowered r_{B1} internal to the UJT in series with R_{B1}. r_{B1} is usually very small (well below 100 Ω). Note that the charge resistance R_1 is usually much larger than the discharge path $r_{B1} + R_{B1}$ so that the discharge time is insignificant relative to the charge time. The period of oscillations is thus approximately, T, the charge time.

The capacitor is discharged up to the point where V_E is low enough to cause the device to turn off. That is, V_E is so low that I_E is reduced below the minimum I_E required to maintain the device in the "on" state. This value of V_E ($V_{E(min)}$) is approximately given by

$$V_{E(min)} \approx V_D + (I_V + I_{B2ON}) \times R_{B1} \qquad (13\text{-}5)$$

where V_D and I_V are the diode drop and the valley current, respectively. I_{B2ON} is often approximated by V_{BB}/r_{B2}.

The period of oscillations as noted is approximately T. This time depends largely on the values of R_1 and C, or more precisely, on the product $R_1 \times C$.[2] The frequency of oscillation can then be approximated by

$$f_{osc} \approx \frac{1}{R_1 C} \qquad (13\text{-}6)$$

This is a very crude approximation that does not account for the particular values

---

[2] The term $R_1 \times C$ is referred to as the "time constant" of the R-C circuit and gives the time it takes for the voltage across the capacitor to reach 63% of the maximum; here, it is 63% of V_{BB}.

of V_{BB} and R_{B1} as well as the particular parameters of the device, all of which affect the frequency. A somewhat better approximation is given by

$$f_{osc} = \frac{1}{KR_1C} \qquad (13\text{-}7)$$

where the value of K is related to the intrinsic stand-off ratio η as shown in Table 13-1.

TABLE 13-1   K VERSUS η

η	K
0.85	1.9
0.8	1.6
0.75	1.4
0.7	1.2
0.65	1.0
0.6	0.9
0.55	0.8
0.5	0.7

This approximation is partially based on the assumptions that $V_{BB} \gg V_{EBI(sat)}$ and $V_{BB} \gg I_{E(ON)} \times R_{B1}$.

The supply voltage is much larger than the emitter-base 1 saturation voltage of the UJT and also much larger than the voltage drop across R_{B1} when the device is "on." ($V_{EBI(sat)} \simeq V_D \simeq 0.5$ V and as long as R_{B1} is small, 100 Ω or less, the drop across R_{B1} is indeed very small.)

The UJT relaxation oscillator has a wide frequency range and is often used as a timing circuit or time delay for long delays (as high as minutes). Its simplicity and low cost make the UJT relaxation oscillator a very popular timing circuit (Fig. 13-7).

The circuit of Fig. 13-7a yields a frequency range of 5 Hz to 1 kHz (period of 0.2 s to 1 ms). This is based on $\eta \approx 0.65$ (typical for the 2N4851 at 30°C). Note that Eq. (13-7) becomes

$$f_{osc} = \frac{1}{R_1C} \qquad (13\text{-}7a)$$

for the value $\eta = 0.65$ for which $K = 1$ (Table 13-1). The highest frequency occurs for $R_1 = 10$ kΩ (the potentiometer set at zero) for which $f_{osc} = 1/(10 \times 10^3 \times 0.1 \times 10^{-6}) = 1$ kHz. The lowest frequency is obtained when R_1 is set to its maximum value 2 MΩ + 10 k$\Omega \approx 2$ MΩ so that $f_{osc} = 1/(2 \times 10^6 \times 0.1 \times 10^{-6}) = 5$ Hz.

It is interesting to note that the waveform at B_1, v_{B1}, consists of short voltage spikes coincident with the discharge time of the capacitor (Fig. 13-7b). These short spikes are often used as triggers for an SCR.

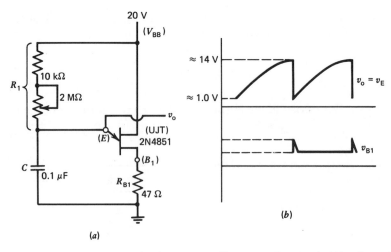

FIGURE 13-7 A practical relaxation oscillator. (a) Circuit. (b) Waveforms V_E, V_{B1}.

13-5 OTHER TYPES OF UJTs

There are available a number of devices, with characteristics similar to the UJT, that offer some specific advantages. For the most part, these devices are structurally very different from the UJT; however, functionally there are great similarities, giving rise to the word UJT in their name.

13-5.1 The Complementary UJT (CUJT)

This device, though it is essentially an integrated circuit (transistorized) device, has characteristics of a complement to the UJT. All its operating voltages and currents have a polarity opposite that of the UJT (Fig. 13-8).

Some of the advantages of the CUJT are its very low I_E leakage, well controlled η (the variations in η are small, compared to the UJT); low voltage operation and

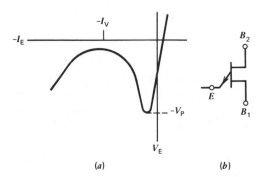

FIGURE 13-8 The complementary UJT (CUJT). (a) Characteristic plot. (b) Symbol.

OTHER TYPES OF UJTs

the fact that it can be used as a complement to the UJT allowing control of both positive and negative voltages.

13-5.2 The Programmable UJT (PUT)

Here again the device construction is not that of a UJT but rather more like the SCR, or a four-layer device. The main feature of the PUT is the fact that r_{BB} (r_{B1} and r_{B2}) can be selected by external components (there are no internal r_{B1} and r_{B2}). That means that both the η as well as the I_{B2B1} (off) current can be arbitrarily selected. This offers a larger range of applications with much tighter design parameters. Figure 13-9 shows the symbol of the PUT and the two external (programming) resistors R_1, R_2 that are essential to the operation of the device. Here,

$$\eta = \frac{R_1}{R_2 + R_1} \qquad (13\text{-}8)$$

The letters A, K, G stand for anode, cathode, and gate (reminiscent of vacuum tube notation). The anode functions much like the emitter of the UJT, while the cathode is the equivalent of B_1 of the UJT. The point marked (B_2) is indeed similar in its function to the base 2 (B_2) of the standard UJT. It is not however a terminal of the device itself.

One of the important differences between the PUT and UJT, besides the programmability of η, is the fact that I_{B1} for the PUT is very small until the device is turned on. For the UJT this current is dependent on the internal resistance r_{BB} (besides external components) that is in the order of 5 kΩ. I_{B1} for the PUT consists largely of leakage currents.

The trigger point (V_P) can be calculated for the PUT much like it has been done for the UJT (Fig. 13-9).

$$V_p = \eta \times V_{BB} + V_D \qquad (13\text{-}9)$$

where $V_D \approx 0.5$ to 0.7 V. No external components are included and

$$\eta = \frac{R_1}{R_2 + R_1}$$

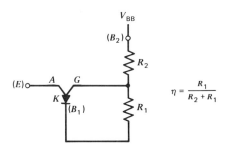

FIGURE 13-9 The PUT.

532   MISCELLANEOUS SEMICONDUCTOR DEVICES

Due to the low currents involved in the operation of the PUT, it is often used for very long time delays. Figure 13-10 demonstrates the delay circuit. The input voltage, a step voltage, charges the capacitor C through R. Until V_E (the voltage across the capacitor) reaches V_P, there is no current in the device; hence, no v_o. At the trigger point, when $V_E = V_P$, a sudden current produces a voltage spike across R_{B1}. The device turns off immediately because the capacitor is discharged by the current I_E.

EXAMPLE 13-2

For the circuit of Fig. 13-10, $V_{BB} = 10$ V, $R_2 = 56$ kΩ, $R_1 = 10$ kΩ, $R_{B1} = 1$ kΩ, $R = 20$ MΩ, $C = 10$ μF, $I_{E(sat)} = 2$ mA, v_{in} is a 10 V step, ($E = 10$ V). Find:
  (a) The time delay, approximately.
  (b) The output waveform (relative to v_{in}).

SOLUTION

The capacitor C is being charged (starting at t_0) up to 10 V, the amplitude of v_{in}. Triggering or turnon will occur when $V_E = V_P$. V_P can be obtained by $V_P = \eta V_{BB} + V_D = R_1/(R_2 + R_1) \times 10 + 0.7 = 10/66 \times 10 + 0.7 = 2.2$ V.
  (a) The time it takes for V_E to reach $V_P = 2.2$ V (the delay time) can be approximated by assuming a linear charge rate, that is, $V_C = V_E \approx t/RC \times E$. (The larger the RC, the longer it takes.) This is a reasonable approximation as long as $V_C \ll E$.
  For $V_E = V_P$

$$V_E = V_P \approx \frac{t}{RC} \times E$$

This can be solved for t, yielding the value of the delay time t_d.

$$t_d = \frac{V_P}{E} \times RC = \frac{2.2}{10} \times 20 \times 10^6 \times 10 \times 10^{-6} = 44 \text{ s}$$

[The exact value for t can be found by solving for t in the exact charge up expression $V_E = V_C = E \times (1 - e^{-t/RC})$ for the condition $V_E = V_P = 2.2$ V. This yields $2.2 = 10(1 - e^{-t/RC})$.]

$$1 - \frac{2.2}{10} = e^{-t/RC}$$

$$t = |\ln 0.78| \times RC = 2.3 |\log 0.78| \times RC$$
$$= 2.3 |\log 0.78| \times 20 \times 10^6 \times 10 \times 10^{-6} = 49.7 \text{ s}$$

  (b) Before turnon $v_o \approx 0$ (v_o consists of a very small leakage current times R_{B1}).

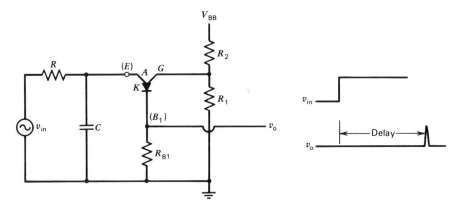

FIGURE 13-10 The PUT circuit.

When "on," $I_{E(sat)} = 2$ mA and $v_o = I_{E(sat)} \times R_{B1} = 2 \times 10^{-3} \times 1 \times 10^3 = 2$ V. The amplitude of $v_o = 2$ V.

The waveforms are shown in Fig. 13-11.

13-5.3 Comparison Among UJTs

The actual structure of the UJT substantially affects its performance. There are three common structures: the bar, cube, and annular structures. The differences between these structures are in the shape of the basic semiconductor structure connecting B_2 to B_1.

The typical characteristics of the three structures are quite different as can be seen from Table 13-2.

TABLE 13-2 COMPARISON OF KEY PARAMETERS FOR THE THREE DIFFERENT UJT STRUCTURES. (Courtesy Motorola Inc.)

Parameter	Typical values for		
	Bar	Cube	Annular
Intrinsic stand-off ratio η	0.6	0.65	0.7
Interbase resistance r_{BB}	7 K	7 K	7 K
Emitter saturation voltage $V_{EB1(SAT)}$	3 V	1.5 V	2.5 V
Peak point current I_p	2 μA	1 μA	0.1 μA
Valley point current I_v	15 mA	10 mA	7 mA
Emitter reverse current I_{EO}	1 μA	0.1 μA	5 nA

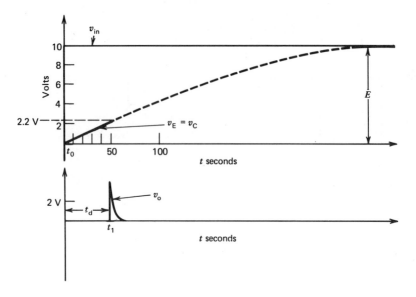

FIGURE 13-11 Voltage waveforms for Ex. 13-2.

Note in particular the very slow leakage current I_{EO} as well as the low I_p for the annular structure.

13-6 PHOTOSENSITIVE DEVICES

There is a wide range of devices that are light sensitive, in which a device parameter, resistance, current, or voltage is dependent on the light intensity incident on the device. Most of these devices are limited to a specific light spectrum, that is, to a specific range of light wavelength or frequency. Note that the visible light, red to violet, for example, ranges in frequency from 4×10^8 MHz to 7.5×10^8 MHz, respectively (the corresponding wavelengths are 0.7×10^{-6} m and 0.4×10^{-6} m, or 0.7 μm and 0.4 μm). This spectral limitation clearly indicates that an important criteria for the suitability of a device is whether the device is operating in the desired spectral range. To simplify matters here, it will always be assumed that the device described is operating with a compatible light source.

Another important parameter is light intensity. The terms that will be used to refer to this quantity are *irradiance*, H, in mW/cm² and total *incident power*, P_i, in mW. The irradiance is the light power per unit area of the device while the incident flux or power is the total light power incident on the device; thus,

$$P_i = A \times H \tag{13-10}$$

where A is the effective (active) area of the light sensitive device. The term radiant flux density is often used in place of irradiance, and the units used for irradiance are sometimes foot-candle (fc) rather than mW/cm². Each fc corresponds to 0.05 mW/cm².[3]

There are four basic types of light sensitive devices.

1. Photoemissive;
2. photoconductive-bulk type;
3. photovoltaic;
4. photoconductive-junction type.

The photoemissive is typified by the photomultiplier vacuum tube (or the simple phototube) where electrons are emitted from a cathode in response to impinging light.

The photoconductive-bulk type is essentially a piece of material, usually cadmium sulfide or cadmium selenide, that changes its resistance as a result of exposure to light.

The photovoltaic is usually a silicon or selenium PN junction that produces a current or voltage proportional to the light incident on the device. This type is a self-generating device and thus requires no external voltage supply for its operation, for example, the solar cell.

The photoconductive-junction type are semiconductor devices, diodes, transistors, etc., that are light sensitive—that is, the current through the device changes with changes in the incident light. (This can be viewed as resistive changes since the current is produced by an externally applied supply voltage.) Only the last two classes of devices will be discussed here.

13-6.1 The Photovoltaic Cell (Photocell)

The photovoltaic cell is a PN junction operated with no external bias and designed to allow light to reach the semiconductor junction (Fig. 13-12). The energy from the light produces an electron current in the diode that can be delivered to a load resistance. Figure 13-13 shows a simple circuit incorporating the photocell. The current I_P, the photocurrent, produces a voltage drop across R_L with a polarity that tends to forward bias the diode. To operate properly, the diode may *not* be forward biased. This sets a limit on the value of R_L, for a particular light intensity.

One of the important characteristics of the device is its *responsivity*, R_ϕ, that is defined as the ratio of current produced in amps to incident light power in watts (or µA per µW). Responsivity is given for a specified light wavelength since, as noted before, the device responds to a limited range of wavelengths. For example, the model PV-040 (made by EG & G Electrooptic Inc.) has a responsivity of 0.47 A/W at 950 nm wavelength and 0.05 A/W at 350 nm. It is clear that in order to obtain a good response (i.e. larger currents per watt of light power), the device

---

[3] Foot-candle (fc) is defined for the visible spectrum and mW/cm² is a measure of total radiation intensity. The conversion is meaningful only if a particular spectral content is specified. The conversion factor shown is for spectral content corresponding to a temperature of 2870°C.

536 MISCELLANEOUS SEMICONDUCTOR DEVICES

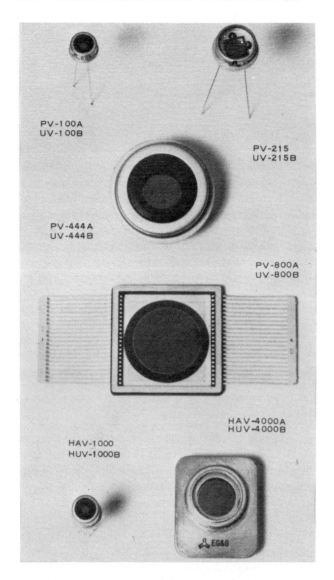

FIGURE 13-12 Various types of photocells (EG & G Electrooptics).

should be used with a source that radiates at 950 nm (0.95 μm). When operating at other wavelengths, the plot shown in Fig. 13-14 can be used to find corresponding responsivity. The current produced is given by

$$I_P = P_i \times R_\phi \qquad (13\text{-}11)$$

I_P — photocurrent in amps
P_i — incident power in watts
R_ϕ — responsivity in A/W

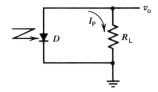

FIGURE 13-13 Photocell circuit.

EXAMPLE 13-3

The PV-040 is used in the circuit shown in Fig. 13-13, with $R_L = 10$ kΩ. The light source is an LED (see Sec. 13-7) that radiates at 670 nm with an irradiance, H (at the photodetector located 1.0 cm from the LED), of 0.5 mW/cm^2. The active area of the PV-040 is given as 0.81 mm^2 and the responsivity is given by Fig. 13-14. Find:
(a) The photocurrent.
(b) The voltage across the diode, v_o.

SOLUTION

(a) From Fig. 13-14 at 670 nm, the responsivity is 0.35 A/W. The incident power P_i can be found from Eq. (13-10). First, however, the active area must be converted to cm^2 (it is given in mm^2), to conform to the units of H.

$$A = 0.81 \times 10^{-2} \text{ cm}^2 \quad (1 \text{ mm}^2 = 10^{-2} \text{ cm}^2)$$
$$P_i = A \times H = 0.81 \times 10^{-2} \times 0.5 = 0.004 \text{ mW} \quad (13\text{-}10)$$

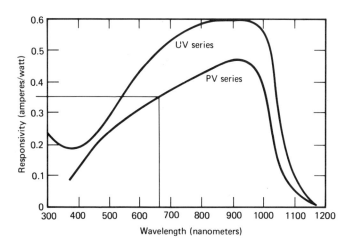

FIGURE 13-14 Typical responsivity versus wavelength for PV and UV models (courtesy EG & G Electrooptics).

(Since H is mW/cm² and A is cm² the answer is in mW.)

$$P_i = 4 \times 10^{-6} \text{ W} \quad (4 \text{ μW})$$

The photocurrent is Eq. (13-11)

$$I_P = P_i \times R_\phi = 4 \times 10^{-6} \times 0.35 = 1.4 \times 10^{-6} \text{ A}$$
$$= \underline{1.4 \text{ μA}}$$

(b) $v_o = I_P \times R_L = 1.4 \times 10^{-6} \times 10 \times 10^3$
$= \underline{0.014 \text{ V} = 14 \text{ mV}}$

The values in Ex. 13-3 are typical and represent the average photovoltaic cell (not the solar cell that is designed specifically for much higher currents with much larger active areas). The very low currents and voltage produced indicate the need for amplification. In systems, for example, designed to measure light intensity, the signal v_o, which represents light intensity, must be substantially larger. An integrated package containing the photocell and an OA is shown in Fig. 13-15 (HAV-1000). It is manufactured by EG & G Inc. The responsivity of the complete package with $R_f = 200$ MΩ is 90 V/μW. Had we used this device in Ex. 13-3, the output voltage would have been

$$90 \text{ V/μW} \times 4 \text{ μW} = 360 \text{ V}$$

This is well beyond saturation for the amplifier. The point is that with the HAV-1000, light powers in the nW (10^{-9} W) range can be measured easily. For a $P_i =$

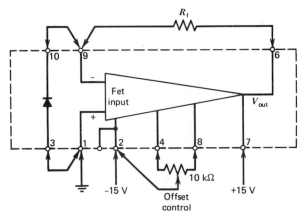

FIGURE 13-15 Photocell—OA combination, model HAV-1000 (Courtesy EG & G Electrooptics Corp.).

10 nW (0.01 μW) the output would be

$$90 \times 0.01 = 0.9 \text{ V}$$

Note also that the responsivity is given in volts per watt rather than A/W; the OA output is in volts, not amperes.

13-6.2 Photodiode

Similar to the photocell, the photodiode is also, as is clear from the name, a PN junction. This time it is operated with a reverse bias applied externally.

There are two types of photodiodes, the *PIN* diode and the avalanche photodiode. (A P-N type is also available, with characteristics much like those of the *PIN*.) The *PIN* diode has a sandwichlike structure. A P material anode and an N material cathode are separated by a thin layer of intrinsic semiconductor material (undoped). The name *PIN* represents the structure P followed by intrinsic followed by N material. This diode is operated in the reverse bias mode, without reaching avalanche. Essentially the light or radiated power incident on the junction (through a "window") changes the leakage current in the diode. The typical responsivity for a high speed diode (the HP 5082-4203) is 0.5 μA/μW, the active area, 2×10^{-3} cm² and the peak response is at about 800 nm.

The photocurrent depends not only on the incident radiation but also on the reverse bias. A plot of this dependence is shown in Fig. 13-16. Typically the leakage current remains constant as long as the diode is reverse biased. Figure 13-16 shows that for P_i, incident power, of 50 μW, $I_P = 16$ μA (as long as reverse bias is applied). Note that Fig. 13-16 also gives the variation in I_P resulting from changes in incident power P_i. For example, for $P_i = 100$ μW, $I_P = 30$ μA, etc. Figure 13-16 does not show what I_P is for $P_i = 0$, no incident power. This current, called

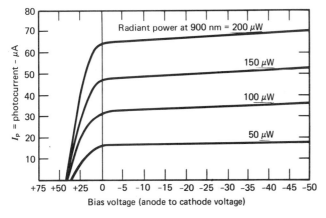

FIGURE 13-16 I_p versus bias current for the *PIN* diode at 900 nm (HP series 5082-4200) (courtesy of HP Corp.).

the diode <u>dark current</u> I_D, is usually given in the data sheet. For the HP 5082-4203 the dark current, I_D, is 2.0 nA maximum (at 25°C). This is very small relative to the operating photocurrent (many μA).

Here again, the currents are very small and must be amplified to become useful in any application. A circuit using an OA is shown in Fig. 13-17. To balance the circuit against voltage offset in the OA it is advisable to have

$$R_1 = R_2$$

(It should really be $R_2 = R_1 \| R_D$, where R_D is the diode resistance. Since R_D is very large, it may be ignored.) The output voltage in Fig. 13-17 is given by

$$v_o = R_1 \times (I_P + I_D) \qquad (13\text{-}12)$$

where $I_P + I_D$ represents the total diode current.

The *avalanche photodiode*, APD, operates in the avalanche region, requiring substantially higher reverse bias voltages. Its responsivity is much higher than that of the PIN so that the APD responds to much lower light levels; its sensitivity is much higher than that of the PIN (or P-N) diode. This higher sensitivity is due to an "avalanche gain." The free electrons (conduction electrons) produced by the impinging light in turn produce many more current electrons (see the description of avalanche in Chapter 2). In some cases this electron multiplication or avalanche gain, m, is as high as 200. This means that very low light levels can be detected.

One of the major drawbacks of the APD is that it requires a high supply voltage (avalanche occurs typically at 170 V for the TISL59 APD made by Texas Instruments). In addition, the avalanche breakdown voltage varies from unit to unit of the same type, as well as with temperature. This requires that the power supply used for the APD be well stabilized and adjustable. (It must be readjusted every time the APD is replaced, even when it is replaced with the same type.)

Some typical characteristics of the TISL59 APD are given in Table 13-3 for operation at 25°C.

The avalanche gain of 200 effectively increases the responsivity by this factor. (Avalanche gain is dependent on the operating conditions, device structure, and doping levels.)

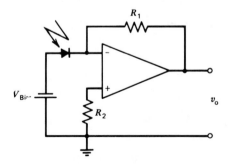

FIGURE 13-17 *PIN* diode with OA.

TABLE 13-3   TISL59 CHARACTERISTICS (PARTIAL LISTING).

Active area	4.6×10^{-3} cm²
Reverse breakdown (avalanche) at $I_R =$ 10 µA	170 V
Dark current ($V_R = 100$ V)	20 pA (20×10^{-12} A)
Responsivity at 907 nm with no avalanche gain ($m = 1$)	0.5 A/W
Gain BW product	80×10^3 MHz (80 Gigahertz)
Avalanche gain m	200

13-6.3 Phototransistor

The phototransistor has all the basic performance characteristics of a regular transistor with the added feature that I_C, the collector current, varies with incident radiation energy. Variation in incident energy has the same effect as changes in I_B. This results in a very high responsivity. The inherent transistor current gain comes into play. The phototransistor can be two or three orders of magnitude (100 or 1000 times) more sensitive than the photodiode.

A set of phototransistor collector characteristics is shown in Fig. 13-18. The plot looks like the standard transistor plot with I_B replaced by H as a parameter. Similar to β_{ac}, a performance parameter called *collector-emitter radiation sensitivity* S_{RCEO}

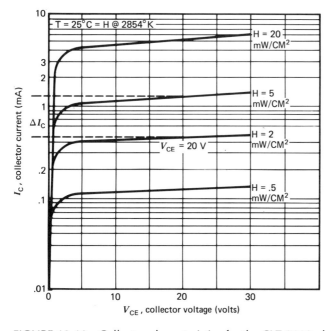

FIGURE 13-18   Collector characteristics for the CLT 3160 phototransistor (courtesy Clairex Corp.).

542 MISCELLANEOUS SEMICONDUCTOR DEVICES

is defined as the changes in I_C, ΔI_C for a given change in irradiation H, ΔH. For the plot of Fig. 13-18 at $V_{CE} = 20$ V, a change in H from 2 mW/cm^2 to 5 mW/cm^2 produces an I_C change from 0.45 to 1.3 mA (note that the I_C scale is logarithmic) and $\Delta I_C = 0.85$ mA.

The sensitivity is

$$S_{RCEO} = \frac{\Delta I_C}{\Delta H} = \frac{0.85 \text{ mA}}{3 \text{ mW/cm}^2}$$

$$= 0.28 \frac{\text{mA}}{\text{mW/cm}^2}$$

The sensitivity is given in terms of irradiance, or radiation energy per unit area, rather than in terms of total incident energy as in the case of the photodiode.

The relation between ΔH, changes in incident radiation, and ΔI_C in the phototransistor is not a linear relation. As a result, the phototransistor is rarely used as a linear light amplifier (where I_C directly represents radiation energy) but rather in switching applications.

The phototransistor can be used with no biasing at all (Fig. 13-19a) or with a base bias circuit that sets up the quiescent operating point (Fig. 13-19b). In Figs. 13-19a and b, v_o can be obtained if H, the input light signal, is known.

EXAMPLE 13-4

In Fig. 13-19a, $R_E = 2.5$ kΩ, $V_{CC} = 30$ V, the transistor is a Motorola MRD 300 (characteristics given in Fig. 13-19c). Find v_o for
 (a) $H = 1.25$ mW/cm^2.
 (b) $H = 7.0$ mW/cm^2.

SOLUTION

We resort to the load line technique. The KVL equation for the output portion is

$$V_{CC} - V_{CE} - I_C R_E = 0$$

$$(I_C \approx I_E)$$

so that

$$V_{CE} = V_{CC} - I_C R_E$$
$$V_{CE} = 30 - I_C \times 2.5 \times 10^3$$

(This equation is plotted in Fig. 13-19c.)
 (a) Point A in Fig. 13-19c gives V_{CE} for $H = 1.25$ mW/cm^2

$$V_{CE} = 27 \text{ V}$$

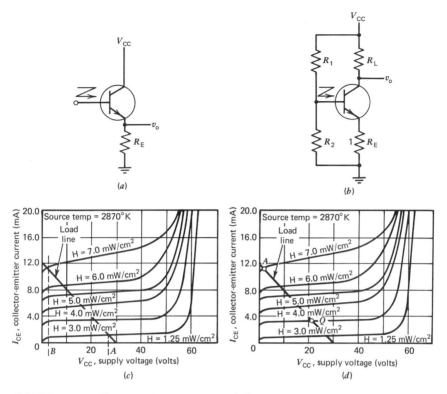

FIGURE 13-19 Phototransistor circuits and characteristics. (a) Circuit with no base bias (base open). (b) Circuit with voltage divider bias. (c) MRD 300 characteristics with load line. (Courtesy Motorola Inc.) (d) MRD 300 characteristics (Ex. 13-5).

Since

$$v_o = I_C R_E = V_{CC} - V_{CE} = 30 - 27 = \underline{3 \text{ V}}$$

(b) From point B for $H = 7$ mW/cm²

$$V_{CE} = 2 \text{ V}$$
$$v_o = V_{CC} - V_{CE} = 30 - 2 = 28 \text{ V}$$

Point B is very near saturation.

EXAMPLE 13-5

In the circuit of Fig. 13-19b, $V_{CC} = 30$ V, $R_1 = 8.2$ kΩ, $R_2 = 1.5$ kΩ, $R_C = 1.5$ kΩ, $R_E = 1$ kΩ, the transistor is MRD300. Find v_o for
(a) $H = 0$.
(b) $H = 4$ mW/cm².

SOLUTION

We draw the load line

$$V_{CC} - I_C R_C - V_{CE} - I_C R_E = 0$$
$$V_{CC} - I_C(R_C + R_E) = V_{CE}$$
$$30 - I_C(2.5 \text{ k}\Omega) = V_{CE}$$

shown in Fig. 13-19d. The quiescent I_C (no light signal) is found by the same methods used in transistor voltage divider biasing.

$$V_B \approx \frac{30 \times 1.5 \text{ k}\Omega}{8.2 \text{ k}\Omega + 1.5 \text{ k}\Omega} = 4.6 \text{ V}$$

$$I_E \approx I_C = \frac{4.6 - 0.7}{1 \text{ k}\Omega} = 3.9 \text{ mA}$$

The Q-point is shown in Fig. 13-19d.

(a) With no light signal $I_C = 3.9$ mA and

$$v_o = V_{CC} - I_C R_C$$
$$= 30 - 3.9 \times 10^{-3} \times 1.5 \times 10^3 = \underline{24.2 \text{ V}}$$

(b) For $H = 4$ mW/cm² the operating point is made to slide up the load line by an additional 4 mW/cm². Note that the added signal gives a total equivalent H of 7 mW/cm², point A, for which $I_C \approx 11.0$ mA so

$$v_o = V_{CC} - I_C R_C$$
$$= 30 - 11 \times 10^{-3} \times 1.5 \times 10^3 = \underline{13.5 \text{ V}}$$

For larger currents and higher sensitivities one may choose to use the photo-Darlington transistor. A set of characteristics and device symbol are shown in Fig. 13-20. Note the relatively low values of H as compared with the characteristics of Fig. 13-18. This implies a much higher collector-emitter radiation sensitivity, S_{RCEO}.

Other photodevices include the photo-FET, the photothyristor, or light activated SCR (LASCR), and so on. Practically every type of semiconductor device is available as a photodevice.

13-6.4 Some Common Characteristics of Photodevices

Besides responsivity (or sensitivity for the phototransistor) there are a number of features that relate to all photodevices discussed.

All are sensitive to the direction from which the light (radiation) comes. Some have a wide angle of acceptance; these are designed to collect the energy from as

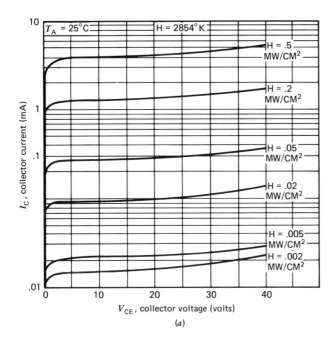

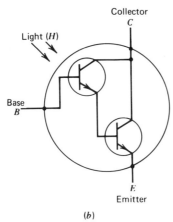

FIGURE 13-20 Photo-Darlington CLR 2050 (Clairex Corp.). (a) Collector characteristics. (b) Circuit.

wide an angle as possible. Others are designed to be directional. Special lenses are often used to obtain the desired angular response.

The various devices have different rise and fall times. The photodiode 5082-4203 (made by HP), for example, when properly reverse biased, has a rise and fall time of 1 ns. The photocurrent reaches its 90% level, corresponding to the incident energy, within 1 ns.

546 MISCELLANEOUS SEMICONDUCTOR DEVICES

The noise characteristics of the devices vary substantially. All produce some internal electrical noise under given conditions. The *PIN* has a relatively low noise operation.

The linearity varies from device to device. The *PIN* diode is probably the most linear, while the phototransistor is the least so.

13-7 SEMICONDUCTOR LIGHT SOURCES

There are two semiconductor light sources: the LED (light-emitting diode) and the LASER (light amplification by stimulated emission of radiation). Both of these devices are basically *PN* junctions that, when sufficiently forward biased, emit light (not necessarily visible light). The physical phenomena producing radiation in the two devices are fundamentally different, so that the performance characteristics are also much different.

The laser is a high efficiency emitter that emits energy over a very narrow range of wavelengths. A typical data sheet for a GaAlAs laser diode model SCW-20, manufactured by Laser Diode Labs Inc., is shown in Fig. 13-21*a*. (The notation GaAlAs stands for Gallium Aluminum Arsenic, which are the elements used in constructing the diode.)

For this particular diode, peak emission occurs typically at 830 nm wavelength. The emission is single color; that is, only an extremely narrow range of wavelengths, $\Delta\lambda = 1$ nm, is produced (as opposed to white light that contains a broad band of wavelengths). It is a very fast diode, reaching 90% of full emission (t_r) in 100 ps (100×10^{-12} s). It emits radiation over a narrow angle, 10° in one direction and 35° in the other. (This is the far field beam divergence specification.) Figure 13-21*b* shows a typical plot of forward current versus radiant power. The diode is in a "lasing" mode only for the range where $I_F > I_{th}$—that is, when the forward current through the diode, I_F, is larger than the threshold current I_{th} (shown in Fig. 13-21*b* for the 30°C plot). Below I_{th} the device behaves like an LED. Note that the radiated power varies very steeply with I_F in the lasing region. A small change in I_F causes a large change in radiated power.

One of the very serious problems in using laser diodes is the fact that I_{th} varies with temperature. As Fig. 13-21*b* shows for 30°C $I_{th} \simeq 75$ mA, while for 80°C $I_{th} = 120$ mA. The problem is further compounded by the heat generated as a result of diode current, which raises junction temperatures. Most laser diodes require special temperature control methods to minimize the effects of temperature variations.

The *LED* radiates at much lower energy levels and over a much wider wavelengths range, typically $\Delta\lambda = 40$ nm. A power versus current plot is shown in Fig. 13-22. While laser power was in mW, LED power is in μW (at similar current levels). The plot is linear, no "knee" as for the laser. The LED radiates at much wider angles (about 35° in all directions) and its rise time is substantially larger, 25 ns typically.

The above discussion of light (radiated energy) sources was particularly oriented toward data transmission applications. Both devices are used in transmitting infor-

SEMICONDUCTOR LIGHT SOURCES

	Symbol	Min.	Typ.	Max.	Units
Total Radiant Flux at rated Io (CW)	Pom	4	7.5		mW
Peak Wavelength	λp		830		
Spectral Width	Δλ		1		nm
Source Size			0.2×7.0		μm
Rise Time of Radiant Flux	Tr		100	800	ps
Far Field Beam Divergence	θpx θn		10×35		degrees
Threshold Current	Ith		60	120	ma
Operating Current	Io		85	Ith+25	ma
Forward Voltage at Io	Vo		2.0		volts
Operating Temperature	To	0		60*	°C
Storage Temperature	Ts	−55		125	°C

*Selections to higher operating temperatures are available.

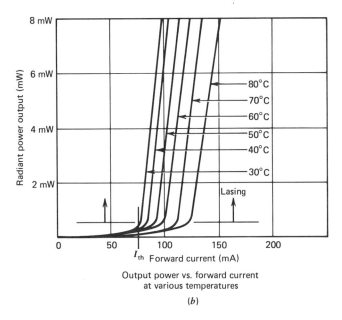

Output power vs. forward current at various temperatures

(b)

FIGURE 13-21 Laser diode SCW-20 (Laser Diode Labs, Inc.). (a) Specifications. (b) Output power versus forward current at various temperatures.

mation much like the broadcast system. Here, the radiated power that is indeed electromagnetic waves (like for the broadcast systems) carries the information. The laser, because it is nonlinear, is useful mostly for digital transmission, on-off type signals, while the LED can be used to transmit analog signals (the light changing linearly with the information signals). Other comparisons between the LED and LASER are shown in Table 13-4.

In the data transmission application, light from the LED or laser transmitter is detected by a photosensitive device at the receiver. The transmission may be through the atmosphere or in speical light carrying glass fibers (optical fibers). A similar but somewhat simpler application using similar principles is the *optoisolator*. A light

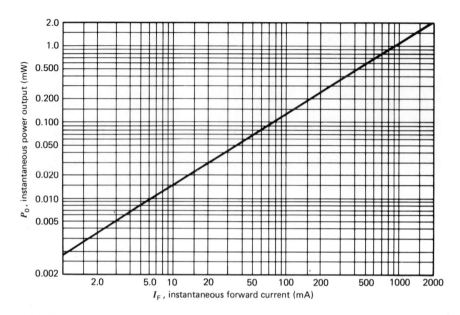

FIGURE 13-22 Power out versus forward current in LED MFOE 102F (courtesy Motorola Inc.).

source (usually an LED) and a photodetector (photodiode, phototransistor, etc.) are housed in a single package so that the radiated energy from the source is detected by the photodetector. In this way the signal input to the light source, LED, is transmitted through to the output, the photodetector, with no electrical connection between them (Fig. 13-23). Note that the input is an electrical signal, while the output may be a varying resistance if a photodiode is used, a varying voltage for a photovoltaic cell, etc. The characteristics of the optoisolator are a composite of the source and detector parameters. The basic purpose of the optoisolator is to provide electrical isolation between two parts of an electrical system.

TABLE 13-4 LASER - LED COMPARISON (Typical Values)

	Maximum radiated power	Spectral widths	Angular radiation	Linearity	Temperature stability	Application
LASER	8 mW	1 nm	10° × 35°	very poor	very poor	high power digital
LED	0.1 mW	40 nm	35°	1%	good	analog, digital

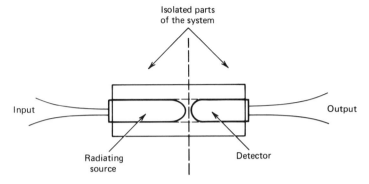

FIGURE 13-23 Optoisolator.

13-8 OTHER SEMICONDUCTOR DEVICES

The following brief descriptions are intended to serve as a survey of a number of other semiconductor devices. Only basic operating principles will be presented with occasional references to possible applications.

13-8.1 The Tunnel Diode

The tunnel diode consists of a heavily doped P-N junction. The heavy doping modifies the behavior of the P-N junction as compared to the regular diode, yielding a V-I curve as shown in Fig. 13-24. The region between I_P and I_V is the negative

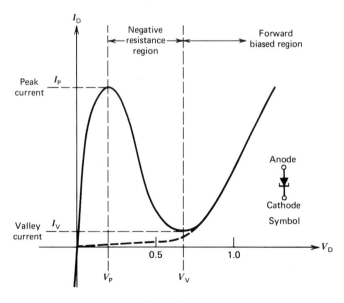

FIGURE 13-24 Tunnel diode characteristic curve (values shown are representative only).

550 MISCELLANEOUS SEMICONDUCTOR DEVICES

resistance region, because here, an increase in voltage produces a decrease in current (see Sec. 13-3).

The dashed line in Fig. 13-24 represents the plot for a standard diode. The tunneling effect (see Chapter 2) occurs for low forward bias, below the normal forward diode drop.

Tunnel diodes may be used as amplifiers and oscillators. Owing to their fast response, they are suitable for operation at very high frequencies (as high as 2×10^{12} Hz).

13-8.2 The Varactor (Variable Voltage Capacitor, Varicap)

In discussing the P-N junction depletion phenomenon (Chapter 2), we noted that a large reverse bias increases that region and effectively increases the separation between the anode and cathode regions. If one considers the anode and cathode as two plates of a capacitor, then the increase in separation will cause a reduction in anode-cathode capacitance. The varactor is a diode designed to take advantage of the capacitance changes with applied reverse voltage.

The varactor is then a P-N junction operated in the region between forward bias and avalanche breakdown (never forward biased). The capacitance of the varactor can be approximated by

$$C_j = \frac{C_o}{\sqrt{1 + 2V_R}} \qquad (13\text{-}13)$$

where C_o is the capacitance at zero bias and V_R is the reverse bias voltage. C_j is the junction capacity. [Equation (13-13) is dependent on the degree of doping and structural details.] In practice the total varactor capacitance C_T includes the fixed capacitance of the contacts and the case in which the junction is housed.

A plot of C_T versus V_R for a number of typical varactors is shown in Fig. 13-25. As expected, from the small size of the junction, C_T is small and rarely exceeds 30 pF. The extent of capacitance change, sometimes called the *tuning ratio*, is usually given as the capacitance near zero bias over the capacitance at maximum allowable reverse voltage. For the MV1870D (top line in Fig. 13-25) the tuning ratio given for C_2/C_{60} is

$$C_2 = 20 \text{ pF, capacitance at } V_R = 2 \text{ V}$$
$$C_{60} = 4.5 \text{ pF, capacitance at } V_R = 60 \text{ V (max } V_R)$$

so the tuning ratio

$$C_2/C_{60} = \frac{20}{4.5} = 4.4$$

A common application of the varactor is in voltage controlled oscillators. The idea is to change the resonance frequency of a tuned circuit that determines the frequency of oscillation of the circuit. Figure 13-26 shows the resonance circuit as part of an oscillator circuit and a method of controlling the frequency by varying

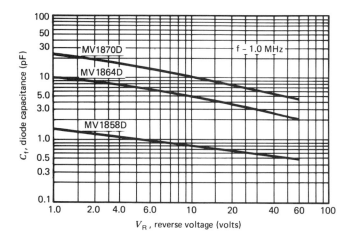

FIGURE 13-25 Capacity versus V_R for varactors (courtesy Motorola, Inc.).

V_C. C_1 is large so that the total tuning capacity is $C + C_T$. As C_T is varied by changing V_C, so does the resonance frequency given approximately by

$$f_o = \frac{1}{2\pi \sqrt{L(C_T + C)}} \qquad (13\text{-}14)$$

Capacitor C is used as a tuning capacitor.

Most varactor specifications relate directly to the standard diode specifications and need not be repeated.

SUMMARY OF IMPORTANT TERMS

Active area The area of a photosensitive device that is exposed to light and contributes to the photocurrent.

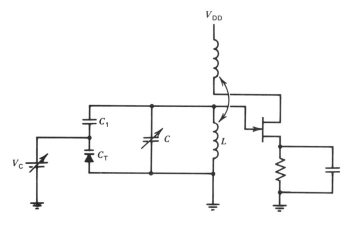

FIGURE 13-26 Voltage controlled oscillator.

APD Avalanche photodiode. A photosensitive diode operating in the avalanche region.

Avalanche gain Current multiplication resulting from the avalanche phenomenon in an APD.

Collector-emitter radiation sensitivity (S_{RCEO}) The ratio of the current (I_C) changes to changes in irradiance in a phototransistor.

CUJT Complementary UJT. An integrated transistor structure that behaves like a UJT with all the voltage polarities reversed (similar to complementary transistors NPN-PNP).

Foot-candle (fc) A unit of light intensity, light power per unit area.

Incident power, P_i Total incident radiated power (in W).

Intrinsic stand-off ratio, η The ratio of $r_{B1}/(r_{B1} + r_{B2})$ in a UJT.

Irradiance, H The radiated energy incident on a surface in power per unit area, mW/cm².

Laser An energy (or light) radiating semiconductor device with relatively high radiant power and narrow spectral response.

LED A radiating semiconductor, low power, high linearity.

Negative resistance A region in the operating range of a device in which an increase in voltage across the device is accompanied by a decrease in current through it.

Optoisolator A light source–photodetector combination that provides electrical isolation.

Photoconduction Electrical conductivity that is dependent on, or results from, incident radiant energy.

Photocurrent Electrical current changes resulting from incident radiant energy.

Photoemission Electrons emitted from certain materials (under specified conditions) when exposed to radiant energy.

Photovoltaic The generation of the voltage or current in response to radiant energy.

PIN A type of photodiode.

PUT Programmable UJT. A UJT in which η is externally determined (r_{B1} and r_{B2} are externally selectable).

Relaxation oscillator A signal generating circuit that produces a triangular wave signal.

Responsivity (R_ϕ) A measure of device sensitivity to radiant energy: the current produced over incident power (A/W).

Spectral response Dependence of a device on the radiant energy wavelength.

Tuning ratio Applies to the varactor. It is a ratio of capacitance at a voltage near zero to that at the maximum allowable voltage.

Tunnel diode A heavily doped P-N junction that exhibits negative resistance in part of its operating region.

UJT Unijunction transistor. A P-N junction with three terminals, two edges of an N bar (B_2, B_1 terminals), and the P region of the junction (E terminal).

Varactor A diode with its capacitance controlled by the applied reverse bias voltage.

nm nanometer = 10^{-9} meter = 1000 μm, a unit of wavelength.

μ or μm 10^{-6} meter, a unit of wavelength.

PROBLEMS

13-1. Which portion of the UJT characteristics is called the negative resistance region and why?

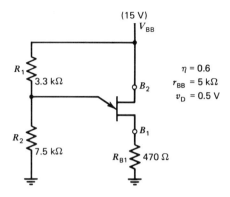

FIGURE 13-27

13-2. A silicon UJT with $\eta = 0.6$ is used with a base-2 to base-1 voltage of 9 V. Find V_P ($V_D = 0.6$ V).

13-3. In Problem 13-2, $r_{BB} = 5.3$ kΩ. Determine the value of r_{B1} and r_{B2}.

13-4. In the circuit of Fig. 13-27 find whether the UJT is "on" (is $V_E > V_P$)?

13-5. Repeat Problem 13-4 for the case in which $V_{BB} = 8$ V.

13-6. Repeat Problem 13-4 with $R_{B1} = 1.5$ kΩ. *Hint:* Find V_{B2B1} for the "off" state. The voltage drop across R_{B1} must be considered when determining whether the device is on.

13-7. In the circuit of Fig. 13-28 with v_{in} amplitude given, V_{BB} is being varied from 10 to 2 V. For what value of V_{BB} will the UJT turn on (for the 3 V input level)?

13-8. A relaxation oscillator uses a UJT with the following specifications: $\eta = 0.7$, $r_{BB} = 7$ kΩ, $V_D = 0.6$ V (Fig. 13-6a). The component values are $R_1 = 1$ MΩ, $C = 10$ μF, $R_{B1} = 82$ Ω. Find the oscillation frequency (use Table 13-1).

13-9. What is the basic difference between the UJT and the PUT?

13-10. What are the four basic types of photosensitive devices? Briefly describe each.

13-11. A photovoltaic cell has a responsivity of 0.5 A/W at a given wavelength. If the active area is 1.2×10^{-2} cm^2 and it is subject to an irradiance of 2 mW/cm^2, find the photocurrent.

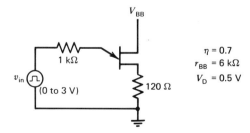

FIGURE 13-28

554 MISCELLANEOUS SEMICONDUCTOR DEVICES

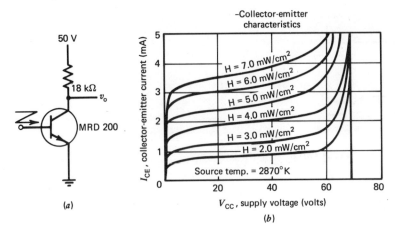

FIGURE 13-29 (a) Circuit. (b) Collector characteristics MRD 200 (courtesy Motorola Inc.).

13-12. Explain what may happen in Problem 13-11 if the wavelength of the irradiating source is substantially changed.

13-13. What are the advantages and disadvantages of the APD relative to the PIN?

13-14. The circuit of Fig. 13-29a uses an MRD 200 with the characteristics given in Fig. 13-29b. Draw the load line and find v_o for H varying between 2.0 to 6 mW/cm².

13-15. What are some of the differences between the laser and the LED?

13-16. An LED has a forward drop of 2.0 V and a maximum allowable current of 25 mA. It is desired to operate the LED at 20 mA in the circuit shown in Fig. 13-30. The LED must be on when S_1 is closed only. Find:

(a) R_C.
(b) R_B.
(You may assume R_2 is an open circuit.)

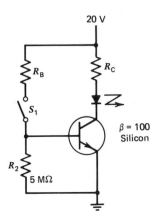

FIGURE 13-30

13-17. In what ways is a tunnel diode different from a regular diode?

13-18. A varicap has a tuning ratio $C_2/C_{10} = 1.5$ (capacitance at 2 V compared to capacitance at 10 V). If used in a tuned L-C circuit, what is the ratio of resonance frequencies that can be achieved ($V_R = 2$ V to $V_R = 10$ V)?

APPENDIX A

FORWARD RESISTANCE OF THE DIODE, r_f

To obtain the expression for r_f from Eq. (2-1b), it is necessary to evaluate the derivative of V_D with respect to I_D and evaluate that quantity at a particular value of I_D. From Eq. (2-1b),

$$I_D = I_s e^{V_D/(kT/q)} \tag{2-1b}$$

we can evaluate dI_D/dV_D and $r_f = 1/(dI_D/dV_D) = dV_D/dI_D$

$$\frac{dI_D}{dV_D} = \frac{1}{kT/q} I_s e^{V_D/(kT/q)} = \frac{1}{kT/q} I_D \text{ }^1$$

and

$$r_f = \frac{1}{\frac{1}{kT/q} \times I_D} = \frac{kT/q}{I_D} \text{ ohms } (\Omega) \tag{2-5}$$

---

[1] From differential calculus we have that for $y = Ie^{ax}$, $dy/dx = a \times I \times e^{ax}$. This results from $d(e^x)/dx = e^x$.

APPENDIX B

DERIVATION OF $S(\beta)$ FOR THE FIXED BIAS CIRCUIT WITH R_E

Equation (3-15a) gives I_B as

$$I_B = \frac{V_{CC} - V_{BE}}{R_B + \beta R_E}$$

Since $I_C = \beta I_B$, we get

$$I_C = \beta I_B = \beta \frac{(V_{CC} - V_{BE})}{R_B + \beta R_E} = \frac{V_{CC} - V_{BE}}{\frac{R_B}{\beta} + R_E}$$

The derivative $dI_C/d\beta$ is then

$$\frac{dI_C}{d\beta} = \left[\frac{V_{CC} - V_{BE}}{\left(\frac{R_B}{\beta} + R_E\right)^2} \right] \times \frac{R_B}{\beta^2}$$

Since

$$S(\beta) = \frac{dI_C/I_C}{d\beta/\beta} = \frac{dI_C}{d\beta} \times \frac{\beta}{I_C}$$

$$= \frac{V_{CC} - V_{BE}}{\left(\dfrac{R_B}{\beta} + R_E\right)^2} \times \frac{R_B}{\beta^2} \times \frac{\beta}{\dfrac{V_{CC} - V_{BE}}{\dfrac{R_B}{\beta} + R_E}}$$

$$= \frac{V_{CC} - V_{BE}}{(R_B/\beta + R_E)^2} \times \frac{R_B}{\beta^2} \times \frac{\beta(R_B/\beta + R_E)}{V_{CC} - V_{BE}}$$

$$= \frac{R_B}{R_B + \beta R_E}$$

$$S(\beta) = \frac{R_B}{R_B + \beta R_E}$$

This last expression shows that $S(\beta) \approx 1$ if R_B is large compared to βR_E (poor stability), and $S(\beta) \ll 1$ if R_B is much less than βR_E (good stability).

APPENDIX C

S(β) FOR THE VOLTAGE DIVIDER BIAS CIRCUIT

In Appendix B we showed that for the fixed bias circuit with R_E

$$S(\beta) = \frac{R_B}{R_B + \beta R_E}$$

As we show in Fig. 3-18c, the voltage divider circuit is essentially a fixed bias circuit in which R_B is replaced by R_{Th} and V_{CC} by V_{Th}.

As a result, the stability $S(\beta)$ for the voltage divider becomes

$$S(\beta) = \frac{R_{Th}}{R_{Th} + \beta R_E}$$

APPENDIX D

THE DERIVATION OF g_m

The transconductance, g_m, is defined by $g_m = \Delta I_D/\Delta V_{GS}$. For small values of ΔV_{GS} ($\Delta V_{GS} \to 0$), this becomes $g_m = dI_D/dV_{GS}$, the derivative of I_D with respect to V_{GS}. This derivative is found directly by the differentiation of Eq. (6-1), Shockley's equation.

$$I_D = I_{DSS}(1 - V_{GS}/V_P)^2 \qquad (6\text{-}1)$$

where I_D and V_{GS} are the variables, and I_{DSS} and V_P are device constants given by the manufacturer.

$$\frac{dI_D}{dV_{GS}} = \frac{-2I_{DSS}}{V_P}\left(1 - \frac{V_{GS}}{V_P}\right) \qquad (\text{D-1})$$

$$g_m = \frac{-2I_{DSS}}{V_P}\left(1 - \frac{V_{GS}}{V_P}\right)$$

APPENDIX

E

EXACT SOLUTION OF FIG. 5-26b

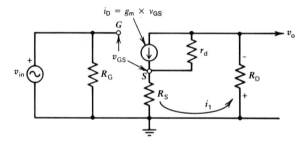

FIGURE E-1 Exact solution for Eq. circuit of Fig. 5-26b

By current division,

$$i_1 = \frac{i_D \times r_d}{R_D + R_S + r_d} = \frac{i_D \times r_d}{R} \quad \text{(E-1)}$$

where $R = R_D + R_S + r_d$

$$v_o = -i_1 \times R_D = -\frac{i_D \times r_d \times R_D}{R} \quad \text{(E-2)}$$

Since

$$i_D = g_m v_{GS} \tag{6-7b}$$

$$v_o = \frac{-g_m v_{GS} \times r_d \times R_D}{R} \tag{E-3}$$

by KVL

$$v_{in} - v_{GS} - v_S = 0$$

Since

$$v_S = i_1 \times R_S = \frac{i_D \times r_d \times R_S}{R}$$

$$= \frac{g_m v_{GS} r_d R_S}{R}$$

we get

$$v_{in} - v_{GS} - \frac{g_m v_{GS} r_d R_S}{R} = 0$$

$$v_{in} = v_{GS} \left[1 + \left(\frac{g_m r_d R_S}{R}\right)\right]$$

$$v_{GS} = \frac{v_{in}}{1 + g_m \dfrac{r_d R_S}{R}} \tag{E-4}$$

Combining Eqs. (E-3) and (E-4) yields

$$v_o = \frac{-g_m \left(\dfrac{v_{in}}{1 + g_m \dfrac{r_d R_s}{R}}\right) R_D \times r_d}{R}$$

$$= \frac{-g_m \times v_{in} R_D r_d}{R\left(1 + g_m \dfrac{r_d R_S}{R}\right)}$$

$$v_o = \frac{-g_m v_{in} \times R_D r_d}{R + g_m r_d R_S}$$

and

$$\frac{v_o}{v_{in}} = \frac{-g_m R_D r_d}{R_D + R_S + r_d + g_m r_d R_S} = A_v$$

If we divide the numerator and denominator by r_d

$$A_v = \frac{-g_m R_D}{1 + g_m R_S + \dfrac{R_D + R_S}{r_d}} \tag{E-5}$$

The same solution may be arrived at by Nodal analysis

$$v_o = \frac{\begin{vmatrix} -i_D & -\dfrac{1}{r_d} \\ i_D & \dfrac{1}{R_S} + \dfrac{1}{r_d} \end{vmatrix}}{\Delta}$$

$$v_S = \frac{\begin{vmatrix} \dfrac{1}{r_d} + \dfrac{1}{R_D} & -i_D \\ -\dfrac{1}{r_d} & i_D \end{vmatrix}}{\Delta}$$

$$\Delta = \begin{vmatrix} \dfrac{1}{r_d} + \dfrac{1}{R_D} & -\dfrac{1}{r_d} \\ -\dfrac{1}{r_d} & \dfrac{1}{R_S} + \dfrac{1}{r_d} \end{vmatrix}$$

and KVL applied to the input yielding Eqs. (E-4) and (E-5). With R_L coupled to the collector, R_D in Eq. (E-5) must be replaced by $R_D \| R_L$ resulting in

$$A_v = \frac{-g_m R_D \| R_L}{1 + g_m R_S + \dfrac{R_D \| R_L + R_S}{r_d}} \tag{E-6}$$

APPENDIX F
THE LOGARITHM (BASE 10)

The logarithm of the number a is defined as that power of 10 that will yield a. Log $a = K$ (The logarithm of a equals K) if $10^K = a$ or $10^{\log a} = a$. Logarithms of numbers may be found in standard tables or by the use of an appropriate calculator. Based on the definition of logarithm we have the following:

$$\log(10^a) = a \tag{F-1}$$

Hence,

$$\log 1 = \log(10^0) = 0$$
$$\log 10 = \log(10^1) = 1$$
$$\log 100 = \log(10^2) = 2$$
$$\log(.1) = \log(10^{-1}) = -1, \text{ etc.}$$

LOGARITHM OF A PRODUCT

$$\log(a \times b) = \log a + \log b$$

If $\log 2 = 0.3010$ (from tables), then

$$\log 20 = \log(2 \times 10) = \log 2 + \log 10 = 0.3010 + 1$$
$$= \underline{1.3010}$$

LOGARITHM OF A FRACTION

$$\log(a/b) = \log a - \log b \qquad \text{(F-2)}$$

Hence,

$$\log 50 = \log\left(\frac{100}{2}\right) = \log 100 - \log 2$$
$$= 2 - 0.3010 = 1.699$$

LOGARITHM OF A POWER

$$\log(a^b) = b \log a \qquad \text{(F-3)}$$

Hence,

$$\log 8 = \log(2^3) = 3 \log 2 = 3 \times .3010 = 0.903$$

To find the antilogarithm, that is, given the logarithm of a number find the number, we may use tables or a calculator. Given that $\log a = 2$,

$$a = 10^{\log a} = 10^2 = 100$$

APPENDIX G
ANTILOGARITHM (BASE 10)

To get the antilogarithm of a number L, we proceed in the reverse direction as that for finding the logarithm.

$$\text{Antilogarithm } (L) = 10^L$$

For example,

$$\text{Antilogarithm } 2 = 10^2 = 100$$
$$\text{Antilogarithm } 1.3010 = 10^{1.3010}$$
$$= 10^1 \times 10^{0.3010} = 20$$

We may use tables or an appropriate calculator.

APPENDIX H

MAXIMUM TRANSISTOR DISSIPATION IN CLASS B CIRCUITS

We calculate here the largest possible power dissipation in the transistors of a class B two-transistor circuit. (Use Fig. 9-16 as reference.)

The power dissipated by both transistors is the difference between the total power delivered to the circuit and the ac output in the load. Total power is

$$P_{tot} = V_{CC1} \times \frac{v_{op}}{\pi R_L} + V_{CC2} \times \frac{v_{op}}{\pi R_L} \qquad \text{(H-1)}$$
$$= \frac{(V_{CC1} + V_{CC2})v_{op}}{\pi R_L}$$

$v_{op}/\pi R_L$ is the average current through Q_1 (or Q_2) with a peak output voltage of v_{op} (average of a half sinewave). The ac power to the load is:

$$P_o = v_{o(p-p)}^2/8R_L \qquad \text{(H-2)}$$

(The rms value of the output voltage is $v_{o(p-p)}/2\sqrt{2}$, and power is expressed by, $P = V^2/R$.) The power dissipated by the *two* transistors is

$$P_{T2} = P_{tot} - P_o \qquad \text{(H-3)}$$
$$= \frac{V_{CC1} + V_{CC2}}{\pi R_L \times 2} \times v_{o(p-p)} - \frac{v_{o(p-p)}^2}{8 \times R_L}$$

In the expression for P_{tot}, $v_{o(p-p)}/2$ has been substituted for v_{op}. To find the value of $v_{o(p\text{-}p)}$ that produces a maximum for P_{T2} (the power dissipated in the transistors), we use the maxima theorem.

$$dP_{2T}/dv_{op} = 0$$
$$dP_{T2}/dv_{op} = (V_{CC1} + V_{CC2})/2\pi R_L - 2v_{o(p\text{-}p)}/8R_L$$

Equating to zero and solving for $v_{o(p\text{-}p)}$ yields

$$0 = \frac{V_{CC1} + V_{CC2}}{2\pi R_L} - \frac{2v_{o(p\text{-}p)}}{8R_L}$$

$$v_{o(p\text{-}p)} = \frac{2}{\pi}(V_{CC1} + V_{CC2}) \tag{H-4}$$

To find the maximum P_{T2} we substitute $v_{o(p\text{-}p)}$ from Eq. (H-4) in Eq. (H-3).

$$\begin{aligned} P_{T2} &= \frac{(V_{CC1} + V_{CC2})}{\pi R_L} \times \frac{\frac{2}{\pi}(V_{CC1} + V_{CC2})}{2} \\ &\quad - \frac{\left[\frac{2}{\pi}(V_{CC1} + V_{CC2})\right]^2}{8R_L} \\ &= \frac{(V_{CC1} + V_{CC2})^2}{\pi^2 R_L} - \frac{(V_{CC1} + V_{CC2})^2}{2\pi^2 R_L} \\ &= \frac{(V_{CC1} + V_{CC2})^2}{2\pi^2 R_L} \end{aligned} \tag{H-5}$$

The last expression gives the maximum power for the two transistors. If we assume there is equal power dissipation in both transistors, the power dissipated in each transistor is

$$P_{T(max)} = \frac{1}{2}\frac{(V_{CC1} + V_{CC2})^2}{2\pi^2 R_L} = \frac{(V_{CC1} + V_{CC2})^2}{4\pi^2 R_L} \tag{9-23}$$

APPENDIX I

INTEGRATOR AND DIFFERENTIATOR GENERAL EXPRESSIONS

INTEGRATOR

The circuit shown is a typical integrator circuit.

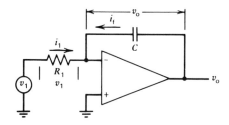

FIGURE I-1

The feedback current i_f is given by

$$i_f = C \frac{dv_o}{dt} \tag{I-1}$$

(v_o is the voltage across C.)

The input current i_1 is

$$i_1 = v_1/R_1 \tag{I-2}$$

(v_1 is across R_1.)

From Eqs. (I-1) and (I-2), and the fact that $i_f + i_1 = 0$ [Eq. (11-8)], we get

$$i_f + i_1 = C\frac{dv_o}{dt} + \frac{v_1}{R_1} = 0 \tag{I-3}$$

Solving Eq. (I-3) for dv_o/dt yields

$$\frac{dv_o}{dt} = -\frac{v_1}{CR_1} \tag{I-4}$$

If we integrate Eq. (I-4),

$$v_o = -\frac{1}{R_1 C}\int v_1\, dt \tag{I-5}$$

Equation (I-5) is the general expression for v_o for any input v_1.

DIFFERENTIATOR

The circuit shown is a typical differentiator.

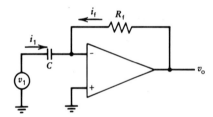

FIGURE I-2

Here,

$$i_f = v_o/R_f \tag{I-6}$$

$$i_1 = C\frac{dv_1}{dt} \tag{I-7}$$

and

$$\frac{v_o}{R_f} + \frac{Cdv_1}{dt} = 0 \tag{I-8}$$

$$v_o = -R_f C\frac{dv_1}{dt} \tag{I-9}$$

The integrator and differentiator circuits can be implemented with inductors instead of capacitors.

INTEGRATOR

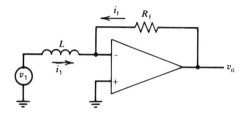

FIGURE I-3

$$i_f = v_o/R_f \tag{I-10}$$

$$i_1 = \frac{1}{L}\int v_1\, dt \tag{I-11}$$

$$i_f + i_1 = \frac{v_o}{R_f} + \frac{1}{L}\int v_1\, dt = 0 \tag{I-12}$$

$$v_o = -\frac{R_f}{L}\int v_1\, dt \tag{I-13}$$

DIFFERENTIATOR

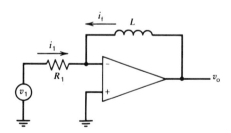

FIGURE I-4

$$i_f = \frac{1}{L}\int v_o\, dt \tag{I-14}$$

$$i_1 = v_1/R_1 \tag{I-15}$$

$$i_f + i_1 = \frac{1}{L}\int v_o\, dt + \frac{v_1}{R_1} = 0 \tag{I-16}$$

$$\int v_o\, dt = -\frac{L}{R_1} v_1 \tag{I-17}$$

Differentiating both sides, we get

$$v_o = -\frac{L}{R_1}\frac{dv_1}{dt} \tag{I-18}$$

The last expression becomes

$$v_o = -\frac{L}{R_1}\frac{\Delta v_1}{\Delta t} \tag{11-32}$$

when the rate of change of v_1 is relatively constant, allowing the use of $\Delta v_1/\Delta t$.

ANSWERS TO MISCELLANEOUS PROBLEMS

Chapter 1

1-1. (a) -0.5 V (at $I = 0.8$ A)
          $+0.7$ A (at $V = 1.4$ V)
      (b) -0.3 V (at $I = 2.3$ A)
          $+0.7$ A (at $V = 2.05$ V)
      (c) Less than 0.1 V
          Less than 0.1 A
1-3. $I_5 = 6$ mA
1-5. $V_o = -20$ V
1-7. (a) $E_{Th} = 7 \sin(10t + 45°)$
          $z_{Th} = z_N = 25\ k\Omega - j5k\Omega\ (25.5\ \underline{/-11.3°})$
          $I_N = 0.27 \sin(10t + 56.3°)$
      (b) $E_{Th} = 10 \sin(100t + 6°)$
          $z_{Th} = z_N = 20\ k\Omega\ \underline{/-3°}\ (\approx 20\ k\Omega)$
          $I_N = 0.5 \sin(100t + 9°)$
      (c) $E_{Th} \approx 10 \sin 1000t$
          $z_{Th} = z_N \approx 20\ k\Omega$
          $I_N \approx 0.5 \sin 1000t$
1-9. $V_3 = 6.6$ V (Polarity $\pm$)
1-11. $I_1 = 1/10\ k\Omega = 0.1$ mA
      $v_o = 15$ V

Chapter 2

2-1. (a) $I_D = 2\ pA \times e^{20} = 0.970$ mA
      (b) $I_D = 2\ pA \times e^{18.8} = 0.29$ mA
      (c) $(0.29 - 0.97)/20 = 0.034$ mA/°C
2-3. (a) $I_D = 0.2$ A
      (b) $I_D = 0.06$ A
      (c) $I_D = 0.055$ A
      (d) $I_D = 0.056$ A

2-5. (a) $R_F = 536\ \Omega$
(b) $r_F = 27\ \Omega$
(c) $R_F = 1793\ \Omega$
(d) $r_F = 90\ \Omega$
2-7. (a) $I_A = 0.8\ \mu A$
(b) $V_R = 0.8\ V$

Chapter 3

3-1.

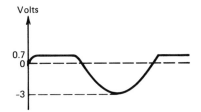

FIGURE P-1

3-3.

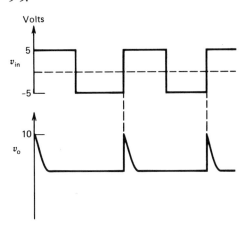

FIGURE P-2

3-5. (a)

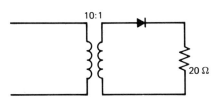

FIGURE P-3

(b)

FIGURE P-4

(c) V_{dc} = 5.2 V
(d) I_{dc} = 0.26 A
(e) I_{peak} = 0.8 A
(f) PIV = 16.2 V

3-7. (a) C = 4000 μF
(b) ϕ_c = 21.6°
(c) I_{peak} = 3.32 A
(d) V_{Srms} = 8.7 V, ratio = 115/8.7 = 13.2

3-9. (a)

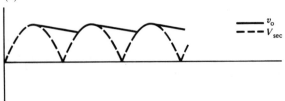

FIGURE P-5

(b) % r = 4.8%
(c) I_{peak} = 0.52 A
(d) E_m = 5.4 V
(e) Turns ratio = 30:1
(f) PIV = 10.8 V

3-11. (a) C = 190 μF
(b) % r = 0.6%

3-13. (a) % Reg. = 16%
(b) % Reg. = 4.4%

3-15. (a)

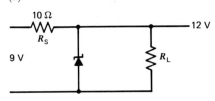

FIGURE P-7

(b) $P_{RS} = 1.1$ W
(c) $P_2 = 4$ W (use 5 W or 10 W zener)

3-17.

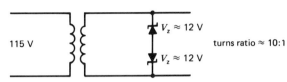

FIGURE P-8

3-19. $V_{dc} = 40$ V

Chapter 4

4-5. (a) CE
      (b) CB
      (c) CC (EF)
4-7. (a) Voltage divider bias with the largest R_E.
4-9. (a) $I_{BQ} = 33$ μA
      (b) $I_{CQ} = 1.7$ mA
      (c) $V_C = 5$ V
4-11. $I_{CQ} = 3.1$ mA
      $V_{CEQ} = 8.6$ V
4-13. (a) $I_{CQ} = 4.1$ mA, $V_E = 0.82$ V
          $R_2 = 2$ kΩ $R_1 = 31$ kΩ
      (b) $I_{CQ} = 2.5$ mA
          $I_{BQ} = 25$ μA
          $R_B = 0.77$ MΩ
4-15. $I_{BQ} = 82$ μA
      $I_{CQ} = 4.1$ mA
      $V_{CQ} = 8.8$ V
4-17. (a) $R_C = 2.36$ kΩ
      (b) $R_B = 112$ kΩ
4-19. $V_E \approx 1.0$ V
      $R_E = 140$ Ω
      $R_2 = 1$ kΩ
      $R_1 = 20$ kΩ

FIGURE P-9

$V_E \approx 1.0$ V
$R_E = 140\ \Omega$
$R_2 = 1$ kΩ
$R_1 = 20$ kΩ

4-21. (a) $I_{BQ} = 36\ \mu A$
(b) $I_{CQ} = -1.8$ mA
(c) $V_{CEQ} = -4.3$ V
(d) $V_{CQ} = 4$ V

Chapter 5

5-1. (a) $A_{vT} = 0.62$
(b) $\dfrac{i_o}{i_{in}} = 68.2$
(c) $A_p = 42.3$

5-3. (a) $v_o = 62\ \mu V$
$P_o = 0.038$ nW $(0.038 \times 10^{-9}$ W$)$
(b) $v_o = 2.2$ mV
$P_o = 47.5$ nW

5-5. $v_o = 0.57$ V
$i_o = 0.47$ mA
$r_{in} = 1.95$ kΩ
$r_o \approx 1.2$ kΩ

5-7. (a)

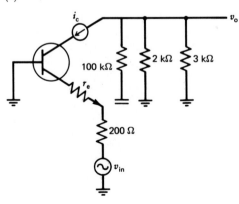

FIGURE P-10

(b) $A_v = 5.7$
$A_i = 0.4$
$r_{in} = 209\ \Omega$
$r_o = 2\ k\Omega$

5-9. $A_v \simeq 1$
$r_{in} \simeq 182\ k\Omega$
$r_o \simeq 15\ \Omega$
$v_{o\ p\text{-}p} = 0.1\ V$

5-11. (a) $R_C = 3\ k\Omega$
$R_{E1} = 200\ \Omega$
$R_{E2} = 200\ \Omega$
$R_2 = 6\ k\Omega$
$R_1 = 60\ k\Omega$
(b) $A_i = -7.7$
$r_{in} = 4.6\ k\Omega$
$r_o = 3\ k\Omega$
(c) $C_E = 78\ \mu F$

5-13. $A_v = -9.5$
$A_i = -43$
$r_{in} = 18.2\ k\Omega$
$r_o = 3.6\ k\Omega$

5-15. $r_{in} = 860\ \Omega$
$r_o = 2.4\ k\Omega$
$A_v = -50$
$A_i = -40$

5-17. (a)

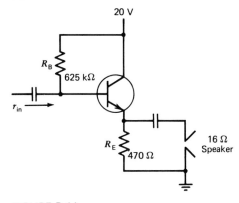

FIGURE P-11

(b) $\beta_{min} = 250$
(c) $R_B = 625\ k\Omega$

5-19. (a) $A_v = 5.5$
(b) $r_{in} = 235\ \Omega$
(c) $A_i = 0.59$
(d) $r_o = 4.7\ k\Omega$

5-21. (a) $R_B = 189$ kΩ
(b) $\beta_{min} = 133$

Chapter 6

6-5. $I_D = 5.3$ mA
$V_D = 9.4$ V
6-7. (a) $V_{GSQ} = -1.7$ V
(b) $I_{DQ} = 3.3$ mA
(c) $V_{DS} = 5.8$ V
(d) $V_S = 3.3$ V
6-9. (a) $V_{GSQ} = -0.6$ V
(b) $I_{DQ} = 7$ mA
(c) $V_{DQ} = 11$ V
6-11. $I_{DQ} = 2.2$ mA
$V_{GSQ} = -2.3$ V
$V_{DQ} = 13.6$ V
6-13. $V_{GSQ} = 2.4$ V
$I_{DQ} = -1.35$ mA
$V_{DQ} = -14.8$ V
6-15. (a) $A_v = -1.95$
(b) $r_{in} = 1$ MΩ
(c) $r_o = 3.9$ kΩ
6-17. (a) $A_v = -7.5$
$v_o = -0.75$ V
(b) $r_{in} = 2$ MΩ
(c) $r_o = 4$ kΩ
6-19. (a) $A_{v1} = -9$ $v_{o1} = -1.8$ V
(b) $A_{v2} = 0.6$ $v_{o2} = 0.12$ V
(c) $r_{o1} = 12$ kΩ
(d) $r_{o2} = 312$ Ω
6-21. (a) $A_v = 0.61$ $r_o = 385$ Ω
(b) $A_v = 0.76$ $r_o = 476$ Ω
(c) $A_v = 0.89$ $r_o = 555$ Ω
6-23. $R_{S1} = 250$ Ω; $R_{S2} = 417$ Ω; $R_D = 4.67$ kΩ; $R_G = 1$ MΩ
6-25. (a) $A_v = -5.2$
(b) $r_{in} = 10$ MΩ
(c) $r_o = 4.7$ kΩ
6-27. (b) $r_{in} = 31.4$ MΩ

Chapter 7

7-1. (a) $r_{in} = 10$ kΩ
(b) $r_o = 100$ Ω
(c) $A_{vT} = 53.6$
7-3. (a) $V_{E1} = 6.8$ V

APPENDIX I   579

        (b) $I_{C1} = 1.4$ mA
        (c) $V_{E2} = 6.1$ V
        (d) $I_{C2} = 6.1$ mA
7-5. (a) $I_{C1} = 1.38$ mA
      (b) $V_{CQ1} = 9.6$ V
      (c) $I_{C2} = 1.8$ mA
      (d) $V_{CQ2} = 8$ V
7-7. (a) $I_{DQ1} = 3.5$ mA $V_{DQ1} = 8.45$ V
      (b) $I_{CQ2} = 23.5$ mA $V_{CEQ2} = 12.25$ V
7-9. (a) $A_{vT} = -5.5$
      (b) $r_{in} = 5.5$ kΩ
      (c) $r_o = 12.3$ Ω
7-11. $A_{vT} = +177$
         $r_{in} = 1.5$ kΩ
         $r_o = 750$ Ω
7-13. (a) $r_{in} = 47$ Ω
      (b) $r_o = 150$ Ω
      (c) $A_{vT} = -158$ (A_{v1} is positive and A_{v2} is negative)
7-15. (a) $r_{in} = 99$ kΩ
      (b) $A_{vT} = 0.86$
      (c) $A_i = 85$
      (d) $r_o = 7$ Ω
7-17. (a) $A_{vT} = 36.4$
      (b) $r_{in} = 1$ MΩ
      (c) $r_o = 3.3$ kΩ
7-19. (a) $I_{DQ1} = 2$ mA $V_{DSQ1} = 13.2$ V
      (b) $I_{CQ2} = 5.4$ mA $V_{CEQ2} = 15.6$ V
      (c) $I_{CQ3} = 49$ mA $V_{CEQ3} = 5.1$ V

Chapter 8

8-1. (a) 3 dB
     (b) -10 dB
     (c) 6 dB
     (d) -17 dB
8-3. (a) 20 dB
     (b) -6 dB
     (c) 40 dB
     (d) 26 dB
     (e) 20 dB
     (f) 60 dB
8-5. $A_{v3} = 22$ dB
8-7. $f_{LC} = 13.3$ Hz
8-9. (a) $f_{LC} = 9.4$ Hz
     (b) $|A_{vmb}| = 2.1$
     (c) A_v at cutoff $= 1.5$

8-11. (a) $f_{LC} = 1.6$ Hz
(b) $|A_{vmb}| = 5.3$
8-13. (a) $f_{LC} = 2$ Hz
(b) $f_{LC} = 1.2$ Hz
(c) $f_{LC} = 18.5$ Hz
8-15. (a) $f_{UC} = 4$ MHz
(b) $f_{UC} = 4$ MHz
8-17. $f_{UC} = 3.6$ MHz
8-19. $f_T = 50$ MHz

Chapter 9

9-1. $r_{e(av)} = 1.8\ \Omega$
9-3. (a) $P_T = 7.5$ W   O.K.
(b) $P_T = 10.4$ W   Exceeds P_D
(c) $P_T = 7.5$ W   O.K.
(d) $P_T = 4.5$ W   O.K.
9-5. (a) $v'_{o(max)} = 150\ V_{p-p}$
(b) $N_1/N_2 = 4.3$
(c) $v_{o(max)} = 35\ V_{p-p}$
9-7. (a) $P_{dc} = 5$ W $P_{ac} = 0.05$ W $\eta = 1\%$
(b) $P_{dc} = 5$ W $P_{ac} = 2.45$ W $\eta = 49\%$
9-9. (a) $\eta = 27.5\%$
(b) $\eta = 55\%$
(c) $v_{in} = 14.3$ V rms
9-11. $r_{in} = 19\ \Omega$
9-13. (a) $P = 2.3$ W
(b) $\theta_{cA} = 2.5°C/W$

Chapter 10

10-1. $v_o = 2 + 6 \times 10^{-3} = 2.006$ V
10-3. (a) CMRR = 50,000 = 94 dB
(b) $v_o = 1 + 2.4 \times 10^{-3} = 1.0024$ V
(c) DA of Problem 10-3 has higher (better) CMRR. (CMRR of Prob. 10-1 is 92 dB.)
10-5. $I_C = 0.66$ mA
10-7. $I_{C2} \approx 50\ \mu A$
10-9. $|v_o/(v_{in1} - v_{in2})| = 173$
10-11. $r_{ind} = 416\ k\Omega$

Chapter 11

11-1. (a) $A_{CL} = 100$
(b) $A_{CL} = 98$
(c) % error = 2%

11-3. (a) $BW = 50$ kHz
(b) $BW = 10$ kHz
(c) $BW = 50$ Hz
11-5. $v_o = -0.54$ V
11-7. $v_o = -0.1$ V rms
11-9. $f_o = 11.3$ kHz
$A_{CL} = -1$
11-11. $v_o = 0$ V
11-13. $v_o = -8.2$ V
11-15. $v_o = -400t$

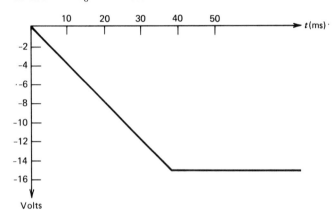

FIGURE P-12

11-17. $v_o = -5 \times 10^5 t$

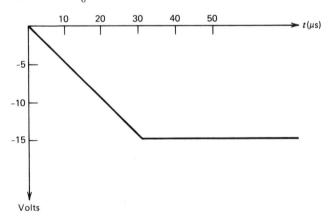

FIGURE P-13

11-19. (a) $A = 952$
(b) % error = 5%

582 APPENDIX I

11-21. $V_{OFF} = 2.0$ mV
11-23. slew rate = 10^5 V/s

Chapter 12

12-3. (a) Place short across SCR.
(b) Disconnect load.
12-5.

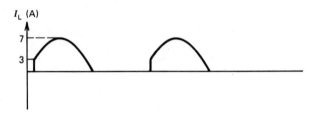

FIGURE P-14

12-7.

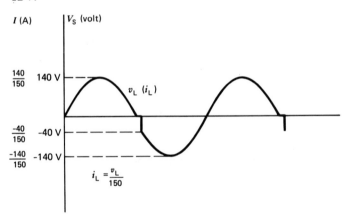

FIGURE P-15

12-9. (a) SCR failed in open state.
(b) Diode D_1 failed—open. (SCR never turns on.)
12-11. (a) Diodes D_1 and/or D_2 are shorted.

Chapter 13

13-2. $V_P = 6$ V
13-3. $r_{B1} = 3.2$ kΩ
$r_{B2} = 2.1$ kΩ
13-5. Trig. level = 5.6 V
$V_E = 5.5$ V $\rightarrow$ OFF

13-7. $V_{BB} = 3.6$ V
13-11. $I_P = 12$ μA
13-13.

	APD	PIN	
SENS.	High	Low	APD Advantage
Sup. volt.	High	Low	APD Disadvantage
Noise	Medium	Low	PPD Disadvantage

13-15.

	Laser	LED
High power	High	Low
Light BW	Narrow	Wide
Driving current	High	Low
slope of light		
Power versus current	Steep	Shallow

13-17. The tunnel diode characteristic has a negative resistance region.

INDEX

Acceptor Atom, 36
Active Region, 131
Adder, 473
Alpha (α), 123, 124
Ambient, 413, 418
Amplifier:
    audio, 397, 450
    bridge, 475
    differential, 430, 431, 456, 474
    feedback, 451
    inverting, 457, 459
    non-inverting, 459, 463, 465
    operational, 450, 456, 480
    power, 407
    VMOS, 407
Avalanche, 50, 99, 127, 540

Bel, 350
Beta (β), 123, 125, 126, 129, 381
Biasing, 319
    circuits, 134
    Darlington pair, 323
    differential amplifier, 430
    with feedback, 136
    fixed, 134, 243
    JFET, 240
    MOSFET, 289
    self, 141, 243
    stability, 132
    for switching, 156
    two supplies, 148
    voltage divider, 143, 147, 246
    see also Transistor biasing; FET biasing
Bipolar Transistor (BJT), 119, 320, 324
Boltzman, 44
Breakover voltage, 504
BW (Bandwidth), 345, 369, 454, 487

Clamper, 64
Class (of Operation), 152, 389
Class A, 391, 416
Class B, 397, 399, 400, 407
Class C, 410, 412
Clipper, 64
Closed loop gain, 451, 453, 478, 479
Collector, 122
Common base, 132, 133, 193
Common collector, 132, 133, 202

INDEX

Common drain (CD), 271
Common emitter, 132, 133, 180
Common mode, 432
Common mode gain, 432
Common mode rejection, 433, 434
Common source (CS), 261
Complementary symmetry, 400, 402, 407, 506
Conduction angle, 87, 88, 505
Conduction band, 31, 32
Coupling, 319, 341
    capacitor, 354
    direct, 332
    R-C, 332
    transformer, 339, 394, 416
Covalent bond, 43
CUJT, 530
Current amplification, 123
Current gain, 168, 186, 201, 205, 269
Current source, 436
Current source mirror, 438
Current source widlar, 439
Cutoff, 124, 131
Cutoff frequency, 454, 477

Darlington pair, 323, 327, 444
Decade, 349
Decibel, 349, 350
Depletion, 39, 126, 291, 286, 287, 291
Derating factor, 414
Differential gain, 432, 443
Differential input, 435, 443, 445
Differential output, 442
Differentiator, 450, 466
Diffusion, 38, 43, 44, 120, 121
Diode, 28, 29, 42, 63, 107
    bilateral, 517
    forward biased, 52
    four layer, 515
    PIN, 539, 546
    reverse bias, 58
    Shockley, 515
    tunnel, 549
    resistance, dynamic, 57
    resistance, static, 57
Distortion, 211, 383, 398, 453, 455
Distortion crossover, 398
Donor, 36
Doping, 35

Drain, 234, 235
Drift, 39, 41, 43
Dynamic Resistance, 175

Early, J.M., 125
Efficiency, 385, 386, 394, 402
Electrical characteristics, 113
Electron volt, 32
Emitter, 122
Emitter follower, see Common collector
Energy band, 30, 31
Energy gap, 30
Enhancement, 284–287, 291, 293

f_B, 349
f_{LC}, 349, 354, 356
f_T, 349, 373
f_{UC}, 349
Falloff, 477
Field effect transistor, see Junction FET
Feedback:
    negative, 481
    positive, 499
Filter:
    active, 450, 477
    R-C, 477
Filtering, 80
Filtering capacitor, 81
Filtering L-C, 81, 91
Filtering, PI, 92
Filtering R-C, 81, 88
Frequency:
    lower cutoff, see f_{LC}
    upper cutoff, see f_{UC}
Frequency compensation, 489
Frequency response, 349, 353, 360, 368, 390, 453, 486

Gain bandwidth, 454
Gate, 234

Heat sink, 417, 418
High pass, 355, 356
Hybrid parameters, 172–174

IGFET, 284
Impurities, 35
Input bias current, 482

Input resistance, 169, 184, 198, 205, 209, 223, 459, 474
Insulator, 32
Integrator, 450, 466
Intrinsic standoff, 522
Irradiance, 534

Junction barrier, 38
Junction FET (JFET), 232, 233, 236, 324, 422
    N channel, 241
    P channel, 241
Junction forward bias, 41
    reverse bias, 41

KCL, 1, 3, 55, 56
Knee, 48
KVL, 1, 5, 55, 56

LASCR, 544
LASER, 546
Leakage, 42, 48
LED, 537, 546
Linearity, 2, 383, 548
Load line, 55, 544
    AC, 216, 395
    DC, 152
Low pass, 477, 478

Majority carriers, 37, 121
Maximum power, 12
Maximum rating, 110, 419
Midband, 353, 365, 372, 375, 452
Miller effect, 349, 362, 363
Minority carriers, 37, 121
MOSFET, 232, 233, 284–298
"Multiplication", 50

Negative resistance, 524, 549
Noise figure, 380
Nonlinear region, 52
Norton's, 2, 7–12, 169
NPN, 122, 157, 159, 439

Octave, 349
Offset current, 484
Offset nulling, 486, 487
Offset voltage, 482, 485
Open Loop, 451, 479

Optoisolator, 547, 548, 552
Output resistance, 170, 187, 202, 206, 270, 274, 459

Passband, 477
Peak current, 75, 77, 78, 88, 96
Peak detector, 70
Peak inverse voltage (PIV), 74, 78, 87, 96
Photoconductive, 535
Photo current, 535
Photo darlington, 544, 545
Photodiode, *see* Diode, photo
Photoemissive, 535
Phototransistor, 541
Photovoltaic cell, 535
PI, *see* Filter, PI
Pinchoff, 235, 237
PNP, 121, 157, 159
Power gain, 171, 188
Power rating, 385, 413
Power supply, 72–76
Push-pull, 404, 407, 409
Put, 631

Q point, 130–156, 232, 242, 244, 254
Quasi complementary, 407

Radiation sensitivity, 541
Rectifier, 73
    bridge, 78
    fullwave, 76–78
    halfwave, 73–76
Regulation, 92, 105
Regulator, voltage, 93, 100, 103
Relaxation oscillator, 526, 528
Resonance, 416
Responsivity, 535
Ripple factor, 83–85
Ripple per cent, 83
Ripple voltage, 82, 86
Rise time, *see* T_r

Saturation, 42, 124
SCR, 499, 500, 506, 511
Shockley, 53, 238, 239, 276
Shockley diode, *see* Diode, Shockley
Sieman, 253

Single ended, 435, 442, 445, 490
Slew rate, 488, 489
Source follower, *see* Common drain
Stopband, 477
Subtractor, 450, 461, 463, 464
Superposition, 14–18, 440
SUS, 517
Switching, 196, 425
Switching, analog, 298
Switching, digital, 300

T_f (fall time), 426
T_r (rise time), 381, 425, 486
Temperature coefficient, 106, 107
Thermal resistance, 413
Thevenin's, 2, 7–12, 144, 150, 169
Thyristor, 499, 515, 517
Transconductance (g_m), 251–258

Transformer, 76, 338, 339
Transistor, 170
Transistor biasing, 130
Triac, 505, 506, 511
Tunnelling, 50

Unijunction transistor (UJT), 522

Varactor, 550
Valence band, 31, 32
Virtual ground, 453
VMOSFET, 301
Voltage doubler, 97
Voltage gain, 168, 194, 203, 261, 452, 456
Voltage regulator, *see* Regulator, voltage

Zener, 50, 98, 106, 408